ICIAM2023 Springer Series

Volume 2

This series aims to publish some of the most relevant results presented at the ICIAM 2023 held in Tokyo in August 2023.

The series is managed by an independent Editorial Board, and will include peer-reviewed content only, including the Invited Speakers volume as well as books resulting from mini-symposia and collateral workshops.

The series is aimed at providing useful reference material to academic and researchers at an international level.

Xiangfeng Wang · Xingju Cai · Deren Han
Editors

Splitting Optimization

Theory, Methodology, and Applications

Springer

Editors
Xiangfeng Wang
School of Computer Science
and Technology
East China Normal University
Shanghai, China

Xingju Cai
School of Mathematical Science
Nanjing Normal University
Nanjing, China

Deren Han
School of Mathematical Sciences
Beihang University
Beijing, China

ISSN 3091-3098 ISSN 3091-3101 (electronic)
ICIAM2023 Springer Series
ISBN 978-981-96-8573-8 ISBN 978-981-96-8574-5 (eBook)
https://doi.org/10.1007/978-981-96-8574-5

Mathematics Subject Classification: 65K05, 65K15, 49M27, 90-08

This Springer imprint is published by the registered company Springer Nature Singapore Pte Ltd.
The registered company address is: 152 Beach Road, #21-01/04 Gateway East, Singapore 189721, Singapore

Preface

We are delighted to present *Splitting Optimization: Theory, Methodology, and Applications*, a comprehensive volume dedicated to the forefront of research in splitting methods within the field of optimization. As optimization problems become increasingly complex and high-dimensional across various scientific and engineering disciplines, the need for efficient and scalable algorithms is more critical than ever. Splitting optimization has emerged as a powerful paradigm, offering robust theoretical frameworks and practical algorithms for decomposing and solving large-scale optimization problems.

This book brings together a collection of pioneering research papers authored by leading experts and scholars. Each contribution explores innovative aspects of splitting optimization, encompassing theoretical advancements, methodological developments, and a diverse array of applications. Our aim is to provide readers with a deep and cohesive understanding of the current state of splitting optimization, as well as to inspire future research and innovations in this dynamic field.

Splitting methods have gained significant attention due to their ability to handle complex optimization problems by breaking them down into simpler subproblems. This decomposition not only facilitates parallel and distributed computing but also allows for the exploitation of problem structures, leading to more efficient algorithms. The versatility of splitting methods makes them applicable to a wide range of problems, including those in machine learning, signal processing, image analysis, and game theory.

"A Trust-Region-Based Splitting Method for Optimization Problems with Linear Constraints" with authors Leyu Hu, Yannan Chen, Xingju Cai, and Deren Han is presented in Chap. 1. This chapter addresses a fundamental challenge in augmented Lagrangian-based splitting methods: the selection of penalty parameters. Fixed parameters can lead to slow convergence or inadequate progress in either the primal or dual variables. The authors introduce a novel trust-region-based approach that adaptively adjusts the trust-region radius during iterations. This method smartly balances the trade-off between primal and dual advancements, ensuring robust global convergence under mild conditions. An $O(1/\epsilon)$ convergence rate is achieved in an ergodic sense. Numerical experiments on medical image recovery and logistic regression problems demonstrate the efficiency and potential of the proposed method, highlighting its applicability in real-world scenarios.

The following chapter contains "Forward-Reflected-Backward Method with Extrapolation and Linesearch for Monotone Inclusion Problems" with authors Tanxing Wang, Heng Zhang, and Xingju Cai. Monotone inclusion problems are central to various applications in optimization and variational analysis. The authors present an enhanced forward-reflected-backward algorithm that incorporates a new extrapolation direction and a novel line-search procedure using locally Lipschitz constants. Unlike existing methods that rely on global Lipschitz constants, this approach allows for larger step sizes, improving convergence speed and computational efficiency. Weak convergence is established under standard assumptions. Extensive numerical experiments on the lasso problem and ℓ_1-regularized logistic regression illustrate the method's superiority over classical algorithms, making it a valuable tool for solving large-scale optimization problems.

"Fast Adaptive ADMM with Gaussian Back Substitution for Multiple Block Linear Constrained Separable Problems" with the author Xiangfeng Wang is presented in the following chapter. The Alternating Direction Method of Multipliers (ADMM) is a popular algorithm for solving separable optimization problems with linear constraints. This chapter introduces the Fast Adaptive ADMM with Gaussian Back Substitution (ADMM-G-V), an innovative framework designed for multiple-block settings. By integrating an adaptive penalty parameter that dynamically adjusts during the iterative process and utilizing Gaussian back substitution, the proposed method enhances both convergence properties and computational efficiency. Theoretical analyses establish global convergence and optimal convergence rates in both ergodic and non-ergodic senses. Numerical experiments on consensus problems over networked agents and distributed logistic regression tasks showcase the algorithm's effectiveness, underscoring its potential in distributed optimization and large-scale machine learning applications.

Chapter 4 contains Inertial "Inertial Alternating Direction Method of Multipliers with Logarithmic-Quadratic Proximal Regularization" with the author Zhongming Wu. Acceleration techniques are crucial for improving the performance of iterative optimization algorithms. This chapter explores the inertial proximal point method applied to the ADMM and symmetric ADMM with logarithmic-quadratic proximal (LQP) regularization. By incorporating inertial terms and appropriate step sizes, the proposed methods accelerate convergence while leveraging the separable structure of the problem. The utilization of LQP regularization transforms constrained subproblems into more manageable unconstrained ones during iterations. Under mild conditions, global convergence is rigorously established, enriching the theoretical understanding of inertial methods in the context of splitting optimization.

"A Class of Augmented-Lagrangian-Type Algorithms for Solving Generalized Nash Equilibrium Problems" with authors Xiaoxi Jia, Shiwei Wang, and Lingling Xu is presented in the following chapter. Generalized Nash Equilibrium Problems (GNEPs) with shared constraints present significant analytical and computational challenges. The authors propose a class of regularized augmented Lagrangian methods that penalize shared linear constraints within each player's augmented Lagrangian function, transforming the original GNEP into a convex Nash Equilibrium subproblem (NEP). Under strong monotonicity and Lipschitz continuity assumptions of the pseudo-gradient, Fejér monotonicity of iterative points is proved with respect to the solution set. By adding a correction step, the authors relax the cocoercivity requirement of the pseudo-gradient, broadening the applicability of their method. Numerical examples validate the effectiveness of the proposed algorithms, demonstrating their potential in economic modeling and resource allocation problems.

The chapters in this volume collectively advance the field of splitting optimization through:—Adaptive Strategies: Introducing adaptive mechanisms for parameter selection, such as trust-region adjustments and adaptive penalty parameters, to enhance convergence and stability without relying on fixed or global constants.—Acceleration Techniques: Employing inertial terms, extrapolation directions, and line-search procedures to accelerate the convergence of iterative methods, making them more practical for large-scale and real-time applications.—Theoretical Advancements: Providing rigorous convergence analyses under less restrictive assumptions, thereby strengthening the theoretical foundations and broadening the applicability of splitting methods.—Diverse Applications: Showcasing the versatility of splitting optimization through practical applications in medical imaging, logistic regression, consensus networks, and game theory, highlighting the impact of these methods across different domains.

Splitting Optimization: Theory, Methodology, and Applications represents a significant milestone in the ongoing evolution of optimization research. As the demand for efficient and scalable algorithms continues to grow, splitting methods will undoubtedly play an increasingly vital role. By presenting these innovative contributions, we hope to foster a deeper understanding of splitting optimization and stimulate further advancements in both theoretical and practical realms. We believe that the insights and methodologies presented in this book will not only enrich the knowledge of readers but also inspire new ideas and collaborations. The challenges addressed and solutions proposed here reflect the dynamic nature of optimization, and we are confident that this volume will serve as a valuable foundation for future exploration and innovation.

We extend our sincere gratitude to all the contributing authors for their outstanding work and dedication. Their commitment to excellence has made this volume possible. We also thank the reviewers for their insightful comments and suggestions, which have greatly enhanced the quality of the chapters. Finally, we express our appreciation to the broader optimization community for their ongoing support and engagement.

Shanghai, China — Xiangfeng Wang
Nanjing, China — Xingju Cai
Beijing, China — Deren Han

Contents

Chapter 1
A Trust-Region-Based Splitting Method for Optimization Problems with Linear Constraints

Leyu Hu, Yannan Chen, Xingju Cai, and Deren Han

Abstract The augmented Lagrangian-based splitting methods have found more and more applications in scientific and engineering computation, such as compressive sensing, covariance selection, image processing, and transportation research. One of the basic difficulties in such algorithms is the selection of the parameter in the augmented Lagrangian function; a larger one may make the primal progress too small, while a smaller one may slow down the dual progress. To overcome this difficulty, in this paper, we propose to solve the splitting subproblems in a trust region manner, and the radius can be adjusted smartly. Under the same mild conditions as those for classical augmented Lagrangian-based splitting methods, we prove the global convergence of the proposed algorithm. Moreover, the $O(1/\epsilon)$ convergence rate is also analyzed in an ergodic sense. We present some preliminary numerical experiments on medical image recovery and logistic regression, which show that the trust region-based splitting method is efficient and promising.

Keywords Trust region · Splitting method · Separable convex optimization · Convergence rate

---

L. Hu
LMIB, School of Mathematical Sciences, Beihang University, Beijing, China
e-mail: huleyu@buaa.edu.cn

Y. Chen
School of Mathematical Sciences, South China Normal University, Guangzhou, China
e-mail: ynchen@scnu.edu.cn

X. Cai
School of Mathematical Sciences, Ministry of Education Key Laboratory of NSLSCS, Nanjing Normal University, Nanjing, China
e-mail: caixingju@njnu.edu.cn

D. Han (✉)
LMIB, School of Mathematical Sciences, Beihang University, Beijing, China
e-mail: handr@buaa.edu.cn

X. Wang et al. (eds.), *Splitting Optimization*, ICIAM2023 Springer Series 2,
https://doi.org/10.1007/978-981-96-8574-5_1

1.1 Introduction

We consider the following separable convex optimization problem with linear constraints:

$$\begin{cases} \min & f(x) + g(y) \\ \text{s.t.} & Ax + By = c, \\ & x \in \mathbb{R}^n, \;\; y \in \mathbb{R}^m, \end{cases} \tag{1.1}$$

where $A \in \mathbb{R}^{\ell \times n}$ and $B \in \mathbb{R}^{\ell \times m}$ are two fixed matrices, the vector $c \in \mathbb{R}^{\ell}$, and f and g are convex and smooth functions with Lipschitz continuous Hessian matrices. Hence, the solution set of (1.1) is convex, and we assume it is nonempty.

Numerous applications from science and engineering fall into the well-structured model (1.1). For example, the sparse solution recovery problem in compressive sensing (basis pursuit) [4, 25]; the matrix completion problem with or without noise [3]; the nonnegative tensor factorization [2]; the constrained total-variation image restoration and reconstruction problems [18]; the estimation of the higher-order generalized diffusion tensor in nuclear magnetic resonance imaging of medical engineering [5]; the ℓ_1-norm penalized log-likelihood covariance selection models [28]; the route based traffic assignment problems in transportation [14]; etc. Boyd et al. [1] show that several problems from statistical and machine learning can be modeled naturally as the separable convex optimization problem (1.1).

One of the efficient numerical algorithms for solving (1.1) is the classical augmented Lagrangian function method [19, 21]. At iteration k with an estimator of multiplier $\lambda^k \in \mathbb{R}^{\ell}$, it minimizes variables x and y simultaneously to get the next iterate

$$(x^{k+1}, y^{k+1}) := \operatorname{argmin}_{x,y} \mathcal{L}_H(x, y, \lambda^k), \tag{1.2}$$

and then update the multiplier

$$\lambda^{k+1} := \lambda^k - H(Ax^{k+1} + By^{k+1} - c). \tag{1.3}$$

Here, the augmented Lagrangian function is defined as

$$\mathcal{L}_H(x, y, \lambda) := f(x) + g(y) - \lambda^{\top}(Ax + By - c) + \frac{1}{2}\|Ax + By - c\|_H^2 \quad \text{for } (x, y, \lambda) \in \mathbb{R}^{n \times m \times \ell}, \tag{1.4}$$

and $H \in \mathbb{R}^{\ell \times \ell}$ is a penalized symmetric positive definite matrix. In many classical literatures, the matrix H is a scalar matrix $H = \beta I$, where β is a positive scalar and I is the identity matrix.

To exploit the inherent separable structure, the alternating direction method of multiplier (ADMM) [10, 12] minimizes the variables x and y separately. Given current iterates y^k and λ^k, it generates the next iterate via the following three steps successively:

$$\begin{cases} x^{k+1} := \text{argmin}_{x\in\mathbb{R}^n} \mathcal{L}_H(x, y^k, \lambda^k), \\ y^{k+1} := \text{argmin}_{y\in\mathbb{R}^m} \mathcal{L}_H(x^{k+1}, y, \lambda^k), \\ \lambda^{k+1} := \lambda^k - H(Ax^{k+1} + By^{k+1} - c). \end{cases} \tag{1.5}$$

That is to say, the large optimization problem (1.2) is divided into two small optimization problems in (1.5). The global convergence of ADMM is extensively studied by many researchers including Gabay [9], Tseng [22], and Eckstein and Bertsekas [7]. Recently, He and Yuan [17] show that ADMM admits the $O(1/\epsilon)$ convergence rate in an ergodic sense.

Both in (1.2) and (1.5), the matrix H plays an important role by penalizing the violation of the linear constraint. Although theoretically the algorithms converge for any positive definite matrix H, the numerical performance varies significantly for different choices. Generally, a larger one may lead to larger progress for the dual variable (it can be viewed as the stepsize for the dual variable λ), while it may cause smaller progress for the primal variables, and vice versa. Hence, how to choose a reasonable one is a difficult task. He et al. [15] propose to add proximal point terms to the subproblems,

$$\begin{cases} x^{k+1} = \text{argmin}_{x\in\mathbb{R}^n} \mathcal{L}_H(x, y^k, \lambda^k) + \dfrac{1}{2}\|x - x^k\|^2_{R_k}, \\ y^{k+1} = \text{argmin}_{y\in\mathbb{R}^m} \mathcal{L}_H(x^{k+1}, y, \lambda^k) + \dfrac{1}{2}\|y - y^k\|^2_{S_k}, \\ \lambda^{k+1} = \lambda^k - H(Ax^{k+1} + By^{k+1} - c), \end{cases} \tag{1.6}$$

where $\{R_k\}$ and $\{S_k\}$ are symmetric positive definite matrices. So the resulting subproblems in (1.6) are strongly convex. Alternatively, Xu [24] and Yuan [27] prefer to use fixed matrices R and S in the proximal point terms. It is worthwhile to note that a special choice of matrices R and S reduces to the split Bregman method [13], which is powerful for problems arising from compressive sensing and image denoising.

In this paper, we propose to solve the subproblems in a trust region framework, since the trust region method [6] is robust and efficient in nonlinear programming. Different from the soft regularization, trust region methods impose an explicit trust region constraint to force the new iterate being in a neighborhood of the current iterate. Formally, we represent the trust region-based splitting method as follows:

$$\begin{cases} x^{k+1} = \text{argmin}_{x\in\mathbb{R}^n} \left\{\widetilde{\mathcal{L}}_H(x, y^k, \lambda^k | x^k, y^k, \lambda^k) \mid \|x - x^k\| \le \Delta_k\right\}, \\ y^{k+1} = \text{argmin}_{y\in\mathbb{R}^m} \left\{\widetilde{\mathcal{L}}_H(x^{k+1}, y, \lambda^k | x^{k+1}, y^k, \lambda^k) \mid \|y - y^k\| \le \Gamma_k\right\}, \\ \lambda^{k+1} = \lambda^k - H(Ax^{k+1} + By^{k+1} - c), \end{cases}$$

where Δ_k and Γ_k are trust region radii for x-subproblem and y-subproblem, respectively, and $\widetilde{\mathcal{L}}_H$ is the quadratic approximation of the augmented Lagrangian function, which is defined as

$$
\begin{aligned}
&\widetilde{\mathcal{L}}_H(x, y, \lambda | x^k, y^k, \lambda^k) \qquad (1.7)\\
&= f(x^k) + \nabla f(x^k)^\top (x - x^k) + \frac{1}{2}(x - x^k)^\top \widetilde{\nabla}^2 f(x^k)(x - x^k)\\
&\quad + g(y^k) + \nabla g(y^k)^\top (y - y^k) + \frac{1}{2}(y - y^k)^\top \widetilde{\nabla}^2 g(y^k)(y - y^k)\\
&\quad - \lambda^k (Ax + By - c) + \frac{1}{2}\|Ax + By - c\|_H^2,
\end{aligned}
$$

where $\widetilde{\nabla}^2 f(x^k)$ and $\widetilde{\nabla}^2 g(y^k)$ are the Hessian-like matrices of f and g at x^k and y^k respectively obtained by certain mechanism such as BFGS. Theoretically, finding a solution to an optimization problem in a trust region is equivalent to solve a quadratic proximal regularized optimization problem with a suitable regularization parameter. However, for the regularization-based method, the selection of the regularization parameter in each iteration is a challenging problem. It is noteworthy that in a trust region framework, we do not need to directly choose the regularization parameter. This is because the concept of the trust region is geometric and more intuitive. According to nonlinear programming [6, 19, 21], dozens of algorithms and codes have been established to solve the trust region subproblem fast and accurately.

The proposed trust region-based splitting method is built in a prediction and correction framework. In the prediction step, we solve the x-subproblem and y-subproblem in an explicit trust region constraint and construct an efficient descent direction. In the correction step, a suitable step length is customized for the descent direction to accelerate the convergence of the proposed algorithm. Combining these two steps, we show that the sequence of iterates generated by the trust region-based splitting method is globally convergent to the optimal solution set. Moreover, an ϵ-approximate optimal solution of the equivalent variational inequality is obtained within the $O(1/\epsilon)$ number of iterations in the worst case. Numerical experiments on the constrained TV-ℓ_2 image deblurring and denoising problem and the ℓ_1 regularized logistic regression problem are performed. The results indicate that the novel trust region-based splitting method is efficient and promising. Significantly, in the image processing example, the involved trust region subproblem is solved efficiently and accurately when the classical fast Fourier transformation is explored.

The outline of this paper is as follows. In Sect. 1.2, we give some useful notations and the variational characterization of the separable convex optimization with linear constraints. The prediction and correction steps of the trust region-based splitting method are described in Sect. 1.3. The global convergence of the proposed algorithm is given in Sect. 1.4.1. And in Sect. 1.4.2, the $O(1/\epsilon)$ convergence rate is analyzed in an ergodic sense. Numerical validation on some real problems is reported in Sect. 1.5. Finally, we complete the paper by drawing some conclusions in Sect. 1.6.

1.2 Variational Characterization

In this section, we give some notations and preliminary results that are useful in the rest of this paper. Let $\lambda \in \mathbb{R}^\ell$ be a multiplier corresponding to the equality constraint $Ax + By - c = 0$ in (1.1). We define

$$u := \begin{pmatrix} x \\ y \end{pmatrix}, w := \begin{pmatrix} u \\ \lambda \end{pmatrix}, \qquad \text{then } w \in \Omega := \mathbb{R}^{n \times m \times \ell}.$$

The presentation of the new splitting method is in the framework of variational inequalities. Now, we give the variational characterization of the separable convex optimization (1.1).

The Lagrangian function of the original optimization problem (1.1) is

$$L(x, y, \lambda) := f(x) + g(y) - \lambda^\top (Ax + By - c), \tag{1.8}$$

where $(x, y, \lambda) \in \Omega$. Suppose $(x^*, y^*, \lambda^*) \in \Omega$ is a primal-dual solution of the convex optimization (1.1). Then, it must be a saddle point of the Lagrangian function, i.e.,

$$L(x^*, y^*, \lambda) \le L(x^*, y^*, \lambda^*) \le L(x, y, \lambda^*).$$

From the viewpoint of variational inequalities, $(x^*, y^*, \lambda^*) \in \Omega$ is a saddle point of the Lagrangian function (1.8) if and only if it satisfies

$$\begin{cases} (x - x^*)^\top (\nabla f(x) - A^\top \lambda^*) \ge 0, \\ (y - y^*)^\top (\nabla g(y) - B^\top \lambda^*) \ge 0, \qquad \forall\, (x, y, \lambda) \in \Omega. \\ (\lambda - \lambda^*)^\top (Ax^* + By^* - c) \ge 0, \end{cases} \tag{1.9}$$

For convenience, we define a mapping F:

$$F(w) := \begin{pmatrix} \nabla f(x) - A^\top \lambda \\ \nabla g(y) - B^\top \lambda \\ Ax + By - c \end{pmatrix} = \begin{pmatrix} 0 & 0 & -A^\top \\ 0 & 0 & -B^\top \\ A & B & 0 \end{pmatrix} w + \begin{pmatrix} \nabla f(x) \\ \nabla g(y) \\ 0 \end{pmatrix} + \begin{pmatrix} 0 \\ 0 \\ c \end{pmatrix}. \tag{1.10}$$

Obviously, the linear mapping F is a monotone operator since the coefficient matrix is skew-symmetric and the convexity of f and g. Then, the variational inequality (VI) (1.9) has a compact form: find $w^* \in \Omega$ such that

$$(w - w^*)^\top F(w^*) \ge 0, \qquad \forall\, w \in \Omega. \tag{1.11}$$

Since we suppose the solution set of the convex optimization (1.1) is nonempty, the solution set Ω^* of the VI (1.11), which is convex, is also nonempty. We remark that the VI (1.11) plays a critical role in the following analysis.

1.3 The Trust-Region-Based Splitting Method

The trust region method is a powerful and robust method in nonlinear programming [6, 19, 21]. At each iteration, the trust region method finds the solution of a subproblem in a special neighborhood of the current iterate, which is called a trust region. In the process of iteration, the size of the trust region could be enlarged or contracted according to some rules.

Our trust region-based splitting method for solving the separable optimization problem (1.1) contains two steps: prediction and correction. We now describe the prediction step in detail.

1.3.1 The Prediction Step

Algorithm 1 (*Prediction Step*) In iteration k, (x^k, y^k, λ^k) is given. we perform the following three steps successively.

$$\begin{aligned} \widetilde{x}^k &:= \operatorname{argmin}_{x\in\mathbb{R}^n} \widetilde{\mathcal{L}}_H(x, y^k, \lambda^k | x^k, y^k, \lambda^k) \\ &\quad \text{s.t. } \|x - x^k\| \le \Delta_k, \end{aligned} \tag{1.12}$$

$$\begin{aligned} \widetilde{y}^k &:= \operatorname{argmin}_{y\in\mathbb{R}^m} \widetilde{\mathcal{L}}_H(\widetilde{x}^k, y, \lambda^k | \widetilde{x}^k, y^k, \lambda^k) \\ &\quad \text{s.t. } \|y - y^k\| \le \Gamma_k, \end{aligned} \tag{1.13}$$

and

$$\widetilde{\lambda}^k := \lambda^k - H(A\widetilde{x}^k + B\widetilde{y}^k - c). \tag{1.14}$$

The quadratic approximation of the augmented Lagrangian function $\widetilde{\mathcal{L}}_H$ is defined in (1.7). The optimal condition for the trust region subproblem (1.12) in the variable x. For the convenience of the following discussion, we rewrite (1.12) as follows, by removing the irrelevant terms and rearranging the terms:

$$\begin{cases} \min\limits_{x\in\mathbb{R}^n} \nabla f(x^k)^\top (x - x^k) + \dfrac{1}{2}(x - x^k)^\top \widetilde{\nabla}^2 f(x^k)(x - x^k) \\ \qquad + \dfrac{1}{2}\|Ax + By^k - c - H^{-1}\lambda^k\|_H^2 \\ \text{s.t. } \|x - x^k\| \le \Delta_k. \end{cases}$$

According to nonlinear programming, the first-order necessary condition for the trust region subproblem (1.12′) is

$$0 \in \nabla f(x^k) + \widetilde{\nabla}^2 f(x^k)(x - x^k) + A^\top H(Ax + By^k - c - H^{-1}\lambda^k) + \mathcal{N}_{\odot(x^k,\Delta_k)}(x)$$

where $\mathcal{N}_{\odot(x^k,\Delta_k)}(x)$ is the normal cone at x of the trust region constraint $\|x - x^k\| \leq \Delta_k$ at x^k was

$$\mathcal{N}_{\odot(x^k,\Delta_k)}(x) := \left\{ v \in \mathbb{R}^n \middle| v = \begin{cases} \delta(x - x^k), & \|x - x^k\| = \Delta_k, \\ 0, & \|x - x^k\| < \Delta_k, \end{cases}, \delta \geq 0 \right\}.$$

Then we have the KKT conditions of (1.12′) as, for $\|\widetilde{x}^k - x^k\| \leq \Delta_k, \delta_k \geq 0$,

$$\begin{cases} \nabla f(x^k) + \widetilde{\nabla}^2 f(x^k)(\widetilde{x}^k - x^k) + A^\top H(A\widetilde{x}^k + By^k - c - H^{-1}\lambda^k) + \delta_k(\widetilde{x}^k - x^k) = 0, \\ \nabla g(y^k) + \widetilde{\nabla}^2 g(y^k)(\widetilde{y}^k - y^k) + B^\top H(A\widetilde{x}^k + B\widetilde{y}^k - c - H^{-1}\lambda^k) + \gamma_k(\widetilde{y}^k - y^k) = 0. \end{cases} \tag{1.15}$$

where δ_k here is the optimal Lagrange multiplier for the trust region constraint. Through this, we could get the optimal δ_k when we solve the trust region subproblem (1.12):

$$\delta_k = \frac{\|\nabla f(x^k) + \widetilde{\nabla}^2 f(x^k)(\widetilde{x}^k - x^k) + A^\top H(A\widetilde{x}^k + By^k - c - H^{-1}\lambda^k)\|}{\|\widetilde{x}^k - x^k\|},$$

when $\widetilde{x}^k \neq x^k$. And we define $\delta_k = 0$ when $\widetilde{x}^k = x^k$. For convenience, we denote

$$\begin{aligned} \nabla_x^k \widetilde{\mathcal{L}} &:= \nabla f(x^k) + A^\top H(Ax^k + By^k - c - H^{-1}\lambda^k), \\ \nabla_y^k \widetilde{\mathcal{L}} &:= \nabla g(y^k) + B^T H(A\widetilde{x}^k + By^k - c - H^{-1}\lambda^k). \end{aligned} \tag{1.16}$$

Then we have

$$\delta_k = \frac{\left\| \nabla_x^k \widetilde{\mathcal{L}} + \left(\widetilde{\nabla}^2 f(x^k) + A^\top HA\right)(\widetilde{x}^k - x^k) \right\|}{\|\widetilde{x}^k - x^k\|}, \tag{1.17}$$

and same for the trust region subproblem (1.13) in the variable y as

$$\gamma_k = \frac{\left\| \nabla_y^k \widetilde{\mathcal{L}} + \left(\widetilde{\nabla}^2 g(y^k) + B^\top HB\right)(\widetilde{y}^k - y^k) \right\|}{\|\widetilde{y}^k - y^k\|}. \tag{1.18}$$

Similarly to (1.15), we could get the KKT conditions of the trust region subproblem (1.13) as, for $\|y - y^k\| \leq \Gamma_k, \gamma_k \geq 0$.

Combining (1.15) and substituting (1.14), we get the following equations:

$$\begin{cases} \nabla f(x^k) + \widetilde{\nabla}^2 f(x^k)(\widetilde{x}^k - x^k) - A^\top \widetilde{\lambda}^k + \delta_k(\widetilde{x}^k - x^k) - HA^\top B(\widetilde{y}^k - y^k) = 0, \\ \nabla g(y^k) + \widetilde{\nabla}^2 g(y^k)(\widetilde{y}^k - y^k) - B^\top \widetilde{\lambda}^k + \gamma_k(\widetilde{y}^k - y^k) = 0, \\ (A\widetilde{x}^k + B\widetilde{y}^k - c) + H^{-1}(\widetilde{\lambda}^k - \lambda^k) = 0. \end{cases} \tag{1.19}$$

Note that by the mean value theorem for integrals, we have for some $\zeta_x^k = x^k + \theta_x^k(\widetilde{x}^k - x^k)$, $\zeta_y^k = y^k + \theta_y^k(\widetilde{y}^k - y^k)$, where $\theta_x^k, \theta_y^k \in (0, 1)$,

$$\begin{aligned} \nabla f(\widetilde{x}^k) - \nabla f(x^k) &= \nabla^2 f(\zeta_x^k)(\widetilde{x}^k - x^k), \\ \nabla g(\widetilde{y}^k) - \nabla g(y^k) &= \nabla^2 g(\zeta_y^k)(\widetilde{y}^k - y^k), \end{aligned} \tag{1.20}$$

Denote

$$\begin{aligned} R_x^k &:= \begin{cases} \widetilde{\nabla}^2 f(x^k) - \nabla^2 f(\zeta_x^k), & \text{if } \widetilde{x}^k \neq x^k, \\ 0, & \text{if } \widetilde{x}^k = x^k, \end{cases} \\ R_y^k &:= \begin{cases} \widetilde{\nabla}^2 g(y^k) - \nabla^2 g(\zeta_y^k), & \text{if } \widetilde{y}^k \neq y^k, \\ 0, & \text{if } \widetilde{y}^k = y^k, \end{cases} \end{aligned} \tag{1.21}$$

$$M_k := \begin{pmatrix} \delta_k I_n + R_x^k & & \\ & B^\top HB + \gamma_k I_m + R_y^k & \\ & & H^{-1} \end{pmatrix}, \tag{1.22}$$

and

$$N := \begin{pmatrix} 0 & -A^\top HB & \\ & -B^\top HB & \\ & & 0 \end{pmatrix}. \tag{1.23}$$

Then, the equations (1.19) can be rewritten as

$$F(\widetilde{w}^k) + (M_k + N)(\widetilde{w}^k - w^k) = 0. \tag{1.24}$$

Lemma 1.1 *Suppose $\widetilde{w}^k$ is generated by Algorithm 1 from an iterate w^k. And suppose $M_k \succeq 0$, then,*

$$\begin{aligned} (w^* - w^k)^\top M_k(\widetilde{w}^k - w^k) &\geq \|\widetilde{w}^k - w^k\|_{M_k}^2 + (\widetilde{\lambda}^k - \lambda^k)B(\widetilde{y}^k - y^k) \\ &\geq \frac{1}{2}\|\widetilde{w}^k - w^k\|_{M_k}^2 \end{aligned} \tag{1.25}$$

where matrices M_k is defined in (1.22).

Proof Since $w^* \in \Omega^*$ and the mapping F (1.10) is monotone, we have

$$(\widetilde{w}^k - w^*)^\top F(\widetilde{w}^k) \geq (\widetilde{w}^k - w^*)^\top F(w^*) = 0. \tag{1.26}$$

Thus, we have

$$\begin{aligned} &(w^* - w^k)^\top M_k(\widetilde{w}^k - w^k) \\ &= \|\widetilde{w}^k - w^k\|_{M_k}^2 + (w^* - \widetilde{w}^k)^\top M_k(\widetilde{w}^k - w^k) \end{aligned}$$

$$
\begin{aligned}
&\overset{(1.24)}{=} \|\widetilde{w}^k - w^k\|_{M_k}^2 + (\widetilde{w}^k - w^*)^\top F(\widetilde{w}^k) + (\widetilde{w}^k - w^*)^\top N(\widetilde{w}^k - w^k) \\
&\overset{(1.26)}{\geq} \|\widetilde{w}^k - w^k\|_{M_k}^2 + (\widetilde{w}^k - w^*)^\top N(\widetilde{w}^k - w^k).
\end{aligned}
$$

According to the definition (1.22) of N, we have

$$
\begin{aligned}
(\widetilde{w}^k - w^*)^\top N(\widetilde{w}^k - w^k) &= \begin{pmatrix} \widetilde{x}^k - x^* \\ \widetilde{y}^k - y^* \end{pmatrix}^\top \begin{pmatrix} -A^\top \\ -B^\top \end{pmatrix} HB(\widetilde{y}^k - y^k) \\
&= -(A\widetilde{x}^k + B\widetilde{y}^k - Ax^* - By^*)^\top HB(\widetilde{y}^k - y^k) \\
&= \left(-H(A\widetilde{x}^k + B\widetilde{y}^k - c)\right)^\top B(\widetilde{y}^k - y^k) \\
&\overset{(1.14)}{=} (\widetilde{\lambda}^k - \lambda^k)B(\widetilde{y}^k - y^k) \\
&\geq -\frac{1}{2}\|\widetilde{\lambda}^k - \lambda^k\|_{H^{-1}}^2 - \frac{1}{2}\|B(\widetilde{y}^k - y^k)\|_H^2 \\
&\geq -\frac{1}{2}\|\widetilde{w}^k - w^k\|_{M_k}^2.
\end{aligned}
$$

Here, $Ax^* + By^* = c$ is used in the second equality. And we prove the inequality (1.25). □

Remark 1.1 Note that if M_k is positive semi-definite, then we could use M_k to construct a merit function to characterize the decline after the prediction step, which is

$$
\begin{aligned}
&\|w^k - w^*\|_{M_k}^2 - \|\widetilde{w}^k - w^*\|_{M_k}^2 \\
&= \|w^k - w^*\|_{M_k}^2 - \|\widetilde{w}^k - w^k + w^k - w^*\|_{M_k}^2 \\
&= 2(w^* - w^k)^\top M_k(\widetilde{w}^k - w^k) - \|\widetilde{w}^k - w^k\|_{M_k}^2 \geq 0.
\end{aligned}
\tag{1.27}
$$

1.3.2 Prediction Step with Trust Region Radii Update

Lemma 1.1 shows that the prediction step is a descent direction of the merit function $\|w - w^*\|_{M_k}^2$. But we need to ensure $(w^* - w^k)^\top M_k(\widetilde{w}^k - w^k) \geq \|\widetilde{w}^k - w^k\|_{M_k}^2 > 0$ to guarantee the descent property, and we need $M_k \succeq 0$ to ensure the norm is valid.

By definition,

$$
\|\widetilde{w}^k - w^k\|_{M_k}^2 = \|\widetilde{x}^k - x^k\|_{\delta_k I_n + R_x^k}^2 + \|\widetilde{y}^k - y^k\|_{\gamma_k I_m + R_y^k + B^\top HB}^2 + \|A\widetilde{x}^k + B\widetilde{y}^k - c\|_H^2.
$$

To ensure $\|\widetilde{w}^k - w^k\|_{M_k}^2 > 0$, it suffices to demonstrate under the following conditions:

(i) Either $\widetilde{x}^k = x^k$ or $\delta_k I_n + R_x^k$ is positive definite;
(ii) Either $\widetilde{y}^k = y^k$ or $\gamma_k I_m + R_y^k + B^\top HB$ is positive definite;

(iii) Either $A\widetilde{x}^k + B\widetilde{y}^k - c = 0$ or H is positive definite;
(iv) Either $\widetilde{x}^k \neq x^k$, $\widetilde{y}^k \neq y^k$, or $A\widetilde{x}^k + B\widetilde{y}^k - c \neq 0$.

Since H is positive definite, the condition (iii) is always satisfied. We first study when the matrix $\delta_k I_n + R_x^k$ and $\gamma_k I_m + R_y^k + B^\top HB$ are positive definite.

Assumption 1.1 The Hessian matrix $\nabla^2 f(x)$ and $\nabla^2 g(y)$ are Lipschitz continuous, which means there exist positive constants L_f and L_g such that

$$\|\nabla^2 f(x) - \nabla^2 f(y)\| \leq L_f \|x - y\|, \quad \|\nabla^2 g(x) - \nabla^2 g(y)\| \leq L_g \|x - y\|. \tag{1.28}$$

Assumption 1.2 The BFGS-like matrix $\widetilde{\nabla}^2 f(x^k)$ and $\widetilde{\nabla}^2 g(y^k)$ are close to the Hessian matrix $\nabla^2 f(x)$ and $\nabla^2 g(y)$, respectively. Specifically, there exists a positive constant ε such that

$$\|\widetilde{\nabla}^2 f(x^k) - \nabla^2 f(x^k)\| \leq \varepsilon, \quad \|\widetilde{\nabla}^2 g(y^k) - \nabla^2 g(y^k)\| \leq \varepsilon. \tag{1.29}$$

Note that, by (1.20), and Lipschitz continuity of the Hessian matrix, and the error bound of the BFGS-like matrix, we have

$$\begin{aligned}
\|R_x^k\| &= \|\nabla^2 f(\zeta_x^k) - \widetilde{\nabla}^2 f(x^k)\| \leq L_f \|\zeta_x^k - x^k\| + \varepsilon \leq L_f \|\widetilde{x}^k - x^k\| + \varepsilon, \\
\|R_y^k\| &= \|\nabla^2 g(\zeta_y^k) - \widetilde{\nabla}^2 g(y^k)\| \leq L_g \|\zeta_y^k - y^k\| + \varepsilon \leq L_g \|\widetilde{y}^k - y^k\| + \varepsilon.
\end{aligned} \tag{1.30}$$

Now we can see if δ_k and γ_k are large enough, we could have the positive definiteness. And this can always be satisfied as long as the trust region radii Δ_k and Γ_k are small enough. We can see this through (1.16). By using the triangle inequality and induced norm, we have

$$\begin{aligned}
\delta_k &\geq \frac{\left|\|\nabla_x^k \widetilde{\mathcal{L}}\| - \|(\widetilde{\nabla}^2 f(x^k) + A^\top HA)(\widetilde{x}^k - x^k)\|\right|}{\|\widetilde{x}^k - x^k\|} \\
&\geq \frac{\|\nabla_x^k \widetilde{\mathcal{L}}\| - \|(\widetilde{\nabla}^2 f(x^k) + A^\top HA)(\widetilde{x}^k - x^k)\|}{\|\widetilde{x}^k - x^k\|} \\
&\geq \frac{\|\nabla_x^k \widetilde{\mathcal{L}}\|}{\|\widetilde{x}^k - x^k\|} - \|\widetilde{\nabla}^2 f(x^k) + A^\top HA\| \\
&\geq \frac{\|\nabla_x^k \widetilde{\mathcal{L}}\|}{\Delta_k} - \|\widetilde{\nabla}^2 f(x^k) + A^\top HA\|,
\end{aligned} \tag{1.31}$$

where the last inequality is due to the trust region constraint $\|\widetilde{x}^k - x^k\| \leq \Delta_k$. This lower bound of δ_k is not tight, but it gives us an approximate lower bound before we solve the trust region subproblem (1.12). Similarly, we also have a lower bound of γ_k, combine the two lower bounds, we have

$$\begin{aligned} \|\nabla_x^k \widetilde{\mathcal{L}}\|/\Delta_k - \|\widetilde{\nabla}^2 f(x^k) + A^\top H A\| &\leq \delta_k, \\ \|\nabla_y^k \widetilde{\mathcal{L}}\|/\Gamma_k - \|\widetilde{\nabla}^2 g(y^k) + B^\top H B\| &\leq \gamma_k. \end{aligned} \tag{1.32}$$

So, it suffices to have

$$\delta_k \geq \|\nabla_x^k \widetilde{\mathcal{L}}\|/\Delta_k - \|\widetilde{\nabla}^2 f(x^k) + A^\top H A\| \geq L_f \|\widetilde{x}^k - x^k\| + \varepsilon > 0,$$

to ensure $\delta_k I_n + R_x^k$ is positive definite. And since then $\delta_k > 0$, the constraint is active, $\|\widetilde{x}^k - x^k\| = \Delta_k$. We can then solve the above inequality to get the necessary condition of Δ_k. When

$$\Delta_k \leq \frac{\sqrt{(\|\widetilde{\nabla}^2 f(x^k) + A^\top H A\| + \varepsilon)^2 + 4L_f \|\nabla_x^k \widetilde{\mathcal{L}}\|} - (\|\widetilde{\nabla}^2 f(x^k) + A^\top H A\| + \varepsilon)}{2L_f}, \tag{1.33}$$

$\delta_k I_n + R_x^k$ is positive definite.

Note that, when $\|\nabla_x^k \widetilde{\mathcal{L}}\|$ approaches to zero, the valid range of Δ_k vanishes. And also note that, $\|\nabla_x^k \widetilde{\mathcal{L}}\| = 0$ means the subproblem (1.12) attains its minimum at x^k. So we could set a tolerance TOL_x to determine whether the subproblem (1.12) is solved to optimality. When $\|\nabla_x^k \widetilde{\mathcal{L}}\| \leq \mathrm{TOL}_x$, we set $\widetilde{x}^k = x^k$ and solve other subproblems or update λ. And if $\|\nabla_x^k \widetilde{\mathcal{L}}$, $\|\nabla_y^k \widetilde{\mathcal{L}}$, and $\|A\widetilde{x}^k + B\widetilde{y}^k - c\|$ are all small enough, then we find the saddle point of the Lagrangian function, and we stop the algorithm. Following this logic, we could have Algorithm 2 below.

Algorithm 2 (*Prediction Step with Trust Region Radii Update*)

1. Set $1 < \eta_1 < \eta_2$, tolerance $\mathrm{TOL}_x, \mathrm{TOL}_y, \mathrm{TOL}_\lambda > 0$. Calculate estimated Hessians $\widetilde{\nabla}^2 f(x^k)$ and $\widetilde{\nabla}^2 g(y^k)$ by a BFGS-like scheme satisfying (1.29) and calculate $\nabla_x^k \widetilde{\mathcal{L}}$ and $\nabla_y^k \widetilde{\mathcal{L}}$ by (1.16);
2. If $\|\nabla_x^k \widetilde{\mathcal{L}}\| \leq \mathrm{TOL}_x$, then set $\widetilde{x}^k = x^k$, $\Delta_{k+1} = \Delta_k$; otherwise, obtain $\widetilde{x}^k$ and update the trust region radii Δ_{k+1} through Step 2′;
3. If $\|\nabla_y^k \widetilde{\mathcal{L}}\| \leq \mathrm{TOL}_y$, then set $\widetilde{y}^k = y^k$, $\Gamma_{k+1} = \Gamma_k$; otherwise, obtain $\widetilde{y}^k$ and update the trust region radii Γ_{k+1} through Step 3′;
4. If $\widetilde{u}^k = u^k$ and $\|A\widetilde{x}^k + B\widetilde{y}^k - c\| \leq \mathrm{TOL}_\lambda$, then stop;

For the x-subproblem, if $\|\nabla_x^k \widetilde{\mathcal{L}}\| > \mathrm{TOL}_x$, we update Δ_{k+1} as follows:

Step 2′(x**-subproblem).** Set $t = 0$, and $\Delta_{k,0} := \Delta_k$.

(a) For given $\Delta_{k,t}$, complete a prediction step of the x-subproblem (1.12) to get $\widetilde{x}^{k,t}$ and evaluate $\delta_{k,t}$ by (1.17);
(b) If

$$\delta_{k,t} \geq L_f \|\widetilde{x}^{k,t} - x^k\| + \varepsilon := r_x^{k,t}, \tag{1.34}$$

then go to step (c), otherwise set $\Delta_{k,t+1} = \|\widetilde{x}^{k,t} - x^k\|/4$, $t := t + 1$ and then go back to step (a);
(c) Set $\widetilde{x}^k = \widetilde{x}^{k,t}$, and

$$\Delta_{k+1} = \begin{cases} \max\left\{\Delta_{k,t}, 1.5\|\widetilde{x}^{k,t} - x^k\|\right\}, & \delta_{k,t} \in [\eta_2 r_x^{k,t}, +\infty), \\ \Delta_{k,t}, & \delta_{k,t} \in [\eta_1 r_x^{k,t}, \eta_2 r_x^{k,t}], \\ \max\left\{0.5\Delta_{k,t}, 0.75\|\widetilde{x}^{k,t} - x^k\|\right\}, & \delta_{k,t} \in [r_x^{k,t}, \eta_1 r_x^{k,t}]. \end{cases} \tag{1.35}$$

Similarly, for the y-subproblem, if $\|\nabla_y^k \widetilde{\mathcal{L}}\| > \text{TOL}_y$, we update Γ_{k+1} as follows:

Step 3′(y**-subproblem).** Set $t = 0$, and $\Gamma_{k,0} := \Gamma_k$.

(a) For given $\Gamma_{k,t}$, complete a prediction step of the y-subproblem (1.13) to get $\widetilde{y}^{k,t}$ and evaluate $\gamma_{k,t}$ by (1.18);
(b) If

$$\gamma_{k,t} \geq L_g \|\widetilde{y}^{k,t} - y^k\| + \varepsilon := r_y^{k,t}, \tag{1.36}$$

then go to step (c), otherwise set $\Gamma_{k,t+1} = \|\widetilde{y}^{k,t} - y^k\|/4$, $t := t + 1$ and then go back to step (a);
(c) Set $\widetilde{y}^k = \widetilde{y}^{k,t}$, and

$$\Gamma_{k+1} = \begin{cases} \max\left\{\Gamma_{k,t}, 1.5\|\widetilde{y}^{k,t} - y^k\|\right\}, & \gamma_{k,t} \in [\eta_2 r_y^{k,t}, +\infty), \\ \Gamma_{k,t}, & \gamma_{k,t} \in [\eta_1 r_y^{k,t}, \eta_2 r_y^{k,t}], \\ \max\left\{0.5\Gamma_{k,t}, 0.75\|\widetilde{y}^{k,t} - y^k\|\right\}, & \gamma_{k,t} \in [r_y^{k,t}, \eta_1 r_y^{k,t}]. \end{cases} \tag{1.37}$$

Following the analysis above, we the lemma below to illustrate how we get $\|\widetilde{w}^k - w^k\|_{M_k}^2 > 0$ and $M_k \succeq 0$.

Lemma 1.2 *Suppose $\widetilde{w}^k$ is generated by Algorithm 2, which means for the x-subproblem, either $\|\nabla_x^k \widetilde{\mathcal{L}}\| \leq TOL_x$ or* (1.34) *holds; for the y-subproblem, either $\|\nabla_y^k \widetilde{\mathcal{L}}\| \leq TOL_y$ or* (1.36) *holds. Then one of the following conditions holds:*

(1) $\|\nabla_x^k \widetilde{\mathcal{L}}\| \le TOL_x$, $\|\nabla_y^k \widetilde{\mathcal{L}}\| \le TOL_y$, *and* $\|A\widetilde{x}^k + B\widetilde{y}^k - c\| \le TOL_\lambda$;
(2) *We have*

$$\|\widetilde{w}^k - w^k\|_{M_k}^2 > 0, \quad M_k \succeq 0. \tag{1.38}$$

Proof If $\|\nabla_x^k \widetilde{\mathcal{L}}\| \le \mathrm{TOL}_x$ and $\|\nabla_y^k \widetilde{\mathcal{L}}\| \le \mathrm{TOL}_y$, then we have $\widetilde{x}^k = x^k$ and $\widetilde{y}^k = y^k$. Then by the definition of $\|\widetilde{w}^k - w^k\|_{M_k}^2$, we have $\|\widetilde{w}^k - w^k\|_{M_k}^2 = \|\widetilde{\lambda} - \lambda^k\|_{H^{-1}}^2 = \|Ax^k + By^k - c\|^2$.

Thus, either $\|A\widetilde{x}^k + B\widetilde{y}^k - c\| \le \mathrm{TOL}_\lambda$ which means statement (1) holds or $\|\widetilde{w}^k - w^k\|_{M_k}^2 \ge \mathrm{TOL}_\lambda^2 > 0$. Since $\widetilde{x}^k = x^k$ and $\widetilde{y}^k = y^k$, we have $\zeta_x^k = x^k$ so that $R_x^k = R_y^k = 0$, which means $M_k \succeq 0$.

If $\|\nabla_x^k \widetilde{\mathcal{L}}\| > \mathrm{TOL}_x$ then

$$\|\widetilde{w}^k - w^k\|_{M_k}^2 \ge \|\widetilde{x}^k - x^k\|_{M_k}^2 \ge (\delta_k - \|R_x^k\|)\|\widetilde{x}^k - x^k\|^2.$$

By (1.30) and (1.34), we have $\|\widetilde{w}^k - w^k\|_{M_k}^2 > 0$ and $\|\delta_k I_n - R_x^k\| \ge 0$. Similarly, if $\|\nabla_y^k \widetilde{\mathcal{L}}\| > \mathrm{TOL}_y$, we also have $\|\widetilde{w}^k - w^k\|_{M_k}^2 > 0$ and $\|\gamma_k I_m - R_y^k\| \ge 0$. Combine the above analysis, we have the statement (2) holds.

Remark 1.2 When the condition (1) in Lemma 1.2 hold, we have $\widetilde{x}^k = x^k$, and $\widetilde{y}^k = y^k$. Thus the condition is equivalent to $\|F(w^k)\| \le \mathrm{TOL}$ with $\mathrm{TOL} = \sqrt{\mathrm{TOL}_x^2 + \mathrm{TOL}_y^2 + \mathrm{TOL}_\lambda^2}$. This means the algorithm stops when the optimality condition (1.11) is satisfied with tolerance TOL.

Next, we will show that given the tolerance, $\|M_k\|$ has an upper bound globally. We could prove it by giving an upper bound of both δ_k and γ_k. And this could be done by showing the lower bound of the trust region radii Δ_k and Γ_k. Note that we could get an upper bound of δ_k by the same way as we get the lower bound in (1.31):

$$\delta_k \le \frac{\|\nabla_x^k \widetilde{\mathcal{L}}\|}{\|\widetilde{x}^k - x^k\|} + \left\|\widetilde{\nabla}^2 f(x^k) + A^\top H A\right\|.$$

If $\|\widetilde{x}^k - x^k\| < \Delta_k$, which means the trust region constraint is inactive, we have $\delta_k = 0$, so that $\delta_k \le \left\|\nabla_x^k \widetilde{\mathcal{L}}\right\| / \Delta_k + \left\|\widetilde{\nabla}^2 f(x^k) + A^\top H A\right\| \ge 0$. And when $\|\widetilde{x}^k - x^k\| = \Delta_k$, this upper bound is also valid. Thus we have

$$\begin{aligned} \delta_k &\le \|\nabla_x^k \widetilde{\mathcal{L}}\| / \Delta_k + \|\widetilde{\nabla}^2 f(x^k) + A^\top H A\|, \\ \gamma_k &\le \|\nabla_y^k \widetilde{\mathcal{L}}\| / \Gamma_k + \|\widetilde{\nabla}^2 g(y^k) + B^\top H B\|. \end{aligned} \tag{1.39}$$

By Assumption 1.1, if the iteration points are bounded, we would have a uniform upper bound of the Hessian matrix $\widetilde{\nabla}^2 f(x^k)$, $\widetilde{\nabla}^2 g(y^k)$, and the gradients $\nabla_x^k \widetilde{\mathcal{L}}$, $\nabla_y^k \widetilde{\mathcal{L}}$. Then we could have a uniform upper bound of δ_k and γ_k in the following lemma.

Lemma 1.3 *Suppose $\widetilde{w}^k$ is generated by the prediction step with trust region radii update above from an iterate w^k. Assume $\|\widetilde{\nabla}^2 f(x^k)\| \le \Sigma_f$, and $\|\widetilde{\nabla}^2 g(y^k)\| \le \Sigma_g$, then we have*

$$\begin{aligned}\Delta_k > \Delta_{min} &:= 0.5 \cdot \frac{1}{4} \cdot \frac{\sqrt{(\Sigma_f+\varepsilon)^2 + 4L_f TOL_x} - (\Sigma_f + \varepsilon)}{2L_f},\\ \Gamma_k > \Gamma_{min} &:= 0.5 \cdot \frac{1}{4} \cdot \frac{\sqrt{(\Sigma_g+\varepsilon)^2 + 4L_g TOL_y} - (\Sigma_g + \varepsilon)}{2L_g}.\end{aligned} \tag{1.40}$$

And further assume $\|\nabla_x^k \widetilde{\mathcal{L}}\| \le \nabla_x^{\max}$, $\|\nabla_y^k \widetilde{\mathcal{L}}\| \le \nabla_y^{\max}$, then there exists a constant $\kappa_1 > 0$ such that for all $k \ge 0$, we have

$$\|M_k\| \le \kappa_1. \tag{1.41}$$

Proof According to (1.32), before we conduct a prediction step, in order to ensure (1.34) holds, it is sufficient to let Δ_k satisfy

$$L_f\|\widetilde{x}^k - x^k\| + \varepsilon = L_f\Delta_k + \varepsilon \le \|\nabla_x^k\widetilde{\mathcal{L}}\|/\Delta_k - \|\widetilde{\nabla}^2 f(x^k) + A^\top H A\| \le \delta_k.$$

So, as long as Δ_k satisfies

$$\Delta_k \ge \frac{\sqrt{(\|\widetilde{\nabla}^2 f(x^k) + A^\top HA\| + \varepsilon)^2 + 4L_f\|\nabla_x^k\widetilde{\mathcal{L}}\|} - (\|\widetilde{\nabla}^2 f(x^k) + A^\top HA\| + \varepsilon)}{2L_f},$$

Since $\|\widetilde{\nabla}^2 f(x^k)\| \le \Sigma_f$, and Δ_k only shrink when we update the trust region radii, which means $\|\nabla_x^k\widetilde{\mathcal{L}}\| > \mathrm{TOL}_x$, we have for all $k \ge 0$,

$$\Delta_k > \Delta_{\min} = 0.5 \cdot \frac{1}{4} \cdot \frac{\sqrt{(\Sigma_f+\varepsilon)^2 + 4L_f \mathrm{TOL}_x} - (\Sigma_f + \varepsilon)}{2L_f}.$$

The other inequality for Γ_k is similar.

Having the uniform lower bound of the trust region radii, the uniform upper bound of the Hessian matrix $\widetilde{\nabla}^2 f(x^k)$, $\widetilde{\nabla}^2 g(y^k)$, and the gradients $\nabla_x^k\widetilde{\mathcal{L}}$, $\nabla_y^k\widetilde{\mathcal{L}}$, we could obtain a uniform upper bound of δ_k and γ_k through (1.39) as

$$\begin{aligned}\delta_k &\le \nabla_x^{\max}/\Delta_{\min} + \Sigma_f + \varepsilon + \|A^\top HA\|,\\ \gamma_k &\le \nabla_y^{\max}/\Gamma_{\min} + \Sigma_g + \varepsilon + \|B^\top HB\|.\end{aligned} \tag{1.42}$$

Thus there exists $\kappa_1 > 0$ such that for all $k \ge 0$, we have (1.41).

Remark 1.3 The x-subproblem and y-subproblem are solved by trust region subproblems. The reason we use trust region subproblems is usually that the actual subproblems are hard to solve exactly. If one of the subproblems is easy to solve or even has a closed-form solution, we could use the exact solution to update the corresponding variable. In this case, we see the variable to be solved exactly as y-subproblem. Since the y-subproblem is solved exactly, we have $R_y^k = 0$ and $\gamma_k = 0$. Then we always have the y-part of M_k is $B^\top HB \succeq 0$.

As a result, this version of the Algorithm still satisfies all the Lemmas above. Specifically, (1.25), (1.38), and (1.41) still hold for this version of Algorithm 2 with Step 3″.

Step 3″(exact y-subproblem). Obtain $\widetilde{y}$ by solving the exact y-subproblem:

$$\widetilde{y}^k = \text{argmin}_{y \in \mathbb{R}^m} \mathcal{L}_H(\widetilde{x}^k, y, \lambda^k), \tag{1.43}$$

1.3.3 The Correction Step

Through Remark 1.1 and Lemma 1.2, we know that the direction achieved by the prediction step is a descent direction of the merit function $\|w - w^*\|^2_{M_k}$. However, M_k is not a constant matrix, so we need a correction step to handle this issue.

Let W be a symmetric positive definite matrix and $w^* \in \Omega^*$. Then, $\|w - w^*\|^2_{W^{-1}}$ could be viewed as a measurement for the iterate w being an optimal solution. Next, we show that $WM_k(\widetilde{w}^k - w^k)$ is a descent direction of the merit function $\|w - w^*\|^2_{W^{-1}}$ at an iterate w^k.

Before we proceed, it is worth noting that $WM_k(\widetilde{w}^k - w^k)$ is calculable although M_k is unknown. Specifically, we have

$$\begin{aligned}
&M_k(\widetilde{w}^k - w^k)\\
&= \begin{pmatrix} (\delta_k I_n + R_x^k)(\widetilde{x}^k - x^k) \\ (B^\top HB + \gamma_k I_m + R_y^k)(\widetilde{y}^k - y^k) \\ A\widetilde{x}^k + B\widetilde{y}^k - c \end{pmatrix}\\
&= \begin{pmatrix} \delta_k(\widetilde{x}^k - x^k) + \widetilde{\nabla}^2 f(x^k)(\widetilde{x}^k - x^k) - \nabla f(\widetilde{x}^k) + \nabla f(x^k) \\ \gamma_k(\widetilde{y}^k - y^k) + \left(\widetilde{\nabla}^2 g(y^k) + B^\top HB\right)(\widetilde{y}^k - y^k) - \nabla g(\widetilde{y}^k) + \nabla g(y^k) \\ A\widetilde{x}^k + B\widetilde{y}^k - c \end{pmatrix}.
\end{aligned} \tag{1.44}$$

So $WM_k(\widetilde{w}^k - w^k)$ is a well-defined descent direction. We also have terms like $\|\widetilde{w}^k - w^k\|^2_{M_k}$ be calculable. Next, we equip the descent direction with a suitable step length.

For convenience, we define

$$\varphi(\widetilde{w}^k, w^k) := \|\widetilde{w}^k - w^k\|^2_{M_k} + (\widetilde{\lambda}^k - \lambda^k)^\top B(\widetilde{y}^k - y^k). \tag{1.45}$$

To obtain the maximal decrease of the merit function $\|w - w^*\|^2_{W^{-1}}$, we define

$$w^{k+1}(\alpha) := w^k + \alpha W M_k(\widetilde{w}^k - w^k)$$

and the maximal following difference

$$\begin{aligned}
&\|w^k - w^*\|_{W^{-1}}^2 - \|w^{k+1}(\alpha) - w^*\|_{W^{-1}}^2 \\
&= \|w^k - w^*\|_{W^{-1}}^2 - \|w^k - w^* + \alpha W M_k(\widetilde{w}^k - w^k)\|_{W^{-1}}^2 \\
&= -2\alpha(w^k - w^*)^\top M_k(\widetilde{w}^k - w^k) - \alpha^2\|W M_k(\widetilde{w}^k - w^k)\|_{W^{-1}}^2 \\
&\geq 2\alpha\varphi(\widetilde{w}^k, w^k) - \alpha^2\|M_k(\widetilde{w}^k - w^k)\|_W^2 \\
&=: \Psi(\alpha),
\end{aligned}$$

It is worthwhile to note that the quadratic function $\Psi(\alpha)$ attains its maximum at

$$\alpha_k = \frac{\varphi(\widetilde{w}^k, w^k)}{\|M_k(\widetilde{w}^k - w^k)\|_W^2}, \tag{1.46}$$

Since $\|w^k - w^*\|^2 - \|w^{k+1}(\alpha) - w^*\|^2 \geq \Psi(\alpha)$, we choose a larger step size $\rho\alpha_k$, where $\rho = [1, 2)$. The resulting correction step is drawn as follows.

Algorithm 3 (*Correction step*) Obtain a prediction step $\widetilde{w}^k$ by Algorithm 2. Compute a step size α_k by (1.46) and generate the new iterate:

$$w^{k+1} := w^k + \rho\alpha_k W M_k(\widetilde{w}^k - w^k). \tag{1.47}$$

From this correction step, we show the decrease of the merit function $\|w - w^*\|_{W^{-1}}^2$ in the next lemma.

Lemma 1.4 *Suppose $\widetilde{w}_k$ and w^{k+1} are generated by Algorithms 2 and 3, respectively. Suppose Lemma 1.2 (2) holds. Then,*

$$\|w^{k+1} - w^*\|_{W^{-1}}^2 \leq \|w^k - w^*\|_{W^{-1}}^2 - \frac{2-\rho}{\rho}\|w^{k+1} - w^k\|_{W^{-1}}^2. \tag{1.48}$$

Proof By Lemma 1.2 (2), we have Lemma 1.1. By Lemma 1.1, (1.45), the definition (1.46) of α_k and the update rule (1.47) for w^{k+1}, we have

$$\begin{aligned}
&\|w^{k+1} - w^*\|_{W^{-1}}^2 - \|w^k - w^*\|_{W^{-1}}^2 \\
&\overset{(1.47)}{=} \|w^k - w^* + \rho\alpha_k W M_k(\widetilde{w}^k - w^k)\|_{W^{-1}}^2 - \|w^k - w^*\|_{W^{-1}}^2 \\
&= \rho^2\alpha_k^2\|W M_k(\widetilde{w}^k - w^k)\|_{W^{-1}}^2 + 2\rho\alpha_k(w^k - w^*)^\top M_k(\widetilde{w}^k - w^k) \\
&\overset{(1.25),(1.45)}{\leq} \rho^2\alpha_k^2\|M_k(\widetilde{w}^k - w^k)\|_W^2 - 2\rho\alpha_k\varphi(\widetilde{w}^k, w^k) \\
&\overset{(1.46)}{=} -(2-\rho)\rho\alpha_k^2\|M_k(\widetilde{w}^k - w^k)\|_W^2
\end{aligned}$$

$$= -(2-\rho)\rho\alpha_k^2\|WM_k(\widetilde{w}^k - w^k)\|_{W^{-1}}^2$$
$$\overset{(1.47)}{=} -(2-\rho)\rho^{-1}\|w^{k+1} - w^k\|_{W^{-1}}^2.$$

1.4 Convergence Analysis

To prove the global convergence of the trust region-based splitting method, we need an additional assumption.

According to (1.41), there exists a positive constant κ such that

$$\|WM_k\| \le \kappa_1\|W\| := \kappa, \qquad \forall\, k. \tag{1.49}$$

Along with Lemma 1.1, the step size α_k defined in (1.46) is obviously bounded away from zero:

$$\alpha_k = \frac{\varphi\left(\widetilde{w}^k, w^k\right)}{\left\|M_k\left(\widetilde{w}^k - w^k\right)\right\|_W^2} \ge \frac{\frac{1}{2}\left\|\widetilde{w}^k - w^k\right\|_{M_k}^2}{\left\|\widetilde{w}^k - w^k\right\|_{M_k^\top W M_k}^2} \ge \frac{1}{2\kappa} \quad \forall k. \tag{1.50}$$

1.4.1 Global Convergence

Now, we present the main convergence result.

Theorem 1.1 *Given an arbitrary initial iterate w^0 and let sequences $\{w^{k+1}\}$ be generated by Algorithm 3. Then, either we have Lemma 1.2 (1) holds some k or*

- $\lim_{k\to\infty}\|w^{k+1} - w^k\|_{W^{-1}} = 0$.
- $\lim_{k\to\infty}\|M_k(\widetilde{w}^k - w^k)\| = 0$.
- *Any accumulation point of $\{\widetilde{w}^k\}$ is a solution of VI (1.11).*
- *$\{w^k\}$ converges to a solution of VI (1.11).*

Proof Once the initial iterate w^0 is given, according to Lemma 1.2, either (1) holds or (2) holds for every k. If (1) holds for some k, then the theorem is proved. Otherwise, we have (2) holds for every k, then by Lemma 1.4, the sequence of merit function $\{\|w^k - w^*\|_{W^{-1}}^2\}$ is decrease. So $\{w^k\}$ is bounded above and $\|w^{k+1} - w^k\|_{W^{-1}}$ must be vanish as $k \to \infty$.

By the update rule (1.47) of w^{k+1} and the low bound (1.50) of the step size α_k, we get the second assertion.

Since the merit function $\{\|w^k - w^*\|_{W^{-1}}^2\}$ is decreasing, $\{w^k\}$ is bounded, so Lemma 1.3 holds. Thus we have (1.41) and then the lower bound (1.50) of α_k. Combining with Lemma 1.1, (1.47), and the definitions (1.22) of matrices M_k and N, we have

$$\begin{aligned}\left\|M_k\left(\widetilde{w}^k - w^k\right)\right\|_W^2 &= \alpha_k \cdot \varphi\left(\widetilde{w}^k, w^k\right)\\ &\geq \frac{1}{2\kappa}\cdot\frac{1}{2}\left\|\widetilde{w}^k - w^k\right\|_{M_k}^2\\ &\geq \frac{1}{4\kappa}\left\|B\left(\widetilde{y}^k - y^k\right)\right\|_H^2.\end{aligned}$$

Since $\|M_k(\widetilde{w}^k - w^k)\|_W \to 0$ as $k \to \infty$, we get $\|B(\widetilde{y}^k - y^k)\|_H \to 0$. So $\|N(\widetilde{w}^k - w^k)\| \to 0$. Recalling the important inequality (1.24), we obtain

$$\|F(\widetilde{w}^k)\| = \|\left(M_k + N\right)\left(\widetilde{w}^k - w^k\right)\| \to 0.$$

So, the third assertion is also valid.

Since $\{w^k\}$ is bounded, it has at least one accumulation point, which is also an accumulation point of $\{\widetilde{w}^k\}$; hence solution point of VI (1.11). The inequality (1.48) implies that $\{w^k\}$ has just one accumulation point, and the last assertion follows immediately. □

1.4.2 Convergence Rate

In this section, we show that the proposed trust region-based splitting method enjoys the $O(1/\epsilon)$ convergence rate in an ergodic sense.

Theorem 1.2 ([Theorem 2.1 in *[17]*) *The optimal solution set* Ω^* *of the VI (1.11) is convex and can be characterized as*

$$\Omega^* = \bigcap_{w\in\Omega}\left\{\widetilde{w} \in \Omega \mid (w - \widetilde{w})^\top F(w) \geq 0\right\}. \tag{1.51}$$

According to this theorem, we say that $\overline{w}_t$ is an ϵ-approximate optimal solution if it satisfies

$$(w - \overline{w}_t)^\top F(w) \geq -\epsilon \qquad \forall\, w \in \Omega, \tag{1.52}$$

where ϵ is a small positive number. To obtain an ϵ, e-approximate optimal solution, we reveal that the worst-case iterative number t satisfies $t = O(1/\epsilon)$.

Next, we introduce a new vector

$$\widehat{w}^k = \begin{pmatrix}\widehat{x}^k\\ \widehat{y}^k\\ \widehat{\lambda}^k\end{pmatrix} := \begin{pmatrix}\widetilde{x}^k\\ \widetilde{y}^k\\ \lambda^k - H(A\widetilde{x}^k + By^k - c)\end{pmatrix}, \tag{1.53}$$

and a new matrix

$$\widehat{M}_k := \begin{pmatrix} \delta_k I_n + R_x^k & & \\ & B^\top HB + \gamma_k I_m + R_y^k & \\ & -B & H^{-1} \end{pmatrix}, \tag{1.54}$$

that will be used in this section. By some calculation, we obtain the following Lemma.

Lemma 1.5 *According to the new notations (1.53) and (1.54), the prediction step produces* $\widehat{w}^k \in \Omega$ *satisfying*

$$(w - \widehat{w}^k)^\top [F(\widehat{w}^k) + \widehat{M}_k(\widehat{w}^k - w^k)] \geq 0, \quad \forall\, w \in \Omega. \tag{1.55}$$

Moreover, the following equalities hold

$$M_k(\widetilde{w}^k - w^k) = \widehat{M}_k(\widehat{w}^k - w^k), \tag{1.56}$$

and

$$\varphi(\widetilde{w}^k, w^k) = (\widehat{w}^k - w^k)^\top \widehat{M}_k^\top (\widehat{w}^k - w^k). \tag{1.57}$$

Proof Based on the new vector $\widehat{w}^k$, (1.19) could be rewritten as

$$\begin{cases} \nabla f(\widehat{x}^k) + \widetilde{\nabla}^2 f(x^k)(\widehat{x}^k - x^k) - A^\top \widehat{\lambda}^k + \delta_k(\widehat{x}^k - x^k) = 0, \\ \nabla g(\widehat{y}^k) + \widetilde{\nabla}^2 g(y^k)(\widehat{y}^k - y^k) - B^\top \widehat{\lambda}^k + (B^\top HB + \gamma_k I_m)(\widehat{y}^k - y^k) = 0, \\ (A\widehat{x}^k + B\widehat{y}^k - c) - B(\widehat{y}^k - y^k) + H^{-1}(\widehat{\lambda}^k - \lambda^k) = 0. \end{cases}$$

In a compact form, the above VI equals to (1.55).

Since $H^{-1}(\widetilde{\lambda}^k - \lambda^k) = -(A\widetilde{x}^k + B\widetilde{y}^k - c) = H^{-1}(\widehat{\lambda}^k - \lambda^k) - B(\widehat{y}^k - y^k)$, the equation (1.56) hold.

From the definition of $\varphi(\widetilde{w}^k, w^k)$ in (1.45), the equality (1.56) and the definitions (1.22) and (1.54) of matrices M_k and $\widehat{M}_k$, we have

$$\begin{aligned} \varphi(\widetilde{w}^k, w^k) &= \|\widetilde{w}^k - w^k\|_{M_k}^2 + (\widetilde{\lambda}^k - \lambda^k)^\top B(\widetilde{y}^k - y^k) \\ &= (\widetilde{w}^k - w^k)^\top \widehat{M}_k(\widehat{w}^k - w^k) + (\widetilde{\lambda}^k - \lambda^k)^\top B(\widetilde{y}^k - y^k) \\ &= (\widetilde{w}^k - w^k)^\top M_k(\widehat{w}^k - w^k) \\ &= (\widehat{w}^k - w^k)^\top M_k(\widetilde{w}^k - w^k) \\ &= (\widehat{w}^k - w^k)^\top \widehat{M}_k(\widehat{w}^k - w^k). \end{aligned}$$

The proof is complete.

Lemma 1.6 *Suppose the sequence* $\{\widehat{w}^k\}$ *is generated by the trust region-based splitting method. Then, we get*

$$(w - \widehat{w}^k)^\top F(w) \geq (\widehat{w}^k - w)^\top \widehat{M}(\widehat{w}^k - w^k), \tag{1.58}$$

Proof Since the mapping F (1.10) is monotone, we have

$$(w-\widehat{w}^k)^\top F(w)-(w-\widehat{w}^k)^\top F(\widehat{w}^k)\geq 0.$$

From the VI (1.55) in Lemma 1.5, we get

$$(w-\widehat{w}^k)^\top F(\widehat{w}^k)\geq (\widehat{w}^k-w)^\top \widehat{M}_k(\widehat{w}^k-w^k)$$

Adding the above two inequalities, we obtain the validation of this lemma.

Lemma 1.7 *Suppose the sequences* $\{\widehat{w}^k\}$ *and* $\{w^{k+1}\}$ *are generated by the prediction and correction steps, respectively. Then, we have*

$$(\widehat{w}^k-w)^\top W^{-1}(w^{k+1}-w^k)\geq \frac{1}{2}\left(\|w-w^{k+1}\|^2_{W^{-1}}-\|w-w^k\|^2_{W^{-1}}\right), \quad (1.59)$$

Proof By some calculation, we have

$$\begin{aligned}&(\widehat{w}^k-w)^\top W^{-1}(w^{k+1}-w^k)\\&=\frac{1}{2}\left(\|w-w^{k+1}\|^2_{W^{-1}}-\|w-w^k\|^2_{W^{-1}}\right)+\frac{1}{2}\left(\|w^k-\widehat{w}^k\|^2_{W^{-1}}-\|w^{k+1}-\widehat{w}^k\|^2_{W^{-1}}\right).\end{aligned}$$

Hence, the rest of the proof is to show that the second term is nonnegative. By the update rule (1.47) of w^{k+1}, the definition (1.46) of α_k, and Lemma 1.5, we get the following equality.

$$\begin{aligned}\left(w^{k+1}-w^k\right)^\top W^{-1}\left(w^k-\widehat{w}^k\right)&=\left[\rho\alpha_k WM_k\left(\widetilde{w}^k-w^k\right)\right]^\top W^{-1}\left(w^k-\widehat{w}^k\right)\\&=-\rho\alpha_k\left(\widehat{w}^k-w^k\right)^\top \widehat{M}_k^\top\left(\widehat{w}^k-w^k\right)\\&=-\rho\alpha_k\varphi\left(\widetilde{w}^k,w^k\right)\\&=-\rho\alpha_k^2\left\|M_k\left(\widetilde{w}^k-w^k\right)\right\|^2_W\\&=-\rho^{-1}\left\|w^{k+1}-w^k\right\|^2_{W^{-1}}.\end{aligned}$$

So, the second term is

$$\begin{aligned}&\|w^{k+1}-\widehat{w}^k\|^2_{W^{-1}}-\|w^k-\widehat{w}^k\|^2_{W^{-1}}\\&=\|w^{k+1}-w^k\|^2_{W^{-1}}+2(w^{k+1}-w^k)^\top W^{-1}(w^k-\widehat{w}^k)\\&=\|w^{k+1}-w^k\|^2_{W^{-1}}-2\rho^{-1}\left\|w^{k+1}-w^k\right\|^2_{W^{-1}}\\&=\frac{\rho-2}{\rho}\|w^{k+1}-w^k\|^2_{W^{-1}}\geq 0.\end{aligned}$$

Therefore, this lemma is proved.

Finally, we give the main theorem.

Theorem 1.3 *For any positive integer t, we define*

$$\overline{w}_t = \frac{\sum_{k=0}^{t} \alpha_k \widehat{w}^k}{\sum_{k=0}^{t} \alpha_k}. \tag{1.60}$$

Then, $\overline{w}_t \in \Omega$ *and*

$$(\overline{w}_t - w)^\top F(w) \le \frac{\kappa}{\rho(t+1)} \|w - w^0\|_{W^{-1}}^2, \tag{1.61}$$

which means that the proposed trust region-based splitting method enjoys the $O(1/\epsilon)$ *convergence rate in an ergodic sense.*

Proof Since $\overline{w}_t$ is a convex combination of iterates $\{\widehat{w}^0, \widehat{w}^1, \ldots, \widehat{w}^t\}$ and the set Ω is convex, we get $\overline{w}_t \in \Omega$.

From Lemmas 1.5, 1.6 and 1.7, we get for all $w \in \Omega$,

$$\begin{aligned} \rho\alpha_k \left((w - \widehat{w}^k)^\top F(w)\right) &\ge (\widehat{w}^k - w)^\top \left(\rho\alpha_k \widehat{M}_k(\widehat{w}^k - w^k)\right) \\ &= (\widehat{w}^k - w)^\top W^{-1}(w^{k+1} - w^k) \\ &\ge \frac{1}{2}\left(\|w - w^{k+1}\|_{W^{-1}}^2 - \|w - w^k\|_{W^{-1}}^2\right) \end{aligned}$$

Then, we sum over the above inequalities for $k = 0, 1, \ldots, t$ and divide $\rho\sum_{k=0}^{t}\alpha_k$,

$$\begin{aligned} \left(w - \frac{\sum_{k=0}^{t}\alpha_k \widehat{w}^k}{\sum_{k=0}^{t}\alpha_k}\right)^\top F(w) &\ge \frac{1}{2\rho\sum_{k=0}^{t}\alpha_k}\left(\|w - w^{t+1}\|_{W^{-1}}^2 - \|w - w^0\|_{W^{-1}}^2\right) \\ &\ge -\frac{1}{2\rho\sum_{k=0}^{t}\alpha_k}\|w - w^0\|_{W^{-1}}^2. \end{aligned}$$

Furthermore, we note that $\alpha_k \ge \frac{1}{2\kappa}$ by (1.50), so we have $2\sum_{k=0}^{t}\alpha_k \ge \frac{1}{\kappa}(t+1)$. Hence, we conclude that

$$(w - \overline{w}_t)^\top F(w) \ge -\frac{\kappa}{\rho(t+1)}\|w - w^0\|_{W^{-1}}^2, \qquad \forall\, w \in \Omega.$$

The proof is complete.

Therefore, for any given initial iterate $w^0 \in \Omega$ and any nonempty compact set $\mathcal{D} \in \Omega$, we denote $d := \max\{\|w - w^0\|_{W^{-1}} \mid w \in \mathcal{D}\}$. Then, for any small positive number ϵ, after

$$t = \left\lfloor \frac{\kappa d^2}{\rho\epsilon} \right\rfloor$$

iterations, the trust region-based splitting method produces a point $\overline{w}_t$ (1.60) such that

$$\sup_{w \in \mathcal{D}} \left\{ (\overline{w}_t - w)^\top F(w) \right\} \le \epsilon^{-1}.$$

In this viewpoint, we approve that the trust region-based splitting method enjoys the $O(1/\epsilon)$ convergence rate in an ergodic sense.

1.5 Numerical Experiments

In this section, we apply the proposed trust region-based splitting method to image recovery problems and logistic regression problems, and compare it with some existing splitting methods.

1.5.1 Constrained TV-ℓ^2 Image Recovery Problems

The image recovery problem is a sort of inverse problem in image analysis. Our interest is in finding the true image z from its degraded observation z_0. The degradation comes from two facts. The first one is blur which is created by an improper lens adjustment or a movement. The other one is random noise b that is assumed to be Gaussian, white, and additive. The linear model of the degradation process is

$$z_0 = \Phi z + b,$$

where $z_0, z, b \in \mathbb{R}^n$, $\Phi \in \mathbb{R}^{n \times n}$ is a linear blur map and n is the number of pixels.

Next, we give the definition of the total variation (TV) of a two-dimensional (2D) image [20]. For a given image z, its discrete derivative ∇z in horizontal and vertical directions are denoted by $\partial_1 z$ and $\partial_2 z$, respectively. Then,

$$\nabla z := \begin{pmatrix} \partial_1 z \\ \partial_2 z \end{pmatrix} \in \mathbb{R}^{2n}. \tag{1.62}$$

When the circular boundary conditions are assumed, these two partial derivatives could be efficiently computed using the 2D discrete Fourier transformation $\mathcal{F}$:

$$\partial_1 z = \mathcal{F}^{-1} D_1 \mathcal{F} z, \quad \text{and} \quad \partial_2 z = \mathcal{F}^{-1} D_2 \mathcal{F} z,$$

where $D_1, D_2 \in \mathbb{C}^{n \times n}$ are two diagonal matrices. Moreover, the blur map is also diagonalized as $\Phi = \mathcal{F}^{-1} K \mathcal{F}$, where $K \in \mathbb{C}^{n \times n}$ is the blur kernel. Then, the TV of z is defined as:

$$\| |\nabla z| \|_1 := \sum_{i=1}^{n} \sqrt{[\partial_1 z]_i^2 + [\partial_2 z]_i^2} \in \mathbb{R}^n. \tag{1.63}$$

In this experiment, we consider the constrained TV-ℓ^2 model for image recovery which is studied in [18]:

$$\begin{cases} \min & \||\nabla z|\|_1 \\ \text{s.t.} & \|\Phi z - z_0\|^2 \le \sigma, \end{cases} \tag{1.64}$$

where σ denotes a noise level which is known in advance. To apply the proposed trust region-based splitting algorithm, we introduce two auxiliary variables $s = \nabla z$, $t = \Phi z - z_0$ and an indicator function of a ball $\odot = \{t \mid \|t\|^2 \le \sigma\}$

$$\pi_{\odot}(t) = \begin{cases} 0 & \text{if } \|t\|^2 \le \sigma, \\ +\infty & \text{otherwise.} \end{cases}$$

Then, the constrained TV-ℓ^2 model (1.64) is equivalent to

$$\begin{cases} \min & \||s|\|_1 + \pi_{\odot}(t) \\ \text{s.t.} & \nabla z - s = 0,\ s \in \mathbb{R}^{2n}, \\ & \Phi z - t = z_0,\ t \in \mathbb{R}^n. \end{cases}$$

Let

$$x = z, \quad y = \begin{pmatrix} s \\ t \end{pmatrix}, \quad \lambda = \begin{pmatrix} \mu \\ \nu \end{pmatrix} \in \mathbb{R}^{2n} \times \mathbb{R}^n,$$
$$f(x) = 0, \quad g(y) = \||s|\|_1 + \pi_{\odot}(t),$$
$$A = \begin{pmatrix} \nabla \\ \Phi \end{pmatrix}, \quad B = \begin{pmatrix} -I_{2n} & 0 \\ 0 & -I_n \end{pmatrix}, \quad c = \begin{pmatrix} 0 \\ z_0 \end{pmatrix}, \quad H = \begin{pmatrix} \xi I_{2n} & 0 \\ 0 & \eta I_n \end{pmatrix}.$$

The augmented Lagrangian function is represented as follows:

$$\begin{aligned} \mathcal{L}_{(\xi,\eta)}(s, t, z, \mu, \nu) = &\ \||s|\|_1 + \pi_{\odot}(t) - \mu^\top(\nabla z - s) - \nu^\top(\Phi z - t - z_0) \\ &+ \frac{\xi}{2}\|\nabla z - s\|^2 + \frac{\eta}{2}\|\Phi z - t - z_0\|^2. \end{aligned}$$

whereafter, we give the initial image $z^0 = z_0$, initial auxiliary variables $s = 0$ and $t = 0$, initial multiplier $\mu^0 = 0$ and $\nu^0 = 0$ and begin the following loops with $k \leftarrow 0$.

- Update the variable $x = z$ in a trust region $\|z - z^k\| \le \Delta_k$.

$$\begin{aligned} \tilde{z}^k &= \text{argmin}_z \left\{ -(\mu^k)^\top \nabla z - (\nu^k)^\top \Phi z + \frac{\xi}{2}\|\nabla z - s^k\|^2 + \frac{\eta}{2}\|\Phi z - t^k - z_0\|^2 \right\} \\ &\quad \text{s.t. } \|z - z^k\| \le \Delta_k \\ &= \text{argmin}_z \left\{ \frac{\xi}{2}\|\nabla z - s^k - \xi^{-1}\mu^k\|^2 + \frac{\eta}{2}\|\Phi z - t^k - z_0 - \eta^{-1}\nu^k\|^2 \right\} \\ &\quad \text{s.t. } \|z - z^k\| \le \Delta_k. \end{aligned}$$

Let $d := z - z^k$. Then we turn to minimize d

$$\begin{cases} \operatorname{argmin}_d \ \frac{\xi}{2}\|\nabla d + \nabla z^k - s^k - \xi^{-1}\mu^k\|^2 + \frac{\eta}{2}\|\Phi d + \Phi z^k - t^k - z_0 - \eta^{-1}\nu^k\|^2 \\ \text{s.t.} \ \ \|d\| \le \Delta_k. \end{cases} \tag{1.65}$$

When the 2D discrete Fourier transformation is employed, we define $\widehat{d} := \mathcal{F}d$. Then, the trust region constraint $\|d\| \le \Delta_k$ is equivalent to $\|\widehat{d}\| \le \sqrt{n}\Delta_k$. Furthermore, let

$$\begin{pmatrix} p \\ q \end{pmatrix} := \nabla z^k - s^k - \xi^{-1}\mu^k, \qquad \text{and} \qquad r := \Phi z^k - t^k - z_0 - \eta^{-1}\nu^k,$$

where $p, q, r \in \mathbb{R}^n$. And we define $\widehat{p} := \mathcal{F}(p)$, $\widehat{q} := \mathcal{F}(q)$, $\widehat{r} := \mathcal{F}(r)$. Then, solving the optimization problem (1.65) is equivalent to minimizing $\widehat{d}$

$$\begin{cases} \operatorname{argmin}_{\widehat{d}} \ \frac{\xi}{2}\left(\|D_1\widehat{d} + \widehat{p}\|^2 + \|D_2\widehat{d} + \widehat{q}\|^2\right) + \frac{\eta}{2}\|H\widehat{d} + \widehat{r}\|^2 \\ \text{s.t.} \ \ \|\widehat{d}\| \le \sqrt{n}\Delta_k. \end{cases} \tag{1.66}$$

This trust region subproblem, whose Hessian is diagonal, could be solved by a dozen of classical algorithm, and the corresponding multiplier δ_k is estimated simultaneously. Here, we prefer to use a nearly exact solver which is introduced in the famous book [19, Algorithm 4.4].

After the solution of the trust region subproblem (1.66) is obtained, we compute

$$\widetilde{z}^k = z^k + \mathcal{F}^{-1}\widehat{d}.$$

- Update $y = (s, t)$. Since variables s and t are not coupled. we could update them separately. For the simplicity of the computation, we set the trust region radius Γ_k of this subproblem is large enough. So, the trust region constraints for variables s and t are inactive and the corresponding multiplier $\gamma_k = 0$.
  First, we consider the variable s.

$$\begin{aligned} \widetilde{s}^k &= \operatorname{argmin}_s \left\{ \||s|\|_1 + (\mu^k)^\top s + \frac{\xi}{2}\|\nabla\widetilde{z}^k - s\|^2 \right\} \\ &= \operatorname{argmin}_s \left\{ \||s|\|_1 + \frac{\xi}{2}\|s - (\nabla\widetilde{z}^k - \xi^{-1}\mu^k)\|^2 \right\}. \end{aligned} \tag{1.67}$$

According to the definition (1.63) of TV, the above optimization problem is separable. Recalling that the auxiliary variable $s(= \nabla z) \in \mathbb{R}^{2n}$ and the special structure (1.62) of ∇z, we correspondingly rewrite

$$s =: \begin{pmatrix} s_1 \\ s_2 \end{pmatrix} \qquad \text{and} \qquad \nabla\widetilde{z}^k - \xi^{-1}\mu^k =: \begin{pmatrix} \upsilon_1 \\ \upsilon_2 \end{pmatrix},$$

where $s_1, s_2, \upsilon_1, \upsilon_2 \in \mathbb{R}^n$. Let

$$\check{s}^i := \begin{pmatrix} [s_1]_i \\ [s_2]_i \end{pmatrix} \in \mathbb{R}^2 \quad \text{and} \quad \check{\upsilon}^i := \begin{pmatrix} [\upsilon_1]_i \\ [\upsilon_2]_i \end{pmatrix} \in \mathbb{R}^2.$$

Then, the separable optimization problem (1.67) is equivalent to the following small problems:

$$\operatorname{argmin}_{\check{s}^i} \left\{ \|\check{s}^i\| + \frac{\xi}{2}\|\check{s}^i - \check{\upsilon}^i\|^2 \right\}, \qquad \text{for } i = 1, \ldots, n.$$

The closed-form solution of this 2D optimization is given by the 2D shrinkage formula [23]:

$$\check{s}^i = \max\left\{ \|\check{\upsilon}^i\| - \frac{1}{\xi}, 0 \right\} \frac{\check{\upsilon}^i}{\|\check{\upsilon}^i\|}, \qquad \text{for } i = 1, \ldots, n,$$

where $0 \cdot (0/0) = 0$ is followed. So $\widetilde{s}^k$ is obtained cheaply.
Second, we turn to the variable t.

$$\begin{aligned} \widetilde{t}^k &= \operatorname{argmin}_t \left\{ \pi_{\odot}(t) + (\nu^k)^\top t + \frac{\eta}{2}\|\Phi\widetilde{z}^k - t - z_0\|^2 \right\} \\ &= \operatorname{argmin}_t \left\{ \pi_{\odot}(t) + \frac{\eta}{2}\|t - (\Phi\widetilde{z}^k - z_0 - \eta^{-1}\nu^k)\|^2 \right\} \\ &= \mathcal{P}_{\odot}(\Phi\widetilde{z}^k - z_0 - \eta^{-1}\nu^k), \end{aligned}$$

where $\mathcal{P}_{\odot}(\cdot)$ is the projective operator onto the ball $\odot$:

$$\mathcal{P}_{\odot}(a) = \begin{cases} a & \text{if } \|a\|^2 \le \sigma, \\ \dfrac{\sqrt{\sigma}}{\|a\|} a & \text{otherwise.} \end{cases}$$

- Update the multiplier $\lambda = (\mu, \nu)$.

$$\begin{aligned} \widetilde{\mu}^k &= \mu^k - \xi(\nabla\widetilde{z}^k - \widetilde{s}^k), \\ \widetilde{\nu}^k &= \nu^k - \eta(\Phi\widetilde{z}^k - \widetilde{t}^k - z_0). \end{aligned}$$

- The correction step is straightforward.

Then, we apply the novel trust region-based splitting method (TRSP) to a T1-weighted magnetic resonance image. Figure 1.1a, which is the true image, shows a horizontal slice of a human brain and Fig. 1.1b illustrates its gradient map

$$\left(\sqrt{[\partial_1 z]_i^2 + [\partial_2 z]_i^2}\right)_{i=1,2,\ldots,n}.$$

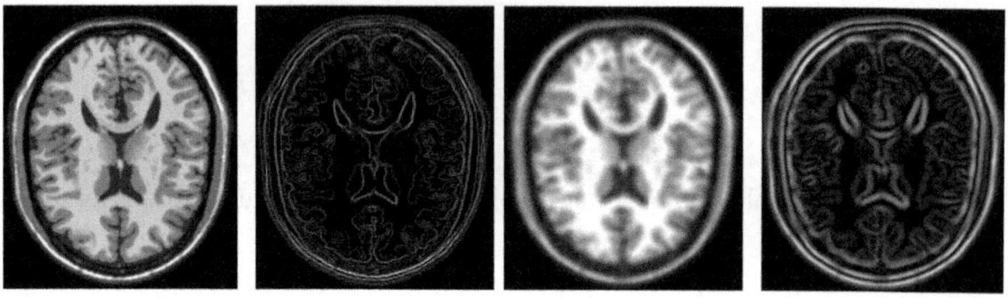

(a) (b) (c) (d)

Fig. 1.1 A T1-weighted magnetic resonance image on a horizontal slice of a human brain. The true image (**a**) and its gradient map (**b**). The blurry and noisy image (**c**) and its gradient map (**d**)

A blurry and noisy image and its gradient are shown in Fig. 1.1c and d. Obviously, the degraded image has thick and blurry edges. The TV-based model is an efficient way to impose sharp edges.

Here, we compare the novel TRSP with the basic ADMM (bADM) and the descent ADMM (dADM) [26]. In the first 50 iterations of TRSP, we allow parameters ξ and η in the penalty matrix adjust adaptively [8, 16]. All these algorithms start from the same initial points, and stop when the relative improvement of the objective TV of the restored image is small enough:

$$\frac{\left| \| |\tilde{z}^k| \| - \| |\tilde{z}^{k-1}| \| \right|}{\| |\tilde{z}^k| \|} \leq 10^{-5}.$$

To measure the quality of restored images, we employ the scalar measurement named the signal-to-noise ratio (SNR):

$$\text{SNR} = 10 \log_{10} \frac{\|z_* - \bar{z}_*\|^2}{\|z_* - z\|^2},$$

where z_* is the true image and $\bar{z}_*$ is its mean intensity value. We draw the SNRs versus iterations for bADM, dADM, and TRSP in Fig. 1.2. While the SNR of images obtained by bADM and dADM are respectively 26.5 and 27.2, TRSP returns an image with largest SNR$=$ 27.5 which means that a higher quality image is reconstructed. We illustrate restored images and corresponding gradient maps in Fig. 1.3. However, all the restored images have little difference in visualization.

Finally, we turn to compare the convergence rate of bADM, dADM, and TRSP. The detailed results on the number of iterations and CPU times are shown in Table 1.1. Compared with the basic ADM, the proposed TRSP saves about 59% iterations and 44% CPU times. These results indicate that the trust region-based splitting algorithm is efficient and promising.

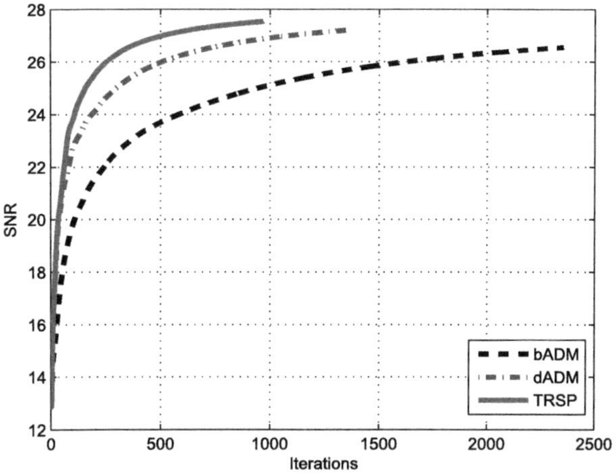

Fig. 1.2 Comparison of SNRs for bADM, dADM and TRSP

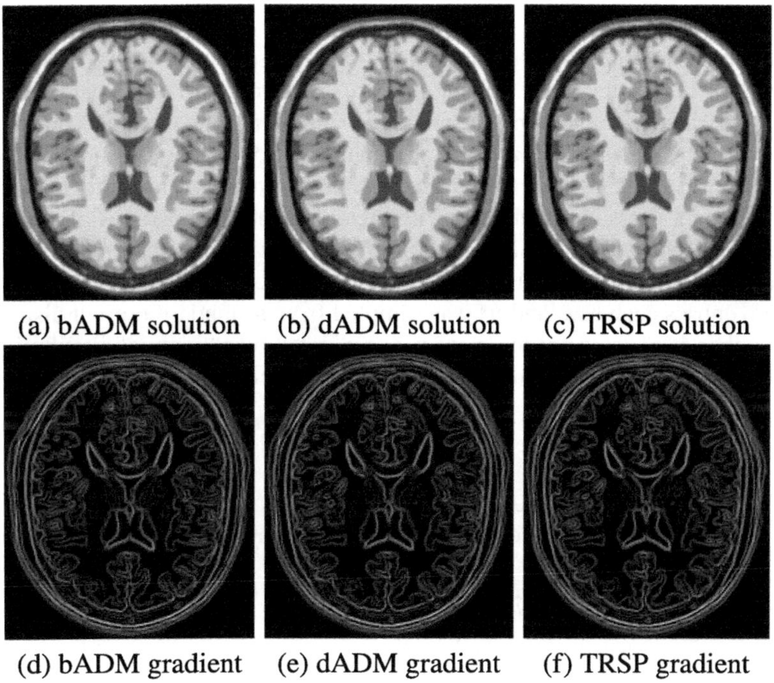

Fig. 1.3 Restored images and their gradient maps

Table 1.1 Iterations and CPU times (second) for the image recovery problem

	bADM	dADM	TRSP
Iterations	2352	1351	969
CPU times	44.7	31.8	24.9

1.5.2 ℓ_1 Regularized Logistic Regression

The image recovery problem has a quadratic objective function, which lacks analysis between TRSP, linearized ADMM (lADM), and majorized ADMM (mADM). In this section, we consider a ℓ_1 regularized logistic regression problem for binary classification in [1, p. 92, (11.1)]. The problem is

$$\min_{w\in\mathbb{R}^n, v\in\mathbb{R}} \sum_{i=1}^{m} \log\left(1+\exp(-b_i(a_i^T w+v))\right)+\mu\|w\|_1, \tag{1.68}$$

with the training set consists of m pairs (a_i, b_i), where $a_i \in \mathbb{R}^n$ is the feature vector and $b_i \in \{-1, 1\}$ is the label. The parameter μ is the regularization parameter.

We generated a problem instance with $m = 2000$ training examples and $n = 200$ features. Each feature vector a_i has a random number of nonzero features which were drawn from a standard normal distribution. The number of nonzero features was generated by a Poisson distribution with mean 40, and the positions of the nonzero features were chosen uniformly at random. We pick a random vector w^* and a random scalar v^* to be the true solution. $w^* \in \mathbb{R}^n$ was generated to have approximately 40 nonzero entries each row, and v^* was sampled from a standard normal distribution. The number of nonzero entries in w^* was generated by a binomial distribution with probability 0.2 with a uniformly random position. The labels b_i were generated by

$$b_i = \operatorname{sign}(a_i^T w^* + v^* + \epsilon_i),$$

where ϵ_i is a random variable drawn from $\mathcal{N}(0, 0.1)$.

As in [1, p. 92, (11.1)], we set $\mu = 0.1\mu_{\max}$, where $\mu_{\max}$ is obtained by $\mu_{\max} = \|A^T\widetilde{b}\|_\infty$, where

$$\widetilde{b}_i = \begin{cases} \theta^{\text{pos}}, & \text{if } b_i = 1, \\ \theta^{\text{neg}}, & \text{if } b_i = -1, \end{cases}$$

where θ^{pos} and θ^{neg} are the fraction of positive and negative labels, respectively.

We fit the ℓ_1 regularized logistic regression model (1.68) into a global consensus problem as

$$\begin{aligned} \min_{x_i,\, z} & \sum_{i=1}^{m} l_i(x_i) + \mu\|z\|_1, \\ \text{s.t.} \quad & x_i = z, \quad i = 1, 2, \ldots, m, \end{aligned} \tag{1.69}$$

where $x_i = (v_i, w_i^T)^T$, $z = (v, w^T)^T$, $v, v_i \in \mathbb{R}$, $w, w_i \in \mathbb{R}^n$, $l_i(x_i) = \log(1 + \exp(-b_i(a_i^T w_i + v_i)))$.

Let

$$x = \begin{pmatrix} x_1 \\ x_2 \\ \vdots \\ x_m \end{pmatrix}, \quad y = z, \quad \lambda = \begin{pmatrix} \lambda_1 \\ \lambda_2 \\ \vdots \\ \lambda_m \end{pmatrix},$$

$$f(x) = \sum_{i=1}^m l_i(x_i), \quad g(y) = \mu \|y\|_1,$$

$$A = I_{mn}, \quad B = \begin{pmatrix} -I_n \\ -I_n \\ \vdots \\ -I_n \end{pmatrix} \in \mathbb{R}^{mn \times n}, \quad c = 0, \quad H = \xi I_{mn},$$

For TRSP, we give initial points $x^0 = 0$, $y^0 = 0$, $\lambda^0 = 0$ and begin the following loops. Before the iteration, we set proper TOL_x, TOL_y and TOL_λ, and give a guess of L_f and approximate error of Hessian matrix ε.

- Update the variable x in a trust region $\|x - x^k\| \le \Delta_k$, due to the separability of the objective function, we could update x_i separately.

$$\widetilde{x}_i^k = \text{argmin}_{\|x_i - x_i^k\| \le \Delta_{k,i}} \left\{ \frac{\xi}{2} \|x_i - y^k - \xi^{-1}\lambda_i^k\|^2 + \widetilde{l}_i^k(x_i) \right\}.$$

where $\widetilde{l}_i^k(x_i)$ is the quadratic approximation of $l_i(x_i)$ at x_i^k. According to Sect. 1.3.1, we have the subproblem is equivalent to (1.15). Then we could solve the subproblem by

$$\widetilde{x}_i^k = (\widetilde{\nabla}^2 l_i(x_i) + (\xi + \delta_{k,i}) I_n)^{-1} (\nabla l_i(x_i) + \xi(x_i^k - y^k - \xi^{-1}\lambda_i^k)). \tag{1.70}$$

This inverse matrix could be computed under an L-BFGS framework, such as in [11]. If $\|\nabla l_i(x_i) + \xi(x_i^k - y^k - \xi^{-1}\lambda_i^k)\| \le \text{TOL}_x$, we skip the subproblem and set $\widetilde{x}_i^k = x_i^k$. Otherwise, we need to ensure $\delta_{k,i} \ge L_f \|\widetilde{x}_i^k - x_i^k\| + \varepsilon$ to get $\widetilde{x}_i^k$. Here we let $\delta_k \le \delta_{k,i}$ for all i such that $\widetilde{x}_i^k \ne x_i^k$. After we get $\widetilde{x}^k$, we update δ_k in a similar way in Algorithm 2, Step 2′. For given boundaries $1 < \eta_1 < \eta_2$, we recall the notation $r_{x,i}^k = L_f \|\widetilde{x}^k - x^k\| + \varepsilon$ and update δ_k by

$$\delta_{k+1,i} = \begin{cases} 0.5 \max(\delta_{k,i}, (1+\eta_1) r_{x,i}^k), & \delta_{k,i} \in [\eta_2 r_{x,i}^k, +\infty), \\ \delta_{k,i}, & \delta_{k,i} \in [\eta_1 r_{x,i}^k, \eta_2 r_{x,i}^k), \\ .5\delta_{k,i}, & \delta_{k,i} \in [r_{x,i}^k, \eta_1 r_{x,i}^k). \end{cases}$$

- Update the variable y in an exact solution as the soft-thresholding operator

$$\widetilde{y}^k = \underset{\mu\xi/m}{\text{shrinkage}} \left(\frac{1}{m} \sum_{i=1}^m (\widetilde{x}_i^k + \xi^{-1}\lambda^k) \right).$$

Solve the y-subproblem exactly is valid in the TRSP scheme as discussed in Remark 1.3.

- Update the multiplier λ by

$$\widetilde{\lambda}^k = \lambda^k - \xi(\widetilde{x}^k - B\widetilde{y}^k).$$

Here we compare the TRSP with the basic ADMM, with the subproblem solved by the L-BFGS method. All the algorithms start from the same initial points, $x = 0$, $y = 0$, $\lambda = 0$. Through Fig. 1.4a, we observe the convergence rate of the residual norm, which is the residual of the consensus constraint, defined as $r_{\text{norm}}^k := \sqrt{\sum i = 1^m \|x_i^k - z^k\|^2}$, for the three algorithms. The TRSP method demonstrates a significantly faster convergence compared to both the basic ADMM and Majorized iPADMM (MADMM), with the residual norm decreasing more rapidly and smoothly. Figure 1.4b shows the dual residual norm, defined as $s_{\text{norm}}^k := \|z^k - z^{k-1}\|_2$. The TRSP method again exhibits superior performance, converging faster and achieving lower dual residual norms compared to the basic ADMM and MADMM. Figure 1.4c presents the comparison of the objective values among the methods. We calculate

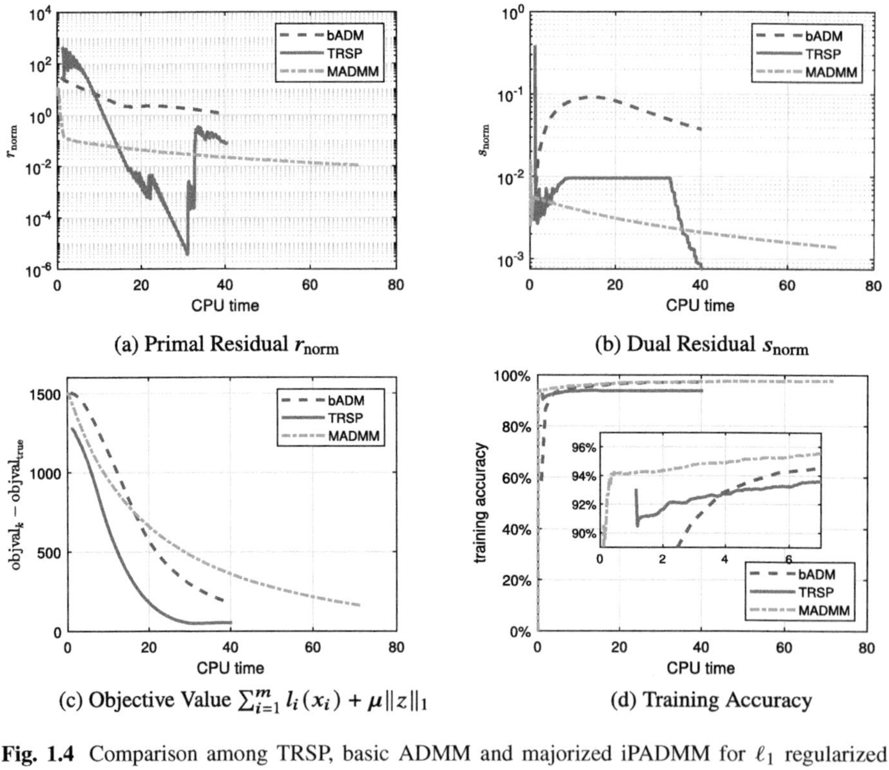

(a) Primal Residual r_{norm} (b) Dual Residual s_{norm}

(c) Objective Value $\sum_{i=1}^m l_i(x_i) + \mu\|z\|_1$ (d) Training Accuracy

Fig. 1.4 Comparison among TRSP, basic ADMM and majorized iPADMM for ℓ_1 regularized logistic regression

the objective value of the k-th iteration and true solution with $x_i^* = z^* = (v^*, w^{*T})^T$ as

$$\text{objval}_k = \sum_{i=1}^{m} l_i(x_i^k) + \mu\|z^k\|_1, \quad \text{objval}_{\text{true}} = \sum_{i=1}^{m} l_i(x_i^k) + \mu\|z^k\|_1$$

and compare $\text{objval}_k - \text{objval}_{\text{true}}$ for all methods. The TRSP method consistently achieves lower objective values faster than both the basic ADMM and MADMM. Lastly, Fig. 1.4d illustrates the training accuracy for all methods. The training accuracy is defined as the percentage of correctly classified training examples, and is calculated as

$$\text{acc}_{\text{train}}^k = \frac{1}{2m}\|b_{\text{pred}}^k - b\|_1,$$

where $b_{\text{pred}}^k = \text{sign}\left[(1 + \exp(-a_i^T w^k - v^k))^{-1} - 0.5\right]$. All algorithms achieve an accuracy over 90%.

In summary, the numerical experiments clearly demonstrate that TRSP outperforms both the basic ADMM and Majorized iPADMM (MADMM) in terms of convergence rate, dual residual norm reduction, objective value optimization, and training accuracy. The use of L-BFGS for solving subproblems in TRSP contributes to its enhanced performance.

1.6 Conclusion

In this paper, we propose to solve the subproblems of a splitting method in an explicit trust region constraint. As a result, we established an efficient trust region-based splitting method for solving a class of separable convex optimization and variational inequalities. We analyzed the global convergence of the new algorithm under some mild assumptions. Moreover, we proved that the new algorithm enjoys the $O(1/\epsilon)$ convergence rate in an ergodic sense. Numerical experiments showed that the novel trust region-based splitting method is competitive with many existing algorithms for image recovery and logistic regression problems.

References

1. Boyd, S., Parikh, N., Chu, E., Peleato, B., Eckstein, J.: Distributed optimization and statistical learning via the alternating direction method of multipliers. Found. Trends Mach. Learn. **3**(1), 1–122 (2010). https://doi.org/10.1561/2200000016
2. Cai, X., Chen, Y., Han, D.: Nonnegative tensor factorizations using an alternating direction method. Front. Math. China **8**(1), 3–18 (2013). https://doi.org/10.1007/s11464-012-0264-8
3. Chen, C., He, B., Yuan, X.: Matrix completion via an alternating direction method. IMA J. Numer. Anal. **32**(1), 227–245 (2012). https://doi.org/10.1093/imanum/drq039

4. Chen, S.S., Donoho, D.L., Saunders, M.A.: Atomic decomposition by basis pursuit. SIAM J. Sci. Comput. **20**(1), 33–61 (1998). https://doi.org/10.1137/S1064827596304010
5. Chen, Y., Dai, Y., Han, D., Sun, W.: Positive semidefinite generalized diffusion tensor imaging via quadratic semidefinite programming. SIAM J. Imaging Sci. **6**(3), 1531–1552 (2013). https://doi.org/10.1137/110843526
6. Conn, A.R., Gould, N.I.M., Toint, P.L.: Trust-Region Methods. SIAM (2000)
7. Eckstein, J., Bertsekas, D.P.: On the Douglas-Rachford splitting method and the proximal point algorithm for maximal monotone operators. Math. Program. **55**(1–3), 293–318 (1992). https://doi.org/10.1007/BF01581204
8. Fu, X., He, B.: Self-adaptive projection-based prediction-correction method for constrained variational inequalities. Front. Math. China **5**(1), 3–21 (2010). https://doi.org/10.1007/s11464-009-0045-1
9. Gabay, D.: Applications of the method of multipliers to variational inequalities. In: Fortin, M., Glowinski, R. (eds.) Augmented Lagrangian Methods: Applications to the Numerical Solution of Boundary-Value Problems. Studies in Mathematics and its Applications, vol. 15, chap. IX, pp. 299–331. Elsevier (1983)
10. Gabay, D., Mercier, B.: A dual algorithm for the solution of nonlinear variational problems via finite element approximation. Comput. Math. Appl. **2**(1), 17–40 (1976). https://doi.org/10.1016/0898-1221(76)90003-1
11. Gao, G., Florez, H., Vink, J.C., Wells, T.J., Saaf, F., Blom, C.P.: Performance analysis of trust region subproblem solvers for limited-memory distributed BFGS optimization method. Front. Appl. Math. Stat. **7**, 673412 (2021). https://doi.org/10.3389/fams.2021.673412
12. Glowinski, R., Marroco, A.: Sur l'approximation, par éléments finis et la résolution par pénalisation-dualité d'une classe de problémes de Dirichlet non linéaires. Rev. Fr. Autom. Inform. Rech. Oper. **9**(R-2), 41–76 (1975). https://doi.org/10.1051/m2an/197509R200411
13. Goldstein, T., Osher, S.: The split Bregman method for L1-regularized problems. SIAM J. Imaging Sci. **2**(2), 323–343 (2009). https://doi.org/10.1137/080725891
14. Han, D., Lo, H.K.: Solving non-additive traffic assignment problems: A descent method for co-coercive variational inequalities. Eur. J. Oper. Res. **159**(3), 529–544 (2004). https://doi.org/10.1016/S0377-2217(03)00423-5
15. He, B., Liao, L.Z., Han, D., Yang, H.: A new inexact alternating directions method for monotone variational inequalities. Math. Program. **92**(1), 103–118 (2002). https://doi.org/10.1007/s101070100280
16. He, B., Yang, H., Wang, S.: Alternating direction method with self-adaptive penalty parameters for monotone variational inequalities. J. Optim. Theory Appl. **106**(2), 337–356 (2000). https://doi.org/10.1023/A:1004603514434
17. He, B., Yuan, X.: On the O(1/n) convergence rate of the Douglas Rachford alternating direction method. SIAM J. Numer. Anal. **52**(2), 700–709 (2012). https://doi.org/10.1137/110836936
18. Ng, M.K., Weiss, P., Yuan, X.: Solving constrained total-variation image restoration and reconstruction problems via alternating direction methods. SIAM J. Sci. Comput. **32**(5), 2710–2736 (2010). https://doi.org/10.1137/090774823
19. Nocedal, J., Wright, S.J.: Numerical Optimization. Science Press, Beijing (2006)
20. Rudin, L.I., Osher, S., Fatemi, E.: Nonlinear total variation based noise removal algorithms. Phys. D Nonlinear Phenom. **60**(1–4), 259–268 (1992). https://doi.org/10.1016/0167-2789(92)90242-F
21. Sun, W., Yuan, Y.X.: Optimization Theory and Methods: Nonlinear Programming. Springer, New York (2006)
22. Tseng, P.: Applications of a splitting algorithm to decomposition in convex programming and variational inequalities. SIAM J. Control Optim. **29**(1), 119–138 (1991). https://doi.org/10.1137/0329006
23. Wang, Y., Yang, J., Yin, W., Zhang, Y.: A new alternating minimization algorithm for total variation image reconstruction. SIAM J. Imaging Sci. **1**(3), 248–272 (2008). https://doi.org/10.1137/080724265

24. Xu, M.: Proximal alternating directions method for structured variational inequalities. J. Optim. Theory Appl. **134**(1), 107–117 (2007). https://doi.org/10.1007/s10957-007-9192-2
25. Yang, J., Zhang, Y.: Alternating direction algorithms for ℓ_1-problems in compressive sensing. SIAM J. Sci. Comput. **31**(1), 250–278 (2011). https://doi.org/10.1137/090777761
26. Ye, C., Yuan, X.: A descent method for structured monotone variational inequalities. Optim. Methods Softw. **22**(2), 329–338 (2007). https://doi.org/10.1080/10556780600552693
27. Yuan, X.: An improved proximal alternating direction method for monotone variational inequalities with separable structure. Comput. Optim. Appl. **49**(1), 17–29 (2011). https://doi.org/10.1007/s10589-009-9293-y
28. Yuan, X.: Alternating direction method for covariance selection models. J. Sci. Comput. **51**(2), 261–273 (2012). https://doi.org/10.1007/s10915-011-9507-1

Chapter 2
Forward-Reflected-Backward Method with Extrapolation and Linesearch for Monotone Inclusion Problems

Tanxing Wang, Heng Zhang, and Xingju Cai

Abstract The extrapolation technique has been widely used to accelerate the forward-reflected-backward method for monotone inclusion problems. This paper considers a new forward-reflected-backward method with extrapolation (FRB_e), which adapts a new extrapolation direction different from the existing acceleration method and uses the latest extrapolation point for the Lipschitz operator. Further, to improve the numerical performance, we propose the linesearch procedure based on the FRB_e by using only the locally Lipschitz constant. Compared to existing methods, our proposed methods not only cover some classical methods, but can also offer a larger stepsize that does not depend on the global Lipschitz constant. We establish the weak convergence of the proposed methods under mild and standard assumptions. In addition, we conduct some numerical experiments on the lasso problem and the ℓ_1 regularized logistic regression problem to demonstrate the advantage of the proposed methods.

Keywords Forward-reflected-backward method · Monotone inclusion · Extrapolation · Linesearch · Weak convergence

2.1 Introduction

In this paper, we propose two methods for finding a zero in the sum of two monotone operators in a real Hilbert space $\mathcal{H}$. Specifically, we consider the monotone inclusion problem:

$$\text{find } x \in \mathcal{H} \text{ such that } 0 \in (A + B)x, \tag{2.1}$$

T. Wang · H. Zhang · X. Cai (✉)
School of Mathematical Sciences, Ministry of Education Key Laboratory of NSLSCS, Nanjing Normal University, Nanjing, China
e-mail: caixingju@njnu.edu.cn

T. Wang
e-mail: wangtanxing2021@163.com

H. Zhang
e-mail: 1174848718@qq.com

X. Wang et al. (eds.), *Splitting Optimization*, ICIAM2023 Springer Series 2,
https://doi.org/10.1007/978-981-96-8574-5_2

where $A: \mathcal{H} \rightrightarrows \mathcal{H}$ and $B: \mathcal{H} \to \mathcal{H}$ are two (maximally) monotone operators with B (locally) Lipschitz continuous such that $(A+B)^{-1}(0) \neq \emptyset$. The monotone inclusion problem (2.1) has drawn much attention because it provides a broad unifying framework for variational inequalities, convex minimization problems, split feasibility problems and equilibrium problems, and has been applied to solve various real-world problems from machine learning, signal processing and image restoration, see [1, 2, 5, 10, 13, 27, 32].

A popular model, which can be formulated under the monotone inclusion, is the following optimization problem of the sum of two functions:

$$\min_{x \in \mathcal{H}} f(x) + g(x), \tag{2.2}$$

where $f: \mathcal{H} \to (-\infty, \infty]$ is proper, closed, convex, and $g: \mathcal{H} \to \mathcal{R}$ is convex with (locally) Lipschitz continuous gradient denoted as ∇g. The solutions to this minimization problem are precisely the points $x \in \mathcal{H}$ which satisfy the first-order optimality condition:

$$0 \in \partial f(x) + \nabla g(x),$$

where ∂f is the subdifferential of f. In this case, the optimization problem is equivalent to the monotone inclusion problem (2.1) with $A = \partial f$ and $B = \nabla g$.

In recent years, there is a growing interest in the design and analysis of splitting algorithms for solving the monotone inclusion problem (2.1). In 1979, Passty [24] and Lions et al. [16] proposed the following forward-backward (FB) method, which combines one forward evaluation of B and one backward evaluation of A in each iteration. More precisely, the method generates a sequence $\{x^k\}_{k\in N}$ according to

$$x^{k+1} = J_{\lambda A}(x^k - \lambda B x^k),$$

where $\lambda \in (0, 2/L)$, L is the Lipschitz continuity modulus of B and $J_{\lambda A} = (I + \lambda A)^{-1}$. Under the assumption that the operator $B: \mathcal{H} \to \mathcal{H}$ is $1/L$-cocoercive, the authors proved that each bounded sequence generated by FB converges weakly to a solution.

It is worthing that coercivity of an operator is a stronger property than Lipschitz continuity and hence can be difficult to satisfy for a general monotone inclusion problem. In 2000, Tseng [33] proposed a modification of the forward-backward method to relax the coercivity assumption, which is known as Tseng's method or the forward-backward-forward (FBF) method. It only requires the Lipschitzness of B at the expense of an additional forward evaluation. Applied to (2.1), Tseng's method generates sequences $\{x^k\}_{k\in N}$ according to

$$\begin{cases} y^k = J_{\lambda A}(x^k - \lambda B x^k), \\ x^{k+1} = y^k - \lambda B y^k + \lambda B x^k, \end{cases}$$

where $\lambda \in (0, 1/L)$ and converges weakly to a solution.

Under the same assumptions as Tseng's method [33], Malitsky and Tam [18] proposed the forward-reflected-backward (FRB) method for solving (2.1), which requires only one forward evaluation per iteration instead of two. The corresponding update scheme reads as

$$x^{k+1} = J_{\lambda A}(x^k - \lambda Bx^k - \lambda(Bx^k - Bx^{k-1})),$$

and converges weakly if B is L-Lipschitz and the stepsize is chosen to satisfy $\lambda < 1/2L$.

As FB, FBF, and FRB are the first-order splitting algorithms, acceleration techniques are of great practical interest to improve the performance of these algorithms. One simple and efficient strategy is to incorporate the inertial technique, which was first introduced by Polyak [25] in 1964. Recently, there are increasing interests in studying inertial type algorithms, such as inertial forward-backward splitting methods for separable optimization problems under the nonconvex setting [20] and strongly convex setting [19], inertial versions of the Douglas-Rachford operator splitting method [4], inertial forward-backward-forward method [3] based on Tseng's approach [33] and general inertial proximal point method for the mixed variational inequality problem [9]. Specially, motivated by the idea of the heavy-ball method [25], Malitsky and Tam [18] proposed an inertial forward-reflected-backward splitting algorithm (iFRB) for solving problem (2.1). The corresponding update scheme reads as

$$\begin{cases} y^k = x^k + \alpha(x^k - x^{k-1}), \\ x^{k+1} = J_{\lambda A}(y^k - \lambda Bx^k - \lambda(Bx^k - Bx^{k-1})), \end{cases} \tag{2.3}$$

where $\alpha \in [0, 1/3)$, $0 < \lambda L < (1 - 3\alpha)/2$. It is worth noting that the iteration scheme (2.3) does not use the latest extrapolation point y^k for the calculation of operator B. In view of this, in this paper, we propose a new forward-reflected-backward method with extrapolation (FRB_e) for solving the monotone inclusion problem (2.1). Further, taking into account that the global Lipschitz constant is not easily obtained and it can often lead to overconservative stepsize, we propose a forward-reflected-backward method with extrapolation and linesearch (FRB_el). Under mild and standard assumptions, we establish the weak convergence of the sequences generated by the proposed methods. Numerical experiments on the lasso problem and the ℓ_1 regularized logistic regression problem to demonstrate the advantage of the proposed methods.

The rest of this paper is organized as follows. In Sect. 2.2, we recall some definitions and results for further analysis. In Sect. 2.3, we introduce FRB_e and FRB_el and investigate their convergence. Some numerical results are reported in Sect. 2.4. Finally, we draw some conclusions in Sect. 2.5.

2.2 Preliminaries

In this section, we summarize some necessary notations and results for further analysis.

Let $\mathcal{H}$ be a real Hilbert space with the inner product $\langle \cdot, \cdot \rangle$ and the induced norm $\| \cdot \|$, and use $\| \cdot \|_1$ to denote the ℓ_1 norm. Let $\{x^k\}_{k \in N}$ be a sequence in $\mathcal{H}$. We write $x^k \rightharpoonup x$ to stand for the weak convergence of the sequence $\{x^k\}_{k \in N}$ to $x \in \mathcal{H}$ as $k \to +\infty$. Let $A : \mathcal{H} \rightrightarrows \mathcal{H}$ be a set-valued operator. The graph of $A : \mathcal{H} \rightrightarrows \mathcal{H}$, denoted by gph$(A)$ is defined by

$$\text{gph}(A) := \{(x, u) \in \mathcal{H} \times \mathcal{H} \mid \S \in \mathcal{H}, \sqcap \in \mathcal{A}\S\}.$$

Let us review some basic definitions and concepts.

Definition 2.1 ([2]) We say that the operator A is

(i) monotone if

$$\langle x - y, u - v \rangle \geqslant 0, \ \forall (x, u), \ (y, v) \in \text{gph}(A);$$

(ii) maximal monotone if it is monotone and gph$(A) \supset$ gph(B) where B is any other monotone operator.

Remark 2.1 An important property of a maximal monotone operator $A : \mathcal{H} \rightrightarrows \mathcal{H}$ is that, if a pair $(x, u) \in \mathcal{H} \times \mathcal{H}$ and $\langle x - y, u - v \rangle \geqslant 0$ for all $(y, v) \in \text{gph}(A)$, then $u \in A(x)$.

For a given maximal monotone operator A, the resolvent

$$J_{\lambda A}(x) = (I + \lambda A)^{-1}(x), \tag{2.4}$$

for $x \in \mathcal{H}$ and $\lambda > 0$ is a single-valued mapping, where I is the identity operator on $\mathcal{H}$. Furthermore, $\|J_{\lambda A}(x) - J_{\lambda A}(y)\| \leqslant \|x - y\|$ for all $x, y \in \mathcal{H}$. A single-valued operator $B : \mathcal{H} \to \mathcal{H}$ is L-Lipschitz continuous if there exists $L > 0$ such that

$$\|Bx - By\| \leqslant L\|x - y\|, \ \forall x, \ y \in \mathcal{H},$$

and B is β-cocoercive if there exists $\beta > 0$ such that

$$\langle Bx - By, x - y \rangle \geqslant \beta \|Bx - By\|^2, \ \forall x, \ y \in \mathcal{H}.$$

Let $C(x, \delta)$ denote the open ball in $\mathcal{H}$ with centra x and radius $\delta > 0$. Recall that $B : \mathcal{H} \to \mathcal{H}$ is called locally Lipschitz provided for each $x \in \mathcal{H}$, there exists $\delta_x > 0$ such that the restriction of B to $C(x, \delta_x)$ is Lipschitz. The following result gives the maximal monotonicity of the sum of two monotone operators.

Lemma 2.1 ([23]) *Suppose $A\colon \mathcal{H} \rightrightarrows \mathcal{H}$ is a maximal monotone operator and $B\colon \mathcal{H} \to \mathcal{H}$ is a Lipschitz continuous and monotone operator. Then $A + B$ is a maximal monotone operator.*

We use the following identities in our convergence analysis.

Lemma 2.2 ([2, 14]) *For any vectors a, b, c and $d \in \mathcal{H}$, the following identity holds*

$$\langle a - b, c - d\rangle = \frac{1}{2}(\|a - d\|^2 - \|a - c\|^2) + \frac{1}{2}(\|b - c\|^2 - \|b - d\|^2).$$

Lemma 2.3 ([2]) *For any vectors x and $y \in \mathcal{H}$, $\alpha \in \mathcal{R}$, then*

$$\|\alpha x + (1 - \alpha)y\|^2 = \alpha\|x\|^2 + (1 - \alpha)\|y\|^2 - \alpha(1 - \alpha)\|x - y\|^2.$$

The lemma below is quite well known and plays an important role in proving weak convergence of sequences in a Hilbert space.

Lemma 2.4 ([21]) *Let K be a nonempty subset of a Hilbert space $\mathcal{H}$ and let $\{x^k\}_{k\in N}$ be bounded sequence in $\mathcal{H}$. Assume the following two conditions are satisfied:*

(i) $\lim_{k\to\infty}\|x - x^k\|$ *exists for each $x \in K$;*
(ii) every weak cluster point of $\{x^k\}_{k\in N}$ belongs to K.

Then $\{x^k\}_{k\in N}$ is weakly converges to a point in K.

Lemma 2.5 ([2]) *Let $A\colon \mathcal{H} \rightrightarrows \mathcal{H}$ be maximally monotone and let $x \in \mathcal{H}$. Set*

$$(\forall\, \lambda \in \mathcal{R}_{++})\; x_\lambda = J_{\lambda A}x.$$

Then $x_\lambda \to P_{\overline{\mathrm{dom} A}}(x)$ as $\lambda \downarrow 0$, and exactly one of the following holds:

(i) $\mathrm{zer} A \neq 0$ *and $x_\lambda \to P_{\mathrm{zer} A}$ as $\lambda \uparrow +\infty$;*
(ii) $\mathrm{zer} A = 0$ *and $||x_\lambda|| \to \infty$ as $\lambda \uparrow +\infty$.*

2.3 Algorithms and Convergence Analysis

In this section, we first introduce a forward-reflected-backward method with extrapolation ($\mathrm{FRB_e}$) for solving the monotone inclusion problem (2.1), described in Algorithm 1 in Sect. 2.3.1 This method adapts a new extrapolation direction different from the existing acceleration method [5, 7, 8, 11, 12, 22, 29–31] and uses the latest extrapolation point for operator B. Then, we establish the convergence results of $\mathrm{FRB_e}$ in Sect. 2.3.2 Furthermore, we incorporate a linesearch procedure into the $\mathrm{FRB_e}$ method, named forward-reflected-backward method with extrapolation and linesearch ($\mathrm{FRB_{el}}$) in Sect. 2.3.3 To this end, we make some necessary assumptions on problem (2.1) below, which will be used throughout this section.

Assumption 2.1 (i) $A\colon \mathcal{H} \rightrightarrows \mathcal{H}$ is a set-valued maximal monotone operator;
(ii) $B\colon \mathcal{H} \to \mathcal{H}$ is a single-valued monotone operator and L-Lipschitz continuous;
(iii) The solution set $(A+B)^{-1}(0)$ of inclusion problem (2.1) is nonempty.

Note that points (i), (ii), and (iii) in Assumption 2.1 are the common assumptions for monotone inclusion problem, which are consistent with the forward-backward-forward method in [33], the forward-reflected-backward method in [18] and the double inertial forward-backward-forward method in [34].

2.3.1 *Proposed* FRB$_{\mathrm{e}}$ *for Solving Inclusion Problem*

Motivated by the idea of forward-reflected-backward method in [18] and the inertial algorithms in [5, 7, 8, 11, 12, 22, 29–31], we propose a forward-reflected-backward method with extrapolation, denoted as FRB$_{\mathrm{e}}$, in Algorithm 1. Compared with existing algorithms, our algorithm provides more flexibility in choosing parameters, which would potentially enhance the algorithm's performance.

Algorithm 1 Forward-reflected-backward method with extrapolation (FRB$_{\mathrm{e}}$)

1: Choose $\lambda_0 > 0$, $\epsilon > 0$ which are close to 0, $\alpha \in [0, 1)$. Set $x^0 = y^{-1} \in \mathcal{H}$.
2: For $k = 0, 1\cdots$, compute

$$\begin{cases} y^k = x^k + \alpha(x^k - y^{k-1}), \\ x^{k+1} = J_{\lambda_k A}(y^k - \lambda_k B y^k - \lambda_{k-1}(Bx^k - By^{k-1})), \end{cases} \tag{2.5}$$

where the stepsize sequence $\{\lambda_k\}_{k\in N}$ is nondecreasing and satisfies

$$\lambda_0 \leqslant \lambda_k \leqslant \frac{1-\alpha-\epsilon}{L(\alpha^2+2\alpha+2)}. \tag{2.6}$$

Remark 2.2 FRB$_{\mathrm{e}}$ can reduce to some classical algorithms when the operators and the parameters take specific values, such as proximal point algorithm [28], Popov's method [26], projected reflected gradient method [17], forward-reflected-backward method [18], and so on. For simplicity, we only consider the fixed stepsize case, i.e., there exists $\lambda \in (0, (1-\alpha)/L(\alpha^2+2\alpha+2))$ such that $\lambda_k = \lambda$ for all k. In this case, FRB$_{\mathrm{e}}$ can be expressed compactly as

$$\begin{cases} y^k = x^k + \alpha(x^k - y^{k-1}), \\ x^{k+1} = J_{\lambda A}(y^k - \lambda B y^k - \lambda(Bx^k - By^{k-1})). \end{cases} \tag{2.7}$$

(i) If $\alpha = 0$, FRB$_{\mathrm{e}}$ reduce to the forward-reflected-backward method [18], that is,

$$x^{k+1} = J_{\lambda A}(x^k - \lambda Bx^k - \lambda(Bx^k - Bx^{k-1})).$$

(ii) If $\alpha = 0$ and $B = 0$, then FRB_e simplifies to the proximal point algorithm [14, 28]. That is,

$$x^{k+1} = J_{\lambda A}(x^k).$$

(iii) If $\alpha = 0$, $A = N_C$ is the normal cone to a set C, and B is an affine operator, then FRB_e can be expressed as

$$x^{k+1} = P_C(x^k - \lambda B(2x^k - x^{k-1})),$$

which coincides with the projected reflected gradient method [17] for VIs.

(iv) If $\alpha = 0$ and $A = N_{\mathcal{H}} = 0$, then FRB_e method (2.7) becomes

$$x^{k+1} = x^k - 2\lambda Bx^k + \lambda Bx^{k-1}. \tag{2.8}$$

It is worth noting that the iteration scheme (2.8) can be expressed as the two-step recursion

$$\begin{cases} y^{k+1} = y^k - \lambda Bx^k, \\ x^{k+1} = y^{k+1} - \lambda Bx^k. \end{cases}$$

This is exactly Popov's algorithm [26] for unconstrained VIs. Furthermore, in the GANs literature, (2.8) is also known to be equivalent to the optimistic gradient method. For details, see the discussion in [15].

2.3.2 *Convergence Analysis of* FRB_e

In this section, we investigate the weak convergence of the proposed FRB_e with the help of Assumption 2.1. To this aim, we consider the metric function

$$\begin{aligned} E_k := \; & \frac{1}{1+\alpha}\|x^\star - y^k\|^2 + 2\lambda_{k-1}\langle x^\star - x^k, Bx^k - By^{k-1}\rangle \\ & + \lambda_{k-1}L(\alpha^2 + \alpha + 1)\|x^k - y^{k-1}\|^2, \end{aligned} \tag{2.9}$$

where $x^\star \in (A + B)^{-1}(0)$, i.e., the element of the solution set. In the following, we show that the sequence $\{E_k\}_{k\in N}$ is monotonically nonincreasing.

Lemma 2.6 *Suppose that Assumption 2.1 holds. Let $\{x^k\}_{k\in N}$ be the sequence generated by Algorithm 1. Then, the sequence $\{E_k\}_{k\in N}$ is monotonically nonincreasing. In particular, for any $k \geqslant 0$, it holds that*

$$E_{k+1} \leqslant E_k - C\|x^{k+1} - y^k\|^2, \tag{2.10}$$

where $C = (1 - \alpha) - \lambda_k L(\alpha^2 + 2\alpha + 2) > 0$.

Proof Using the definition of the resolvent in (2.4) and the x^{k+1} in (2.5), we obtain

$$x^{k+1} - y^k + \lambda_k By^k + \lambda_{k-1}(Bx^k - By^{k-1}) \in -\lambda_k Ax^{k+1}. \tag{2.11}$$

By the monotonicity of A, we have

$$\langle x^\star - x^{k+1}, \lambda_k Ax^\star - \lambda_k Ax^{k+1}\rangle \geqslant 0. \tag{2.12}$$

Pick $x^\star \in (A+B)^{-1}(0)$, then

$$-\lambda_k Bx^\star \in \lambda_k Ax^\star. \tag{2.13}$$

Plugging (2.11) and (2.13) into (2.12), we have

$$\langle x^\star - x^{k+1}, x^{k+1} - y^k + \lambda_k By^k + \lambda_{k-1}(Bx^k - By^{k-1}) - \lambda_k Bx^\star\rangle \geqslant 0.$$

Rearranging the above inequality gives

$$\begin{aligned}
0 \leqslant & 2\langle x^\star - x^{k+1}, x^{k+1} - y^k\rangle + 2\lambda_k\langle x^\star - x^{k+1}, By^k - Bx^\star\rangle \\
& + 2\lambda_{k-1}\langle x^\star - x^{k+1}, Bx^k - By^{k-1}\rangle \\
= & \|x^\star - y^k\|^2 - \|x^\star - x^{k+1}\|^2 - \|x^{k+1} - y^k\|^2 \\
& + 2\lambda_k\langle x^\star - x^{k+1}, By^k - Bx^{k+1}\rangle + 2\lambda_k\langle x^\star - x^{k+1}, Bx^{k+1} - Bx^\star\rangle \\
& + 2\lambda_{k-1}\langle x^\star - x^k, Bx^k - By^{k-1}\rangle + 2\lambda_{k-1}\langle x^k - x^{k+1}, Bx^k - By^{k-1}\rangle \\
\leqslant & \|x^\star - y^k\|^2 - \|x^\star - x^{k+1}\|^2 - \|x^{k+1} - y^k\|^2 \\
& + 2\lambda_k\langle x^\star - x^{k+1}, By^k - Bx^{k+1}\rangle + 2\lambda_{k-1}\langle x^\star - x^k, Bx^k - By^{k-1}\rangle \\
& + \lambda_{k-1}L\|x^{k+1} - x^k\|^2 + \lambda_{k-1}L\|x^k - y^{k-1}\|^2,
\end{aligned} \tag{2.14}$$

where the equality follows from Lemma 2.2 and the second inequality holds by using the monotonicity of B, the Lipschitz continuity of B, and the Cauchy inequality $2ab \leqslant a^2 + b^2$ with $a = \|x^k - x^{k+1}\|$ and $b = \|x^k - y^{k-1}\|$.
Therefore, we have

$$\begin{aligned}
& \|x^\star - x^{k+1}\|^2 + 2\lambda_k\langle x^\star - x^{k+1}, Bx^{k+1} - By^k\rangle + \|x^{k+1} - y^k\|^2 \\
\leqslant & \|x^\star - y^k\|^2 + 2\lambda_{k-1}\langle x^\star - x^k, Bx^k - By^{k-1}\rangle + \lambda_{k-1}L\|x^{k+1} - x^k\|^2 \\
& + \lambda_{k-1}L\|x^k - y^{k-1}\|^2.
\end{aligned} \tag{2.15}$$

From (2.5), we obtain

$$x^{k+1} = \frac{1}{1+\alpha}y^{k+1} + \frac{\alpha}{1+\alpha}y^k, \tag{2.16}$$

$$y^{k+1} - y^k = (1+\alpha)(x^{k+1} - y^k), \tag{2.17}$$

$$y^k - x^k = \alpha(x^k - y^{k-1}). \tag{2.18}$$

Thus, it follows from (2.16), (2.17), (2.18) and Lemma 2.3, we have

$$\begin{aligned}
&\|x^\star - x^{k+1}\|^2 \\
=&\left\| x^\star - \left(\frac{1}{1+\alpha} y^{k+1} + \frac{\alpha}{1+\alpha} y^k\right)\right\|^2 \\
=&\left\| \frac{1}{1+\alpha}(x^\star - y^{k+1}) + \frac{\alpha}{1+\alpha}(x^\star - y^k)\right\|^2 \\
=&\frac{1}{1+\alpha}\left\|x^\star - y^{k+1}\right\|^2 + \frac{\alpha}{1+\alpha}\left\|x^\star - y^k\right\|^2 - \frac{\alpha}{(1+\alpha)^2}\left\|y^{k+1} - y^k\right\|^2 \\
=&\frac{1}{1+\alpha}\left\|x^\star - y^{k+1}\right\|^2 + \frac{\alpha}{1+\alpha}\left\|x^\star - y^k\right\|^2 - \alpha\left\|x^{k+1} - y^k\right\|^2,
\end{aligned} \tag{2.19}$$

and

$$\begin{aligned}
&\|x^{k+1} - x^k\|^2 \\
=&\|x^{k+1} - y^k + y^k - x^k\|^2 = \|x^{k+1} - y^k + \alpha(x^k - y^{k-1})\|^2 \\
=&\|x^{k+1} - y^k\|^2 + \alpha^2\|x^k - y^{k-1}\|^2 + 2\alpha\langle x^{k+1} - y^k, x^k - y^{k-1}\rangle \\
\leqslant&(1+\alpha)\|x^{k+1} - y^k\|^2 + (\alpha^2 + \alpha)\|x^k - y^{k-1}\|^2,
\end{aligned} \tag{2.20}$$

where the last inequality follows from Cauchy-Schwarz and Young inequalities.

Substituting (2.19) and (2.20) into (2.15), we get

$$\begin{aligned}
&\frac{1}{1+\alpha}\|x^\star - y^{k+1}\|^2 + 2\lambda_k\langle x^\star - x^{k+1}, Bx^{k+1} - By^k\rangle + (1-\alpha)\|x^{k+1} - y^k\|^2 \\
\leqslant&\frac{1}{1+\alpha}\|x^\star - y^k\|^2 + 2\lambda_{k-1}\langle x^\star - x^k, Bx^k - By^{k-1}\rangle + \lambda_{k-1}L\|x^k - y^{k-1}\|^2 \\
&+ \lambda_{k-1}L\|x^{k+1} - x^k\|^2 \\
\leqslant&\frac{1}{1+\alpha}\|x^\star - y^k\|^2 + 2\lambda_{k-1}\langle x^\star - x^k, Bx^k - By^{k-1}\rangle + \lambda_{k-1}L\|x^k - y^{k-1}\|^2 \\
&+ \lambda_{k-1}L(1+\alpha)\|x^{k+1} - y^k\|^2 + \lambda_{k-1}L(\alpha^2+\alpha)\|x^k - y^{k-1}\|^2.
\end{aligned} \tag{2.21}$$

Combining the definition of $\{E_k\}_{k\in N}$ in (2.9) and (2.21), we obtain

$$\begin{aligned}
&E_{k+1} \\
\leqslant&E_k - ((1-\alpha) - \lambda_k L(\alpha^2+\alpha+1) - \lambda_{k-1}L(\alpha+1))\|x^{k+1} - y^k\|^2 \\
\leqslant&E_k - ((1-\alpha) - \lambda_k L(\alpha^2+2\alpha+2))\|x^{k+1} - y^k\|^2 \\
=&E_k - C\|x^{k+1} - y^k\|^2,
\end{aligned}$$

where the second inequality holds follows from the stepsize sequence $\{\lambda_k\}_{k\in N}$ is nondecreasing. Therefore, we get assertion (2.10). In addition, it follows from (2.6) that

$$C = (1-\alpha) - \lambda_k L(\alpha^2 + 2\alpha + 2) > 0.$$

Therefore, the sequence $\{E_k\}_{k\in N}$ is monotonically nonincreasing. □

Theorem 2.1 *The sequence $\{x^k\}_{k\in N}$ generated by Algorithm 1 converges weakly to a point in $(A+B)^{-1}(0)$ when Assumption 2.1 is satisfied.*

Proof We first show that sequence $\{E_k\}_{k\in N}$ is bounded. It follows from (2.10) that

$$\begin{aligned} E_{k+1} \leqslant & E_k - C\|x^{k+1} - y^k\|^2 \\ \leqslant & E_k \leqslant \ldots \leqslant E_0 = \frac{1}{1+\alpha}\left\|x^\star - y^0\right\|^2 < +\infty. \end{aligned} \tag{2.22}$$

Therefore, the sequence $\{E_k\}_{k\in N}$ has an upper bound. And we will next prove that it has a lower bound. Indeed, we can find that

$$\begin{aligned} & 2\lambda_k\langle x^\star - x^{k+1}, Bx^{k+1} - By^k\rangle \\ \geqslant & -2\lambda_k L\|x^\star - x^{k+1}\|\|x^{k+1} - y^k\| \\ \geqslant & -\lambda_k L(\|x^\star - x^{k+1}\|^2 + \|x^{k+1} - y^k\|^2) \\ = & -\lambda_k L\left(\frac{1}{1+\alpha}\left\|x^\star - y^{k+1}\right\|^2 + \frac{\alpha}{1+\alpha}\left\|x^\star - y^k\right\|^2 + (1-\alpha)\left\|x^{k+1} - y^k\right\|^2\right) \\ = & -\lambda_k L\left(\frac{1}{1+\alpha}\|x^\star - y^{k+1}\|^2 + \frac{\alpha}{1+\alpha}\|x^\star - y^{k+1} + y^{k+1} - y^k\|^2\right) \\ & -\lambda_k L(1-\alpha)\|x^{k+1} - y^k\|^2 \\ \geqslant & -\lambda_k L\left(\frac{1}{1+\alpha}\|x^\star - y^{k+1}\|^2 + \frac{3\alpha}{1+\alpha}\|x^\star - y^{k+1}\|^2 + \frac{3\alpha}{2(1+\alpha)}\|y^{k+1} - y^k\|^2\right) \\ & -\lambda_k L(1-\alpha)\|x^{k+1} - y^k\|^2 \\ = & -\lambda_k L\left(\frac{1}{1+\alpha}\|x^\star - y^{k+1}\|^2 + \frac{3\alpha}{1+\alpha}\|x^\star - y^{k+1}\|^2\right) \\ & -\lambda_k L\left(\frac{3\alpha(1+\alpha)}{2}\|x^{k+1} - y^k\|^2 + (1-\alpha)\|x^{k+1} - y^k\|^2\right) \\ = & -\lambda_k L\frac{1+3\alpha}{1+\alpha}\|x^\star - y^{k+1}\|^2 - \lambda_k L\left(\frac{3\alpha^2}{2} + \frac{\alpha}{2} + 1\right)\|x^{k+1} - y^k\|^2, \end{aligned} \tag{2.23}$$

where the first inequality follows from the Cauchy-Schwarz inequality and the Lipschitz continuity of B, the second inequality follows from the Young inequality, the third inequality follows from the Cauchy-Schwarz inequality $ab \leqslant \frac{\beta}{2}a^2 + \frac{1}{2\beta}b^2$, $\beta > 0$ with $a = \|x^\star - y^{k+1}\|$, $b = \|y^{k+1} - y^k\|$ and $\beta = 2$, and the first equality and the third equality follows from (2.19) and (2.17), respectively.

We then obtain from (2.6) and (2.23) that

$$\begin{aligned} &E_{k+1} \\ =&\frac{1}{1+\alpha}\|x^{\star}-y^{k+1}\|^2+2\lambda_k\langle x^{\star}-x^{k+1}, Bx^{k+1}-By^k\rangle \\ &+\lambda_k L(\alpha^2+\alpha+1)\|x^{k+1}-y^k\|^2 \\ \geqslant&\left(\frac{1}{1+\alpha}-\lambda_k L\frac{1+3\alpha}{1+\alpha}\right)\|x^{\star}-y^{k+1}\|^2-\lambda_k L\frac{\alpha(\alpha-1)}{2}\|x^{k+1}-y^k\|^2\geqslant 0, \end{aligned} \tag{2.24}$$

we know that the sequence $\{E_k\}_{k\in N}$ is lower bounded. So we can get $\{E_k\}_{k\in N}$ is bounded. Then, summing up (2.10) from $k=0,\cdots,N$, it yields

$$\sum_{k=0}^{N} C\|x^{k+1}-y^k\|^2\leqslant \sum_{k=0}^{N}(E_k-E_{k+1})=E_0-E_{N+1}\leqslant E_0<+\infty,$$

Therefore, we can deduce that

$$\lim_{k\to+\infty}\|x^{k+1}-y^k\|=0. \tag{2.25}$$

One can then obtain from (2.22) and (2.24) that $\{y^k\}_{k\in N}$ is bounded. Similarly, by the definition of x^{k+1} in (2.16), we have that $\{x^k\}_{k\in N}$ is also bounded. Then $\{x^k\}_{k\in N}$ has a weakly convergent subsequence $\{x^{k_j}\}_{j\in N}$ such that $\{x^{k_j}\}_{j\in N}$ converges weakly to $x^{\infty}\in\mathcal{H}$.

Multiplying both sides by $-1/\lambda_k$ and adding Bx^{k+1} to both sides in (2.11) and choosing $k=k_j-1$, we have

$$-\frac{1}{\lambda_{k_j-1}}(x^{k_j}-y^{k_j-1})+Bx^{k_j}-By^{k_j-1}-\frac{\lambda_{k_j-2}}{\lambda_{k_j-1}}(Bx^{k_j-1}-By^{k_j-2})\in(A+B)x^{k_j}. \tag{2.26}$$

Since $\lim_{k\to+\infty}\|x^{k+1}-y^k\|=0$ and B is Lipschitz continuous, we have $Bx^{k_j}-By^{k_j-1}\to 0$, as $j\to+\infty$. By Lemma 2.1, we have that $A+B$ is maximal monotone. Therefore, the graph of $A+B$ is demiclosed. From (2.6), we obtain $\{\lambda_k\}_{k\in N}$ is bounded and far away from zero. Further, passing to the limit in (2.26), we have $0\in(A+B)x^{\infty}$. To show that $\{x^k\}_{k\in N}$ is weakly convergent, first note that, by the boundedness of $\{E_k\}_{k\in N}$ and (2.10), we deduce the existence of the limit

$$\begin{aligned} \lim_{k\to+\infty}&\frac{1}{1+\alpha}\|x^{\star}-y^{k+1}\|^2+2\lambda_k\langle x^{\star}-x^{k+1}, Bx^{k+1}-By^k\rangle \\ &+\lambda_k(\alpha^2+\alpha+1)\|x^{k+1}-y^k\|^2. \end{aligned}$$

Further, it follows from (2.5), we know

$$\lim_{k\to+\infty} \frac{1}{1+\alpha}\|x^\star - x^{k+1}\|^2 + \frac{\alpha^2}{1+\alpha}\|x^{k+1} - y^k\|^2$$
$$- \frac{2\alpha}{1+\alpha}\langle x^\star - x^{k+1}, x^{k+1} - y^k\rangle + 2\lambda_k\langle x^\star - x^{k+1}, Bx^{k+1} - By^k\rangle$$
$$+ \lambda_k(\alpha^2 + \alpha + 1)\|x^{k+1} - y^k\|^2$$

exists. Let

$$F_k = -2\lambda_k\langle x^\star - x^{k+1}, Bx^{k+1} - By^k\rangle - \lambda_k(\alpha^2 + \alpha + 1)\|x^{k+1} - y^k\|^2,$$

and

$$G_k = \frac{\alpha^2}{1+\alpha}\|x^{k+1} - y^k\|^2 - \frac{2\alpha}{1+\alpha}\langle x^\star - x^{k+1}, Bx^{k+1} - By^k\rangle.$$

Since $Bx^{k+1} - By^k \to 0$, as $k \to +\infty$, $x^{k+1} - y^k \to 0$, as $k \to +\infty$ and $\{x^k\}_{k\in N}$ and $\{\lambda_k\}_{k\in N}$ are bounded, we have that $F_k \to 0, k \to +\infty$ and $G_k \to 0, k \to +\infty$. Then

$$\frac{1}{1+\alpha}\|x^\star - x^{k+1}\|^2 = E_k - G_k + F_k.$$

Since $\lim_{k\to+\infty} E_k$ exists and $\frac{1}{1+\alpha} > 0$, we then have that $\lim_{k\to+\infty}\|x^\star - x^k\|$ exists. By Lemma 2.4, we conclude that $\{x^k\}_{k\in N}$ converges weakly to a point in $(A + B)^{-1}(0)$. The proof is complete. □

Theorem 2.1 has an immediate result when the steps of the sequence $\{\lambda_k\}_{k\in N}$ are constant, and we obtain the following corollary.

Corollary 2.1 *Let $A\colon \mathcal{H} \rightrightarrows \mathcal{H}$ be maximally monotone, let $B\colon \mathcal{H} \to \mathcal{H}$ be monotone and L-Lipschitz and suppose that $(A + B)^{-1}(0) \neq \emptyset$. Choose $\lambda \in \left(0, \frac{1-\alpha}{(\alpha^2+\alpha+2)L}\right)$. Givens $x^0, y^{-1} \in \mathcal{H}$, the sequence $\{x^k\}_{k\in N}$ generated by*

$$\begin{cases} y^k = x^k + \alpha(x^k - y^{k-1}), \\ x^{k+1} = J_{\lambda A}(y^k - \lambda By^k - \lambda(Bx^k - By^{k-1})). \end{cases}$$

Then $\{x^k\}_{k\in N}$ converges weakly to a point contained in $(A + B)^{-1}(0)$.

2.3.3 Forward-Reflected-Backward Method with Extrapolation and Linesearch

The algorithm presented in Sect. 2.3.1 requires information about the single-valued operator's Lipschitz constant in order to choose an appropriate stepsize. But for many practical problems, this requirement is difficult to meet. On the one hand, obtaining the global Lipschitz constant of a single-valued operator requires a significant cost in many cases and leading to poor numerical performance. On the other hand, we can only obtain a locally Lipschitz constant in practical problems. In view of these, we propose a forward-reflected-backward method with a linesearch procedure based on FRB_e (FRB_el), which converges whenever the single-valued operator is locally Lipschitz.

Algorithm 2 Forward-reflected-backward method with extrapolation and linesearch (FRB_el)

1: Choose $x^0 = y^{-1} \in \mathcal{H}, \lambda_0, \lambda_{-1} > 0, \delta \in \left(0, \frac{2(1-\alpha)}{\alpha^2+2\alpha+2}\right), \alpha \in [0, 1), \sigma \in (0, 1), \rho \in \{1, \sigma^{-1}\}$.
2: For $k = 0, 1 \cdots$, compute

$$\begin{cases} y^k = x^k + \alpha(x^k - y^{k-1}), \\ x^{k+1} = J_{\lambda_k A}(y^k - \lambda_k B y^k - \lambda_{k-1}(Bx^k - By^{k-1})), \end{cases}$$

where $\lambda_k = \rho\lambda_{k-1}\sigma^i$, with i being the smallest nonnegative integer satisfying

$$\lambda_k \| Bx^{k+1} - By^k \| \leqslant \frac{\delta}{2} \| x^{k+1} - y^k \|. \tag{2.27}$$

Remark 2.3 As discussed at the beginning of this section, FRB_el combines an inertial step with a linesearch step and a resolvent step. It is worth noting that it is not necessary to estimate the Lipschitz constant of the Lipschitz continuous monotone operator B in our proposed Algorithm 2.

The following lemma shows that the linesearch procedure described in Algorithm 2 is well-defined when operator B is locally Lipschitz continuous.

Lemma 2.7 *Suppose that operator $B\colon \mathcal{H} \to \mathcal{H}$ is locally Lipschitz. Then the linesearch criterion (2.27) is well defined, which means that it will be satisfied after a finite number of iterations.*

Proof Let $x^{k+1}(\lambda) := J_{\lambda A}(y^k - \lambda B y^k - \lambda_{k-1}(Bx^k - By^{k-1}))$. According to the Lemma 2.5, we obtain that $J_{\lambda A}(x^{k+1}(0)) \to P_{\overline{\text{dom} A}}(x^{k+1}(0))$ when $\lambda \downarrow 0$. Since operator A is maximal monotone, we know that $J_{\lambda A}$ is nonexpansive. Therefore, we have

$$\begin{aligned} &\| x^{k+1}(\lambda) - P_{\overline{\text{dom} A}}(x^{k+1}(0)) \| \\ \leqslant &\| x^{k+1}(\lambda) - J_{\lambda A}(x^{k+1}(0)) \| + \| J_{\lambda A}(x^{k+1}(0)) - P_{\overline{\text{dom} A}}(x^{k+1}(0)) \| \\ \leqslant &\| \lambda B y^k \| + \| J_{\lambda A}(x^{k+1}(0)) - P_{\overline{\text{dom} A}}(x^{k+1}(0)) \|. \end{aligned}$$

By taking the limit as $\lambda \downarrow 0$,

$$x^{k+1}(\lambda) \to P_{\overline{\text{dom} A}}(x^{k+1}(0)).$$

Assuming that linesearch finds λ at the kth iteration failed, then for all $\lambda = \rho\lambda_{k-1}\sigma^i, i = 0, 1 \ldots$, it implies

$$\rho\lambda_{k-1}\sigma^i \|Bx^{k+1}(\lambda) - By^k\| > \frac{\delta}{2}\|x^{k+1}(\lambda) - y^k\|.$$

Since operator B is locally Lipschitz, there exists $L > 0$ when i is large enough. And we have

$$\rho\lambda_{k-1}\sigma^i \|Bx^{k+1}(\lambda) - By^k\| > \frac{\delta}{2}\|x^{k+1}(\lambda) - y^k\| \geqslant \frac{\delta}{2L}\|Bx^{k+1}(\lambda) - By^k\|.$$

Therefore, it implies

$$\rho\lambda_{k-1}\sigma^i > \frac{\delta}{2L}.$$

Since $\sigma^i \to 0$ as $i \to \infty$, this inequality gives a contradiction, which completes the proof. □

The next Lemma 2.8 is a direct extension of Lemma 2.6.

Lemma 2.8 *Let $\{x^k\}_{k\in N}$ be the sequence generated by Algorithm 2. Then, the sequence $\{E_k\}_{k\in N}$ is monotonically nonincreasing. In particular, for any $k \geqslant 0$, it holds that*

$$E_{k+1} \leqslant E_k - C\|x^{k+1} - y^k\|^2,$$

where $C = (1-\alpha) - \frac{\delta}{2}(\alpha^2 + 2\alpha + 2) > 0$.

Proof The proof is similar to Lemma 2.6. We use inequality (2.27), which is well-defined due to Lemma 2.7, instead of the Lipschitzness of B to get the inequalities (2.14) and (2.23). □

Theorem 2.2 *Let $\mathcal{H}$ be finite dimensional, $A\colon \mathcal{H} \rightrightarrows \mathcal{H}$ be maximally monotone, and $B\colon \mathcal{H} \to \mathcal{H}$ be monotone and locally Lipschitz continuous, and suppose that $(A+B)^{-1}(0)$ is nonempty. Then the sequence $\{x^k\}_{k\in N}$ generated by Algorithm 2 converges to a point contained in $(A+B)^{-1}(0)$.*

Proof It is similar to Theorem 2.1 but using Lemma 2.8 instead of Lemma 2.6, and (2.27) instead of the Lipschitzness of B. We take $\lambda_k L = \delta/2$ for all $k \in N$ in (2.22) and (2.24), so we can deduce $\{y^k\}_{k\in N}$ is bounded. Since (2.16) and (2.25), we know that $\{x^k\}_{k\in N}$ is bounded and $\|x^{k+1} - y^k\| \to 0$. As a locally Lipschitz operator on a finite-dimensional space, B is Lipschitz on bounded sets. Thus, since $\{x^k\}_{k\in N}$ is bounded, there exists a constant $L > 0$ such that

$$\|Bx^{k+1} - By^k\| \leqslant L\|x^{k+1} - y^k\|. \tag{2.28}$$

Combining (2.27) and (2.28), we see that $\{\lambda_k\}_{k\in N}$ is bounded away from zero. The remainder of the proof is similar to Theorem 2.1. □

2.4 Numerical Experiment

In this section, to demonstrate the effectiveness of the proposed FRB_e and FRB_{el}, we apply them to solve the lasso problem and the ℓ_1 regularized logistic regression problem. All experiments are performed in MATLAB R2021a on a PC with Intel Core i7-13700H and 16.0 GB of RAM.

2.4.1 *Lasso Problem*

In this subsection, we consider the lasso problem:

$$\min_{x \in R^n} F(x) := \frac{1}{2}\|Dx - b\|^2 + \mu\|x\|_1, \tag{2.29}$$

where $D \in R^{m \times n}$, $b \in R^m$ and $\mu > 0$. We observe that (2.29) is in the form of (2.2) with $f(x) = \mu\|x\|_1$ and $g(x) = \frac{1}{2}\|Dx - b\|^2$. Therefore, the minimization problem (2.29) is equivalent to the following monotone inclusion problem

$$\text{find } x \in R^n \text{ such that } 0 \in (A + B)x,$$

where $A = \partial(\mu\|x\|_1)$ and $B = D^{\mathrm{T}}(Dx - b)$.

It is clear that g has a Lipschitz continuous gradient and $f + g$ has compact lower level sets. Thus, Assumption 2.1 is satisfied for (2.29), we can apply Algorithms 1 and 2 to solve (2.29). In addition, it is not hard to show that ∇g has a Lipschitz continuity modulus of $\lambda_{\max}(D^{\mathrm{T}} D)$. In view of this, in the experiment below, we take $L = \lambda_{\max}(D^{\mathrm{T}} D)$ for iFRB and FRB_e.

Now we perform numerical experiments to study the performance of FRB_e and FRB_{el}. We choose $\mu = 1$ and $\mu = 3$ in (2.29), initialize all algorithms at the origin, and use the duality gap of the primal problem (2.29) and its dual problem to terminate the algorithms as in [35]. Specifically, we define

$$u^k = \min\left\{1, \frac{\mu}{\|D^{\mathrm{T}} h(Ax^k)\|_\infty}\right\} \nabla h(Ax^k),$$

and terminate the algorithms when the duality gap is small, i.e.,

$$\frac{|g(x^k) + f(x^k) - d_{ls}(u^k)|}{\max\{f(x^k) + g(x^k), 1\}} \le 10^{-6},$$

where $g(x) = h(Ax) = \frac{1}{2}\|Ax - b\|^2$, and $d_{ls}(u)$ is the optimal value of dual problem.

Table 2.1 Numerical comparisons of different algorithms for solving the lasso problem. In this paper, we use the notation (m, n) to denote the corresponding choices of m and n ($\mu = 1$)

	(100, 200)		(200, 1000)		(400, 2000)		(600, 3000)	
	Iter	Time	Iter	Time	Iter	Time	Iter	Time
iFRB	1900	0.058	11924	0.677	19513	2.148	28368	8.285
FRB_e	1210	0.036	7579	0.620	12401	1.967	18028	7.475
FRB_l	399	0.017	1762	0.138	3040	0.429	4572	1.578
FRB_{el}	**341**	**0.013**	**1390**	**0.120**	**2325**	**0.375**	**3389**	**1.521**

Table 2.2 Numerical comparisons of different algorithms for solving the lasso problem. In this paper, we use the notation (m, n) to denote the corresponding choices of m and n ($\mu = 3$)

	(100, 200)		(200, 1000)		(400, 2000)		(600, 3000)	
	Iter	Time	Iter	Time	Iter	Time	Iter	Time
iFRB	1122	0.028	4800	0.313	7343	0.897	10272	3.100
FRB_e	715	0.025	3053	0.297	4668	0.839	6530	2.987
FRB_l	204	0.011	661	0.052	1463	0.159	1470	0.542
FRB_{el}	**183**	**0.009**	**573**	**0.045**	**1172**	**0.151**	**1175**	**0.507**

The problems used in our experiments are generated as follows. For each m and n, we generate an $m \times n$ matrix D with i.i.d. standard Gaussian entries. We then choose a support set T of size s uniformly at random, and generate an s-sparse vector $\hat{x}$ supported on T with i.i.d. standard Gaussian entries. The vector b is then generated as $b = D\hat{x} + 0.01\tilde{e}$, where $\tilde{e}$ has standard i.i.d. Gaussian entries.

To illustrate the effectiveness of FRB_e and FRB_{el}, we compare them with iFRB [18] and FRB_l [18]. In the experiment, we choose the values of stepsize and inertial parameter to achieve the optimal performance for each algorithm. As outlined in [18], we specify the parameters as follows: $\alpha = 0.2$, $\lambda = 0.99 \times \frac{1}{5L}$ for iFRB; $\alpha = 0$, $\delta = 0.99$, $\sigma = 0.7$, $\rho = \sigma^{-1}$, and the stepsize sequence $\{\lambda_k\}_{k \in N}$ satisfy (2.27) with $y^k = x^k$ for FRB_l. As for FRB_e and FRB_{el}, we choose: $\alpha = 0.2$, $\lambda = 0.99 \times \frac{2}{13L}$ for FRB_e; $\alpha = 0.3$, $\delta = 0.99 \times \dfrac{2(1-\alpha)}{\alpha^2 + 2\alpha + 2}$, $\sigma = 0.7$, $\rho = \sigma^{-1}$ and the stepsize sequence $\{\lambda_k\}_{k \in N}$ satisfy (2.27) for FRB_{el}.

The computational results with $\mu = 1$ and $\mu = 3$ averaged over 50 instances for a range of choices of m and n are presented in Tables 2.1 and 2.2. From these two tables, we can see that FRB_e outperforms the iFRB, and FRB_{el} outperforms the FRB_l in terms of the number of iterations and CPU time for all scenarios. To further observe the convergence of the tested algorithms, we plot the evolutions of $F(x^k) - F^\star$ with respect to the iteration numbers for each algorithm in Figs. 2.1 and 2.2, where $F^\star$ represents the minimum of the objective function values obtained by all tested algorithms. The results show that the FRB_e converges faster than the iFRB,

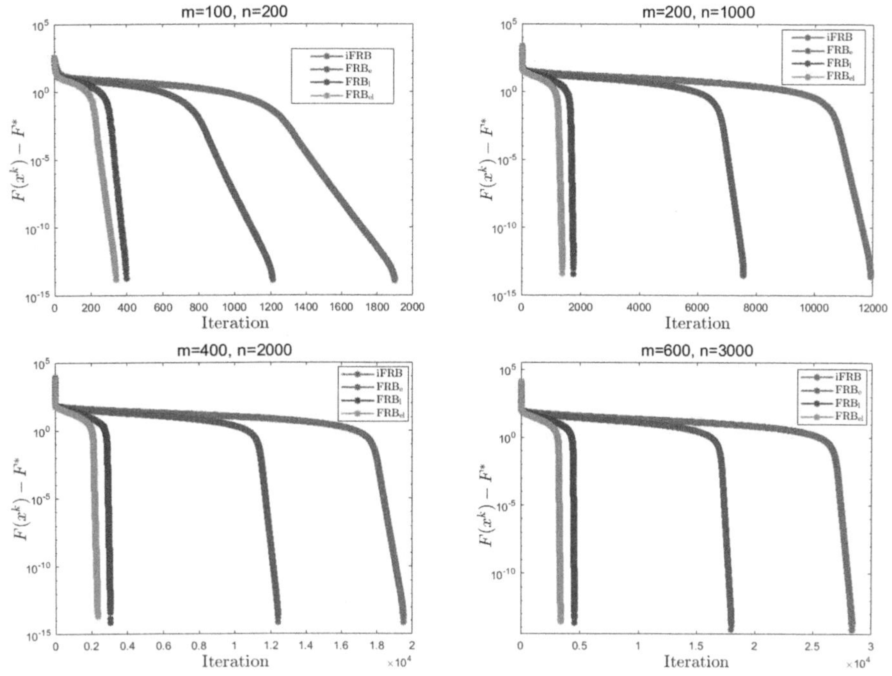

Fig. 2.1 Evolutions of $F(x^k) - F^\star$ with respect to the number of iterations for solving the lasso problem with $\mu = 1$

and the FRB$_{\rm el}$ converges faster than the FRB$_{\rm l}$, demonstrating the advantage of the FRB$_{\rm e}$ and FRB$_{\rm el}$.

2.4.2 ℓ_1 Regularized Logistic Regression Problem

In this subsection, we consider the ℓ_1 regularized logistic regression problem,

$$\min_{\tilde{x} \in R^n,\, x^0 \in R} \; F(x) := \sum_{i=1}^{m} \log(1 + e^{-b_i (a_i^{\rm T} \tilde{x} + x^0)}) + \gamma \|\tilde{x}\|_1, \tag{2.30}$$

where $a_i \in R^n$, $b_i \in \{-1, 1\}$, $i = 1, \cdots, m$, with b_i not all the same, $m < n$, and $\gamma > 0$ is the regularization parameter. So we can apply our methods for solving the problem (2.2) in case $f(\tilde{x}) = \gamma \|\tilde{x}\|_1$, $g(\tilde{x}) = \sum_{i=1}^{m} \log(1 + e^{-b_i (P\tilde{x})_i})$, where $x := (\tilde{x}, x^0) \in R^{n+1}$, and P is the matrix whose ith row is given by $(a_i^{\rm T}\ 1)$. Moreover, one can show that ∇g is Lipschitz continuous with modulus $0.25\lambda_{\max}(P^{\rm T} P)$ [35]. Thus, in our experiments below we take $L = 0.25\lambda_{\max}(P^{\rm T} P)$.

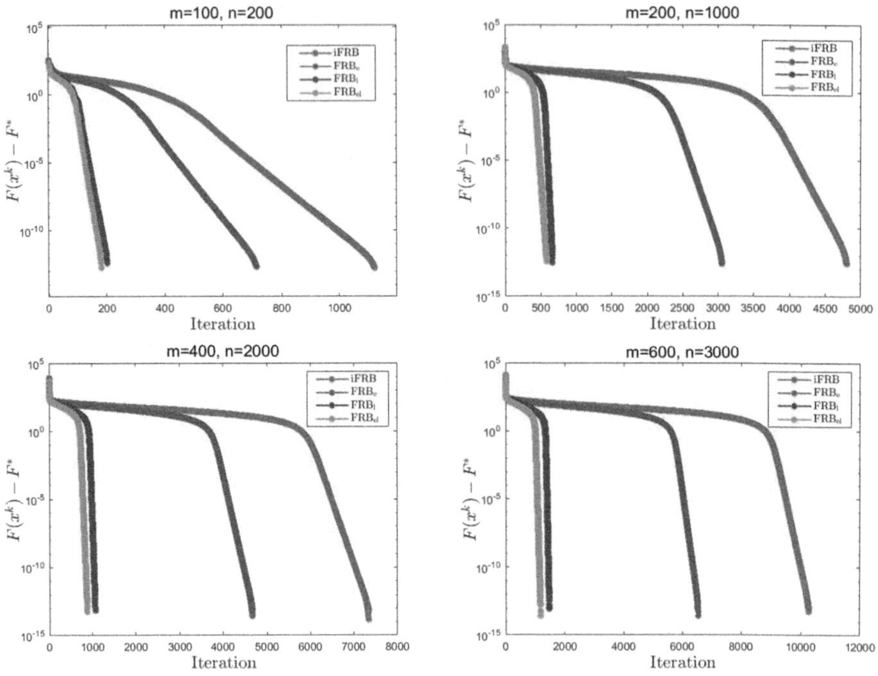

Fig. 2.2 Evolutions of $F(x^k) - F^\star$ with respect to the number of iterations for solving the lasso problem with $\mu = 3$

As that in [35], Assumption 2.1 is satisfied for (2.30). Thus, Algorithm 1 and Algorithm 2 are applicable. In the experiments below, we choose $\gamma = 1$ and $\gamma = 3$ in (2.30). We initialize all algorithms at the origin, and terminate the algorithms as in [35].

We consider random instances for our experiments. For each m and n, we generate an $m \times n$ matrix A with i.i.d. standard Gaussian entries, where A is the matrix whose ith row is a_i^{T}. We then choose a support set T of size s uniformly at random and generate an s-sparse vector $\hat{x}$ supported on T with i.i.d. standard Gaussian entries. The vector b is then generated as $b = \mathrm{sign}(A\hat{x} + ce)$, where c is chosen uniformly at random from $[0, 1]$.

We now perform numerical experiments to verify the efficiency of FRB$_\mathrm{e}$ and FRB$_\mathrm{el}$. In the experiment, we choose the values of stepsize and inertial parameter to achieve the optimal performance for each algorithm. As outlined in [18], we specify the parameters as follows: $\alpha = 0.2$, $\lambda = 0.99 \times \frac{1}{5L}$ for iFRB; $\alpha = 0$, $\delta = 0.99$, $\sigma = 0.8$, $\rho = 1$ and the stepsize sequence $\{\lambda_k\}_{k \in N}$ satisfy (2.27) with $y^k = x^k$ for FRB$_\mathrm{l}$. As for FRB$_\mathrm{e}$ and FRB$_\mathrm{el}$, we choose: $\alpha = 0.2$, $\lambda = 0.99 \times \frac{2}{13L}$ for FRB$_\mathrm{e}$; $\alpha = 0.3$, $\delta = 0.99 \times \dfrac{2(1-\alpha)}{\alpha^2 + 2\alpha + 2}$, $\sigma = 0.8$, $\rho = 1$ and the stepsize sequence $\{\lambda_k\}_{k \in N}$ satisfy (2.27) for FRB$_\mathrm{el}$.

Table 2.3 Numerical comparisons of different algorithms for solving the logistic regression problem. In this paper, we use the notation (m, n) to denote the corresponding choices of m and n $(\gamma = 1)$

	(200, 400)		(400, 1000)		(600, 2000)		(1000, 3000)	
	Iter	Time	Iter	Time	Iter	Time	Iter	Time
iFRB	21861	12.108	67358	257.697	119334	1097.831	166041	3727.979
FRB_e	13890	7.616	42803	163.784	75829	670.698	105511	2353.380
FRB_l	650	0.696	1253	10.355	1545	30.246	1918	99.612
FRB_{el}	**478**	**0.496**	**914**	**7.923**	**1126**	**23.103**	**1204**	**61.330**

Table 2.4 Numerical comparisons of different algorithms for solving the logistic regression problem. In this paper, we use the notation (m, n) to denote the corresponding choices of m and n $(\gamma = 3)$

	(200, 400)		(400, 1000)		(600, 2000)		(1000, 3000)	
	Iter	Time	Iter	Time	Iter	Time	Iter	Time
iFRB	6178	3.348	13008	50.094	26691	243.296	37269	863.537
FRB_e	3928	2.160	8269	33.829	16963	157.555	23683	564.829
FRB_l	378	0.409	541	4.786	845	16.440	1032	53.490
FRB_{el}	**319**	**0.342**	**412**	**3.599**	**668**	**13.675**	**791**	**41.608**

The computational results averaged over 50 instances for a range of choices of m and n are presented in Tables 2.3 and 2.4. From these two tables, we can see that FRB_e outperforms the iFRB, and FRB_{el} outperforms the FRB_l in terms of the number of iterations and CPU time for all scenarios. The results show that the FRB_e converges faster than the iFRB, and the FRB_{el} converges faster than the FRB_l, demonstrating the advantage of the FRB_e and FRB_{el}.

2.5 Conclusion

In this paper, we propose two forward-reflected-backward type methods, named FRB_e and FRB_{el}, for monotone inclusion problems. The proposed methods not only unify some well-known operator splitting methods based on the inertial and linesearch techniques, but can also leads to improved numerical performance. We establish the weak convergence of the proposed methods under mild and standard assumptions. Further, we apply FRB_e and FRB_{el} to solve the lasso problem and the ℓ_1 regularized logistic regression problem. The numerical results on these two important problems verify the efficiency of our proposed methods. In the future, we shall investigate how to combine the relaxation technique and two different extrapolation steps in order to have better performance.

Acknowledgements The work of Xingju Cai was supported by the National Natural Science Foundation of China (NSFC) under Grant 12131004. The work of Tanxing Wang was supported by Postgraduate Research & Practice Innovation Program of Jiangsu Province under Grant 181200003024152.

References

1. Abubakar, J., Kumam, P., Hassan Ibrahim, A., Padcharoen, A.: Relaxed inertial Tseng's type method for solving the inclusion problem with application to image restoration. Mathematics **8**, 1–19 (2020)
2. Bauschke, H.H., Combettes, P.L.: Convex Analysis and Monotone Operator Theory in Hilbert Spaces. Springer, Berln (2011)
3. Boţ, R.I., Csetnek, E.R.: An inertial Tseng's type proximal algorithm for nonsmooth and nonconvex optimization problems. J. Optim. Theory Appl. **171**, 600–616 (2016)
4. Boţ, R.I., Csetnek, E.R., Hendrich, C.: Inertial Douglas-Rachford splitting for monotone inclusion problems. Appl. Math. Comput. **256**, 472–487 (2015)
5. Cai, G., Dong, Q.L., Peng, Y.: Strong convergence theorems for inertial Tseng's extragradient method for solving variational inequality problems and fixed point problems. Optim. Lett. **15**, 1457–1474 (2021)
6. Cai, X.J., Guo, K., Jiang, F., Wang, K., Wu, Z.M., Han, D.R.: The developments of proximal point algorithms. J. Oper. Res. Soc. China. **10**, 197–239 (2022)
7. Cholamjiak, P., Hieu, D.V., Cho, Y.J.: Relaxed forward-backward splitting methods for solving variational inclusions and applications. J. Sci. Comput. **88**, 1–23 (2021)
8. Cholamjiak, P., Hieu, D.V., Muu, L.D.: Inertial splitting methods without prior constants for solving variational inclusions of two operators. Bull. Iran. Math. Soc. **48**, 3019–3045 (2022)
9. Chen, C.H., Ma, S.Q., Yang, J.F.: A general inertial proximal point algorithm for mixed variational inequality problem. SIAM J. Optim. **25**, 2120–2142 (2015)
10. Combettes, P.L., Wajs, V.R.: Signal recovery by proximal forward-backward splitting. Multiscale Model. Simul. **4**, 1168–1200 (2005)
11. Çopur, A.K., Hacioğlu, E., Gursoy, F., Ertürk, M.: An efficient inertial type iterative algorithm to approximate the solutions of quasi variational inequalities in real Hilbert spaces. J. Sci. Comput. **89**, 1–28 (2021)
12. Dong, Q., Jiang, D., Cholamjiak, P., Shehu, Y.: A strong convergence result involving an inertial forward-backward algorithm for monotone inclusions. J. Fixed Point Theory Appl. **19**, 3097–3118 (2017)
13. Duchi, J., Singer, Y.: Efficient online and batch learning using forward-backward splitting. J. Mach. Learn. Res. **10**, 2899–2934 (2009)
14. He, B.S., Yuan, X.M.: Convergence analysis of primal-dual algorithms for a saddle-point problem: from contraction perspective. SIAM J. Imaging Sci. **5**, 119–149 (2012)
15. Hsieh, Y.G., Iutzeler, F., Malick, J., Mertikopoulos, P.: On the convergence of single-call stochastic extra-gradient methods. Adv. Neural Inf. Process. Syst. **32**, (2019)
16. Lions, P.L., Mercier, B.: Splitting algorithms for the sum of two nonlinear operators. SIAM J. Numer. Anal. **16**, 964–979 (1979)
17. Malitsky, Y.: Projected reflected gradient methods for monotone variational inequalities. SIAM J. Optim. **25**, 502–520 (2015)
18. Malitsky, Y., Tam, M.K.: A forward-backward splitting method for monotone inclusions without cocoercivity. SIAM J. Optim. **30**, 1451–1472 (2020)
19. Ochs, P., Brox, T., Pock, T.: iPiasco: inertial proximal algorithm for strongly convex optimization. J. Math. Imaging Vision. **53**, 171–181 (2015)

20. Ochs, P., Chen, Y.J., Brox, T., Pock, T.: iPiano: inertial proximal algorithm for nonconvex optimization. SIAM J. Imaging Sci. **7**, 1388–1419 (2014)
21. Opial, Z.: Weak convergence of the sequence of successive approximations for nonexpansive mappings. Bull. Am. Math. Soc. **73**, 591–597 (1967)
22. Padcharoen, A., Kitkuan, D., Kumam, W., Kumam, P.: Tseng methods with inertial for solving inclusion problems and application to image deblurring and image recovery problems. Comput. Math. Methods **3**, 1–14 (2020)
23. Pascall, D., Sburlan, S.: Nonlinear Mappings of Monotone Type. Springer, Dordrecht (1978)
24. Passty, G.B.: Ergodic convergence to a zero of the sum of monotone operators in Hilbert space. J. Math. Anal. Appl. **72**, 383–390 (1979)
25. Polyak, B.T.: Some methods of speeding up the convergence of iteration methods. Comput. Math. Math. Phys. **4**, 1–17 (1964)
26. Popov, L.D.: A modification of the Arrow-Hurwicz method for finding saddle points. Math. Notes **28**, 845–848 (1980)
27. Raguet, H., Fadili, J., Peyré, G.: A generalized forward-backward splitting. SIAM J. Imaging Sci. **6**, 1199–1226 (2013)
28. Rockafellar, R.T.: Monotone operators and the proximal point algorithm. SIAM J. Control Optim. **14**, 877–898 (1976)
29. Shehu, Y., Iyiola, O.S., Reich, S.: A modified inertial subgradient extragradient method for solving variational inequalities. Optim. Eng. **23**, 421–449 (2021)
30. Tan, B., Cho, S.Y.: Strong convergence of inertial forward-backward methods for solving monotone inclusions. Appl. Anal. **101**, 5386–5414 (2022)
31. Thong, D.V., Vinh, N.T., Cho, Y.J.: A strong convergence theorem for Tseng's extragradient method for solving variational inequality problems. Optim. Lett. **14**, 1157–1175 (2020)
32. Thong, D.V., Yang, J., Cho, Y.J., Rassias, T.M.: Explicit extragradient-like method with adaptive stepsizes for pseudomonotone variational inequalities. Optim. Lett. **15**, 2181–2199 (2021)
33. Tseng, P.: A modified forward-backward splitting method for maximal monotone mapping. SIAM J. Control. Optim. **38**, 431–446 (2000)
34. Wang, Z.B., Lei, Z.Y., Long, X., Chen, Z.Y.: A modified Tseng splitting method with double inertial steps for solving monotone inclusion problems. J. Sci. Comput. **96**, 1–29 (2023)
35. Wen, B., Chen, X.J., Pong, T.K.: Linear convergence of proximal gradient algorithm with extrapolation for a class of nonconvex nonsmooth minimization problems. SIAM J. Optim. **27**, 124–145 (2017)

Chapter 3
Fast Adaptive ADMM with Gaussian Back Substitution for Multiple Block Linear Constrained Separable Problems

Xiangfeng Wang

Abstract This work presents a novel algorithmic framework, called *Fast Adaptive Alternating Direction Method of Multipliers with Gaussian Back Substitution* (ADMM-G-V), tailored for solving multiple block linear constrained separable problems. The proposed method extends the classical multi-block ADMM by incorporating an adaptive penalty parameter, which is dynamically adjusted during the iterative process to enhance convergence properties and computational efficiency. A comprehensive theoretical analysis is provided by establishing the global convergence and worst-case convergence rate of the algorithm in both ergodic and non-ergodic senses. We demonstrate the effectiveness of our method through numerical experiments on consensus problems over networked agents and distributed logistic regression tasks.

Keywords ADMM · Self-adaptive · Gaussian back substitution · Multiple-block

3.1 Introduction and Motivation

In this paper, we consider a general structured multiple block minimization model, i.e.,

$$\min_{\{x_i \in \mathcal{E}_i\}} \left\{ \sum_{i=1}^{m} \theta_i(x_i) \mid \sum_{i=1}^{m} \mathcal{A}_i(x_i) = b \right\}, \tag{3.1}$$

where each $\theta_i : \mathcal{E}_i \to \mathbb{R}$ is a convex real-valued function, each $\mathcal{A}_i : \mathcal{E}_i \to \mathbb{R}^\ell$ denotes a linear operator with $b \in \mathbb{R}^\ell$ be a given vector. Throughout, the solution set $X^\star$ of (3.1) is assumed to be nonempty. This general formulation (3.1) includes the popular two-block problem as a special case, which has proved to be a reasonable and effective model in plenty of applications [1]. The motivation for considering this

X. Wang (✉)
School of Computer Science and Technology and Key Laboratory of Mathematics and Engineering Applications MoE, East China Normal University, Shanghai, China
e-mail: xfwang@cs.ecnu.edu.cn

X. Wang et al. (eds.), *Splitting Optimization*, ICIAM2023 Springer Series 2,
https://doi.org/10.1007/978-981-96-8574-5_3

Table 3.1 $\mathcal{A}$, $\mathcal{B}$, C, C and D for RASL and TILT: $\mathcal{I}$ denotes the identity transform; $\mathcal{J}$, $\mathcal{J}_i$, ϵ_i, M and τ are all given matrix or vector; $\circ$ denotes a given linear operation

Model	$\mathcal{A}$	$\mathcal{B}$	$C(C)$	C	D
RASL	$\mathcal{I}$	$\mathcal{I}$	$-\sum_{i=1}^{n} \mathcal{J}_i \Delta\tau(\epsilon_i \epsilon_i^{\mathrm{T}})$	$\Delta\tau$	$M \circ \tau$
TILT	$\mathcal{I}$	$\mathcal{I}$	$-\mathcal{J}\Delta\tau$	$\Delta\tau$	$M \circ \tau$

particular, more general structured problem (3.1) is that for many interesting applications, the goal (objective) and environment (constraint) become more complicated, especially as more and more application-driven distinctive structures or information are introduced to guarantee better performance. In the following, we propose some practical applications to draw forth our work.

Motivation applications (Low-rank or sparse structure-based image processing problems): The low-rank structure is popularly used to model the invariant property or great relevance between images, while the sparse structure is a byproduct of modeling uncorrelated information. A general formulation can be summarized as

$$\begin{aligned} \min_{L,S,\Delta} \quad & \|L\|_\star + \gamma \|S\|_1 \\ \text{s.t.} \quad & \mathcal{A}(L) + \mathcal{B}(S) + C(C) = D, \end{aligned} \tag{3.2}$$

where the matrix $L \in \mathbb{R}^{m \times n}$ denotes the low-rank part, the matrix $S \in \mathbb{R}^{m \times n}$ represents the sparse part, while Δ usually additional parameters driven by the application task. Concretely, we evoke two popular models, e.g., *robust alignment by sparse and low-rank decomposition* (RASL) [2] and *transform invariant low-rank textures* (TILT) [3]. In order to solve these two models, some subproblem has to be efficiently calculated, which can be included into (3.2) as in Table 3.1.

The unified formulation (3.2) can be contained in the general model (3.1); as a result, the requirement of efficiently computing impels us to design a structured algorithm framework for (3.1). For algorithmic-design purposes, we may need to consider using these two functions individually as well because they usually have different properties/structures. Define the augmented Lagrange function of (3.1) for the case $m = 2$ be

$$\mathcal{L}_\beta(x_1, x_2, \lambda) = \theta_1(x_1) + \theta_2(x_2) - \lambda^{\mathrm{T}}(A_1x_1 + A_2x_2 - b) + \frac{\beta}{2}\|A_1x_1 + A_2x_2 - b\|^2, \tag{3.3}$$

with $\lambda \in \mathbb{R}^m$ denotes the Lagrange multiplier and $\beta > 0$ the penalty parameter. The alternating direction method of multipliers (ADMM) was first proposed by Glowinski and Marrocco in [4] for solving specific nonlinear elliptic equations, and its iterative scheme for (3.1) reads as

$$x^{k+1} = \arg\min\left\{\mathcal{L}_\beta\left(x, y^k, \lambda^k\right) \mid x \in \mathcal{X}\right\}; \tag{3.4a}$$

$$y^{k+1} = \arg\min\left\{\mathcal{L}_\beta\left(x^{k+1}, y, \lambda^k\right) \mid y \in \mathcal{Y}\right\}; \tag{3.4b}$$

$$\lambda^{k+1} = \lambda^k - \beta\left(Ax^{k+1} + By^{k+1} - b\right). \tag{3.4c}$$

Recently, ADMM has found many applications arising in different areas such as image processing, statistical learning, computer vision, and wireless communication network. We refer to [1, 5, 6] for some review papers of ADMM. The convergence of ADMM has been well analyzed in some earlier references such as [7, 8]. For the convergence rate, according to the known explanation in [7, 9] that the ADMM is a special case of the proximal point algorithm (PPA) in [10] and the convergence results of PPA in [11], it is easy to perceive that the ADMM scheme (3.4), without further assumptions such as the strong convexity or special structure assumptions on one or both the functions in the objective, some restrictions on the penalty parameter β, or some error bounds or metric subregularity assumptions on the solution set, is sublinear. In [12, 13], the authors established the sublinear convergence rate for the ADMM scheme (3.4), in sense of both the egrodic and non-ergodic worst-case $\mathcal{O}\left(\frac{1}{n}\right)$ natures respectively measured by the iteration complexity where n is the iteration counter.

The classical ADMM (3.4) has been comparatively thoroughly analyzed and understood. In this paper, we mainly focus on the multiple-block case, which also has popular applications in image processing, machine learning, etc. In order to take advantage of each θ_i's properties individually, a natural idea for solving the general case of (3.1) with $m \geq 3$ is to extend the classical ADMM scheme (3.4) straightforwardly—yielding the following scheme for iteration $k+1$:

$$\begin{cases} x_1^{k+1} = \arg\min\left\{\mathcal{L}_\beta\left(x_1, x_2^k, \ldots, x_m^k, \lambda^k\right) \mid x_1 \in \mathcal{X}_1\right\}, \\ \quad\cdots \\ x_i^{k+1} = \arg\min\left\{\mathcal{L}_\beta\left(x_1^{k+1}, \ldots, x_{i-1}^{k+1}, x_i, x_{i+1}^k, \ldots, x_m^k, \lambda^k\right) \mid x_i \in \mathcal{X}_i\right\}, \\ \quad\cdots \\ x_m^{k+1} = \arg\min\left\{\mathcal{L}_\beta\left(x_1^{k+1}, \ldots, x_{m-1}^{k+1}, x_m, \lambda^k\right) \mid x_m \in \mathcal{X}_m\right\}, \\ \lambda^{k+1} = \lambda^k - \beta\left(\sum_{i=1}^m A_i x_i^{k+1} - b\right), \end{cases} \tag{3.5}$$

with the introduction of the augmented Lagrangian function as follows

$$\mathcal{L}_\beta\left(x_1, \ldots, x_m, \lambda\right) = \sum_{i=1}^m \theta_i(x_i) - \lambda^{\mathrm{T}}\left(\sum_{i=1}^m A_i x_i - b\right) + \frac{\beta}{2}\left\|\sum_{i=1}^m A_i x_i - b\right\|^2. \tag{3.6}$$

Just as the original ADMM scheme (3.4), the iterative scheme (3.5) can be easily derived by decomposing the augmented Lagrangian function of (3.1) in the Gauss-Seidel fashion. In (3.5), the variables x_i's are minimized in alternating order, and the decomposed subproblems are much easier than the original problem (3.1) since only one function θ_i is involved in its x_i-subproblem. Then, the step of updating the Lagrange multiplier coordinates all these solutions to local small subproblems to find a solution to a global large problem. Note that the direct extension of the

ADMM scheme (3.5) reduces to the augmented Lagrangian method (ALM) [14] and the standard ADMM scheme when $m = 1$ and $m = 2$ in (3.1), respectively.

The convergence of the scheme (3.5), however, had perplexed authors for a long time. On one hand, the scheme (3.5) empirically works well for some applications, see [2, 15]. On the other hand, in the literature, its convergence could be shown only under some further assumptions. For example, in [16–18], the convergence of (3.5) was shown under some additional strongly convex conditions, and the penalty parameter β should be chosen judiciously within a certain interval. Moreover, when each function θ_i in (3.1) is of particular structure and the update of λ^{k+1} in (3.5) is required to adopt a new step size rather than β, i.e.,

$$\lambda^{k+1} = \lambda^k - \tau\beta \left(\sum_{i=1}^{m} A_i x_i^{k+1} - b \right), \tag{3.7}$$

where $\tau > 0$ is sufficiently small to fulfill certain error bound conditions, the resulting scheme was proved to be convergent in [19], together with some linear convergent results. In fact, the scheme (3.5) but with the new update of λ in (3.7) can be regarded as an implementation of the dual ascent method to the dual of (3.1) with a shrunk step size. A counterexample was given in [20] showing that the scheme (3.5) is not necessarily convergent without further assumptions, and a sufficient condition ensuring the convergence of (3.5) was given therein.

In general, it is not easy to verify whether the step size τ in (3.7) is small enough to satisfy the desired error bound. Thus it should be stick to the direct extension of ADMM (3.5) where the step size for updating λ is taken as the same as the penalty parameter (thus it is not necessarily very small) and the function θ_i's in (3.5) are only assumed to be generic nonsmooth convex functions, and study in which way the convergence of (3.5) can be derived. In [21], the authors have shown that the resulting sequence is convergent if the output of (3.5) is further corrected by a Gaussian back substitution procedure. The numerical efficiency of the Gaussian back substitution procedure, together with its superiority to some other relevant work based on (3.5), has been illustrated numerically in [15, 22]. The detailed algorithm framework of ADMM with Gaussian back substitution (ADMM-G) is presented as follows, in which the direct extension of ADMM (3.5) can be considered as the first prediction step:

$$\begin{cases} \tilde{x}_1^k = \arg\min \left\{ \mathcal{L}_\beta \left(x_1, x_2^k, \ldots, x_m^k, \lambda^k \right) \mid x_1 \in \mathcal{X}_1 \right\}, \\ \cdots \\ \tilde{x}_i^k = \arg\min \left\{ \mathcal{L}_\beta \left(\tilde{x}_1^k, \ldots, \tilde{x}_{i-1}^k, x_i, x_{i+1}^k, \ldots, x_m^k, \lambda^k \right) \mid x_i \in \mathcal{X}_i \right\}, \\ \cdots \\ \tilde{x}_m^k = \arg\min \left\{ \mathcal{L}_\beta \left(\tilde{x}_1^k, \ldots, \tilde{x}_{m-1}^k, x_m, \lambda^k \right) \mid x_m \in \mathcal{X}_m \right\}, \\ \tilde{\lambda}^k = \lambda^k - \beta \left(\sum_{i=1}^{m} A_i \tilde{x}_i^k - b \right), \end{cases} \tag{3.8a}$$

$$\begin{cases} \lambda^{k+1} = \lambda^k + \alpha\left(\tilde{\lambda}^k - \lambda^k\right), \\ x_m^{k+1} = x_m^k + \alpha\left(\tilde{x}_m^k - x_m^k\right), \\ \quad \cdots \\ x_i^{k+1} = x_i^k + \alpha\left(\tilde{x}_i^k - x_i^k\right) - \sum\limits_{j=i+1}^{m} \left(A_i^{\mathrm{T}} A_i\right)^{-1}\left(A_i^{\mathrm{T}} A_j\right)\left(x_j^{k+1} - x_j^k\right), \\ \quad \cdots \\ x_1^{k+1} = \tilde{x}_1^k, \end{cases} \tag{3.8b}$$

where we need to assume that all A_is have full column rank. The above Gaussian back substitution procedure (3.8) still requires to compute the inverses of $A_i^{\mathrm{T}} A_i$ for $i = 2, \ldots, m-1$, which could be computationally expensive for generic A_i's arising in some image processing applications. The computing procedure of ADMM-G can be graphically explained through the following Fig. 3.1. If we choose $\alpha = 1$, it is obvious that the direct extension of ADMM with Gaussian back substitution can reduce to the classical original ADMM (3.4a)–(3.4c) for (3.1) with $m = 2$. In classical ADMM and ADMM-G, the penalty parameter β can be an arbitrary positive scalar to theoretically guarantee the convergence. This is a nice property, and it provides flexibility in choosing different values when the ADMM is implemented for various specific applications. Meanwhile, it is well known that the efficiency of ADMM highly depends on the appropriate choice of the penalty parameter, and the efficiency can vary dramatically with different values of the penalty parameter for the same problem. Moreover, choosing a good value of the penalty parameter is application-sensitive; a value working well for one application might be completely ineffective for another one. Indeed, so far, this is no general theory for how to choose the "optimal" value of β for the ADMM-type scheme (3.4) or (3.8). This concern forces practitioners to tune the penalty parameter a priori for the specific application under discussion when the ADMM-type scheme (3.4) or (3.8) is implemented. There is a large volume of literature mentioning this issue.

On the other hand, the penalty parameter β can be adjusted dynamically subject to certain rules and it results in the ADMM-type methods with variable penalty parameter. The prior searching for an appropriate value of β can thus be avoided. Indeed, the convergence of ADMM with variable penalty parameter can be theoretically carried over if the rules are meticulously chosen; see, e.g., [21, 23–25], for some relatively earlier literature for the convergence analysis. Some important properties for establishing the convergence of ADMM with variable penalty parameter can be

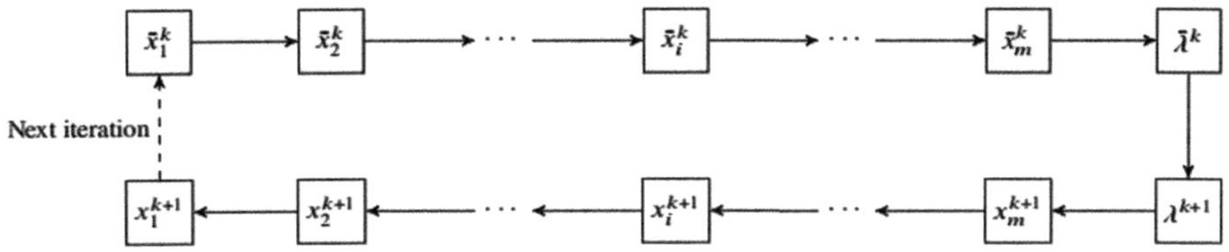

Fig. 3.1 Computing procedure of ADMM-G

found in [24]. Empirically, it is arguable to say which one is preferred: tuning the ADMM-type scheme a prior to find a suitable choice of β or iteratively adjusting the parameters in the iterative process. We believe there is no conclusive assertion in this regard, and both methodologies are equally important from either the mathematical or empirical perspective. Depending on different computation loads in different phases, each of the strategies can find supportive applications in the very large set of literature. Recently, [26] extends the self-adaptive penalty parameter scheme to the consensus problem, which is in a special form of (3.1). Another adaptive penalty scheme is proposed in [27], which also can be included in the framework of [21]. [27] introduces an adaptive relaxed ADMM, which is a popular practical variant of ADMM, and proves the convergence result for the adaptive penalty parameter case. In applications, the rule initiated in [21] for adjusting the penalty parameter has been well verified in many applications such as machine learning [28, 29], computer vision [30], computer graphics [31], smart grid [32], geophysical image processing [33, 34] and elucidated in [1].

Nevertheless, until now, there is no result about adaptive ADMM-G (3.8) with variable penalty parameters in which the penalty parameters are iteratively adjusted. In this paper, we will propose an adaptive ADMM-G algorithm framework, while further establishing the global convergence and worst-case convergence rate of the ADMM with variable penalty parameters in both the ergodic and non-ergodic senses. Before presenting the algorithm framework, we mention that we have to compute all inverses of $A_i^{\mathrm{T}} A_i$ $(i = 1, \ldots, m-1)$ in ADMM-G, which maybe heavy computational complexity. In this paper, we introduce a new variable $u(w)$ which will help modify ADMM-G to avoid calculating the inverses, and some concrete related definitions are as follows

$$x = \begin{pmatrix} x_1 \\ x_2 \\ \vdots \\ x_m \end{pmatrix}, \quad w = \begin{pmatrix} x_1 \\ x_2 \\ \vdots \\ x_m \\ \lambda \end{pmatrix}, \quad u = \begin{pmatrix} x_2 \\ \vdots \\ x_m \\ \lambda \end{pmatrix}, \tag{3.9a}$$

$$v(w) = \begin{pmatrix} v(w)_1 \\ v(w)_2 \\ \vdots \\ v(w)_m \\ v(w)_{m+1} \end{pmatrix} := \begin{pmatrix} A_1 x_1 \\ A_2 x_2 \\ \vdots \\ A_m x_m \\ \lambda \end{pmatrix}, \quad u(w) := \begin{pmatrix} A_2 x_2 \\ \vdots \\ A_m x_m \\ \lambda \end{pmatrix}, \tag{3.9b}$$

and we further rewrite the augmented Lagrangian function (3.6) into the following equivalent reformulation, e.g.,

$$\begin{aligned}\mathcal{L}_\beta(x_1,\dots,x_m,\lambda) &= \sum_{i=1}^{m}\theta_i(x_i) - \lambda^{\mathrm{T}}\left(\sum_{i=1}^{m}A_i x_i - b\right) + \frac{\beta}{2}\left\|\sum_{i=1}^{m}A_i x_i - b\right\|^2 \\ &= \sum_{i=1}^{m}\theta_i(x_i) - \left[v(w)_{m+1}\right]^{\mathrm{T}}\left(\sum_{i=1}^{m}v(w)_i - b\right) + \frac{\beta}{2}\left\|\sum_{i=1}^{m}v(w)_i - b\right\|^2.\end{aligned}$$

We re-record the above equation when only focusing on x_i as

$$\begin{aligned}&\hat{\mathcal{L}}_\beta\left(v(w)_1,\dots,v(w)_{i-1},x_i,v(w)_{i+1},\dots,v(w)_m,v(w)_{m+1}\right) \\ =&\theta_i(x_i) - \left[v(w)_{m+1}\right]^{\mathrm{T}}\left(A_i x_i + \sum_{j=1,j\neq i}^{m}v(w)_j - b\right) + \frac{\beta}{2}\left\|A_i x_i + \sum_{j=1,j\neq i}^{m}v(w)_j - b\right\|^2.\end{aligned} \tag{3.10}$$

As a result, each subproblem with respect to x_i is determined only by

$$\begin{aligned}v(w)_{-i} &:= \left[v(w)_{[1,i-1]}, v(w)_{[i+1,m+1]}\right] \\ &= \left[v(w)_1,\dots,v(w)_{i-1},v(w)_{i+1},\dots,v(w)_m,v(w)_{m+1}\right],\end{aligned}$$

which means in iteration k+1 of (3.8), we have

$$\begin{aligned}\tilde{x}_i^k &= \operatorname{argmin}\left\{\mathcal{L}_\beta\left(\tilde{x}_1^k,\dots,\tilde{x}_{i-1}^k,x_i,x_{i+1}^k,\dots,x_m^k,\lambda^k\right) \mid x_i\in\mathcal{X}_i\right\} \\ &= \arg\min_{x_i\in\mathcal{X}_i}\left\{\theta_i(x_i) + \frac{\beta}{2}\left\|A_i x_i + \sum_{j=1}^{i-1}v(\tilde{w}^k)_j + \sum_{j=i+1}^{m}v(w^k)_j - b - \frac{\lambda^k}{\beta}\right\|^2\right\} \\ &= \arg\min_{x_i\in\mathcal{X}_i}\left\{\hat{\mathcal{L}}_\beta\left(v(\tilde{w}^k)_1,\dots,v(\tilde{w}^k)_{i-1},x_i,v(w^k)_{i+1},\dots,v(w^k)_m,v(w^k)_{m+1}\right)\right\}.\end{aligned} \tag{3.11}$$

Thus, the recursion of ADMM-G with adaptive variable penalty parameter (ADMM-spsGspsV) for the structured convex optimization (3.1) can be written as follows in Algorithm 1. We give some further explanations of Algorithm 1 (ADMMspsGspsV) as follows:

- The matrix $\mathcal{M}_k$ in each iteration is defined as

$$\mathcal{M}_k = \begin{pmatrix} I_\ell & -I_\ell & 0 & \cdots & \cdots & 0 \\ 0 & I_\ell & -I_\ell & \ddots & \ddots & \vdots \\ \vdots & \ddots & \ddots & \ddots & \ddots & \vdots \\ \vdots & \ddots & \ddots & \ddots & -I_\ell & 0 \\ 0 & \cdots & \cdots & 0 & I_\ell & 0 \\ \ell & -\beta_k I_\ell & \cdots & \cdots & -\beta_k I_\ell & I_\ell \end{pmatrix} \in \mathbb{R}^{(m\ell)\times(m\ell)};$$

Algorithm 1 ADMM-G with adaptive variable penalty parameters (ADMM-G-V)

1: Require $u\left(w^0\right) \in \underbrace{\mathbb{R}^\ell \times \cdots \times \mathbb{R}^\ell}_{m}$, $\beta_0 > 0$, $\alpha \in (0, 1)$;
2: Calculate a new adaptive variable penalty parameter β_k;
3: while not converged do
4: For given $u\left(w^k\right)$, $\tilde{w}^k := \left(\tilde{x}_1^k, \ldots, \tilde{x}_m^k, \tilde{\lambda}^k\right)$ are calculated according to (3.8a), e.g.,

$$\begin{cases} \tilde{x}_1^k = \arg\min\left\{\hat{\mathcal{L}}_{\beta_k}\left(x_1, v\left(w^k\right)_2, \ldots, v\left(w^k\right)_m, v\left(w^k\right)_{m+1}\right) \mid x_1 \in \mathcal{X}_1\right\}, \\ \vdots \\ \tilde{x}_i^k = \arg\min\left\{\hat{\mathcal{L}}_{\beta_k}\left(v\left(\tilde{w}^k\right)_1, \ldots, v\left(\tilde{w}^k\right)_{i-1}, x_i, v\left(w^k\right)_{i+1}, \ldots, v\left(w^k\right)_m, v\left(w^k\right)_{m+1}\right) \mid x_i \in \mathcal{X}_i\right\}, \\ \vdots \\ \tilde{x}_m^k = \arg\min\left\{\hat{\mathcal{L}}_{\beta_k}\left(v\left(\tilde{w}^k\right)_1, \ldots, v\left(\tilde{w}^k\right)_m, v\left(w^k\right)_{m+1}\right) \mid x_i \in \mathcal{X}_i\right\}, \\ \tilde{\lambda}^k = \lambda^k - \beta_k\left(A_1\tilde{x}_1^k + \sum_{i=2}^{m} A_i x_i^k - b\right); \end{cases} \tag{3.12}$$

5: Generate the new iterate u^{k+1} and x_1^{k+1} by correcting $\tilde{w}^k$ in the backward order, e.g.,

$$u(w^{k+1}) = u(w^k) - \alpha\mathcal{M}_k\left[u(w^k) - u(\tilde{w}^k)\right]; \tag{3.13}$$

6: end while=0

- Two sequences are iteratively computed in ADMM-G, e.g., $\left\{\tilde{w}^k\right\}$ and $\left\{u\left(w^k\right)\right\}$. $\left\{u\left(w^k\right)\right\}$ denotes the truly iterative sequence with $\left\{\tilde{w}^k\right\}$ being an intermediate iterative sequence. As a result, we will consider the theoretical results on both sequences.

The computing procedure of ADMM-G can be graphically explained through the following Fig. 3.2.

This algorithm framework, ADMM-G-V, obviously includes ADMM-G if the penalty parameter is set to be a constant β. In the following, we will consider this new algorithm in a general prediction-contraction framework form. We would like to mention that for succinctness, we skip the rapidly increasing references appearing in the literature for discussing sharper convergence rate results of the ADMM-type methods from various perspectives under stronger assumptions. Instead, we stick

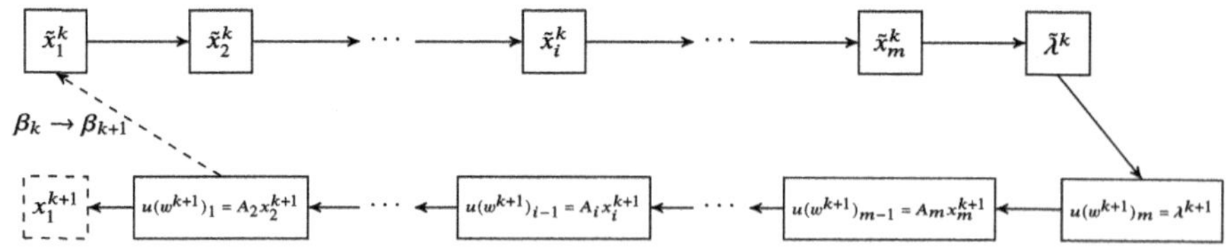

Fig. 3.2 Computing procedure of ADMM-G

to the most general model setting in (3.1) and the iterative scheme in Algorithm 1; and only discuss the global convergence and sublinear convergence rate. The results in this plain context can immediately boost further analysis for deriving sharper convergence rates under various stronger assumptions for the ADMM with variable penalty parameters.

The rest of the paper is organized as follows. In Sect. 3.2, we will first propose some preliminaries and notations. Discussions about the adaptive variable penalty parameter β_k will be given in Sect. 3.3. A prediction-correction reformulation framework will be given in Sect. 3.4. In Sect. 3.5, we will prove the global convergence result and will establish the convergence rate analysis in both ergodic and non-ergodic sense. Section 3.6 contains some numerical experiments to prove the efficiency of variable penalty parameters and the well-established theoretical results. Conclusions are presented in Sect. 3.7.

3.2 Preliminaries and Notations

In this section, we summarize some preliminaries that will be used later in our analysis. Recall we have defined λ as the Lagrange multiplier of the linear constraints in (3.1). Then, the Lagrangian function of this problem is

$$\mathcal{L}(x_1, \ldots, x_m, \lambda) = \sum_{i=1}^{m} \theta_i(x_i) - \lambda^{\mathrm{T}} \left(\sum_{i=1}^{m} A_i x_i - b \right),$$

which is defined on $\mathcal{W} = \mathcal{X}_1 \times \cdots \times \mathcal{X}_m \times \mathbb{R}^\ell$. Let $\left(x_1^*, \ldots, x_m^*, \lambda^*\right) \in \mathcal{W}$ be an saddle point of the Lagrangian function, then as discussed in [35] that finding a saddle point of $\mathcal{L}(x_1, \ldots, x_m, \lambda)$ is equivalent to finding a vector

$$w^* = \begin{pmatrix} x_1^* \\ x_2^* \\ \vdots \\ x_m^* \\ \lambda^* \end{pmatrix} \text{ such that } \begin{cases} \theta_1(x_1) - \theta_1(x_1^*) + \left(x_1 - x_1^*\right)^{\mathrm{T}} \left(-A_1^{\mathrm{T}}\lambda^*\right) \geq 0, \forall x_1 \in \mathcal{X}_1, \\ \theta_2(x_2) - \theta_2(x_2^*) + \left(x_2 - x_2^*\right)^{\mathrm{T}} \left(-A_2^{\mathrm{T}}\lambda^*\right) \geq 0, \forall x_2 \in \mathcal{X}_2, \\ \vdots \\ \theta_m(x_m) - \theta_m(x_m^*) + \left(x_m - x_m^*\right)^{\mathrm{T}} \left(-A_m^{\mathrm{T}}\lambda^*\right) \geq 0, \forall x_m \in \mathcal{X}_m, \\ (\lambda - \lambda^*)^{\mathrm{T}} \left(\sum\limits_{i=1}^{m} A_i x_i^* - b \right) \geq 0, \forall \lambda \in \mathbb{R}^\ell. \end{cases} \tag{3.14}$$

More compactly, the inequalities in (3.14) can be equivalently rewritten as the following variational inequality (VI), e.g.,

$$\mathrm{VI}(\mathcal{W}, F, \theta): \; w^* \in \mathcal{W}, \; \theta(x) - \theta(x^*) + \left(w - w^*\right)^{\mathrm{T}} F\left(w^*\right) \geq 0, \; \forall\, w \in \mathcal{W}, \tag{3.15a}$$

where

$$\theta(x) = \sum_{i=1}^{m} \theta_i(x_i), \quad F(w) = \begin{pmatrix} -A_1^{\mathrm{T}}\lambda \\ -A_2^{\mathrm{T}}\lambda \\ \vdots \\ -A_m^{\mathrm{T}}\lambda \\ \sum_{i=1}^{m} A_i x_i - b \end{pmatrix}. \tag{3.15b}$$

Note that the operator $F(w)$ defined in (3.15b) is monotone because it is affine with a skew-symmetric matrix. Let $\mathcal{W}^*$ be the solution set of VI($\mathcal{W}$, F, θ). Since the solution set of (3.1) is assumed to be nonempty, so that $\mathcal{W}^*$ is also nonempty. In the convergence rate analysis, we shall use a characterization of the solution set $\mathcal{W}^*$ of VI (3.15a). We present it as the following theorem, and its proof can be found in [36, Theorem 2.3.5] or [12, Theorem 2.1].

Theorem 3.1 *The solution set of VI($\mathcal{W}$, F, θ) is convex and it can be characterized as*

$$\mathcal{W}^* = \bigcap_{w \in \mathcal{W}} \left\{ \tilde{w} \in \mathcal{W} : \theta(x) - \theta(\tilde{x}) + (w - \tilde{w})^{\mathrm{T}} F(w) \geq 0 \right\}. \tag{3.16}$$

In the following, we summarize some notations which will be used in later analysis. These notations will make the presentation of our theoretical analysis in later sections more compact. Also, these notations are based on previous work [35, 37], and you can find more details in these references. Define $\mathcal{V} = \mathcal{X}_2 \times \cdots \times \mathcal{X}_m \times \mathbb{R}^{\ell}$, and accordingly we also use the notation

$$\mathcal{V}^* = \left\{ (x_2^*, \ldots, x_m^*, \lambda^*) \mid (x_1^*, x_2^*, \ldots, x_m^*, \lambda^*) \in \mathcal{W}^* \right\}.$$

Further define a matrix sequence $\left\{ H_k = \beta_k I_\ell \in \mathbb{R}^{\ell \times \ell} \right\}$ and a new variable sequence $\{u_k\}$ based on the variable w and the matrix H_k, e.g.,

$$u_k(w) = \begin{pmatrix} u_k(x_2) \\ \vdots \\ u_k(x_m) \\ u_k(\lambda) \end{pmatrix} = \begin{pmatrix} \sqrt{H_k} A_2 x_2 \\ \vdots \\ \sqrt{H_k} A_m x_m \\ \sqrt{H_k}^{-1} \lambda \end{pmatrix} \in \mathbb{R}^{m\ell}, \tag{3.17}$$

from which the notation $u_k(w^k) = \left\{ u_k(x_2^k), \ldots, u_k(x_m^k), u_k(\lambda^k) \right\}$ is also clear. For some concrete applications of (3.1) such as those in [22, 37], the matrices $A_2, \cdots A_m$ are all identity matrices. For these cases, $u_k(w)$ reduces to v if all β_k equal to 1. With these notations, the scheme (1) can be summarized as generating the new iteration $u_k\left(w^{k+1}\right)$ with the input $u_k\left(w^k\right)$. Further based on some notations in [35], we define

$$\mathcal{A}_k = \operatorname{diag}\left(\sqrt{H_k}A_2, \sqrt{H_k}A_3, \ldots, \sqrt{H_k}A_m, \sqrt{H_k}^{-1} I_\ell\right) \in \mathbb{R}^{m\ell \times \left(\sum_{i=2}^{m} n_i + \ell\right)}, \quad (3.18)$$

which indicates that $u_k(w) = \mathcal{A}_k z$. Then, we introduce several matrices in block-wise form as follows:

$$Q = \begin{pmatrix} I_\ell & 0 & \cdots & \cdots & 0 \\ I_\ell & I_\ell & \ddots & \ddots & \vdots \\ \vdots & \ddots & \ddots & \ddots & \vdots \\ I_\ell & \cdots & I_\ell & I_\ell & 0 \\ -I_\ell & -I_\ell & \cdots & -I_\ell & I_\ell \end{pmatrix} \in \mathbb{R}^{m\ell \times m\ell}, \quad \mathcal{M} = \begin{pmatrix} I_\ell & -I_\ell & 0 & 0 & 0 \\ 0 & I_\ell & \ddots & \ddots & \vdots \\ \vdots & \ddots & \ddots & -I_\ell & 0 \\ 0 & \cdots & 0 & I_\ell & 0 \\ -I_\ell & -I_\ell & \cdots & -I_\ell & I_\ell \end{pmatrix} \in \mathbb{R}^{m\ell \times m\ell}, \quad (3.19)$$

which will be used in the substitution procedures to be proposed. Note that Q and $\mathcal{M}$ are all well-structured block lower triangular matrices, and for (3.1) with $m = 2$, we have

$$Q = \mathcal{M} = \begin{pmatrix} I_\ell & 0 \\ -I_\ell & I_\ell \end{pmatrix} \in \mathbb{R}^{2\ell \times 2\ell}.$$

3.3 Adaptive Variable Penalty Parameter β_k

In this section, we will discuss the adaptive updating scheme of the important penalty parameter β_k. The main concern is not only to efficiently accelerate our algorithm framework, but also to guarantee better theoretical results, which include global convergence and iteration complexity for general convex problems. At the beginning, we present the following generic property that the sequence $\{\beta_k\}$ needs to satisfy, and our theoretical analysis is based on the following property.

Summable Bounded Property: there exists another non-negative sequence $\{\eta_k\}$ such that $\{\beta_k, \eta_k\}$ satisfies

$$\frac{1}{1+\eta_k}\beta_k \le \beta_{k+1} \le (1+\eta_k)\beta_k, \qquad \sum_{k=0}^{+\infty} \eta_k < +\infty. \quad (3.20)$$

This summable bounded property indicates that although the penalty parameter β_k can be dynamically adjusted, each pair β_k and β_{k+1} should be "close." They have at most $1 + \eta_k$ times the gap.

Motivated by [21], in which residual balancing scheme is applied to adaptively tune the parameter sequence $\{\beta_k\}$, we define the primal and dual "residuals" with respect to iterations $u\left(w^{k-1}\right)$ and $u\left(w^k\right)$ as follows

$$p_k := \mathcal{A}\left[u\left(w^k\right) - u\left(w^{k-1}\right)\right],$$

$$d_k := \sum_{i=1}^{m} A_i x_i^k - b,$$

where $\mathcal{A} := \operatorname{diag}(A_2, \cdots, A_m) \in \mathbb{R}^{(m-1)\ell \times \sum_{i=2}^{m} n_i}$. According to the convergence results in the following sections, we guarantee that both of these two residuals approach zero as the iterates become more accurate. We observe that increasing β_k strengthens the penalty term, yielding larger primal residual but smaller dual residual; conversely, decreasing β_k leads to smaller primal residual but larger dual residual. Both residuals must be small enough at convergence, it makes sense to "balance" them, i.e., tune β_k to keep both residuals of similar magnitude, i.e., $\|p_k\| \approx \|d_k\|$.

Given a constant $\mu \in (0, 1)$ and a non-negative sequence $\{\eta_k\}$ that satisfies

$$\sum_{k=0}^{+\infty} \eta_k < +\infty, \tag{3.21}$$

we consider the following three strategies, where $t \in \mathbb{N}_+$:

1. $\{\beta_k\}$ is monotonically nondecreasing:

$$\beta_{k+1} = \begin{cases} \beta_k (1 + \eta_k), & \text{if } \|p_k\| < \mu \|d_k\|, \\ \beta_k, & \text{otherwise}, \end{cases} \tag{3.22}$$

   with $\eta_k = 0$ if $(k \bmod t) \neq 0$;
2. $\{\beta_k\}$ is monotonically nonincreasing:

$$\beta_{k+1} = \begin{cases} \frac{\beta_k}{1+\eta_k}, & \text{if } \|d_k\| < \mu \|p_k\|, \\ \beta_k, & \text{otherwise}, \end{cases} \tag{3.23}$$

   with $\eta_k = 0$ if $(k \bmod t) \neq 0$;
3. $\{\beta_k\}$ is self-adaptive:

$$\beta_{k+1} = \begin{cases} \beta_k (1 + \eta_k), & \text{if } \|p_k\| < \mu \|d_k\|, \\ \frac{\beta_k}{1+\eta_k}, & \text{if } \|d_k\| < \mu \|p_k\|, \\ \beta_k, & \text{otherwise}, \end{cases} \tag{3.24}$$

   with $\eta_k = 0$ if $(k \bmod t) \neq 0$.

All the above three cases indicate that we will adjust β every t iterations if possible and we always have

$$\beta_{k+1} = \beta_k (1 + \eta_k), \text{ or } \beta_{k+1} = \frac{\beta_k}{1 + \eta_k}, \text{ or } \beta_{k+1} = \beta_k,$$

which demonstrates that these strategies all guarantee the **summable bounded property**, i.e., (3.20). Furthermore, under condition (3.21), we can obtain that

$$\prod_{i=1}^{\infty}(1+\eta_k) < +\infty.$$

Hence, the sequence $\{\beta_k\}$ is both upper bounded and bounded below away from zero; that is, we have

$$\inf_k \{\beta_k\} > 0, \quad \sup_k \{\beta_k\} < +\infty.$$

3.4 Prediction-Correction Reformulation Framework

In this section, we reformulate the proposed Algorithm 1 as a prediction-correction framework; some lemmas are proved accordingly. Note that this prediction-correction framework is simply a theoretical revisit to the scheme (3.8) that is more convenient for the coming convergence analysis; it is not necessary to obey this framework for implementing the scheme (3.12). In the following, we propose a key lemma of this paper, which immediately enables us to reformulate the Algorithm 1 with adaptive variable penalty parameters as a prediction-correction framework.

Lemma 3.1 *Let u^{k+1} and $\tilde{w}^k$ are generated by Algorithm 1 with given u^k, together with $u_k(w^k) = \mathcal{A}_k z^k$ and $u_k(\tilde{w}^k) = \mathcal{A}_k \tilde{z}^k$. Then we have*

1. *Prediction step: for all $w \in \mathcal{W}$*

$$\theta(x) - \theta(\tilde{x}^k) + \left(w - \tilde{w}^k\right)^{\mathrm{T}} F(\tilde{w}^k) \geq \left(u_k(w) - u_k(\tilde{w}^k)\right)^{\mathrm{T}} Q\left(u_k(w^k) - u_k(\tilde{w}^k)\right); \quad (3.25)$$

2. *Correction-step: relationship between $u_k(w^{k+1})$ and $u_k(\tilde{w}^k)$ with $\alpha \in (0, 1)$*

$$u_k(w^{k+1}) = u_k(w^k) - \alpha \mathcal{M}\left(u_k(w^k) - u_k(\tilde{w}^k)\right); \quad (3.26)$$

3. *Based on matrices Q and $\mathcal{M}$ in (3.19), we define matrix $\mathcal{H}$ and $\mathcal{G}$*

$$\mathcal{H} = Q\mathcal{M}^{-1} \in \mathbb{R}^{m\ell\times m\ell}, \quad \mathcal{G} = Q^{\mathrm{T}} + Q - \mathcal{M}^{\mathrm{T}}\mathcal{H}\mathcal{M} \in \mathbb{R}^{m\ell\times m\ell}, \quad (3.27)$$

where matrix $\mathcal{H}$ is symmetric and positive definite, while matrix $\mathcal{G}$ is also symmetric but positive semi-definite.

Proof The *prediction-step* (3.25) can be obtained through the same inference as [35, Theorem 3.3] except that $u_k(w)$ is based on the adaptive penalty parameter β_k. So that we ignore the detailed proof here, please check [35, Theorem 3.3] for more details.

As for the *correction-step*, it can be directly obtained by using the notations $u_k(w^k)$, $u_k(\tilde{w}^k)$, $u_k(w^{k+1})$ and $\mathcal{N}$. Also, the positive semi-definiteness of $\mathcal{H}$ and $\mathcal{G}$ can be obtained through some basic matrix calculations. Almost the same technique has been used in [12, 35, 37]. □

Moreover, based on the above Lemma 3.1 and more structure-driven properties of the above-defined notations, we prove two contraction-type convergence properties.

Lemma 3.2 *Let u^{k+1} and $\tilde{w}^k$ are generated by Algorithm 1 with given u^k, together with $u_k(w^k) = \mathcal{A}_k z^k$ and $u_k(\tilde{w}^k) = \mathcal{A}_k \tilde{z}^k$. Then we have*

1. *Basic-contraction-function: for all $w \in \mathcal{W}$*

$$\begin{aligned}&\theta(x) - \theta(\tilde{x}^k) + \left(w - \tilde{w}^k\right)^{\mathrm{T}} F\left(\tilde{w}^k\right)\\ &\geq \frac{1}{2\alpha}\left(\left\|u_k(w) - u_k(w^{k+1})\right\|_{\mathcal{H}}^2 - \left\|u_k(w) - u_k(w^k)\right\|_{\mathcal{H}}^2\right) + \frac{1-\alpha}{2\alpha^2}\left\|u_k(w^k) - u_k(w^{k+1})\right\|_{\mathcal{H}}^2;\end{aligned} \tag{3.28}$$

2. *Strictly-contraction-function: for all $w^* \in \mathcal{W}^*$*

$$\left\|u_k(w^{k+1}) - u_k(w^*)\right\|_{\mathcal{H}}^2 \leq \left\|u_k(w^k) - u_k(w^*)\right\|_{\mathcal{H}}^2 - \frac{1-\alpha}{\alpha}\left\|u_k(w^k) - u_k(w^{k+1})\right\|_{\mathcal{H}}^2. \tag{3.29}$$

Proof For the first *basic-contraction-function*, recall (3.25) and the definition of $\mathcal{H}$, we have for all $w \in \mathcal{W}$

$$\theta(x) - \theta(\tilde{x}^k) + \left(w - \tilde{w}^k\right)^{\mathrm{T}} F\left(\tilde{w}^k\right) \geq \frac{1}{\alpha}\left(u_k(w) - u_k(\tilde{w}^k)\right)^{\mathrm{T}} \mathcal{H}\left(u_k(w^k) - u_k(w^{k+1})\right).$$

Further applying the identity

$$(a-b)^{\mathrm{T}} \mathcal{H}(c-d) = \frac{1}{2}\left\{\|a-d\|_{\mathcal{H}}^2 - \|a-c\|_{\mathcal{H}}^2\right\} + \frac{1}{2}\left\{\|c-b\|_{\mathcal{H}}^2 - \|d-b\|_{\mathcal{H}}^2\right\},$$

to the right-hand side of the above inequality with

$$a = u_k(w),\ b = u_k(\tilde{w}^k),\ c = u_k(w^k),\ d = u_k(w^{k+1}),$$

we thus obtain

$$\begin{aligned}&\left(u_k(w) - u_k(\tilde{w}^k)\right)^{\mathrm{T}} \mathcal{H}\left(u_k(w^k) - u_k(w^{k+1})\right)\\ &= \frac{1}{2\alpha}\left(\left\|u_k(w) - u_k(w^{k+1})\right\|_{\mathcal{H}}^2 - \left\|u_k(w) - u_k(w^k)\right\|_{\mathcal{H}}^2\right)\\ &\quad + \frac{1}{2\alpha}\left(\left\|u_k(w^k) - u_k(\tilde{w}^k)\right\|_{\mathcal{H}}^2 - \left\|u_k(w^{k+1}) - u_k(\tilde{w}^k)\right\|_{\mathcal{H}}^2\right).\end{aligned}$$

As for the last term, we can prove that

$$
\begin{aligned}
&\left\|u_k(w^k)-u_k(\tilde{w}^k)\right\|_{\mathcal{H}}^2-\left\|u_k(w^{k+1})-u_k(\tilde{w}^k)\right\|_{\mathcal{H}}^2\\
&=\left\|u_k(w^k)-u_k(\tilde{w}^k)\right\|_{\mathcal{H}}^2-\left\|\left(u_k(w^k)-u_k(\tilde{w}^k)\right)-\left(u_k(w^k)-u_k(w^{k+1})\right)\right\|_{\mathcal{H}}^2\\
&\overset{(3.2)}{=}\left\|u_k(w^k)-u_k(\tilde{w}^k)\right\|_{\mathcal{H}}^2-\left\|\left(u_k(w^k)-u_k(\tilde{w}^k)\right)-\alpha\mathcal{M}\left(u_k(w^k)-u_k(\tilde{w}^k)\right)\right\|_{\mathcal{H}}^2\\
&=2\alpha\left(u_k(w^k)-u_k(\tilde{w}^k)\right)^{\mathrm{T}}\mathcal{H}\mathcal{M}\left(u_k(w^k)-u_k(\tilde{w}^k)\right)-\alpha^2\left\|u_k(w^k)-u_k(\tilde{w}^k)\right\|_{\mathcal{M}^{\mathrm{T}}\mathcal{H}\mathcal{M}}^2\\
&=\alpha\left(u_k(w^k)-u_k(\tilde{w}^k)\right)^{\mathrm{T}}(Q^{\mathrm{T}}+Q-\mathcal{M}^{\mathrm{T}}\mathcal{H}\mathcal{M})\left(u_k(w^k)-u_k(\tilde{w}^k)\right)\\
&\qquad+\alpha(1-\alpha)\left(u_k(w^k)-u_k(\tilde{w}^k)\right)^{\mathrm{T}}\mathcal{M}^{\mathrm{T}}\mathcal{H}\mathcal{M}\left(u_k(w^k)-u_k(\tilde{w}^k)\right)\\
&\overset{(3.2)}{=}\alpha\left\|u_k(w^k)-u_k(\tilde{w}^k)\right\|_{\mathcal{G}}^2+\frac{1-\alpha}{\alpha}\left\|u_k(w^k)-u_k(w^{k+1})\right\|_{\mathcal{H}}^2\\
&\geq\frac{1-\alpha}{\alpha}\left\|u_k(w^k)-u_k(w^{k+1})\right\|_{\mathcal{H}}^2,
\end{aligned}
\tag{3.30}
$$

with $\alpha\in(0,1)$ and further the assertion of the first part in this theorem is proved. In the following, recall (3.28) and set $w=w^*$, we get

$$
\begin{aligned}
&\|u_k(w^k)-u_k(w^*)\|_{\mathcal{H}}^2-\|u_k(w^{k+1})-u_k(w^*)\|_{\mathcal{H}}^2\\
&\geq\frac{1-\alpha}{\alpha}\|u_k(w^k)-u_k(w^{k+1})\|_{\mathcal{H}}^2+2\alpha\left\{\theta(\tilde{x}^k)-\theta(x^*)+\left(\tilde{w}^k-w^*\right)^{\mathrm{T}}F\left(\tilde{w}^k\right)\right\}\\
&\geq\frac{1-\alpha}{\alpha}\|u_k(w^k)-u_k(w^{k+1})\|_{\mathcal{H}}^2+2\alpha\left\{\theta(\tilde{x}^k)-\theta(x^*)+\left(\tilde{w}^k-w^*\right)^{\mathrm{T}}F\left(w^*\right)\right\}\\
&\geq\frac{1-\alpha}{\alpha}\|u_k(w^k)-u_k(w^{k+1})\|_{\mathcal{H}}^2,
\end{aligned}
$$

where the second last inequality is based on the monotonicity of $F(w)$ and the last inequality is obtained by using the optimality of w^*. Thus, the assertion (3.29) follows directly. □

3.5 Theoretical Analysis

In this section, we will establish the theoretical results for ADMM-G-V, which includes the global convergence and the worst case $O\left(\frac{1}{k}\right)$ convergence rate in both ergodic and non-ergodic senses.

3.5.1 Global Convergence

The proved lemmas are adequate for establishing the global convergence of the proposed ADMM-G-V algorithm, and the analytic framework is standard in the context of contractive-type methods. Before presenting the main theorem, we introduce the variable u, which has no H_k after comparing with $u_k(w)$. Actually, $\{u^k\}$ is the sequence that we should pay more attention to, which is iteratively computed in ADMM-G-V.

Theorem 3.2 *Let $\{u(w^k)\}$ and $\{\tilde{w}^k\}$ be the sequence generated by the proposed ADMM-G-V. Then, we have*

1. $\|u(w^k) - u(w^{k+1})\|_{\mathcal{H}} \overset{k\to\infty}{\longrightarrow} 0$, $\left\|\sum_{i=1}^{m} A_i\tilde{x}_i^k - b\right\|_2 \overset{k\to\infty}{\longrightarrow} 0$;[1]
2. $\theta(\tilde{x}^k) - \theta(x^*) \overset{k\to\infty}{\longrightarrow} 0$ *for any given* $x^* \in \mathcal{X}^*$. *where* $\mathcal{A}_k = diag(A_2, A_3, \ldots, A_m, I_\ell)$.

Proof Recall the *strictly-contraction-function* property, e.g., for all $w^* \in \mathcal{W}^*$ we have

$$\left\|u_k(w^{k+1}) - u_k(w^*)\right\|_{\mathcal{H}}^2 \le \left\|u_k(w^k) - u_k(w^*)\right\|_{\mathcal{H}}^2 - \frac{1-\alpha}{\alpha}\left\|u_k(w^k) - u_k(w^{k+1})\right\|_{\mathcal{H}}^2.$$

Mention again the step-size property (3.20), we have

$$\left\|u_k(w) - u_k(w^k)\right\|_{\mathcal{H}}^2 \le (1+\eta_{k-1}) \cdot \left\|u_{k-1}(w) - u_{k-1}(w^k)\right\|_{\mathcal{H}}^2.$$

By applying this property in the above contractive property with $w = w^*$, we can obtain

$$\begin{aligned}
\frac{1}{\gamma_1^{k+1}}\left\|u_{k+1}(w^{k+1}) - u_{k+1}(w^*)\right\|_{\mathcal{H}}^2 &\le \frac{1+\eta_k}{\gamma_1^{k+1}}\left\|u_k(w^{k+1}) - u_k(w^*)\right\|_{\mathcal{H}}^2 \\
&\le \frac{1}{\gamma_1^k}\left\|u_k(w^k) - u_k(w^*)\right\|_{\mathcal{H}}^2 - \frac{1-\alpha}{\alpha\gamma_1^k}\left\|u_k(w^k) - u_k(w^{k+1})\right\|_{\mathcal{H}}^2,
\end{aligned}$$

where

$$\gamma_k^{t-1} = \prod_{j=k}^{t-1}(1+\eta_j), \quad \gamma_t^{t-1} = 1.$$

[1] $\|u(w^k) - u(w^{k+1})\|_{\mathcal{H}}$ has no relationship with the adaptive β_k, and is only determined by the iterative sequence $\{u(w^k)\}$.

Summing the above inequality from $k = 1$ to $k = t$, we have

$$\begin{aligned}
&\frac{1}{\gamma_1^{t+1}} \left\| u_{t+1}(w^{t+1}) - u_{t+1}(w^*) \right\|_{\mathcal{H}}^2 \\
&\le \frac{1}{\gamma_1^1} \left\| u_1(w^1) - u_1(w^*) \right\|_{\mathcal{H}}^2 - \sum_{k=1}^{t} \frac{1-\alpha}{\alpha \gamma_1^k} \left\| u_k(w^k) - u_k(w^{k+1}) \right\|_{\mathcal{H}}^2 \\
&\le \frac{1}{\gamma_1^1} \left\| u_1(w^1) - u_1(w^*) \right\|_{\mathcal{H}}^2 - \sum_{k=1}^{t} \frac{1-\alpha}{\alpha C_p} \left\| u_k(w^k) - u_k(w^{k+1}) \right\|_{\mathcal{H}}^2 .
\end{aligned} \tag{3.31}$$

As a result, we obtain that for all iterations k

$$\begin{aligned}
\sum_{k=1}^{t} \frac{(1-\alpha)\min\left\{\beta_k, \frac{1}{\beta_k}\right\}}{\alpha C_p} \left\| u\left(w^k\right) - u\left(w^{k+1}\right) \right\|_{\mathcal{H}}^2 &\le \sum_{k=1}^{t} \frac{1-\alpha}{\alpha C_p} \left\| u_k(w^k) - u_k(w^{k+1}) \right\|_{\mathcal{H}}^2 \\
&\le \frac{1}{\gamma_1^1} \left\| u_1(w^1) - u_1(w^*) \right\|_{\mathcal{H}}^2 < \infty.
\end{aligned} \tag{3.32}$$

Further more with the unified boundedness of the sequence $\left\{ \frac{(1-\alpha)\min\left\{\beta_k, \frac{1}{\beta_k}\right\}}{\alpha C_p} \right\}_k$, we can guarantee that

$$\left\| u\left(w^k\right) - u\left(w^{k+1}\right) \right\|_{\mathcal{H}}^2 \overset{k\to\infty}{\longrightarrow} 0. \tag{3.33}$$

As follows, we have $\left\| \lambda^k - \lambda^{k+1} \right\| \overset{k\to\infty}{\longrightarrow} 0$, which indicates

$$\left\| \sum_{i=1}^{m} A_i \tilde{x}_i^k - b \right\|_2 \overset{k\to\infty}{\longrightarrow} 0.$$

Recall the optimality condition with respect to $\tilde{x}_i^k$ and x_i^* respectively, we have

$$\begin{gathered}
\theta_i(x_i) - \theta_i(\tilde{x}_i) + \left(x_i - \tilde{x}_i^k\right)^{\mathrm{T}} \left\{ -A_i^{\mathrm{T}} \tilde{\lambda}^k + \beta A_i^{\mathrm{T}} \left(\sum_{j=2}^{i} A_j^{\mathrm{T}} \left(\tilde{x}_j^k - x_j^k \right) \right) \right\} \ge 0, \\
\theta_i(x_i) - \theta_i\left(x_i^*\right) + \left(x_i - x_i^*\right)^{\mathrm{T}} \left(-A_i^{\mathrm{T}} \lambda^* \right) \ge 0.
\end{gathered}$$

Then by subscribing x_i^* and $\tilde{x}_i^k$ into the above two inequalities respectively, we can obtain

$$\begin{aligned}
&\left(\tilde{x}_i^k - x_i^*\right)^{\mathrm{T}} \left(A_i^{\mathrm{T}} \lambda^*\right) \le \theta_i\left(\tilde{x}_i^k\right) - \theta_i\left(x_i^*\right) \\
&\qquad \le \left(\tilde{x}_i^k - x_i^*\right)^{\mathrm{T}} \left(A_i^{\mathrm{T}} \tilde{\lambda}^k\right) + \left(x_i^* - \tilde{x}_i^k\right)^{\mathrm{T}} \left[\beta_k A_i^{\mathrm{T}} \left(\sum_{j=2}^{i} A_j \left(\tilde{x}_j^k - x_j^k \right) \right) \right].
\end{aligned} \tag{3.34}$$

Summing the above inequality from $i = 1$ to $i = m$, we can further obtain

$$\begin{aligned}
&\frac{1}{\beta_k}\left(\sum_{i=1}^{m} A_i\tilde{x}_i^k - b\right)^{\mathrm{T}} \lambda^* \leq \theta\left(\tilde{x}^k\right) - \theta\left(x^*\right) \\
&\leq \frac{1}{\beta_k}\left(\sum_{i=1}^{m} A_i\tilde{x}_i^k - b\right)^{\mathrm{T}} \tilde{\lambda}^k + \sum_{i=1}^{m}\left\{\left(x_i^* - \tilde{x}_i^k\right)^{\mathrm{T}}\left[\beta_k A_i^{\mathrm{T}}\left(\sum_{j=2}^{i} A_j\left(\tilde{x}_j^k - x_j^k\right)\right)\right]\right\} \\
&= \frac{1}{\beta_k}\left(\sum_{i=1}^{m} A_i\tilde{x}_i^k - b\right)^{\mathrm{T}} \tilde{\lambda}^k + \beta_k\left[u\left(w^*\right) - u\left(\tilde{w}^k\right)\right]^{\mathrm{T}} \mathcal{L}\left[u\left(\tilde{w}^k\right) - u\left(w^k\right)\right],
\end{aligned} \tag{3.35}$$

where matrix $\mathcal{L}$ is defined as

$$\mathcal{L} = \begin{pmatrix} I_\ell & 0 & \cdots\cdots & 0 \\ I_\ell & I_\ell & \ddots \quad \ddots & \vdots \\ \vdots & \ddots & \ddots \quad \ddots & \vdots \\ I_\ell & \cdots\cdots & I_\ell & 0 \\ 0 & \cdots\cdots & \cdots & 0 \end{pmatrix}.$$

We know that

$$\left\|\sum_{i=1}^{m} A_i\tilde{x}_i^k - b\right\|_2 \stackrel{k\to\infty}{\longrightarrow} 0, \quad \left\|u\left(\tilde{w}^k\right) - u\left(w^k\right)\right\| \stackrel{k\to\infty}{\longrightarrow} 0,$$

together the boundedness of sequence $u\left(w^k\right)$, $u\left(\tilde{w}^k\right)$ and $u\left(w^*\right)$ (strictly-contraction-function property), we can conclude that both sides of (3.35) converge to 0. As a result, we guarantee that

$$\theta(\tilde{x}^k) - \theta(x^*) \stackrel{k\to\infty}{\longrightarrow} 0, \quad \forall x^* \in \mathcal{X}^*.$$

This finishes the proof of this theorem. □

Besides the global convergence of ADMM-G-V, the convergence rate is also another focus of attention. In the following two subsections, we will establish the worst-case iteration complexity in both the ergodic sense and the non-ergodic sense.

3.5.2 Convergence Rate in the Ergodic Sense

In Theorem 3.1, we can prove the equivalence of the following two conditions (see [36, Theorem 2.3.5] or [12, Theorem 2.1]), e.g., for all $w \in \mathcal{W}$

$$\tilde{w} \in \mathcal{W}, \quad \theta(x) - \theta(\tilde{x}) + (w - \tilde{w})^{\mathrm{T}} F(\tilde{w}) \geq 0,$$

which is equivalent to

$$\theta(x) - \theta(\tilde{x}) + (w - \tilde{w})^{\mathrm{T}} F(w) \geq 0.$$

We use the late one to define the approximate solution of VI. Namely, for given $\epsilon > 0$, $\tilde{w} \in \mathcal{W}$ is called an ϵ-approximate solution of VI $(\mathcal{W}, F, \theta)$, if it satisfies for all $w \in \mathcal{W}_{(\tilde{w})} := \left\{ w \in \mathcal{W} \;\middle|\; \|w - \tilde{w}\| \leq 1 \right\}$

$$\theta(x) - \theta(\tilde{x}) + (w - \tilde{w})^{\mathrm{T}} F(w) \geq -\epsilon.$$

We need to show that for given $\epsilon > 0$, after some iterations, it can offer a $\tilde{w} \in \mathcal{W}$, such that

$$\tilde{w} \in \mathcal{W}, \quad \sup_{w \in \mathcal{W}_{(\tilde{w})}} \left\{ \theta(\tilde{u}) - \theta(u) + (\tilde{w} - w)^{\mathrm{T}} F(w) \right\} \leq \epsilon. \tag{3.36}$$

Lemma 3.2 is the base for the convergence rate proof. Using the monotonicity of F, we have

$$\left(w - \tilde{w}^k\right)^{\mathrm{T}} F(w) \geq \left(w - \tilde{w}^k\right)^{\mathrm{T}} F\left(\tilde{w}^k\right).$$

Substituting it in (3.28), we obtain for all $w \in \mathcal{W}$

$$\theta(x) - \theta(\tilde{x}^k) + \left(w - \tilde{w}^k\right)^{\mathrm{T}} F(w) + \frac{1}{2\alpha} \left\| u_k(w) - u_k(w^k) \right\|_{\mathcal{H}}^2 \geq \frac{1}{2\alpha} \left\| u_k(w) - u_k(w^{k+1}) \right\|_{\mathcal{H}}^2. \tag{3.37}$$

Lemma 3.3 *Let $\{w^k\}$ be the sequence generated by* **Algorithm-G-V** *for the problem (3.1) and $\tilde{w}^k$ is obtained in the $k+1$-th iteration, together with $u_k(w^k) = \mathcal{A}_k v^k$ and $u_k(\tilde{w}^k) = \mathcal{A}_k \tilde{v}^k$. Then, we have*

$$\sum_{k=0}^{t} \gamma_k^{t-1} \theta(\tilde{x}^k) - \sum_{k=0}^{t} \gamma_k^{t-1} \theta(x) + \left(\sum_{k=0}^{t} \gamma_k^{t-1} \tilde{w}^k - \sum_{k=0}^{t} \gamma_k^{t-1} w \right)^{\mathrm{T}} F(w) \leq \frac{\gamma_0^{t-1}}{2\alpha} \|u_0(w) - u_0(w^0)\|_{\mathcal{H}}^2, \tag{3.38}$$

where

$$\gamma_k^{t-1} = \prod_{j=k}^{t-1} (1 + \eta_j), \quad \gamma_t^{t-1} = 1. \tag{3.39}$$

Proof First, it holds that $\tilde{w}^k \in \mathcal{W}$ for all $k \geq 0$ and for all $w \in \mathcal{W}$

$$\theta(x) - \theta(\tilde{x}^k) + \left(w - \tilde{w}^k\right)^{\mathrm{T}} F(w) + \frac{1}{2\alpha}\left\|u_k(w) - u_k(w^k)\right\|_{\mathcal{H}}^2 \geq \frac{1}{2\alpha}\left\|u_k(w) - u_k(w^{k+1})\right\|_{\mathcal{H}}^2.$$

According to (3.20), we have

$$\left\|u_k(w) - u_k(w^k)\right\|_{\mathcal{H}}^2 \leq (1 + \eta_{k-1}) \cdot \left\|u_{k-1}(w) - u_{k-1}(w^k)\right\|_{\mathcal{H}}^2, \tag{3.40}$$

consequently, it follows that for all $w \in \mathcal{W}$

$$\begin{aligned} &\theta(\tilde{x}^k) - \theta(x) + \left(\tilde{w}^k - w\right)^{\mathrm{T}} F(w) + \frac{1}{2\alpha}\left\|u_k(w) - u_k(w^{k+1})\right\|_{\mathcal{H}}^2 \\ &\leq \frac{1}{2\alpha}\left\|u_k(w) - u_k(w^k)\right\|_{\mathcal{H}}^2 \leq \frac{1}{2\alpha}(1 + \eta_{k-1})\left\|u_{k-1}(w) - u_{k-1}(w^k)\right\|_{\mathcal{H}}^2. \end{aligned} \tag{3.41}$$

Therefore, we have the following sequential inequalities

$$\begin{cases} \theta(\tilde{x}^0) - \theta(x) + \left(\tilde{w}^0 - w\right)^{\mathrm{T}} F(w) + \frac{1}{2\alpha}\left\|u_0(w) - u_0(w^1)\right\|_{\mathcal{H}}^2 \leq \frac{1}{2\alpha}\left\|u_0(w) - u_0(w^0)\right\|_{\mathcal{H}}^2, \\ \theta(\tilde{x}^1) - \theta(x) + \left(\tilde{w}^1 - w\right)^{\mathrm{T}} F(w) + \frac{1}{2\alpha}\left\|u_1(w) - u_1(w^2)\right\|_{\mathcal{H}}^2 \leq \frac{1}{2\alpha}(1 + \eta_0)\left\|u_0(w) - u_0(w^1)\right\|_{\mathcal{H}}^2, \\ \qquad \vdots \qquad\qquad \vdots \\ \theta(\tilde{x}^k) - \theta(x) + \left(\tilde{w}^k - w\right)^{\mathrm{T}} F(w) + \frac{1}{2\alpha}\left\|u_k(w) - u_k(w^{k+1})\right\|_{\mathcal{H}}^2 \leq \frac{1}{2\alpha}(1 + \eta_{k-1})\left\|u_{k-1}(w) - u_{k-1}(w^k)\right\|_{\mathcal{H}}^2, \\ \qquad \vdots \qquad\qquad \vdots \\ \theta(\tilde{x}^t) - \theta(x) + \left(\tilde{w}^t - w\right)^{\mathrm{T}} F(w) + \frac{1}{2\alpha}\left\|u_t(w) - u_t(w^{t+1})\right\|_{\mathcal{H}}^2 \leq \frac{1}{2\alpha}(1 + \eta_{t-1})\left\|u_{t-1}(w) - u_{t-1}(w^t)\right\|_{\mathcal{H}}^2. \end{cases}$$

For the (k+1)-th inequality, we multiply it by a desirable factor $\gamma_k^{t-1} = \prod_{j=k}^{t-1}(1 + \eta_j)$, we can get

$$\begin{cases} \gamma_0^{t-1}\left\{\theta(\tilde{x}^0) - \theta(x) + \left(\tilde{w}^0 - w\right)^{\mathrm{T}} F(w) + \frac{1}{2\alpha}\left\|u_0(w) - u_0(w^1)\right\|_{\mathcal{H}}^2\right\} \leq \frac{1}{2\alpha}\gamma_0^{t-1}\left\|u_0(w) - u_0(w^0)\right\|_{\mathcal{H}}^2, \\ \gamma_1^{t-1}\left\{\theta(\tilde{x}^1) - \theta(x) + \left(\tilde{w}^1 - w\right)^{\mathrm{T}} F(w) + \frac{1}{2\alpha}\left\|u_1(w) - u_1(w^2)\right\|_{\mathcal{H}}^2\right\} \leq \frac{1}{2\alpha}(1 + \eta_0)\gamma_1^{t-1}\left\|u_0(w) - u_0(w^1)\right\|_{\mathcal{H}}^2, \\ \qquad \vdots \\ \gamma_k^{t-1}\left\{\theta(\tilde{x}^k) - \theta(x) + \left(\tilde{w}^k - w\right)^{\mathrm{T}} F(w) + \frac{1}{2\alpha}\left\|u_k(w) - u_k(w^{k+1})\right\|_{\mathcal{H}}^2\right\} \leq \frac{1}{2\alpha}(1 + \eta_{k-1})\gamma_k^{t-1}\left\|u_{k-1}(w) - u_{k-1}(w^k)\right\|_{\mathcal{H}}^2, \\ \qquad \vdots \\ \gamma_t^{t-1}\left\{\theta(\tilde{x}^t) - \theta(x) + \left(\tilde{w}^t - w\right)^{\mathrm{T}} F(w) + \frac{1}{2\alpha}\left\|u_t(w) - u_t(w^{t+1})\right\|_{\mathcal{H}}^2\right\} \leq \frac{1}{2\alpha}(1 + \eta_{t-1})\gamma_t^{t-1}\left\|u_{t-1}(w) - u_{t-1}(w^t)\right\|_{\mathcal{H}}^2. \end{cases}$$

Using the notations in (3.39), the above inequalities can be rewritten as

$$\begin{cases} \gamma_0^{t-1}\left\{\theta(\tilde{x}^0)-\theta(x)+\left(\tilde{w}^0-w\right)^{\mathrm{T}}F(w)\right\}+\frac{\gamma_0^{t-1}}{2\alpha}\left\|u_0(w)-u_0(w^1)\right\|_{\mathcal{H}}^2\le\frac{\gamma_0^{t-1}}{2\alpha}\left\|u_0(w)-u_0(w^0)\right\|_{\mathcal{H}}^2, \\ \gamma_1^{t-1}\left\{\theta(\tilde{x}^1)-\theta(x)+\left(\tilde{w}^1-w\right)^{\mathrm{T}}F(w)\right\}+\frac{\gamma_1^{t-1}}{2\alpha}\left\|u_1(w)-u_1(w^2)\right\|_{\mathcal{H}}^2\le\frac{\gamma_0^{t-1}}{2\alpha}\left\|u_0(w)-u_0(w^1)\right\|_{\mathcal{H}}^2, \\ \qquad\vdots\qquad\vdots \\ \gamma_k^{t-1}\left\{\theta(\tilde{x}^k)-\theta(x)+\left(\tilde{w}^k-w\right)^{\mathrm{T}}F(w)\right\}+\frac{\gamma_k^{t-1}}{2\alpha}\left\|u_k(w)-u_k(w^{k+1})\right\|_{\mathcal{H}}^2\le\frac{\gamma_{k-1}^{t-1}}{2\alpha}\left\|u_{k-1}(w)-u_{k-1}(w^k)\right\|_{\mathcal{H}}^2, \\ \qquad\vdots\qquad\vdots \\ \gamma_t^{t-1}\left\{\theta(\tilde{x}^t)-\theta(x)+\left(\tilde{w}^t-w\right)^{\mathrm{T}}F(w)\right\}+\frac{\gamma_t^{t-1}}{2\alpha}\left\|u_t(w)-u_t(w^{t+1})\right\|_{\mathcal{H}}^2\le\frac{\gamma_{t-1}^{t-1}}{2\alpha}\left\|u_{t-1}(w)-u_{t-1}(w^t)\right\|_{\mathcal{H}}^2. \end{cases}$$

Adding all the above inequalities together, we get (3.38) and the lemma is proved. □

Theorem 3.3 *Let* $\{w^k\}$ *be the sequence generated by* **Algorithm-G-V** *for the problem (3.1) and* $\tilde{w}^k$ *is obtained in the* $k+1$*-th iteration, together with* $u_k(w^k)=\mathcal{A}_k v^k$ *and* $u_k(\tilde{w}^k)=\mathcal{A}_k\tilde{v}^k$*. Then, for any integer number* $t>0$*, we have for all* $w\in\mathcal{W}$

$$\theta(\tilde{x}_t)-\theta(x)+(\tilde{w}_t-w)^{\mathrm{T}}F(w)\le\frac{\gamma_0^{t-1}}{2\alpha\Upsilon_t}\left\|u_0(w)-u_0(w^1)\right\|_{\mathcal{H}}^2, \tag{3.42}$$

where

$$\tilde{w}_t=\frac{1}{\Upsilon_t}\sum_{k=0}^{t}\gamma_k^{t-1}\tilde{w}^k,\qquad \Upsilon_t=\sum_{k=0}^{t}\gamma_k^{t-1}, \tag{3.43}$$

and γ_k^{t-1} *is defined in* (3.39).

Proof Use the notation of $\tilde{w}_t$, (3.38) can be written as for all $w\in\mathcal{W}$

$$\frac{1}{\Upsilon_t}\sum_{k=0}^{t}\gamma_k^{t-1}\theta(\tilde{x}^k)-\theta(x)+(\tilde{w}_t-w)^{\mathrm{T}}F(w)\le\frac{\gamma_0^{t-1}}{2\alpha\Upsilon_t}\left\|u_0(w)-u_0(w^1)\right\|_{\mathcal{H}}^2. \tag{3.44}$$

Since $\theta(u)$ is convex and

$$\tilde{x}_t=\frac{1}{\Upsilon_t}\sum_{k=0}^{t}\gamma_k^{t-1}\tilde{x}^k,$$

we have that

$$\theta(\tilde{x}_t)\le\frac{1}{\Upsilon_t}\sum_{k=0}^{t}\gamma_k^{t-1}\theta(\tilde{x}^k).$$

Substituting it in (3.44), the assertion of this theorem follows directly. □

Further since

$$\gamma_0^{t-1}=\prod_{j=0}^{t-1}(1+\eta_j)\le C_p\qquad\text{and}\qquad\Upsilon_t=\sum_{k=0}^{t}\gamma_k^{t-1}\ge t+1,$$

and it follows from (3.42) that for all $w \in \mathcal{W}$

$$\theta(\tilde{x}_t) - \theta(u) + (\tilde{w}_t - w)^{\mathrm{T}} F(w) \le \frac{C_p}{2\alpha(t+1)} \left\| u_0(w) - u_0(w^1) \right\|_{\mathcal{H}}^2 \sim O\left(\frac{1}{t}\right).$$

3.5.3 Convergence Rate in a Non-ergodic Sense

At the beginning, we first discuss that the quantity $\left\| u_k(w^k) - u_k(w^{k+1}) \right\|_{\mathcal{H}}^2$ can be used to measure the accuracy of the iterate $\tilde{w}^k$ to a solution point of $VI(\mathcal{W}, F, \theta)$. More specifically, since $\mathcal{H}$ is positive definite, we conclude that $u_k(w^k) - u_k(w^{k+1}) = 0$ and $u_k(w^k) - u_k(\tilde{w}^k) = 0$ if $\left\| u_k(w^k) - u_k(w^{k+1}) \right\|_{\mathcal{H}}^2 = 0$. In other words, recall (3.25), we can obtain the following variational inequality characterization

$$\tilde{w}^k \in \mathcal{W},\ \theta(x) - \theta(\tilde{x}^k) + \left(w - \tilde{w}^k\right)^{\mathrm{T}} F\left(\tilde{w}^k\right) \ge 0, \quad \forall w \in \mathcal{W},$$

which indicates $\tilde{w}^k$ is a solution of $VI(\mathcal{W}, F, \theta)$. Therefore, $\left\| u_k(w^k) - u_k(w^{k+1}) \right\|_{\mathcal{H}}^2$ can be considered as an error measurement after k iterates of the ADMM-G-V scheme, and it is reasonable to seek an upper bound of $\left\| u_k(w^k) - u_k(w^{k+1}) \right\|_{\mathcal{H}}^2$ in term of the quantity of $O\left(\frac{1}{k}\right)$ for the purpose of investigating the convergence rate of ADMM-G-V. Based on this fact, we have the following theorem.

Theorem 3.4 *Let* $\{w^k\}$ *be the sequence generated by* **Algorithm-G-V** *for the problem (3.1) and* $\tilde{w}^k$ *is obtained in the* $k+1$*-th iteration, together with* $u_k(w^k) = \mathcal{A}_k v^k$ *and* $u_k(\tilde{w}^k) = \mathcal{A}_k \tilde{v}^k$. *Then, for any integer number* $t > 0$, *we have for all* $w \in \mathcal{W}$

$$\min_{1 \le k \le t} \left\{ \left\| u(w^k) - u(w^{k+1}) \right\|_{\mathcal{H}}^2 \right\} \le \frac{\alpha C_p^2 \beta_0}{(1-\alpha)\gamma_1^1 \cdot t} \left\| u_1(w^1) - u_1(w^*) \right\|_{\mathcal{H}}^2 \sim O\left(\frac{1}{t}\right). \tag{3.45}$$

Proof Recall the *strictly-contraction-function* property in Lemma 3.2, for any $w^* \in \mathcal{W}^*$

$$\left\| u_k(w^{k+1}) - u_k(w^*) \right\|_{\mathcal{H}}^2 \le \left\| u_k(w^k) - u_k(w^*) \right\|_{\mathcal{H}}^2 - \frac{1-\alpha}{\alpha} \left\| u_k(w^k) - u_k(w^{k+1}) \right\|_{\mathcal{H}}^2.$$

Mention again the step-size property (3.40)

$$\left\| u_k(w) - u_k(w^k) \right\|_{\mathcal{H}}^2 \le (1 + \eta_{k-1}) \cdot \left\| u_{k-1}(w) - u_{k-1}(w^k) \right\|_{\mathcal{H}}^2.$$

By applying this property in the above contractive property with $w = w^*$, we can obtain

$$
\begin{aligned}
\frac{1}{\gamma_1^{k+1}} \left\| u_{k+1}(w^{k+1}) - u_{k+1}(w^*) \right\|_{\mathcal{H}}^2 &\le \frac{1+\eta_k}{\gamma_1^{k+1}} \left\| u_k(w^{k+1}) - u_k(w^*) \right\|_{\mathcal{H}}^2 \\
&\le \frac{1}{\gamma_1^k} \left\| u_k(w^k) - u_k(w^*) \right\|_{\mathcal{H}}^2 - \frac{1-\alpha}{\alpha \gamma_1^k} \left\| u_k(w^k) - u_k(w^{k+1}) \right\|_{\mathcal{H}}^2 .
\end{aligned}
$$

Summing the above inequality from $k = 1$ to $k = t$, we have

$$
\begin{aligned}
\frac{1}{\gamma_1^{t+1}} \| u_{t+1}(w^{t+1}) - u_{t+1}(w^*) \|_{\mathcal{H}}^2 &\le \frac{1}{\gamma_1^1} \| u_1(w^1) - u_1(w^*) \|_{\mathcal{H}}^2 - \sum_{k=1}^{t} \frac{1-\alpha}{\alpha \gamma_1^k} \left\| u_k(w^k) - u_k(w^{k+1}) \right\|_{\mathcal{H}}^2 \\
&\le \frac{1}{\gamma_1^1} \| u_1(w^1) - u_1(w^*) \|_{\mathcal{H}}^2 - \sum_{k=1}^{t} \frac{1-\alpha}{\alpha C_p} \left\| u_k(w^k) - u_k(w^{k+1}) \right\|_{\mathcal{H}}^2 ,
\end{aligned}
$$

further we have

$$
\begin{aligned}
& t \cdot \min \left\{ \frac{1-\alpha}{\alpha C_p^2 \beta_0} \left\| u(w^k) - u(w^{k+1}) \right\|_{\mathcal{H}}^2 \right\} \\
\le\; & t \cdot \min \left\{ \frac{(1-\alpha) \min \left\{ \beta_k, \frac{1}{\beta_k} \right\}}{\alpha C_p} \left\| u(w^k) - u(w^{k+1}) \right\|_{\mathcal{H}}^2 \right\} \\
\le\; & t \cdot \min \left\{ \frac{1-\alpha}{\alpha C_p} \left\| u_k(w^k) - u_k(w^{k+1}) \right\|_{\mathcal{H}}^2 \right\} \\
\le\; & \sum_{k=1}^{t} \frac{1-\alpha}{\alpha C_p} \left\| u_k(w^k) - u_k(w^{k+1}) \right\|_{\mathcal{H}}^2 \\
\le\; & \frac{1}{\gamma_1^1} \left\| u_1(w^1) - u_1(w^*) \right\|_{\mathcal{H}}^2 ,
\end{aligned}
$$

which directly supports the above assertion (3.45). □

3.6 Numerical Experiments

We consider the consensus problem over a network $\mathcal{N}$ as follows

$$
\min_{x \in \mathbb{R}^n} f(x) := \sum_{i=1}^{m} f_i(x). \tag{3.46}
$$

This multi-agent system with m agents and the set of agents is denoted by $\mathcal{V} = \{1, \cdots, m\}$. The agents communicate with each other via this undirected and connected graph topology $\mathcal{G} = (\mathcal{V}, \mathcal{E})$ where $\mathcal{E}$ is the edge set with $\mathcal{E} \subset \mathcal{V} \times \mathcal{V}$. Each agent i has a local objective function $f_i(x)$. The agents coordinate with each other to minimize the global objective function $f(x)$ defined above. In order to apply our proposed algorithm framework ADMM-G-V, we first equivalently reformulate (3.46) into the following linear constrained separable problem, i.e.,

$$\begin{aligned} \min_{x_i \in \mathbb{R}^n} \ & \sum_{i=1}^{m} f_i(x_i), \\ s.t. \ & x_i = x_j, (i, j) \in \mathcal{E}, i < j. \end{aligned} \tag{3.47}$$

In this reformulation, each edge is considered only once; as a result, the $i < j$ constraint is additionally added.

3.6.1 Undirected and Connected Graph $\mathcal{G}$

Equation (3.47) can be considered as an m-blocks separable problem which is a special case of the general problem (3.1). However, if we take into consideration the network structure, (3.47) could be formulated as a multiple block separable problem with fewer blocks. By taking the following three specific graph examples in Fig. 3.3, we obtain the following observations:

1. **Tree**: When the network structure is a tree graph as in Fig. 3.3a, we can separate the nodes into two groups, e.g., $\mathcal{V}_1 = \{1, 4, 5, 6, 7\}$ and $\mathcal{V}_2 = \{2, 3, 8, 9\}$ as in Fig. 3.3d. It is obvious that the nodes in $\mathcal{V}_1$ and $\mathcal{V}_2$ only have a relationship with the nodes in another group, not with the nodes in their own group. In other words, problem (3.47) can be formulated into a two-block separable problem, not a multiple block problem.
2. **Bipartite Graph**: When the network structure is a bipartite graph as in Fig. 3.3b, we can separate the nodes into two groups, e.g., $\mathcal{V}_1 = \{1, 2, 3, 4, 5\}$ and $\mathcal{V}_2 = \{6, 7, 8, 9\}$ as in Fig. 3.3e. It is obvious that the nodes in $\mathcal{V}_1$ and $\mathcal{V}_2$ only have a relationship with the nodes in another group, not with the nodes in their own group. In other words, problem (3.47) can also be formulated into a two-block separable problem, not a multiple block problem;
3. **General Graph**: When the network structure is a general graph as in Fig. 3.3c with 6 nodes, we can separate the nodes into three groups, e.g., $\mathcal{V}_1 = \{1, 6\}$, $\mathcal{V}_2 = \{2, 3\}$ and $\mathcal{V}_3 = \{4, 5\}$ as in 3.3f. It is obvious that the nodes in $\mathcal{V}_1$, $\mathcal{V}_2$, and $\mathcal{V}_3$ only have relationships with the nodes in other groups, not with the nodes in their own group. In other words, problem (3.47) can be formulated into a three-block separable problem while $\{x_1, x_6\}$, $\{x_2, x_3\}$ and $\{x_4, x_5\}$ are considered as three variable blocks.

All these examples show that with proper rearrangement (3.47) can be considered as a multiple block problem with much less block number than the node number in the constraint graph $\mathcal{G}$. The greedy algorithm can be used to determine the groups in

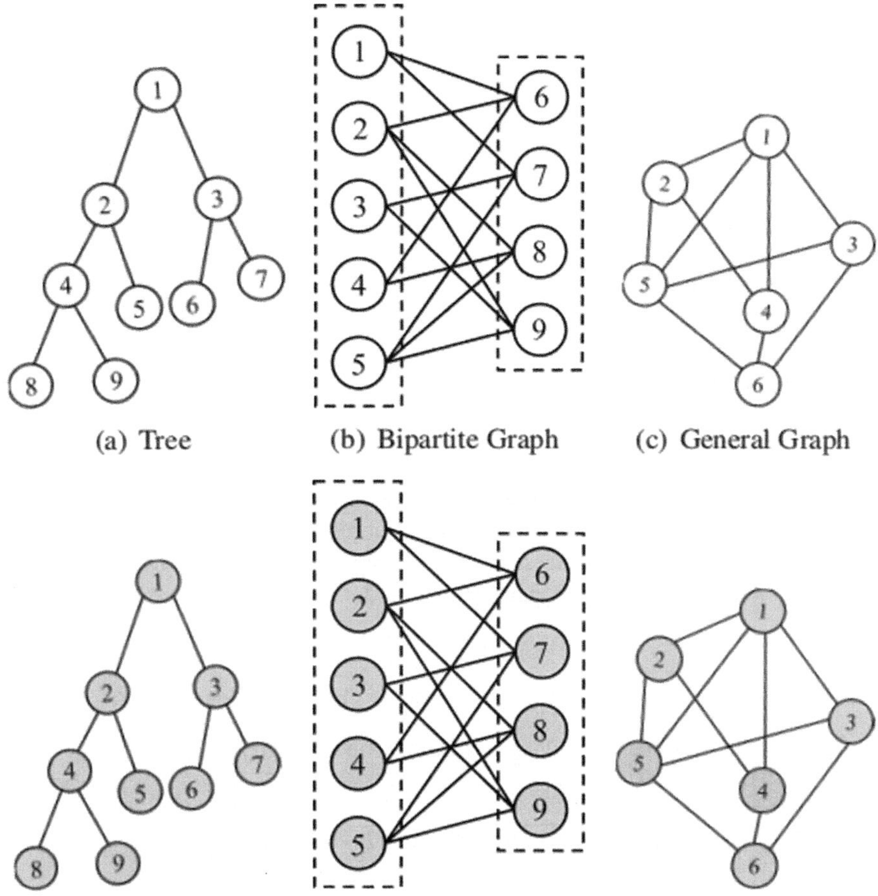

(a) Tree (b) Bipartite Graph (c) General Graph

(d) Tree (e) Bipartite Graph (f) General Graph

Fig. 3.3 Undirected and connected graph examples $\mathcal{G}$

$\mathcal{G}$ by considering this problem as the graph coloring problem [38]. We assume that the nodes $\mathcal{V}$ in graph $\mathcal{G}$ can be separated into m_r groups, i.e., $\{\mathcal{V}_1, \ldots, \mathcal{V}_{m_r}\}$, as a result problem (3.47) can be further reformulated as

$$\min_{x_i \in \mathbb{R}^n} \sum_{i=1}^{m_r} \left\{ \sum_{j \in \mathcal{V}_i} f_j(x_j) \right\}, \quad s.t. \quad \sum_{i=1}^{m_r} \left\{ \sum_{j \in \mathcal{V}_i} \mathcal{A}_j x_j \right\} = 0, \tag{3.48}$$

where $\mathcal{A}_j \in \mathbb{R}^{mn \times n}$ and

$$\mathcal{A}_j = \begin{bmatrix} A_{1j} \\ \vdots \\ A_{mj} \end{bmatrix}, \quad A_{\ell j} \in \mathbb{R}^{n\times n}$$

and

$$A_{\ell j} = \begin{cases} -\mathcal{I}_{n\times n} & \text{if } (\ell, j) \in \mathcal{E},\ \ell < j, \\ \mathcal{I}_{n\times n} & \text{if } (\ell, j) \in \mathcal{E},\ \ell > j, \\ \mathcal{O}_{n\times n} & \text{otherwise,} \end{cases} \qquad \forall\, 1 \le \ell \le n, 1 \le j \le n.$$

3.6.2 Distributed Logistic Regression

Logistic regression (LR) is a popular classification method that has been widely introduced to lots of machine learning applications, including computer vision, natural language processing and etc. The distributed Logistic regression problem considered in this paper implies that the datasets are separated in different nodes; however, we need to learn the overall LR parameter in order to guarantee a complete model. Given m distributed datasets $\{(\mathbf{A_i}, \mathbf{b_i})\}_{i=1}^m$, where $\mathbf{A_i} = \left\{\mathbf{a_i^j}\right\}_{j=1}^{N_i}$ denotes the data samples distributed in node i with $\mathbf{a_i^j} \in \mathbb{R}^{\mathbf{n}}$ and $\mathbf{b_i} = \left\{\mathbf{b_i^j}\right\}_{j=1}^{N_i}$ denotes the label set with $\mathbf{b_i^j} \in \{+\mathbf{1}, -\mathbf{1}\}$. We have the following formulation of each f_i, i.e.,

$$f_i(x_i) = \frac{1}{N_i} \sum_{j=1}^{N_i} \log\left(1 + \exp\left(-\mathbf{b_i^j}\left(\mathbf{x_i^T a_i^j}\right)\right)\right),$$

where $x_i \in \mathbb{R}^n$ denotes the parameter with respect to the i-th dataset.

3.6.3 Datasets, Network Structure and Computing Environment

We apply the model (3.47) on two real datasets. The first is the RCV1 dataset, which is used for text binary classification research [39]. The other dataset is the large-scale URL dataset, which is used for identifying suspicious URLs [40]. The detailed sample number and feature dimension are stated in the following Table 3.2. We notice that both the sample number and feature dimension are huge for the URL dataset, but because of the limitation of processing capacity in each node, we only randomly choose 160,000 data samples for this numerical experiment.

After establishing the graph $\mathcal{G}$, we will separate the datasets into the nodes in $\mathcal{G}$.

Table 3.2 Description of real datasetsDatasets

Dataset	RCV1	URL
#Examples	20000	2.4M
#Features	47236	3.2M
Label ratio +1: -1	1:1	1:2

The connected network $\mathcal{G}$ is randomly generated with $m = 100$ nodes and connectivity ratio $r = 0.2$. Both datasets are randomly and uniformly separated into m nodes, i.g., $N_i = 200$ and $N_i = 24,000$ for each dataset, respectively. By applying the greedy algorithm, we can dye the generated graph with κ colors (without loss of generality, we assume $\kappa > 2$). Further, we can consider the guaranteed problem (3.47) as a κ-block linear constrained separable problem ($\kappa = 17$ for our generated problem). Our algorithm framework will be employed to solve this distributed Logistic regression problem on the generated network $\mathcal{E}$.

All the numerical experiments were implemented on a Laptop with Intel(R) Core(TM) i5-6300U CPU@ 2.40GHz 2.50GHz and 8.00 GB Memory. All the codes were written in MATLAB2016A. We compare with two other algorithms: direct extension of multiple block ADMM (**D-ADMM**) and Algorithm 1 with constant β (**Constant-beta**). All the three β adjustment schemes (3.22), (3.23), (3.24) will the compared, which all start from $\beta = 1$. The stopping criteria for all the algorithms are set to be

$$\max\left\{\|p_k\|_2\,,\ \|d_k\|_2\right\} \leq 10^{-6},$$

while the iteration number ("Iter. #"), objective function value ("Objective"), the constraint violation value $\|d_k\|_2$ ("Constraint") and the computation time ("Time(s)") will be presented in the following tables.

3.6.4 *Numerical Performance*

In the above Table 3.3, generally **Algorithm 1** with self-adaptive β schemes perform better than multiple block ADMM with Gaussian back substitution technique, even better the direct extension of multiple block ADMM. In Table 3.4, we compare **Algorithm 1** for β schemes (3.22) (3.23) and (3.24) with respect to different t. We can find that although for some cases (like (3.22) with $t = 2$ and (3.23) with $t = 2$), a less self-adaptive frequency can guarantee faster convergence. However, overall the self-adaptive β scheme is not computationally consuming for our experiment, so that performing the self-adaptive scheme in each iteration ($t = 1$) will accelerate the algorithm framework.

Table 3.3 Numerical comparison of **D-ADMM** and **Algorithm 1** for different β scheme

Dataset	Algorithm	β_k	Iter. #	Objective	Constraint	Time(s)
RCV1	D-ADMM	10	721	230.0599	9.76e-07	551.15
		1	337	230.0599	9.71e-07	283.14
		0.1	2000	229.8048	2.46e-06	1565.57
	Algorithm 1	10	811	230.0602	9.35e-07	606.42
		1	382	230.0600	9.37e-07	327.93
		0.1	1839	230.0600	9.76e-07	1695.98
		(3.22)	223	230.0600	9.29e-07	236.93
		(3.23)	331	230.0600	9.46e-07	276.69
		(3.24)	136	230.0600	9.82e-07	153.21
URL	D-ADMM	10	889	478.4982	9.78e-07	1122.43
		1	529	478.4974	9.65e-07	632.14
		0.1	2000	477.9725	1.95e-05	2756.87
	Algorithm 1	10	1188	478.4980	9.33e-07	1671.57
		1	654	478.4978	9.45e-07	776.42
		0.1	1976	478.4878	9.69e-07	3097.53
		(3.22)	373	478.4978	9.65e-07	527.82
		(3.23)	362	478.4978	9.57e-07	496.45
		(3.24)	228	478.4978	9.54e-07	332.39

Table 3.4 Numerical comparison of **Algorithm 1** for β schemes (3.22), (3.23) and (3.24) with respect to different t

Dataset	Algorithm	β_k	t	Iter. #	Objective	Constraint	Time(s)
RCV1	**Algorithm 1**	(3.22)	1	223	230.0600	9.29e-07	236.93
			2	246	230.0600	9.67e-07	224.38
			5	313	230.0600	9.31e-07	245.29
		(3.23)	1	331	230.0600	9.46e-07	276.69
			2	347	230.0600	9.49e-07	280.16
			5	362	230.0600	9.37e-07	311.84
		(3.24)	1	136	230.0600	9.82e-07	153.21
			2	175	230.0600	9.75e-07	176.87
			5	224	230.0600	9.81e-07	226.25
URL	**Algorithm 1**	(3.22)	1	373	478.4978	9.65e-07	527.82
			2	423	478.4978	9.57e-07	578.57
			5	469	478.4978	9.69e-07	620.67
		(3.23)	1	362	478.4978	9.57e-07	496.45
			2	396	478.4978	9.29e-07	503.54
			5	423	478.4978	9.42e-07	559.58
		(3.24)	1	228	478.4978	9.54e-07	332.39
			2	247	478.4978	9.46e-07	329.63
			5	284	478.4978	9.39e-07	393.54

3.7 Conclusion and Future Work

In this paper, we propose a theoretical analysis of the alternating direction method of multipliers with a Gaussian back substitution step for the variable penalty parameters case. Both non-ergodic and ergodic worst-case convergence rates are established. Specifically, classical two-block ADMM with variable penalty parameters can be included in the framework. A better penalty parameter tuning mechanism should be further analyzed and considered.

Acknowledgements This work was supported by the NSFC with No. 12071145.

References

1. Boyd, S., Parikh, N., Chu, E., Peleato, B., Eckstein, J.: Distributed optimization and statistical learning via the alternating direction method of multipliers. Found. Trends Mach. Learn. **3**(1), 1–122 (2011)
2. Peng, Y.G., Ganesh, A., Wright, J., Wenli, X., Ma, Y.: RASL: robust alignment by sparse and low-rank decomposition for linearly correlated images. IEEE Trans. Pattern Anal. Mach. Intell. **34**(11), 2233–2246 (2012)
3. Zhang, Z., Ganesh, A., Liang, X., Ma, Y.: TILT: transform invariant low-rank textures. Int. J. Comput. Vis. **99**(1), 1–24 (2012)
4. Glowinski, R., Marrocco, A.: Approximation par éléments finis d'ordre un et résolution par pénalisation-dualité d'une classe de problémes non linéaires. R.A.I.R.O. R2 **60**(8), 41–76 (1975)
5. Eckstein, J., Yao, W.: Understanding the convergence of the alternating direction method of multipliers: theoretical and computational perspectives. Pac. J. Optim. **11**, 619–644 (2015)
6. Glowinski, R.: On alternating direction methods of multipliers: a historical perspective. In: Modeling, Simulation and Optimization for Science and Technology, pp. 59–82 (2014)
7. Glowinski, R., Marrocco, A.: Chapter IX applications of the method of multipliers to variational inequalities. Stud. Math. Appl. **15**, 299–331 (1983)
8. Glowinski, R., Oden, J.T.: Numerical methods for nonlinear variational problems. J. Appl. Mech. **52**, 739 (1985)
9. Eckstein, J., Bertsekas, D.P.: On the Douglas-Rachford splitting method and the proximal point algorithm for maximal monotone operators. Math. Program. **55**(1), 293–318 (1992)
10. Martinet, B.: Brève communication. régularisation d'inéquations variationnelles par approximations successives. ESAIM: Mathematical Modelling and Numerical Analysis—Modélisation Mathématique et Analyse Numérique **4**(R3), 154–158 (1970)
11. Rockafellar, R.T.: Monotone operators and the proximal point algorithm. SIAM J. Control. Optim. **14**(5), 877–898 (1976)
12. He, B.S., Yuan, X.M.: On the $O(1/n)$ convergence rate of the Douglas-Rachford alternating direction method. SIAM J. Numer. Anal. **50**(2), 700–709 (2012)
13. He, B.S., Yuan, X.M.: On non-ergodic convergence rate of Douglas-Rachford alternating direction method of multipliers. Numerische Mathematik **130**(3), 567–577 (2015)
14. Hestenes, M.R.: Multiplier and gradient methods. J. Optim. Theory Appl. **4**(5), 303–320 (1969)
15. Tao, M., Yuan, X.M.: Recovering low-rank and sparse components of matrices from incomplete and noisy observations. SIAM J. Optim. **21**(1), 57–81 (2011)
16. Han, D.R., Cai, X.J., Yuan, X.M.: On the convergence of the direct extension of ADMM for three-block separable convex minimization models with one strongly convex function. Comput. Optim. Appl. **66**(1), 39–73 (2017)

17. Han, D.R., Yuan, X.M.: A note on the alternating direction method of multipliers. J. Optim. Theory Appl. **155**(1), 227–238 (2012)
18. Sun, D.F., Li, M., Toh, K.-C.: A convergent 3-block semi-proximal ADMM for convex minimization problems with one strongly convex block. Asia Pac. J. Oper. Res. **32**(4), 1550024 (2015)
19. Hong, M.Y., Luo, Z.-Q.: On the linear convergence of the alternating direction method of multipliers. Math. Program. **162**(1), 165–199 (2017)
20. Yuan, X., Ye, Y., He, B., Chen, C.: The direct extension of ADMM for multi-block convex minimization problems is not necessarily convergent. Math. Program. **155**(1–2), 57–79 (2016)
21. Yang, H., He, B.S., Wang, S.L.: Alternating direction method with self-adaptive penalty parameters for monotone variational inequalitiest. J. Optim. Theory Appl. **106**(2), 337–356 (2000)
22. Yuan, X.M., Ng, M.K., Zhang, W.X.: Coupled variational image decomposition and restoration model for blurred cartoon-plus-texture images with missing pixels. IEEE Trans. Image Process. **22**(6), 2233–2246 (2013)
23. He, B., Liao, L.-Z., Han, D., Yang, H.: A new inexact alternating directions method for monotone variational inequalities. Math. Program. **92**(1), 103–118 (2002)
24. He, B.S., Yang, H.: Some convergence properties of a method of multipliers for linearly constrained monotone variational inequalities. Oper. Res. Lett. **23**(3), 151–161 (1998)
25. Kontogiorgis, S., Meyer, R.R.: A variable-penalty alternating directions method for convex optimization. Math. Program. **83**(1–3), 29–153 (1998)
26. Yoon, S., Song, C., Pavlovic, V.: Fast ADMM algorithm for distributed optimization with adaptive penalty. In: AAAI, pp. 753–759 (2016)
27. Figueiredo, M., Xu, Z., Goldstein, T.: Adaptive ADMM with spectral penalty parameter selection. In: AISTATS, pp. 718–727 (2017)
28. Ye, J., Li, X., Hu, Y., Zhang, D., He, X.: Fast and accurate matrix completion via truncated nuclear norm regularization. IEEE Trans. Pattern Anal. Mach. Intell. **35**(9), 2117–2130 (2013)
29. Meshi, O., Globerson, A.: An alternating direction method for dual MAP LP relaxation. In: Machine Learning and Knowledge Discovery in Databases, pp. 470–483 (2011)
30. Wohlberg, B.: Efficient algorithms for convolutional sparse representations. IEEE Trans. Image Process. **25**(1), 301–315 (2016)
31. Guibas, L., Solomon, J., Rustamov, R., Butscher, A.: Earth mover's distances on discrete surfaces. ACM Trans. Graph. **33**(4), 67 (2014)
32. Zhang, B., Sun, H., Zheng, W., Wu, W., Liu, Y.: A fully distributed reactive power optimization and control method for active distribution networks. IEEE Trans. Smart Grid **7**(2), 1021–1033 (2016)
33. Bioucas-Dias, J.M., Iordache, M.D., Plaza, A.: Collaborative sparse regression for hyperspectral unmixing. IEEE Trans. Geosci. Remote. Sens. **52**(1), 341–354 (2014)
34. Wu, Y., Tang, W., Shi, Z., Zhang, C.: Sparse unmixing of hyperspectral data using spectral a priori information. IEEE Trans. Geosci. Remote. Sens. **53**(2), 770–783 (2015)
35. Tao, M., He, B.S., Yuan, X.M.: Convergence rate analysis for the alternating direction method of multipliers with a substitution procedure for separable convex programming. Math. Oper. Res. **42**(3), 662–691 (2017)
36. Facchinei, F., Pang, J.-S.: Finite-Dimensional Variational Inequalities and Complementarity Problems, vol. 1. Springer Science & Business Media (2007)
37. Tao, M., He, B.S., Yuan, X.M.: Alternating direction method with Gaussian back substitution for separable convex programming. SIAM J. Optim. **22**(2), 313–340 (2012)
38. West, D.B.: Introduction to Graph Theory, 2nd edn. Prentice Hall (1996)
39. Rose, T.G., Lewis, D.D., Yang, Y., Li, F.: Rcv1: a new benchmark collection for text categorization research. J. Mach. Learn. Res. **5**, 361–397 (2004)
40. Savage, S., Ma, J., Saul, L.K., Voelker, G.M.: Identifying suspicious URLs: an application of large-scale online learning. In: ICML, pp. 681–688 (2009)

Chapter 4
Inertial Alternating Direction Method of Multipliers with Logarithmic-Quadratic Proximal Regularization

Zhongming Wu

Abstract It has been demonstrated that the alternating direction method of multipliers (ADMM) with logarithmic-quadratic proximal (LQP) regularization is efficient in solving a specific class of separable convex optimization problems. This method capitalizes on the individual separable properties and transforms the constrained subproblems into more manageable unconstrained subproblems during the iterative process. In this paper, we investigate the application of the inertial proximal point method and focus on studying the inertial ADMM and symmetric ADMM with LQP regularization for solving constrained separable convex optimization problems. These approaches employ ADMM or symmetric ADMM on extrapolated points with appropriate step sizes to accelerate the convergence rate. Under some mild conditions, we establish the global convergence of the proposed methods.

Keywords Alternating direction method of multipliers · Logarithmicquadratic proximal regularization · Inertial · Convergence analysis

4.1 Introduction

In this paper, we consider the following separable constrained optimization problem

$$\min\left\{\theta_1(x)+\theta_2(y) \mid Ax+By=b, x\in\mathbb{R}^n_+, y\in\mathbb{R}^p_+\right\}, \qquad (4.1)$$

where $\theta_1:\mathbb{R}^n_+\to\mathbb{R}$ and $\theta_2:\mathbb{R}^p_+\to\mathbb{R}$ are closed convex functions, $A\in\mathbb{R}^{m\times n}$, $B\in\mathbb{R}^{m\times p}$ and $b\in\mathbb{R}^m$. The solution set of (4.1) is assumed to be nonempty throughout our discussions in this paper. The augmented Lagrangian function of the problem (4.1) is defined by

Z. Wu (✉)
School of Management Science and Engineering, Nanjing University of Information Science and Technology, Nanjing, Jiangsu, China
e-mail: wuzm@nuist.edu.cn

X. Wang et al. (eds.), *Splitting Optimization*, ICIAM2023 Springer Series 2,
https://doi.org/10.1007/978-981-96-8574-5_4

$$\mathcal{L}_\beta(x, y, \lambda) = \theta_1(x) + \theta_2(y) - \lambda^\top(Ax + By - b) + \frac{\beta}{2}\|Ax + By - b\|^2, \tag{4.2}$$

where $\lambda \in \mathbb{R}^m$ is the Lagrange multiplier associated with the linear constraint in (4.1) and $\beta > 0$ is a penalty parameter. Problem (4.1) captures many applications in various fields such as financial portfolio optimization [34] and traffic management [40, 47].

The alternating direction method of multipliers (ADMM), originally proposed in [22, 23], is widely recognized for its efficient solution to (4.1). This method leverages the individual properties of θ_1 and θ_2. For a comprehensive understanding of ADMM's theoretical analysis and diverse applications, we refer readers to [11, 13, 20, 26, 29, 30, 46] and the relevant references therein. In this paper, we introduce a cyclically equivalent form of ADMM, with further details on this cyclic equivalence provided in [12, 14]. The iterative scheme of ADMM in "$x \to \lambda \to y$" order for (4.1) can be read as

$$\begin{cases} x^{k+1} = \arg\min\{\mathcal{L}_\beta(x, y^k, \lambda^k) \mid x \in \mathbb{R}^n_+\}, \\ \lambda^{k+1} = \lambda^k - \beta(Ax^{k+1} + By^k - b), \\ y^{k+1} = \arg\min\{\mathcal{L}_\beta(x^{k+1}, y, \lambda^{k+1}) \mid y \in \mathbb{R}^p_+\}. \end{cases} \tag{4.3}$$

Another influential and important method for solving problem (4.1) is the Peaceman-Rachford splitting (PRS) method, initially proposed in [37, 41]. When applying the PRS method to the dual of the considered problem, we obtain another alternating method, known as symmetric ADMM [18, 24, 28, 33, 49, 50]. The iterative scheme of symmetric ADMM for solving (4.1) can be read as follows:

$$\begin{cases} x^{k+1} = \arg\min\{\mathcal{L}_\beta(x, y^k, \lambda^k) \mid x \in \mathbb{R}^n_+\}, \\ \lambda^{k+1/2} = \lambda^k - \beta(Ax^{k+1} + By^k - b), \\ y^{k+1} = \arg\min\{\mathcal{L}_\beta(x^{k+1}, y, \lambda^{k+1/2}) \mid y \in \mathbb{R}^p_+\}, \\ \lambda^{k+1} = \lambda^{k+1/2} - \beta(Ax^{k+1} + By^{k+1} - b). \end{cases} \tag{4.4}$$

The symmetric ADMM (4.4) differs from ADMM due to the additional Lagrange multiplier update step, as analyzed in [22]. It has been observed that the symmetric ADMM converges faster than ADMM if it does converge [21, 27]. However, symmetric ADMM (4.4) may fail to converge without further assumptions. To address this limitation, He et al. [27] introduced an underdetermined relaxation factor $\gamma \in (0, 1)$ to both Lagrange multiplier updating steps of (4.4). The parameter $\gamma \in (0, 1)$ plays a crucial role in ensuring the strict contractiveness of the generated sequence. Under this condition, they established convergence results and demonstrated the same sublinear convergence rate as that of ADMM [29].

For the generic convex objective function in (4.1), the constrained subproblems in (4.3) and (4.4) typically require iterative and approximate solutions. However, due to the specific nature of the constraints in these (symmetric) ADMM subproblems, several studies [6–8, 36, 38, 47, 51] propose appropriate regularization of the objective functions to strictly confine the solutions within the interiors of $\mathbb{R}^n_+$ and $\mathbb{R}^p_+$. This regularization allows the constrained subproblems in (4.3) and (4.4) to be converted into unconstrained subproblems. An excellent choice for such regularization is the logarithmic-quadratic proximal (LQP) regularization, initially proposed in [6], and extensively studied in various articles such as [4, 5, 51]. The iterative scheme of ADMM with LQP regularization is as follows:

$$\begin{cases} x^{k+1} = \arg\min\{\mathcal{L}_\beta(x, y^k, \lambda^k) + rd(x, x^k) \mid x \in \mathbb{R}^n_+\}, \\ \lambda^{k+1} = \lambda^k - \beta(Ax^{k+1} + By^k - b), \\ y^{k+1} = \arg\min\{\mathcal{L}_\beta(x^{k+1}, y, \lambda^{k+1}) + sd(y, y^k) \mid y \in \mathbb{R}^p_+\}, \end{cases} \tag{4.5}$$

and the iterative scheme of symmetric ADMM with LQP regularization is

$$\begin{cases} x^{k+1} = \arg\min\{\mathcal{L}_\beta(x, y^k, \lambda^k) + rd(x, x^k) \mid x \in \mathbb{R}^n_+\}, \\ \lambda^{k+1/2} = \lambda^k - \beta(Ax^{k+1} + By^k - b), \\ y^{k+1} = \arg\min\{\mathcal{L}_\beta(x^{k+1}, y, \lambda^{k+1/2}) + sd(y, y^k) \mid y \in \mathbb{R}^p_+\}, \\ \lambda^{k+1} = \lambda^{k+1/2} - \beta(Ax^{k+1} + By^{k+1} - b), \end{cases} \tag{4.6}$$

where $d(\cdot, \cdot)$ is the LQP function defined by

$$d(z', z) := \begin{cases} \sum_{j=1}^{N} \left[\frac{1}{2}(z'_j - z_j)^2 + \mu\left(z_j^2 \log \frac{z_j}{z'_j} + z'_j z_j - z_j^2\right)\right], & \text{if } z' \in \mathbb{R}^N_{++}; \\ +\infty, & \text{otherwise}, \end{cases} \tag{4.7}$$

for any $z \in \mathbb{R}^N_{++}$, r and s are two positive scalars. Indeed, the LQP regularization terms $rd(x, x^k)$ and $sd(y, y^k)$ inherently constrain x^{k+1} and y^{k+1} to belong to $\mathbb{R}^{n_1}_{++}$ and $\mathbb{R}^{n_2}_{++}$, respectively, rendering the constraints $x \in \mathbb{R}^{n_1}_+$ and $y \in \mathbb{R}^{n_2}_+$ inactive. Consequently, only two unconstrained minimization subproblems are involved in (4.5). The convergence and the rate of convergence of these two methods have been widely discussed in the literature [15, 35, 36, 44].

Proximal point algorithm (PPA) [43, 52] is a closely related method with the existing splitting methods such as ADMM and Douglas-Rachford splitting method [19], which is to minimize a differentiable function $\psi : \mathbb{R}^n \to \mathbb{R}$, it can be interpreted as an implicit one-step discretization method for the ordinary differential equation $w' + \nabla\psi(w) = 0$, where $w : \mathbb{R} \to \mathbb{R}^n$ is differentiable, w' denotes its derivative and $\nabla\psi$ denotes the gradient of ψ. To accelerate the convergence speed of the PPA, one can adopt the multi-step methods, which

$$w^{k+1} = (I + \sigma\nabla\psi)^{-1}(w^k + \alpha(w^k - w^{k-1})), \tag{4.8}$$

where $\sigma = h^2/(1 + \tau h), \alpha = 1/(1 + \tau h)$ and I represents the identity operator. Note that the iterative scheme (4.8) can be regarded as applying the PPA to the extrapolated point $w^k + \alpha(w^k - w^{k-1})$, which is usually the so-called inertial PPA [1, 16]. Indeed, this inertial technique can be traced back to [42], and recently there has been extensive work in studying inertial-type algorithms, including the inertial forward-backward splitting method [2, 10, 39, 48], inertial Douglas-Rachford splitting method [9], and so forth.

Inspired by the inertial PPA aforementioned, Chen et al. [14] introduced an inertial proximal ADMM to enhance the convergence speed of ADMM (4.3). This approach can be seen as applying ADMM (4.3) to an extrapolated point with an appropriate step size. Subsequently, the inertial ADMM has been further investigated in [3], where the authors proposed an inertial proximal ADMM with an appropriate adjustment of the viscosity and proximal parameters. The fast convergence properties, as well as the convergence of the iterates to saddle points of the Lagrangian function, can be guaranteed. Recently, the inertial acceleration techniques have been extended to handle nonconvex optimization problems with linear constraints [17, 31, 32, 45]. However, the resulting method still requires iterative and approximate solutions for the constrained subproblems.

In this paper, we apply the inertial technique to ADMM and symmetric ADMM with LQP regularization for solving the linearly constrained separable convex optimization problem (4.1). These proposed methods not only enforce the solutions of the constrained subproblems in (4.3) and (4.4) to remain strictly within the interiors of $\mathbb{R}^n_+$ and $\mathbb{R}^p_+$, facilitating their conversion into easier unconstrained subproblems, but also accelerate the convergence speed of the iterative schemes (4.5) and (4.6) through extrapolation-based steps. Under mild assumptions, we establish the global convergence of the proposed methods.

The remainder of this paper is organized as follows. Section 4.2 provides a summary of relevant preliminary results and introduces the definition of LQP regularization. In Sects. 4.3 and 4.4, we present the inertial ADMM and symmetric ADMM with LQP regularization, respectively, and analyze their convergence properties. Finally, Sect. 4.5 concludes this paper.

4.2 Preliminaries

We first summarize some useful preliminaries known in the literature and present some simple conclusions for further analysis.

4.2.1 The Logarithmic-Quadratic Proximal Regularization

In this subsection, we give some basic knowledge about the LQP regularization. More details are also provided in [6]. Let us define

$$\varphi(c) := \begin{cases} \frac{1}{2}(c-1)^2 + \mu(c - \log c - 1), & \text{if } c > 0; \\ +\infty, & \text{otherwise,} \end{cases}$$

where $\mu \in (0, 1)$ is a given constant. Associated with φ, we define

$$\Phi'(z, z') := (z_1\varphi'(z'_1/z_1), \ldots, z_N\varphi'(z'_N/z_N))^\top, \quad \forall z, z' \in \mathbb{R}^N_{++},$$

where

$$\varphi'(z'_j/z_j) = z'_j/z_j - 1 + \mu(1 - z'_j/z_j), \quad j = 1, 2, \ldots, N.$$

For any $z', z \in \mathbb{R}^N_{++}$, we have $d(z', z) \geq \|z' - z\|^2/2$ and $d(z', z) = 0$ if and only if $z' = z$. Moreover, the function $d(\cdot, \cdot)$ defined in (4.7) can be rewritten as

$$d(z', z) = \sum_{j=1}^{N} z_j^2 \varphi(z'_j/z_j), \quad \forall z', z \in \mathbb{R}^N_{++},$$

and then we have

$$\Phi'(z, z') = \nabla_{z'} d(z', z) = (z' - z) + \mu[z - Z^2(z')^{-1}],$$

where $Z := \text{diag}(z_1, z_2, \ldots, z_N) \in \mathbb{R}^{N\times N}$, $(z')^{-1} \in \mathbb{R}^N$ is a vector whose j-th element is $1/z'_j$.

We now summarize an important lemma, proven in [36], which will be useful for the convergence analysis in subsequent sections.

Lemma 4.1 *Let* $P := \text{diag}(p_1, p_2, \ldots, p_N) \in \mathbb{R}^{N\times N}$ *be a positive definite diagonal matrix,* $q(z) \in \mathbb{R}^N$ *be a monotone mapping of* z *with respect to* $\mathbb{R}^N_+$, *and* $\vartheta : \mathbb{R}^N \to \mathbb{R}$. *Let* $\mu \in (0, 1)$ *be a constant. For given* $\bar{z}, z \in \mathbb{R}^N_{++}$, *we define* $\bar{Z} := \text{diag}(\bar{z}_1, \bar{z}_2, \ldots, \bar{z}^N)$, $z^{-1} := (1/z_1, \ldots, 1/z_N)^\top$ *and*

$$\Phi'(\bar{z}, z) := (z - \bar{z}) + \mu(\bar{z} - \bar{Z}^2 z^{-1}).$$

Then, the variational inequality

$$\vartheta(z') - \vartheta(z) + (z' - z)^\top[q(z) + P\Phi'(\bar{z}, z)] \geq 0, \quad \forall z' \in \mathbb{R}^N_+,$$

has the unique positive solution z. *Besides, for this positive solution* $z \in \mathbb{R}^N_{++}$ *and any* $z' \in \mathbb{R}^N_+$, *we have*

$$\vartheta(z') - \vartheta(z) + (z' - z)^\top q(z) \geq (1 + \mu)(\bar{z} - z)^\top P(z' - z) - \mu\|\bar{z} - z\|_P^2. \quad (4.9)$$

4.2.2 Variational Reformulation

In this subsection, we use a variational inequality (VI) reformulation of the model (4.1) and a characterization of its solution set. Let the Lagrangian function of (4.1) be

$$\mathcal{L}(x, y, \lambda) = \theta_1(x) + \theta_2(y) - \lambda^\top(Ax + By - b),$$

where $x \in \mathbb{R}^n_+$, $y \in \mathbb{R}^p_+$ and $\lambda \in \mathbb{R}^m$ be the Lagrange multiplier. Then finding a saddle point of $\mathcal{L}(x, y, \lambda)$ is to find $(x^*, y^*, \lambda^*) \in \mathbb{R}^n_+ \times \mathbb{R}^p_+ \times \mathbb{R}^m$ such that

$$\mathcal{L}(x^*, y^*, \lambda) \le \mathcal{L}(x^*, y^*, \lambda^*) \le \mathcal{L}(x, y, \lambda^*), \quad \forall x^* \in \mathbb{R}^n_+, \ y^* \in \mathbb{R}^p_+, \ \lambda^* \in \mathbb{R}^m.$$

Therefore, solving (4.1) is equivalent to finding $w^* = (x^*, y^*, \lambda^*) \in \Omega := \mathbb{R}^n_+ \times \mathbb{R}^p_+ \times \mathbb{R}^m$ such that

$$\mathrm{VI}(\Omega, \mathrm{F}, \theta): \ \theta(u) - \theta(u^*) + (w - w^*)^\top F(w^*) \ge 0, \ \ \forall w \in \Omega, \tag{4.10}$$

where

$$\theta(u) = \theta_1(x) + \theta_2(y), \ u = \begin{pmatrix} x \\ y \end{pmatrix}, \ w = \begin{pmatrix} x \\ y \\ \lambda \end{pmatrix} \text{ and } F(w) = \begin{pmatrix} -A^\top \lambda \\ -B^\top \lambda \\ Ax + By - b \end{pmatrix}. \tag{4.11}$$

Since the mapping $F(w)$ defined in (4.11) is affine with a skew-symmetric matrix, it is monotone. We denote by Ω^* the solution set of VI(Ω, F, θ), and suppose that it is nonempty.

4.2.3 *Some Notations*

For notational simplicity, we first define two matrices which will simplify our further analysis in the following sections. Specifically, let

$$G := \begin{pmatrix} (1+\mu)R & 0 & 0 \\ 0 & \beta B^\top B + (1+\mu)S & -B^\top \\ 0 & -B & \frac{1}{\beta} I_m \end{pmatrix}, \quad N := \begin{pmatrix} \mu R & 0 & 0 \\ 0 & \mu S & 0 \\ 0 & 0 & 0 \end{pmatrix}, \tag{4.12}$$

and

$$M := \begin{pmatrix} (1+\mu)R & 0 & 0 \\ 0 & \frac{\gamma+\rho-\gamma\rho}{\gamma+\rho}\beta B^\top B + (1+\mu)S & -\frac{\gamma}{\gamma+\rho} B^\top \\ 0 & -\frac{\gamma}{\gamma+\rho} B & \frac{1}{(\gamma+\rho)\beta} I_m \end{pmatrix}, \tag{4.13}$$

where $R := rI_n$, $S := sI_p$ and γ, ρ are two positive constants. Indeed, it is not difficult to verify that the matrices G, N, and M are positive symmetric definite when $r, s > 0$ and $\gamma, \rho \in (0, 1)$.

4.3 Inertial ADMM with LQP Regularization

In this section, we first introduce the inertial ADMM with LQP regularization as a solution approach for problem (4.1). We then analyze its convergence using a variational inequality framework.

4.3.1 Inertial ADMM with LQP Regularization

We now present the inertial ADMM with LQP regularization. The method first extrapolates the current point in the direction of the last movement and then applies the ADMM with LQP regularization to the extrapolated point. Below, we summarize the overall algorithm framework.

Algorithm 1 Inertial ADMM with LQP regularization

Initialization: Choose the constants $r, s > 0$ and $\beta > 0$, and a sequence of nonnegative parameters $\{\alpha_k\}_{k=0}^{\infty}$. Let $(x^{-1}, y^{-1}, \lambda^{-1}) := (x^0, y^0, \lambda^0) \in \mathbb{R}^n_{++} \times \mathbb{R}^p_{++} \times \mathbb{R}^m$.
for $k = 0, 1, 2, \ldots$ do
Step 1. Update

$$\tilde{w}^k = w^k + \alpha_k(w^k - w^{k-1}), \tag{4.14}$$

ensuring that $\tilde{w}^k \in \mathbb{R}^n_{++} \times \mathbb{R}^p_{++} \times \mathbb{R}^m$.
Step 2. Update the new iterate w^{k+1} by

$$\begin{cases} x^{k+1} = \arg\min\{\mathcal{L}_\beta(x, \tilde{y}^k, \tilde{\lambda}^k) + rd(x, \tilde{x}^k) \mid x \in \mathbb{R}^n_+\}, \\ \lambda^{k+1} = \tilde{\lambda}^k - \beta(Ax^{k+1} + B\tilde{y}^k - b), \\ y^{k+1} = \arg\min\{\mathcal{L}_\beta(x^{k+1}, y, \lambda^{k+1}) + sd(y, \tilde{y}^k) \mid y \in \mathbb{R}^p_+\}. \end{cases} \tag{4.15}$$

end
Return: w^{k+1}.

Remark 4.1 Note that Lemma 4.1 guarantees the existence of a unique solution, denoted as $x^{k+1} \in \mathbb{R}^n_{++}$ for subproblem (4.15), and $y^{k+1} \in \mathbb{R}^p_{++}$ for subproblem (4.15). Additionally, the condition $\tilde{w}^k \in \mathbb{R}^n_{++} \times \mathbb{R}^p_{++} \times \mathbb{R}^m$ can be satisfied by choosing $\alpha_k = \max\left\{0, \min_i \left\{\frac{w_i^{k-1} - w_i^k}{w_i^k}\right\}\right\}$. Moreover, the effectiveness of LQP regularization has been showcased in [47] for a traffic management problem, and the benefits of the inertial technique have been extensively confirmed in various real-world applications, as seen in, for instance, [9, 10, 14]. Hence, we exclude the numerical experiments in this paper.

We now introduce the following assumption for the sequence of parameters $\{\alpha_k\}_{k=0}^{\infty}$.

Assumption 4.1 Suppose that the sequence of parameters $\{\alpha_k\}_{k=0}^{\infty}$ is chosen such that

(i) for all $k \geq 0$, $0 \leq \alpha_k \leq \alpha$ for some $\alpha \in [0, 1)$,
(ii) the sequence $\{w^k\}_{k=0}^{\infty}$ generated by (4.15) satisfies

$$\sum_{k=0}^{\infty} \alpha_k \|w^k - w^{k-1}\|_G^2 < \infty. \tag{4.16}$$

Note that in order to ensure Assumption 4.1, we can choose $\{\alpha_k\}_{k=0}^{\infty}$ adaptively based on the historical iterative information in practice such that (4.16) holds. Alternatively, it is simultaneously satisfied if $\{\alpha_k\}_{k=0}^{\infty}$ satisfies some further conditions; see, e.g., [1, Prop. 2.1].

4.3.2 Convergence Analysis

Below, we start our convergence analysis by proving some properties for the sequence $\{w^k\}$ according to the first-order optimality condition.

Lemma 4.2 *Let the sequence $\{w^k\}$ be generated by the iterative scheme of the inertial ADMM with LQP regularization given in Algorithm 1, and G be defined in* (4.12). *Then, for any $w \in \Omega$, we have $w^{k+1} \in \Omega$ and*

$$\theta(u) - \theta(u^{k+1}) + (w - w^{k+1})^\top [F(w^{k+1}) + G(w^{k+1} - \tilde{w}^k)] + \|w^{k+1} - \tilde{w}^k\|_N^2 \geq 0. \quad (4.17)$$

Proof From the first-order optimality condition of the x-subproblem in (4.15), we have $x^{k+1} \in \mathbb{R}^n_+$, and for any $x \in \mathbb{R}^n_+$, it holds that

$$\begin{aligned}&\theta_1(x) - \theta_1(x^{k+1}) + (x - x^{k+1})^\top \\ &\{-A^\top[\tilde{\lambda}^k - \beta(Ax^{k+1} + B\tilde{y}^k - b)] + r\Phi'(\tilde{x}^k, x^{k+1})\} \geq 0.\end{aligned}$$

Applying Lemma 4.1 to above inequality with $P = R$, $\bar{z} = \tilde{x}^k$, $z = x^{k+1}$, $z' = x$, $\vartheta = \theta_1$ and

$$q(z) = -A^\top[\tilde{\lambda}^k - \beta(Ax^{k+1} + B\tilde{y}^k - b)]$$

in (4.9), then for any $x \in \mathbb{R}^n_+$, we have $x^{k+1} \in \mathbb{R}^n_{++}$ and

$$\begin{aligned}&\theta_1(x) - \theta_1(x^{k+1}) + (x - x^{k+1})^\top(-A^\top\lambda^{k+1}) \\ &\geq (1+\mu)(x - x^{k+1})R(\tilde{x}^k - x^{k+1}) - \mu\|x^{k+1} - \tilde{x}^k\|_R^2.\end{aligned}$$

Rearranging the inequality, we have

$$\begin{aligned}&\theta_1(x) - \theta_1(x^{k+1}) + (x - x^{k+1})^\top[-A^\top\lambda^{k+1} + (1+\mu)R(x^{k+1} - \tilde{x}^k)] \\ &+ \mu\|x^{k+1} - \tilde{x}^k\|_R^2 \geq 0.\end{aligned} \quad (4.18)$$

Similarly, for the y-subproblem (4.15), we have $y^{k+1} \in \mathbb{R}^p_{++}$, and for any $y \in \mathbb{R}^p_+$, it holds that

$$\begin{aligned}\theta_2(y) - \theta_2(y^{k+1}) + (y - y^{k+1})^\top\{&-B^\top[\lambda^{k+1} - \beta(Ax^{k+1} + By^{k+1} - b)] \\ &+ (1+\mu)S(y^{k+1} - \tilde{y}^k)\} + \mu\|y^{k+1} - \tilde{y}^k\|_S^2 \geq 0.\end{aligned} \quad (4.19)$$

Moreover, it follows from (4.15) that

$$\beta(Ax^{k+1} + By^{k+1} - b) = \tilde{\lambda}^k - \lambda^{k+1} + \beta B(y^{k+1} - \tilde{y}^k) = 0, \tag{4.20}$$

then the inequality (4.19) can be rewritten as

$$\begin{aligned}\theta_2(y) - \theta_2(y^{k+1}) + (y - y^{k+1})^\top \Big\{ - B^\top \lambda^{k+1} + [\beta B^\top B + (1+\mu)S](y^{k+1} - \tilde{y}^k) \\ - B^\top(\lambda^{k+1} - \tilde{\lambda}^k)\Big\} + \mu \|y^{k+1} - \tilde{y}^k\|_S^2 \geq 0.\end{aligned} \tag{4.21}$$

Combining (4.18), (4.20) and (4.21), and using the notations in (4.11) and (4.12), we obtain the assertion (4.17) immediately. This completes the proof.

□

Theorem 4.1 *Suppose that* $\{\alpha_k\}_{k=0}^\infty$ *satisfies Assumption 4.1. Let the sequence* $\{w^k\}_{k=0}^\infty$ *be generated by the inertial ADMM with LQP regularization given in Algorithm 1. Then, it holds that*

(i) for any $w^* \in \Omega^*$, $\lim\limits_{k\to\infty} \|w^k - w^*\|_G$ *exists,*
(ii) $\sum_{k=0}^\infty \|w^{k+1} - \tilde{w}^k\|_G^2 < \infty$.

Proof (i) Setting $w = w^*$ in (4.17) and $w = w^{k+1}$ in (4.10), adding these two inequalities and using the monotonicity of F, we obtain

$$(w^{k+1} - w^*)^\top G(w^{k+1} - \tilde{w}^k) - \|w^{k+1} - \tilde{w}^k\|_N^2 \leq 0. \tag{4.22}$$

It follows from (4.14) that

$$(w^{k+1} - w^*)^\top G(w^{k+1} - \tilde{w}^k) = (w^{k+1} - w^*)^\top G[w^{k+1} - w^k - \alpha_k(w^k - w^{k-1})]. \tag{4.23}$$

Define $\varphi_k = \frac{1}{2}\|w^k - w^*\|_G^2$, then we have

$$(w^{k+1} - w^*)^\top G(w^{k+1} - w^k) = \varphi_{k+1} - \varphi_k + \frac{1}{2}\|w^{k+1} - w^k\|_G^2,$$

and

$$\begin{aligned}&(w^{k+1} - w^*)^\top G(w^k - w^{k-1}) \\ &= \varphi_k - \varphi_{k-1} + \frac{1}{2}\|w^k - w^{k-1}\|_G^2 + (w^{k+1} - w^k)^\top G(w^k - w^{k-1}).\end{aligned}$$

Together with (4.22) and (4.23), we have

$$\varphi_{k+1} - \varphi_k - \alpha_k(\varphi_k - \varphi_{k-1}) \leq -\frac{1}{2}\|w^{k+1} - w^k - \alpha_k(w^k - w^{k-1})\|_G^2 + \frac{1}{2}(\alpha_k + \alpha_k^2)\|w^k - w^{k-1}\|_G^2 + \|w^{k+1} - \tilde{w}^k\|_N^2.$$

It follows from (4.14) and $\frac{1}{2}(\alpha_k + \alpha_k^2) \leq \alpha_k$ for any $\alpha_k \in [0, 1)$ that

$$\begin{aligned} &\varphi_{k+1} - \varphi_k - \alpha_k(\varphi_k - \varphi_{k-1}) \\ &\leq -\frac{1}{2}\|w^{k+1} - \tilde{w}^k\|_G^2 + \alpha_k\|w^k - w^{k-1}\|_G^2 + \|w^{k+1} - \tilde{w}^k\|_N^2 \\ &\leq \alpha_k\|w^k - w^{k-1}\|_G^2, \end{aligned} \tag{4.24}$$

where the second inequality follows from $G \succeq 2N$ since $\mu \in (0, 1)$, G and N are defined in (4.12).

Define $\nu_k := \varphi_k - \varphi_{k-1}$ and $\delta_k := \alpha_k\|w^k - w^{k-1}\|_G^2$. Then, the inequality (4.24) implies that $\nu_{k+1} \leq \alpha_k\nu_k + \delta_k \leq \alpha[\nu_k]_+ + \delta_k$, where $[t]_+ = \max\{t, 0\}$ for $t \in \mathbb{R}$. Also, we have

$$[\nu_{k+1}]_+ \leq \alpha[\nu_k]_+ + \delta_k \leq \cdots \leq \alpha^{k+1}[\nu_0]_+ + \sum_{j=0}^{k}\alpha^j\delta_{k-j}.$$

Note that $w^0 = w^{-1}$, which implies that $\nu_0 = [\nu_0]_+ = 0, \delta_0 = 0$. Thus, it follows from the above inequalities and Assumption 4.1 (ii) that

$$\sum_{k=0}^{\infty}[\nu_{k+1}]_+ \leq \frac{1}{1-\alpha}\sum_{k=1}^{\infty}\delta_k < \infty. \tag{4.25}$$

Let $\sigma_k := \varphi_k - \sum_{j=1}^{k}[\nu_j]_+$. Then, it follows from (4.25) and $\varphi_k \geq 0$ that σ_k is bounded from below. Besides, we have

$$\begin{aligned} \sigma_{k+1} &= \varphi_{k+1} - [\nu_{k+1}]_+ - \sum_{j=1}^{k}[\nu_j]_+ \\ &\leq \varphi_{k+1} - \nu_{k+1} - \sum_{j=1}^{k}[\nu_j]_+ \leq \varphi_k - \sum_{j=1}^{k}[\nu_j]_+ = \sigma_k, \end{aligned}$$

thus σ_k is nonincreasing. Consequently, $\{\sigma_k\}_{k=0}^{\infty}$ converges as $k \to \infty$, and the limit

$$\lim_{k\to\infty}\varphi_k = \lim_{k\to\infty}\left(\sigma_k + \sum_{j=1}^{k}[\nu_j]_+\right) = \lim_{k\to\infty}\sigma_k + \sum_{k=1}^{\infty}[\nu_k]_+$$

exists. This completes the proof of assertion (i).

(ii) Applying the identity

$$\|a\|_G^2 - \|b\|_G^2 = \|a-b\|_G^2 + 2b^\top G(a-b),$$

and setting

$$a := w^* - \tilde{w}^k, \quad b := w^* - w^{k+1},$$

we have

$$(w^{k+1} - w^*)^\top G(w^{k+1} - \tilde{w}^k) = \frac{1}{2}\|w^{k+1} - \tilde{w}^k\|_G^2 + \frac{1}{2}\Big(\|w^{k+1} - w^*\|_G^2 - \|\tilde{w}^k - w^*\|_G^2\Big).$$

Together with (4.22), we obtain

$$\|\tilde{w}^k - w^*\|_G^2 - \|w^{k+1} - w^*\|_G^2 \geq \|w^{k+1} - \tilde{w}^k\|_G^2 - 2\|w^{k+1} - \tilde{w}^k\|_N^2.$$

Recalling the definition of G and N in (4.12) and $\mu \in (0, 1)$, there exists a positive constant $0 < c \leq \frac{1-\mu}{1+\mu} < 1$ such that $G - 2N \succeq cG$. Then, the above inequality can be written as

$$\|\tilde{w}^k - w^*\|_G^2 - \|w^{k+1} - w^*\|_G^2 \geq c\|w^{k+1} - \tilde{w}^k\|_G^2. \tag{4.26}$$

Since

$$(w^k - w^*)^\top G(w^k - w^{k-1}) = \frac{1}{2}\|w^k - w^*\|_G^2 - \frac{1}{2}\|w^{k-1} - w^*\|_G^2 + \frac{1}{2}\|w^k - w^{k-1}\|_G^2. \tag{4.27}$$

Combining (4.26), (4.27) and (4.14), we obtain

$$\begin{aligned}
&\|w^{k+1} - \tilde{w}^k\|_G^2 \\
&\leq \frac{1}{c}\Big[\|w^k - w^*\|_G^2 - \|w^{k+1} - w^*\|_G^2 + \alpha_k \nu_k + (\alpha_k + \alpha_k^2)\|w^k - w^{k-1}\|_G^2\Big] \\
&\leq \frac{1}{c}\Big[\|w^k - w^*\|_G^2 - \|w^{k+1} - w^*\|_G^2 + \alpha[\nu_k]_+ + 2\alpha_k\|w^k - w^{k-1}\|_G^2\Big].
\end{aligned}$$

By taking sum over k, we obtain

$$\sum_{k=1}^{\infty} \|w^{k+1} - \tilde{w}^k\|_G^2 \leq \frac{1}{c}\Big[\varphi_1 + \sum_{k=1}^{\infty}\Big([\alpha\nu_k]_+ + 2\delta_k\Big)\Big].$$

Together with (4.25) and Assumption 4.1 (ii), we obtain $\sum_{k=0}^{\infty} \|w^{k+1} - \tilde{w}^k\|_G^2 < \infty$ immediately. □

The following lemma presents the feasibility and convergence of the objective function value for the inertial ADMM with LQP regularization, and the proof can refer to [14, Theorem 4.3]. Here, we omit the details.

Lemma 4.3 *Suppose that* $\{\alpha_k\}_{k=0}^{\infty}$ *satisfies Assumption 4.1. Let the sequence* $\{w_k\}_{k=0}^{\infty}$ *be generated by the inertial ADMM with LQP regularization given in Algorithm 1. Then, we have*

(i) $\sum_{k=0}^{\infty} \|Ax^k + By^k - b\|^2 < \infty$, *and hence* $\lim_{k\to\infty} \|Ax^k + By^k - b\| = 0.$

(ii) The objective function value $f(x^k) + g(y^k)$ *converges to the optimal value of* (4.1) *as* $k \to \infty$.

Note that the results presented in Theorem 4.1 does not ensure the convergence of $\{w^k\}_{k=0}^{\infty}$. In the following theorem, we present the convergence analysis for the sequence $\{w^k\}_{k=0}^{\infty}$ generated by inertial ADMM with LQP regularization scheme (4.15).

Theorem 4.2 *Suppose that* $\{\alpha_k\}_{k=0}^{\infty}$ *satisfies Assumption 4.1. Let the sequence* $\{w^k\}_{k=0}^{\infty}$ *be generated by the inertial ADMM with LQP regularization given in Algorithm 1. Then,* $\{w^k\}_{k=0}^{\infty}$ *converges to a point* $w^* \in \Omega^*$ *as* $k \to \infty$.

Proof It follows from Theorem 4.1 that $\lim_{k\to\infty} \|w^k - w^*\|_G$ exists for any $w^* \in \Omega^*$, thus the sequence $\{Gw^k\}_{k=0}^{\infty}$ is bounded. By the definition of G in (4.12), we have that the sequences $\{Rx^k\}_{k=0}^{\infty}$, $\{Sy^k\}_{k=0}^{\infty}$ and $\{By^k - \frac{\lambda^k}{\beta}\}_{k=0}^{\infty}$ are bounded, and thus $\{x^k\}_{k=0}^{\infty}$, $\{y^k\}_{k=0}^{\infty}$ are bounded by the definition of R and S, respectively. Besides, the assertion of Lemma 4.3 (i) implies that $\{Ax^k + \frac{\lambda^k}{\beta}\}_{k=0}^{\infty}$ is bounded. On the other hand, setting $x = x^*$ in (4.18), we obtain

$$\theta_1(x^*) - \theta_1(x^{k+1}) + (x^* - x^{k+1})^\top[-A^\top\lambda^{k+1} + (1+\mu)R(x^{k+1} - \tilde{x}^k)] \\ + \mu\|x^{k+1} - \tilde{x}^k\|_R^2 \geq 0.$$

Since $w^* = (x^*, y^*, \lambda^*) \in \Omega^*$, thus setting $x = x^{k+1}$, $y = y^*$ and $\lambda = \lambda^*$ in (4.10), we have

$$\theta_1(x^{k+1}) - \theta_1(x^*) + (x^{k+1} - x^*)^\top(-A^\top\lambda^*) \geq 0.$$

Adding the above two inequalities, it holds that

$$\langle A(x^{k+1} - x^*), \lambda^{k+1} - \lambda^*\rangle \geq \langle x^{k+1} - x^*, (1+\mu)R(x^{k+1} - \tilde{x}^k)\rangle - \mu\|x^{k+1} - \tilde{x}^k\|_R^2.$$

From the above inequality, we have

$$\langle A(x^k - x^*), \lambda^k - \lambda^*\rangle \geq -\frac{1+\mu}{2}\left(\|x^k - x^*\|_R^2 + \|x^k - \tilde{x}^{k-1}\|_R^2\right) - \mu\|x^k - \tilde{x}^{k-1}\|_R^2.$$

By the boundedness of $\{Rx^k\}_{k=0}^{\infty}$ and the assertion in Theorem 4.1 (ii), we know that $\langle A(x^k - x^*), \lambda^k - \lambda^* \rangle$ is bounded from below for $k \geq 1$. Then, according to

$$\|A(x^k - x^*\|^2 + \|(\lambda^k - \lambda^*)/\beta\|^2 = \|(Ax^k + \lambda^k/\beta) - (Ax^* + \lambda^*/\beta)\|^2 - \frac{2}{\beta}\langle A(x^k - x^*), \lambda^k - \lambda^* \rangle.$$

It holds that $\{Ax^k\}_{k=0}^{\infty}$ and $\{\lambda^k\}_{k=0}^{\infty}$ are bounded. Therefore, the sequence $\{w^k\}_{k=0}^{\infty}$ is bounded and at least has a limit point. Let w^* be any limit point of $\{w^k\}_{k=0}^{\infty}$ and thus there exists a sequence $w^{k_j} \to w^*$ as $j \to \infty$. First, we have $w^* \in \Omega$ since Ω is closed. Furthermore, taking limits over $k = k_j \to \infty$ in (4.17) and note that $G(w^{k_j} - \tilde{w}^{k_j - 1}) \to 0$, we have

$$\theta(w) - \theta(w^*) + \langle w - w^*, F(w^*) \rangle \geq 0.$$

Since we can choose $w \in \Omega$ vary arbitrarily, thus $w^* \in \Omega^*$. Next, it is routine to prove the uniqueness of the limit points of $\{w^k\}_{k=0}^{\infty}$ based on the results presented in [1, Lemma 2.1] and [14, Theorem 4.7]. Therefore, the sequence $\{w_k\}_{k=0}^{\infty}$ converges to a point $w^* \in \Omega^*$ as $k \to \infty$. □

Remark 4.2 Note that the convergence result does not need any assumptions for the matrices A and B; the result also holds when $A = 0$ or/and $B = 0$ in problem (4.1). Compared with the algorithm presented in [14], the inertial ADMM with LQP regularization not only can force the subproblems in (4.15) converted to unconstrained, but also needs less assumptions for deriving the convergence results.

4.4 Inertial Symmetric ADMM with LQP Regularization

In this section, we propose the inertial symmetric ADMM with LQP regularization for solving the problem (4.1), and analyze its convergence in a similar way as that in Sect. 4.3.

4.4.1 Inertial Symmetric ADMM with LQP Regularization

We now present the inertial symmetric ADMM with LQP regularization, which differs from Algorithm 1 with the additional Lagrange multiplier updating step. Besides, we add two contractive factors to the dual step sizes to guarantee the convergence as that in [27, 36]. More details can be found in the following algorithm framework.

Algorithm 2 Inertial symmetric ADMM with LQP regularization

Initialization: Choose the constants $r, s > 0$, $\gamma, \rho \in (0, 1)$ and $\beta > 0$, and a sequence of nonnegative parameters $\{\alpha_k\}_{k=0}^{\infty}$. Let $(x^{-1}, y^{-1}, \lambda^{-1}) := (x^0, y^0, \lambda^0) \in \mathbb{R}^n_{++} \times \mathbb{R}^p_{++} \times \mathbb{R}^m$.
for $k = 0, 1, 2, \ldots$ do
Step 1. Update

$$\tilde{w}^k = w^k + \alpha_k(w^k - w^{k-1}), \tag{4.28}$$

ensuring that $\tilde{w}^k \in \mathbb{R}^n_{++} \times \mathbb{R}^p_{++} \times \mathbb{R}^m$.
Step 2. Update the new iterate w^{k+1} by

$$\begin{cases} x^{k+1} = \arg\min\{\mathcal{L}_\beta(x, \tilde{y}^k, \tilde{\lambda}^k) + rd(x, \tilde{x}^k) \mid x \in \mathbb{R}^n_+\}, \\ \lambda^{k+1/2} = \tilde{\lambda}^k - \gamma\beta(Ax^{k+1} + B\tilde{y}^k - b), \\ y^{k+1} = \arg\min\{\mathcal{L}_\beta(x^{k+1}, y, \lambda^{k+1/2}) + sd(y, \tilde{y}^k) \mid y \in \mathbb{R}^p_+\}, \\ \lambda^{k+1} = \lambda^{k+1/2} - \rho\beta(Ax^{k+1} + By^{k+1} - b). \end{cases} \tag{4.29}$$

end
return w^{k+1}.

Remark 4.3 Note that there exists a unique solution $x^{k+1} \in \mathbb{R}^n_{++}$ and $y^{k+1} \in \mathbb{R}^p_{++}$ in subproblems (4.28) and (4.29) respectively as that in Algorithm 1. Besides, the Assumption 4.1 also should be satisfied for Algorithm 2. In addition, when choosing $\alpha_k = 0$, the algorithm reduces to the method studied in [36]. Note that Algorithm 2 cannot include Algorithm 1 as a special case, since the step sizes of the Lagrange multiplier update are constrained to $(0, 1)$. This implies that we need to conduct a separate convergence analysis for both algorithms.

4.4.2 Convergence Analysis

Similar to Lemma 4.2 and [25, Lemma 3.1], we give the following lemma which is mainly based on the first-order optimality conditions of (4.28) and (4.29).

Lemma 4.4 *Let the sequence $\{w^k\}$ be generated by the iterative scheme of the inertial symmetric ADMM with LQP regularization given in Algorithm 2, N and M are defined in* (4.12) *and* (4.13), *respectively. Then, for any $w \in \Omega$, we have $w^{k+1} \in \Omega$ and*

$$\begin{aligned} &\theta(u) - \theta(u^{k+1}) + (w - w^{k+1})^\top [F(w^{k+1}) + M(w^{k+1} - \tilde{w}^k)] + \|w^{k+1} - \tilde{w}^k\|_N^2 \\ &\qquad \geq (w^{k+1} - w)^\top \begin{pmatrix} A^\top \\ B^\top \\ 0 \end{pmatrix} [(1-\gamma)\beta(\tilde{y}^k - y^{k+1}) + (1 - \gamma - \rho)\beta r^{k+1}], \end{aligned} \tag{4.30}$$

where $r^{k+1} := Ax^{k+1} + By^{k+1} - b$.

Proof From the first-order optimality condition of (4.29), and applying the Lemma 4.1, we have $x^{k+1} \in \mathbb{R}^n_{++}$, and for any $x \in \mathbb{R}^n_+$, it holds that

$$\theta_1(x) - \theta_1(x^{k+1}) + (x - x^{k+1})^\top \Big\{ - A^\top \tilde{\lambda}^k + \beta A^\top r^{k+1} + \beta A^\top B(\tilde{y}^k - y^{k+1}) \\ + (1+\mu)R(x^{k+1} - \tilde{x}^k) \Big\} + \mu \|x^{k+1} - \tilde{x}^k\|_R^2 \geq 0. \tag{4.31}$$

Similarly, from (4.29), for any $y \in \mathbb{R}^p_+$, we have $y^{k+1} \in \mathbb{R}^p_{++}$ and

$$\theta_2(y) - \theta_2(y^{k+1}) + (y - y^{k+1})^\top \Big\{ - B^\top \lambda^{k+1/2} + \beta B^\top r^{k+1} \\ + (1+\mu)S(y^{k+1} - \tilde{y}^k) \Big\} + \mu \|y^{k+1} - \tilde{y}^k\|_S^2 \geq 0. \tag{4.32}$$

It follows from (4.29) that

$$\lambda^{k+1/2} = \lambda^{k+1} + \rho\beta r^{k+1}$$

and

$$\tilde{\lambda}^k = \lambda^{k+1} + (\gamma + \rho)\beta r^{k+1} + \gamma\beta B(\tilde{y}^k - y^{k+1}). \tag{4.33}$$

Substituting the above equalities into (4.31) and (4.32), we have

$$\theta_1(x) - \theta_1(x^{k+1}) + (x - x^{k+1})^\top \Big\{ - A^\top \lambda^{k+1} + (1 - \gamma - \rho)\beta A^\top r^{k+1} \\ + (1-\gamma)\beta A^\top B(\tilde{y}^k - y^{k+1}) + (1+\mu)R(x^{k+1} - \tilde{x}^k) \Big\} + \mu \|x^{k+1} - \tilde{x}^k\|_R^2 \geq 0. \tag{4.34}$$

and

$$\theta_2(y) - \theta_2(y^{k+1}) + (y - y^{k+1})^\top \Big\{ - B^\top \lambda^{k+1} + (1 - \rho)\beta B^\top r^{k+1} \\ + (1+\mu)S(y^{k+1} - \tilde{y}^k) \Big\} + \mu \|y^{k+1} - \tilde{y}^k\|_S^2 \geq 0. \tag{4.35}$$

On the other hand, it follows from (4.33) that

$$r^{k+1} - \frac{\gamma}{\gamma + \rho} B(y^{k+1} - \tilde{y}^k) + \frac{1}{(\gamma + \rho)\beta} (\lambda^{k+1} - \tilde{\lambda}^k) = 0. \tag{4.36}$$

Combining (4.34), (4.35) and (4.36), the assertion (4.30) is obtained immediately. □

Theorem 4.3 *Suppose that* $\{\alpha_k\}_{k=0}^\infty$ *satisfies Assumption 4.1. Let the sequence* $\{w^k\}_{k=0}^\infty$ *be generated by the inertial symmetric ADMM with LQP regularization given in Algorithm 2. Then, we have*

(i) for any $w^* \in \Omega^*$, $\lim_{k\to\infty} \|w^k - w^*\|_M$ *exists,*

(ii) $\sum_{k=0}^{\infty} \|w^{k+1} - \tilde{w}^k\|_M^2 < \infty$.

Proof (i) Setting $w = w^*$ in (4.30) and $w = w^{k+1}$ in (4.10), adding these two inequalities, using the monotonicity of F and the fact $Ax^* + By^* - b = 0$, we obtain

$$(w^{k+1} - w^*)^\top M(w^{k+1} - \tilde{w}^k) \\ \le -(1-\gamma)\beta(r^{k+1})^\top(\tilde{y}^k - y^{k+1}) - (1-\gamma-\rho)\beta\|r^{k+1}\|^2 + \|w^{k+1} - \tilde{w}^k\|_N^2. \tag{4.37}$$

Let

$$H := \frac{1}{\gamma+\rho}\begin{pmatrix} (\gamma+\rho-\gamma\rho)\beta B^\top B & -\gamma B^\top \\ -\gamma B & \frac{1}{\beta} I_m \end{pmatrix} \text{ and } C := \begin{pmatrix} I_p & 0 \\ \gamma\beta B & (\gamma+\rho)\beta I_m \end{pmatrix}. \tag{4.38}$$

Then, it is easy to verify that H is positive semidefinite if $0 \le \gamma \le 1$ and $\rho > 0$, and

$$C^\top HC := \frac{1}{\gamma+\rho}\begin{pmatrix} (1-\gamma)\beta B^\top B & 0 \\ 0 & (\gamma+\rho)\beta I_m \end{pmatrix}. \tag{4.39}$$

Let $v := (y, \lambda)$, together with (4.33) and (4.38), we have

$$\tilde{v}^k - v^{k+1} = \begin{pmatrix} \tilde{y}^k - y^{k+1} \\ \tilde{\lambda}^k - \lambda^{k+1} \end{pmatrix} = C\begin{pmatrix} \tilde{y}^k - y^{k+1} \\ r^{k+1} \end{pmatrix}. \tag{4.40}$$

With (4.39) and (4.40), we obtain

$$\|\tilde{v}^k - v^{k+1}\|_H^2 = (1-\gamma)\beta\|B(\tilde{y}^k - y^{k+1})\|^2 + (\gamma+\rho)\beta\|r^{k+1}\|^2. \tag{4.41}$$

Applying the inequality $2a^\top b \le \|a\|^2 + \|b\|^2$, we have

$$-(1-\gamma)\beta(r^{k+1})^\top(\tilde{y}^k - y^{k+1}) - (1-\gamma-\rho)\beta\|r^{k+1}\|^2 \\ \le \frac{\gamma+2\rho-1}{2}\beta\|r^{k+1}\|^2 + \frac{1-\gamma}{2}\beta\|B(\tilde{y}^k - y^{k+1})\|^2. \tag{4.42}$$

Combining (4.41), (4.42) and $\gamma \in [0, 1)$, we conclude

$$-(1-\gamma)\beta(r^{k+1})^\top(\tilde{y}^k - y^{k+1}) - (1-\gamma-\rho)\beta\|r^{k+1}\|^2 \le \frac{1}{2}\|v^{k+1} - \tilde{v}^k\|_H.$$

It follows from (4.37) that

$$(w^{k+1} - w^*)^\top M(w^{k+1} - \tilde{w}^k) \le \frac{1}{2}\|w^{k+1} - \tilde{w}^k\|_{\widetilde{H}+2N}^2, \tag{4.43}$$

where

$$\widetilde{H} := \begin{pmatrix} 0 & 0 \\ 0 & H \end{pmatrix}.$$

Then, we can derive the assertion (i) similar to the proof of Theorem 4.1 (i) and omit it here.

(ii) Applying the identity

$$\|a\|_G - \|b\|_G = \|a - b\|_G + 2b^\top G(a - b),$$

and setting

$$a := w^* - \tilde{w}^k, \quad b := w^* - w^{k+1},$$

we have

$$(w^{k+1} - w^*)^\top M(w^{k+1} - \tilde{w}^k) = \frac{1}{2}\|w^{k+1} - \tilde{w}^k\|_M^2 + \frac{1}{2}\Big(\|w^{k+1} - w^*\|_M^2 - \|\tilde{w}^k - w^*\|_M^2\Big).$$

Together with (4.37), we obtain

$$\begin{aligned}
&\|\tilde{w}^k - w^*\|_M^2 - \|w^{k+1} - w^*\|_M^2 \\
&\geq \|w^{k+1} - \tilde{w}^k\|_M^2 + 2(1-\gamma)\beta(r^{k+1})^\top B(\tilde{y}^k - y^{k+1}) \\
&\quad + 2(1-\gamma-\rho)\beta\|r^{k+1}\|^2 - 2\|w^{k+1} - \tilde{w}^k\|_N^2 \\
&= (1-\mu)\|w^{k+1} - \tilde{w}^k\|_N^2 + (1-\gamma)\beta\|B(\tilde{y}^k - y^{k+1})\|^2 \\
&\quad + 2(1-\gamma)\beta(r^{k+1})^\top B(\tilde{y}^k - y^{k+1}) + (2-\gamma-\rho)\beta\|r^{k+1}\|^2 \\
&= (1-\mu)\|w^{k+1} - \tilde{w}^k\|_N^2 + (\tilde{v}^k - v^{k+1})^\top C^{-T} K C(\tilde{v}^k - v^{k+1}),
\end{aligned} \tag{4.44}$$

the last equality follows from (4.40), where

$$K := \begin{pmatrix} (1-\gamma)\beta B^\top B & (1-\gamma)\beta B^\top \\ (1-\gamma)\beta B & (2-\gamma-\rho)\beta I_m \end{pmatrix},$$

and C is defined in (4.38). It is easy to verify that K is positive definite if $0 \leq \gamma < 1$ and $0 < \rho < 1$. Moreover, by some simple calculations, we have

$$K \succeq \tilde{c} C^\top H C$$

when taking $\tilde{c} := \frac{1-\sqrt{1-\gamma+\rho(1-\rho)}}{\gamma+\rho} \in (0, 1)$. On the other hand, if $0 < \bar{c} \leq \frac{1-\mu}{1+\mu} < 1$, we have $(1-\mu)R \succeq \bar{c}(1+\mu)R$ and $(1-\mu)S \succeq \bar{c}(1+\mu)S$. Therefore, taking $\eta = \min\{\tilde{c}, \bar{c}\}$, by the definition of M in (4.13), and together with (4.44), we have

$$\|\tilde{w}^k - w^*\|_M^2 - \|w^{k+1} - w^*\|_M^2 \geq \eta\|w^{k+1} - \tilde{w}^k\|_M^2.$$

Then, it is easy to obtain the assertion (ii) similar to the proof of Theorem 4.1 (ii). This completes the proof. □

Similar to Lemma 4.3, we first obtain the results of the feasibility and the convergence of the objective function value for the inertial symmetric ADMM with LQP regularization.

Lemma 4.5 *Suppose that* $\{\alpha_k\}_{k=0}^{\infty}$ *satisfies Assumption 4.1. Let the sequence* $\{w^k\}_{k=0}^{\infty}$ *be generated by the inertial symmetric ADMM with LQP regularization given in Algorithm 2. Then, we have*

(i) $\sum\limits_{k=0}^{\infty} \|Ax^k + By^k - b\|^2 < \infty$, *and hence* $\lim\limits_{k\to\infty} \|Ax^k + By^k - b\| = 0.$

(ii) The objective function value $f(x^k) + g(y^k)$ *converges to the optimal value of* (4.1) *as* $k \to \infty$.

The following theorem presents the convergence analysis for the sequence $\{w^k\}_{k=0}^{\infty}$ generated by inertial symmetric ADMM with LQP regularization scheme (4.29).

Theorem 4.4 *Suppose that* $\{\alpha_k\}_{k=0}^{\infty}$ *satisfies Assumption 4.1. Let the sequence* $\{w^k\}_{k=0}^{\infty}$ *be generated by the inertial symmetric ADMM with LQP regularization given in Algorithm 2. Then,* $\{w^k\}_{k=0}^{\infty}$ *converges to a point* $w^* \in \Omega^*$ *as* $k \to \infty$.

Proof We prove it in a similar way as that in Theorem 4.2. It follows from Theorem 4.3 that $\lim\limits_{k\to\infty} \|w^k - w^*\|_M$ exists for any $w^* \in \Omega^*$, thus the sequence $\{Mw^k\}_{k=0}^{\infty}$ is bounded. By the definition of M in (4.13), we have that the sequences $\{Rx^k\}_{k=0}^{\infty}$, $\{Sy^k\}_{k=0}^{\infty}$ and $\{\gamma By^k - \frac{\lambda^k}{\beta}\}_{k=0}^{\infty}$ are bounded, and thus $\{x^k\}_{k=0}^{\infty}$, $\{y^k\}_{k=0}^{\infty}$ are bounded by the definition of R and S, respectively. Besides, the assertion of Lemma 4.5 (i) implies that $\{\gamma Ax^k + \frac{\lambda^k}{\beta}\}_{k=0}^{\infty}$ is bounded. On the other hand, setting $x = x^*$ in (4.34), we obtain

$$
\begin{aligned}
&\theta_1(x^*) - \theta_1(x^{k+1}) + (x^* - x^{k+1})^\top \Big\{ - A^\top \lambda^{k+1} + (1 - \gamma - \rho)\beta A^\top r^{k+1} \\
&\quad + (1 - \gamma)\beta A^\top B(\tilde{y}^k - y^{k+1}) + (1 + \mu)R(x^{k+1} - \tilde{x}^k)\Big\} + \mu\|x^{k+1} - \tilde{x}^k\|_R^2 \ge 0.
\end{aligned}
$$

Since $w^* = (x^*, y^*, \lambda^*) \in \Omega^*$, thus setting $x = x^{k+1}$, $y = y^*$ and $\lambda = \lambda^*$ in (4.10), we have

$$
\theta_1(x^{k+1}) - \theta_1(x^*) + (x^{k+1} - x^*)^\top(-A^\top \lambda^*) \ge 0.
$$

Adding the above two inequalities, and let $k := k - 1$, it holds that

$$
\begin{aligned}
&\langle A(x^k - x^*), \lambda^k - \lambda^* + (\gamma + \rho - 1)\beta r^k + (1 - \gamma)\beta B(y^k - \tilde{y}^{k-1})\rangle \\
&\ge \langle x^k - x^*, (1 + \mu)R(x^k - \tilde{x}^{k-1})\rangle - \mu\|x^k - \tilde{x}^{k-1}\|_R^2 \\
&\ge -\frac{1 + \mu}{2}\Big(\|x^k - x^*\|_R^2 + \|x^k - \tilde{x}^{k-1}\|_R^2\Big) - \mu\|x^k - \tilde{x}^{k-1}\|_R^2.
\end{aligned}
$$

By the boundedness of $\{Rx^k\}_{k=0}^{\infty}$ and the assertion in Theorem 4.3 (ii), we know that

$$\langle A(x^k - x^*), \lambda^k - \lambda^* + (\gamma + \rho - 1)\beta A^\top r^k + (1-\gamma)\beta A^\top B(y^k - \tilde{y}^{k-1})\rangle$$

is bounded from below for $k \geq 1$. On the other hand, we have

$$\begin{aligned}
&\|\gamma A(x^k - x^*)\|^2 + \|(\lambda^k - \lambda^*)/\beta + (\gamma + \rho - 1)\beta A^\top r^k + (1-\gamma)\beta A^\top B(y^k - \tilde{y}^{k-1})\|^2 \\
&\leq 2\|(\gamma A x^k + \lambda^k/\beta) - (\gamma A x^* + \lambda^*/\beta)\|^2 \\
&\quad + 2\|(\gamma + \rho - 1)\beta r^k + (1-\gamma)\beta B(y^k - \tilde{y}^{k-1})\|^2 \\
&\quad - \frac{2\gamma}{\beta}\langle A(x^k - x^*), \lambda^k - \lambda^* + (\gamma + \rho - 1)\beta^2 r^k + (1-\gamma)\beta^2 B(y^k - \tilde{y}^{k-1})\rangle.
\end{aligned} \tag{4.45}$$

It follows from Theorem 4.3 (ii) that $\lim_{k\to\infty} \|v^k - \tilde{v}^{k-1}\|_H^2 = 0$, and it also have $\lim_{k\to\infty} r^k = 0$ follows from Lemma 4.5 (i), thus it holds that $\lim_{k\to\infty} B(y^k - \tilde{y}^{k-1}) = 0$, and

$$\lim_{k\to\infty} 2\|(\gamma + \rho - 1)\beta r^k + (1-\gamma)\beta B(y^k - \tilde{y}^{k-1})\|^2 = 0.$$

Together with the boundedness of $\{\gamma A x^k + \frac{\lambda^k}{\beta}\}_{k=0}^{\infty}$ and (4.45), it holds that $\{Ax^k\}_{k=0}^{\infty}$ and $\{\lambda^k\}_{k=0}^{\infty}$ are bounded. Therefore, the sequence $\{w^k\}_{k=0}^{\infty}$ is bounded and at least has a limit point. Let w^* be any limit point of $\{w^k\}_{k=0}^{\infty}$ and thus exists a sequence $w^{k_j} \to w^*$ as $j \to \infty$. First, we have $w^* \in \Omega$ since Ω is closed. Furthermore, taking limits over $k = k_j \to \infty$ in (4.17) and note that $M(w^{k_j} - \tilde{w}^{k_j - 1}) \to 0$, we have

$$\theta(w) - \theta(w^*) + \langle w - w^*, F(w^*)\rangle \geq 0.$$

Since we can choose $w \in \Omega$ vary arbitrarily, thus $w^* \in \Omega^*$. Next, it is routine to prove the uniqueness of the limit points of $\{w^k\}_{k=0}^{\infty}$ based on the results presented in [1, Lemma 2.1] and [14, Theorem 4.7]. Therefore, the sequence $\{w^k\}_{k=0}^{\infty}$ converges to a point $w^* \in \Omega^*$ as $k \to \infty$. □

4.5 Conclusion

We revisited the alternating direction method of multipliers (ADMM) with logarithmic- quadratic proximal (LQP) regularization, which can be utilized to solve a class of convex optimization problems with linear constraints. By incorporating the inertial extrapolation step, we proposed an inertial ADMM and an inertial symmetric ADMM with LQP regularization, respectively. Additionally, the convergence of the proposed methods was established under the framework of variational inequality.

Acknowledgements This work was supported by the National Natural Science Foundation of China grants 12471291, 12001286, and the China Postdoctoral Science Foundation grants 2022M7116

Competing Interests The author declares that he has no known competing financial interests or personal relationships that could have appeared to influence the work reported in this paper.

References

1. Alvarez, F., Attouch, H.: An inertial proximal method for maximal monotone operators via discretization of a nonlinear oscillator with damping. Set-Valued Anal. **9**, 3–11 (2001)
2. Attouch, H., Peypouquet, J., Redont, P.: A dynamical approach to an inertial forward-backward algorithm for convex minimization. SIAM J. Optim. **24**(1), 232–256 (2014)
3. Attouch, H., Chbani, Z., Fadili, J., Riahi, H.: Fast convergence of dynamical ADMM via time scaling of damped inertial dynamics. J. Optim. Theory Appl. **193**(1), 704–736 (2022)
4. Auslender, A., Teboulle, M.: Entropic proximal decomposition methods for convex programs and variational inequalities. Math. Program. **91**, 33–47 (2001)
5. Auslender, A., Teboulle, M.: The log-quadratic proximal methodology in convex optimization algorithms and variational inequalities. In: Equilibrium Problems and Variational Models, pp. 19–52 (2003)
6. Auslender, A., Teboulle, M., Ben-Tiba, S.: A logarithmic-quadratic proximal method for variational inequalities. In: Computational Optimization: A Tribute to Olvi Mangasarian, vol. I, pp. 31–40 (1999)
7. Bai, J., Ma, Y., Sun, H., Zhang, M.: Iteration complexity analysis of a partial LQP-based alternating direction method of multipliers. Appl. Numer. Math. **165**, 500–518 (2021)
8. Bnouhachem, A.: A self-adaptive descent LQP alternating direction method for the structured variational inequalities. Numer. Algorithms **86**, 303–324 (2021)
9. Bot, R.I., Csetnek, E.R., Hendrich, C.: Inertial Douglas-Rachford splitting for monotone inclusion problems. Appl. Math. Comput. **256**, 472–487 (2015)
10. Bot, R.I., Csetnek, E.R., László, S.C.: An inertial forward-backward algorithm for the minimization of the sum of two nonconvex functions. EURO J. Comput. Optim. **4**, 3–25 (2016)
11. Boyd, S., Parikh, N., Chu, E., Peleato, B., Eckstein, J.: Distributed optimization and statistical learning via the alternating direction method of multipliers. Found. Trends Mach. Learn. **3**(1), 1–122 (2011)
12. Cai, X., Gu, G., He, B., Yuan, X.: A proximal point algorithm revisit on the alternating direction method of multipliers. Sci. China Math. **56**, 2179–2186 (2013)
13. Cai, X., Guo, K., Jiang, F., Wang, K., Wu, Z., Han, D.: The developments of proximal point algorithms. J. Operat. Res. Soc. China **10**(2), 197–239 (2022)
14. Chen, C., Chan, R.H., Ma, S., Yang, J.: Inertial proximal ADMM for linearly constrained separable convex optimization. SIAM J. Imag. Sci. **8**(4), 2239–2267 (2015)
15. Chen, C., Li, M., Yuan, X.: Further study on the convergence rate of alternating direction method of multipliers with logarithmic-quadratic proximal regularization. J. Optim. Theory Appl. **166**, 906–929 (2015)
16. Chen, C., Ma, S., Yang, J.: A general inertial proximal point algorithm for mixed variational inequality problem. SIAM J. Optim. **25**(4), 2120–2142 (2015)
17. Dang, Y., Chen, L., Gao, Y.: Multi-block relaxed-dual linear inertial ADMM algorithm for nonconvex and nonsmooth problems with nonseparable structures. Numer. Algorithms 1–35 (2024)
18. Deng, Z., Han, D.: Multi-step inertial strictly contractive PRSM algorithms for convex programming problems with applications. J. Comput. Appl. Math. **437**, 115469 (2024)
19. Douglas, J., Rachford, H.H.: On the numerical solution of heat conduction problems in two and three space variables. Trans. Am. Math. Soc. **82**(2), 421–439 (1956)

20. Eckstein, J., Yao, W.: Understanding the convergence of the alternating direction method of multipliers: theoretical and computational perspectives. Pacific J. Optim. **11**(4), 619–644 (2015)
21. Gabay, D.: Applications of the method of multipliers to variational inequalities. In: Fortinand, M., Glowinski, R. (eds.) Augmented Lagrange Methods: Applications to the Solution of Boundary-Valued Problems, pp. 299–331. North-Holland, Amsterdam (1983)
22. Gabay, D., Mercier, B.: A dual algorithm for the solution of nonlinear variational problems via finite element approximation. Comput. Math. Appl. **2**(1), 17–40 (1976)
23. Glowinski, R., Marroco, A.: Sur lapproximation, parlments nis dordre un, et la rsolution, par penalisation-dualit, dune classe de problmes de Dirichlet non linaires. Revue Francaise d'automatique, Informatique, Recherche Operationnelle. Analyse Numerique **9**(R2), 41–76 (1975)
24. Goldfarb, D., Ma, S., Scheinberg, K.: Fast alternating linearization methods for minimizing the sum of two convex functions. Math. Program. **141**(1), 349–382 (2013)
25. Gu, Y., Jiang, B., Han, D.: A semi-proximal-based strictly contractive Peaceman-Rachford splitting method, pp. 1–20 (2015). arXiv:1506.02221
26. Han, D.: A survey on some recent developments of alternating direction method of multipliers. J. Oper. Res. Soc. China 1–52 (2022)
27. He, B., Liu, H., Wang, Z., Yuan, X.: A strictly contractive Peaceman-Rachford splitting method for convex programming. SIAM J. Optim. **24**(3), 1011–1040 (2014)
28. He, B., Ma, F., Yuan, X.: Convergence study on the symmetric version of ADMM with larger step sizes. SIAM J. Imag. Sci. **9**(3), 1467–1501 (2016)
29. He, B., Yuan, X.: On the O(1/n) convergence rate of the Douglas-Rachford alternating direction method. SIAM J. Numer. Anal. **50**(2), 700–709 (2012)
30. He, B., Yuan, X.: On non-ergodic convergence rate of Douglas-Rachford alternating direction method of multipliers. Numer. Math. **130**(3), 567–577 (2015)
31. Hien, L.T.K., Papadimitriou, D.: An inertial ADMM for a class of nonconvex composite optimization with nonlinear coupling constraints. J. Glob. Optim. 1–22 (2024)
32. Hien, L., Phan, D.N., Gillis, N.: Inertial alternating direction method of multipliers for non-convex non-smooth optimization. Comput. Optim. Appl. **83**(1), 247–285 (2022)
33. Jiang, F., Wu, Z., Cai, X.: Generalized ADMM with optimal indefinite proximal term for linearly constrained convex optimization. J. Ind. Manag. Optim. **16**(2), 835–856 (2020)
34. Lai, Z.R., Yang, P.Y., Fang, L., Wu, X.: Short-term sparse portfolio optimization based on alternating direction method of multipliers. J. Mach. Learn. Res. **19**(63), 1–28 (2018)
35. Li, M., Li, X., Yuan, X.: Convergence analysis of the generalized alternating direction method of multipliers with logarithmic-quadratic proximal regularization. J. Optim. Theory Appl. **164**, 218–233 (2015)
36. Li, M., Yuan, X.: A strictly contractive Peaceman-Rachford splitting method with logarithmic-quadratic proximal regularization for convex programming. Math. Oper. Res. **40**(4), 842–858 (2015)
37. Lions, P.L., Mercier, B.: Splitting algorithms for the sum of two nonlinear operators. SIAM J. Numer. Anal. **16**(6), 964–979 (1979)
38. Liu, Y., Guo, K., Yang, M.: Convergence study on the logarithmic-quadratic proximal regularization of strictly contractive Peaceman-Rachford splitting method with larger step-size. Int. J. Comput. Math. **97**(8), 1744–1766 (2020)
39. Lorenz, D.A., Pock, T.: An inertial forward-backward algorithm for monotone inclusions. J. Math. Imaging Vis. **51**, 311–325 (2015)
40. Nagurney, A., Zhang, D.: Projected Dynamical Systems and Variational Inequalities with Applications, vol. 2. Springer Science & Business Media (2012)
41. Peaceman, D.W., Rachford, H.H., Jr.: The numerical solution of parabolic and elliptic differential equations. J. Soc. Ind. Appl. Math. **3**(1), 28–41 (1955)
42. Polyak, B.T.: Some methods of speeding up the convergence of iteration methods. USSR Comput. Math. Math. Phys. **4**(5), 1–17 (1964)
43. Rockafellar, R.T.: Monotone operators and the proximal point algorithm. SIAM J. Control. Optim. **14**(5), 877–898 (1976)

44. Tao, M., Yuan, X.: On the O(1/t) convergence rate of alternating direction method with logarithmic-quadratic proximal regularization. SIAM J. Optim. **22**(4), 1431–1448 (2012)
45. Wang, X., Shao, H., Liu, P., Wu, T.: An inertial proximal partially symmetric ADMM-based algorithm for linearly constrained multi-block nonconvex optimization problems with applications. J. Comput. Appl. Math. **420**, 114821 (2023)
46. Wu, Z., Cai, X., Han, D.: Linearized block-wise alternating direction method of multipliers for multiple-convex programming. J. Ind. Manag. Optim. **14**(3), 833–855 (2018)
47. Wu, Z., Li, M.: An LQP-based symmetric alternating direction method of multipliers with larger step sizes. J. Oper. Res. Soc. China **7**, 365–383 (2019)
48. Wu, Z., Li, M.: General inertial proximal gradient method for a class of nonconvex nonsmooth optimization problems. Comput. Optim. Appl. **73**, 129–158 (2019)
49. Wu, Z., Li, M., Wang, D.Z., Han, D.: A symmetric alternating direction method of multipliers for separable nonconvex minimization problems. Asia-Pacific J. Oper. Res. **34**(06), 1750030 (2017)
50. Wu, Z., Liu, F., Li, M.: A proximal Peaceman-Rachford splitting method for solving the multi-block separable convex minimization problems. Int. J. Comput. Math. **96**(4), 708–728 (2019)
51. Yuan, X., Li, M.: An LQP-based decomposition method for solving a class of variational inequalities. SIAM J. Optim. **21**(4), 1309–1318 (2011)
52. Zaslavski A.J.: Proximal Point Algorithm. Springer International Publishing (2016)

Chapter 5
A Class of Augmented-Lagrangian-Type Algorithms for Solving Generalized Nash Equilibrium Problems

Xiaoxi Jia, Shiwei Wang, and Lingling Xu

Abstract In this paper, we study the solution of convex generalized Nash equilibrium problems (GNEP) with shared linear constraints, and propose a class of regularized augmented Lagrangian methods. The idea is to penalize the shared linear constraints into the augmented Lagrangian function of each player, so as to construct a convex Nash equilibrium subproblem (NEP). Under the strong monotonicity and Lipschitz continuity assumptions of pseudo-gradient, we prove the Fejér monotonicity of iterative points with respect to the set of solutions. Under the cocoercivity assumption of pseudo-gradient, the iterative scheme of the algorithm is equivalent to the forward-backward splitting algorithm for solving the zero of an operator. If one more correction step is added, the cocoercivity hypothesis of pseudo-gradient can be weakened to the Lipschitz continuity hypothesis. Some numerical examples are given to verify the effectiveness of the algorithm.

Keywords Generalized Nash equilibrium problem · Augmented Lagrangian method · Forward-backward splitting · Convergence analysis

5.1 Introduction

We consider the GNEP of N players with common linear equality constraints, and the optimization problem of each player is as follows

X. Jia
Department of Mathematics and Computer Science, Saarland University, Saarbrcken, Germany
e-mail: xiaoxijia@math.uni-sb.de

S. Wang
School of Mathematical Sciences, Ministry of Education Key Laboratory for NSLSCS, Nanjing Normal University, Nanjing, China

L. Xu (✉)
School of Mathematical Sciences, Ministry of Education Key Laboratory for NSLSCS, Nanjing Normal University, Nanjing, China
e-mail: xulingling@njnu.edu.cn

X. Wang et al. (eds.), *Splitting Optimization*, ICIAM2023 Springer Series 2,
https://doi.org/10.1007/978-981-96-8574-5_5

$$\min_{x_\nu \in \mathcal{X}_\nu} \theta_\nu(x_\nu, x_{-\nu}) \quad \text{s.t.} \quad \sum_{\nu=1}^{N} A_\nu x_\nu = b. \tag{5.1}$$

More generally, common linear inequality constraints are also considered

$$\min_{x_\nu \in \mathcal{X}_\nu} \theta_\nu(x_\nu, x_{-\nu}) \quad \text{s.t.} \quad \sum_{\nu=1}^{N} A_\nu x_\nu \leq b, \tag{5.2}$$

where $x_\nu \in \mathbb{R}^{n_\nu}$ denotes the decision set vector of player v, $\nu = 1, 2, \ldots, N, \sum_{\nu=1}^{N} n_\nu = n$. Write $x := (x_\nu, x_{-\nu})$, where $x_{-\nu}$ represents all decision vectors except x_ν. This notation is used to emphasize the role of the sub-vector block x_ν. Therefore, we have $x = (x_\nu, x_{-\nu}) = (x_1, x_2, \ldots, x_N)$, $\quad (y_\nu, x_{-\nu}) = (x_1, \ldots, x_{\nu-1}, y_\nu, x_{\nu+1}, \ldots, x_N)$. $\mathcal{X}_\nu \subseteq \mathbb{R}^{n_\nu}$ is a non-empty closed convex set. $\theta_\nu : \mathbb{R}^{n_1} \times \cdots \times \mathbb{R}^{n_N} \to \mathbb{R}$ is a continuous differentiable function. For the fixed $x_{-\nu}$, $\theta_\nu(\cdot, x_{-\nu})$ is convex. $A_\nu \in \mathbb{R}^{m \times n_\nu}$ and $b \in \mathbb{R}^m$ are constraint matrix and constraint vector, respectively.

For convenience, we introduce notation

$$\mathcal{X} := \mathcal{X}_1 \times \ldots \times \mathcal{X}_N \subseteq \mathbb{R}^n, \quad x := (x_1, x_2, \ldots, x_N) \in \mathbb{R}^n,$$
$$A := (A_1, A_2, \ldots, A_N) \in \mathbb{R}^{m \times n}, \quad \mathcal{F} := \{x \in \mathcal{X} \mid Ax = b\}.$$

The general GNEP is expressed as (5.3).

$$\min_{x_\nu \in \mathcal{X}_\nu} \theta_\nu(x_\nu, x_{-\nu}) \quad \text{s.t.} \quad x_\nu \in X_\nu(x_{-\nu}), \tag{5.3}$$

where X_ν represents the feasible set mapping of the ν-th player. After the other players' strategies $x_{-\nu}$ are given, the ν-th player's feasible strategy set can be obtained through X_ν. Generally, the solution of GNEP is to find a strategy vector $x^* = (x_1^*, \ldots, x_N^*)$, such that for any $\nu = 1, \ldots, N$, we have

$$\theta_\nu(x^*) \leq \theta_\nu(x_\nu, x_{-\nu}^*) \quad \forall x_\nu \in X_\nu(x_{-\nu}^*).$$

As early as 1955, Nikaido and Isoda [1] proposed the NI function to reformulate the general GNEP in a constrained quasi-optimization problem, where the objective function can be regarded as an evaluation function constructed from the NI function. Here, "quasi-optimization" means that the problem is not a standard optimization problem because the set of constraints is closely related to the decision variables. In 1974, Bensoussan [2] transformed convex GNEP into a quasi-variational inequality problem (QVI for short). Correspondingly, it is also obtained that the convex NEP can be equivalent to a certain variational inequality (VI for short) problem. But the solution for QVI is also very difficult. Rosen [3] considered the jointly convex case in 1965. At this time, the constraint expressions of each player are characterized as the same non-positive convex function. The idea still needs to minimize the evaluation function, so the solution to this problem is still quite difficult. Under the assumption of joint convexity, a solution of QVI, which is the so-called variational equilibrium or regular equilibrium, is the definite solution of GNEP. In addition to the algorithms for solving the GNEP problem, research on the existence, uniqueness, stability, and KKT-type conditions of its solutions is also being followed up. For details, please refer to the two review papers in 2010 and 2014 [4, 5].

The algorithms of convex GNEP include augmented Lagrangian methods (ALM), penalty function methods, Newton methods, split methods, etc. In GNEP, when some players have common constraints, inherent singularities arise when solving GNEP, see [6] for details, which causes much difficulty in designing proper second-order methods where Hessian or Hessian approximation is necessary, such as Newton-type methods [6, 7]. This difficulty inspired us to start looking for algorithms with at least good information about differentiation. Penalty function algorithms are one of them. The first penalty function algorithm for GNEP was proposed by Fukushima [8], and the literature [9, 10] also proposed similar penalty algorithms. But these algorithms have a common defect. Their subproblems are non-smooth NEP, so it is difficult to solve numerically. However, this disadvantage can be overcome by introducing the multiplier, that is, applying the ALM-type methods.

The ALM methods (or multiplier penalty function methods) are the classic methods for solving constrained optimization problems. This type of method is often applied to GNEP. For details, please refer to the review literature [4, 5]. The subproblems of ALM-type methods are generally highly smooth and therefore easier to solve. Pang and Fukushima [11] used ALM-type methods to solve quasi-variational inequality (QVI) problems, and GNEP is a special class of QVI problems. [12] proposed an improved ALM algorithm for QVI. Kanzow in [13] also proposed an ALM algorithm directly for GNEP.

The splitting-type approach can be applied to NEP with relative ease. For example, applying the Gauss-Seidel-type algorithms to NEP, i.e., for participants $\nu = 1, 2, \ldots, N$, do

$$x_\nu^{k+1} = \arg\min_{x_\nu \in X_\nu} \theta_\nu(x_1^{k+1}, \ldots, x_{\nu-1}^{k+1}, x_\nu, x_{\nu-1}^k, \ldots, x_N^k).$$

But its generalization on GNEP becomes more complicated because the feasible set of each participant in GNEP is influenced by the decision variables of the remaining participants. The literature [14] applied this type of splitting method to a class of potential game problems and obtained the corresponding convergence analysis, where the inner semi-continuity of feasible set-valued maps is required, which, however, is usually not easy to verify. The so-called ADMM is also proposed in the literature [15, 16] where the (multiplier) penalty term was employed to get rid of the assumption of inner semi-continuity.

The ADMM method is a well-known algorithm for large-scale optimization problems with block structures. It is efficient for the problems with two-block structures, but for problems with three or more blocks, the classical iterative schemes should be properly corrected or more restricted assumptions should be addicted. For details, please refer to the literature [17–25]. Recently, Kanzow et al. proposed Jacobi-ADMM-type methods [26] and Gauss-Seidel-ADMM-type methods [27] for GNEP with common linear constraints in an infinite-dimensional space, referring to ADMM-type algorithms for solving classical optimization problems.

The Jacobi-ADMM method minimizes the augmented Lagrangian function of the linearized objective function for each participant with an additional regularization term, and then solves it in parallel. Its algorithm is shown in Algorithm 1. Börgens and Kanzow [26] prove that its iterative format (5.4), (5.5) is equivalent to a forward-backward splitting algorithm for solving the zero point of an operator. And the convergence of iterative schemes (5.4) and (5.5) can be obtained by the related theory of operator splitting.

Algorithm 1 Regularized linearized Jacobi-type ADMM method

1: Choose a starting point $(x^0, \lambda^0) \in \mathcal{X} \times \mathbb{R}^m$, parameters $\beta,\ \gamma > 0$.
2: If a suitable termination criterion is satisfied: STOP.
3: For $\nu = 1, 2, 3, \ldots, N$, compute

$$x_\nu^{k+1} := \arg\min_{x_\nu \in \mathcal{X}_\nu} \Big\{ \left\langle \nabla_{x_\nu} \theta_\nu(x^k), x_\nu - x_\nu^k \right\rangle + \langle \lambda^k, A_\nu x_\nu \rangle + \frac{\beta}{2} \big(\|A_\nu x_\nu + \sum_{i \neq \nu} A_i x_i^k - b\|^2 + \gamma \|x_\nu - x_\nu^k\|^2 \big) \Big\}. \tag{5.4}$$

4: Compute

$$\lambda^{k+1} := \lambda^k + \beta \left(\sum_{\nu=1}^{N} A_\nu x_\nu^{k+1} - b \right). \tag{5.5}$$

5: Set $k := k + 1$, and go to Step 1.

The Gauss-Seidel-ADMM-type method is shown in Algorithm 2. Different from Algorithms 1 and 2 only adds a regular term to the augmented Lagrangian function, and then solves the subproblems one by one in sequence. Under strong assumptions (such as strong monotonicity of pseudo-gradients, etc.), Fejér monotonicity of iterates can be obtained.

Algorithm 2 Regularized Gauss-Seidel-type ADMM method

1: Choose a starting point $(x^0, \lambda^0) \in \mathcal{X} \times \mathbb{R}^m$, parameters $\beta,\ \gamma > 0$.
2: If a suitable termination criterion is satisfied: STOP.
3: For $\nu = 1, 2, 3, \ldots, N$, compute

$$\begin{aligned} x_\nu^{k+1} := \arg\min_{x_\nu \in \mathcal{X}_\nu} \big\{ & \theta_\nu(x_1^{k+1}, \ldots, x_{\nu-1}^{k+1}, x_\nu, x_{\nu+1}^k, \ldots, x_N^k) \\ & + \langle \lambda^k, A_\nu x_\nu \rangle + \frac{\gamma}{2} \left\| x_\nu - x_\nu^k \right\|^2 \\ & + \frac{\beta}{2} \| A_\nu x_\nu + \sum_{i=1}^{\nu-1} A_i x_i^{k+1} + \sum_{i=\nu+1}^{N} A_i x_i^k - b \|^2 \big\}. \end{aligned} \tag{5.6}$$

4: Compute

$$\lambda^{k+1} = \lambda^k + \beta \left(\sum_{\nu=1}^{N} A_\nu x_\nu^{k+1} - b \right). \tag{5.7}$$

5: Set $k := k + 1$, and go to Step 1.

Based on these two ADMM-type methods, we propose the corresponding augmented Lagrangian method in this paper. The difference is that we do not solve the subproblems with a block-by-block strategy like the two ADMM-type algorithms, but solve a convex NEP. The connection between the proposed algorithms and operator splitting methods is discussed. We prove the convergence of the algorithms under mild assumptions.

This paper contains four sections. The first section describes the GNEP with a linear equality constraint and introduces several existing related algorithms. In Sect. 5.2 we recall some basic knowledge and notation. In the third section, several types of augmented Lagrangian-type algorithms are proposed, namely, the regularized linearized augmented Lagrangian algorithm, the regularized augmented Lagrangian algorithm, and the corrected regularized linearized augmented Lagrangian algorithm, where a correction step is added to the regularized linearized augmented Lagrangian algorithm. Under the mild assumptions, their Fejér monotonicity of the iterative point sequence is obtained. Section 5.4 carries out the numerical experiments. Seven numerical examples of GNEP are solved by several augmented Lagrangian-type algorithms.

5.2 Preliminaries

In this section, some basic definitions and notations are listed first. Then, a reformulation of the GNEP is given, which aims to find the zero point of the corresponding operator. Finally, the splitting algorithm is briefly introduced for tracking the zero point.

5.2.1 Basic Definitions

Let $\mathbb{R}^n$ be a real n dimensional vector space, $\langle\cdot,\cdot\rangle$ is the inner product defined on $\mathbb{R}^n$, i.e. $\langle x, y\rangle = x^T y$. Define $\|x\| = \sqrt{x^T x}$. Given a symmetric positive definite matrix Q, we define the inner product induced by Q as

$$\langle x, y\rangle_Q := \langle x, Qy\rangle = x^T Qy.$$

The norm induced by the inner product is denoted as $\|x\|_Q = \sqrt{x^T Qx}$. When the sign of the norm has no subscript, it still defaults to 2-norm.

For the theoretical proof, we need to recall some knowledge of operator theory. Let $\mathcal{H}$ be a Hilbert space. For the set-valued operator $T : \mathcal{H} \to 2^{\mathcal{H}}$, its domain is defined as $dom\ T := \{x \in \mathcal{H} \,|\, T(x) \neq \emptyset\}$. The graph of the operator T is defined as $grap\ T := \{(x, u) \in \mathcal{H} \times \mathcal{H} \,|\, u \in T(x)\}$. Its zero point set is defined as $zer\ T := \{x \in \mathcal{H} \,|\, 0 \in T(x)\}$.

An operator T is called monotonic if it satisfies

$$\langle u - v, x - y\rangle \geq 0 \quad \forall (x, u), (y, v) \in grap\ T.$$

On the basis of monotonicity, we give the definition of a maximal monotone operator.

Definition 5.1 Let the operator $A : \mathcal{H} \to 2^{\mathcal{H}}$ be a monotone operator. A is called a maximal monotone operator if there is no monotone operator $B : \mathcal{H} \to 2^{\mathcal{H}}$ that satisfies $grap\ A \subsetneqq grap\ B$. That is, for all $(x, u) \in \mathcal{H} \times \mathcal{H}$, we have

$$(x, u) \in grap\ A \iff \langle x - y, u - v\rangle \geq 0 \quad \forall (y, v) \in grap\ A. \tag{5.8}$$

The set-valued operator T is called strongly monotonic, if there exists a constant $\rho > 0$, such that

$$\langle u - v, x - y\rangle \geq \rho\|x - y\|^2 \quad \forall (x, u), (y, v) \in grap\ T.$$

The subdifferential of a convex function $f : \mathcal{H} \to (-\infty, \infty]$ at a point x_0 is defined as

$$\partial f(x_0) := \{g \in \mathcal{H} \,|\, f(x) - f(x_0) \geq \langle g, x - x_0\rangle, \forall x \in \mathcal{H}\}.$$

It is easy to know that the subdifferential operator of a closed, convex function is a maximal monotone operator.

The single-valued operator $T : \mathcal{H} \to \mathcal{H}$ is called α-cocoercive, if for $\alpha > 0$, we have

$$\langle T(x) - T(y), x - y\rangle \geq \alpha\|T(x) - T(y)\|^2 \quad \forall x, y \in \mathcal{H}.$$

By the Cauchy-Schwarz inequality, it is easy to prove that the α-cocoercive operator is also $1/\alpha$-Lipschitz continuous. And it is easy to prove that an operator satisfying ρ-strong monotonicity and L-Lipschitz continuity is ρ/L^2-cocoercive. On the other hand, cocoercivity cannot infer strong monotonicity. In the GNEP for inequality constraints proposed later, after the introduction of new variables, the gradient of the objective function will no longer maintain strong monotonicity, but the cocoercivity can be retained.

Given a set $\mathcal{X} \subseteq \mathcal{H}$, its normal cone is defined as

$$N_{\mathcal{X}}(x) := \begin{cases} \{s \in \mathcal{H} \mid \langle s, y - x\rangle \le 0, \forall y \in \mathcal{X}\} & \text{if } x \in \mathcal{X}, \\ \emptyset \quad \text{if } x \notin \mathcal{X}. \end{cases}$$

Now we introduce an important concept about iterative sequences.

Definition 5.2 (*Fejér monotonicity*) Let C be a non-empty subset of $\mathcal{H}$, $\{x_k\}_{k\in\mathbb{N}}$ is the point sequence in $\mathcal{H}$, we call that $\{x_k\}_{k\in\mathbb{N}}$ is Fejér monotone about the set C, if

$$\|x^{k+1} - x\| \le \|x^k - x\| \quad \forall k \in \mathbb{N},\ x \in C.$$

There is an important result about Fejér monotone sequences. See the proof in [28].

Lemma 5.1 *Let $\{x_k\}_{k\in\mathbb{N}}$ be the sequence in $\mathcal{H}$, and C is a non-empty subset of $\mathcal{H}$. If $\{x_k\}_{k\in\mathbb{N}}$ is Fejér monotone about the set C, and each cluster point of the sequence is in the set C, then $\{x_k\}_{k\in\mathbb{N}}$ converges to a certain point in C.*

Some common basic inequalities, such as the Cauchy-Schwarz inequality and the following Young inequality, will also be used in the theoretical analysis.

Lemma 5.2 (Young's inequality) *For all $a, b \in \mathbb{R}$, $\varepsilon > 0$, we have $|a \cdot b| \le \frac{\varepsilon}{2}a^2 + \frac{1}{2\varepsilon}b^2$.*

5.2.2 Reformulation of GNEP

In this subsection, we will reformulate the GNEP (5.1) considered in this paper as the problem of finding the zeros of a maximal monotone operator. For this purpose, two special types of solutions to GNEP (5.1) need to be considered.

First, two classes of convexity assumptions about general GNEP are introduced. For the general GNEP (5.3), the constraint $X_\nu(x_{-\nu}) = \{x_\nu \in \mathcal{X}_\nu \mid g_\nu(x_\nu, x_{-\nu}) \le 0\}$, then GNEP (5.3) can be reformulated as

$$\min_{x_\nu \in \mathcal{X}_\nu} \theta_\nu(x_\nu, x_{-\nu}) \qquad \text{s.t.} \quad g_\nu(x_\nu, x_{-\nu}) \le 0, \tag{5.9}$$

where $g_\nu : \mathcal{H} \to \mathbb{R}^{m_\nu}$ is the constraint function of the ν-th player, and the objective function is continuously differentiable. The problem is said to satisfy the player convexity, if for fixed $x_{-\nu}$, $\theta_\nu(\cdot, x_{-\nu})$ is convex, and the feasible set $X_\nu(x_{-\nu})$ is also convex. Under the framework of linear equality or inequality constraints in this paper, the player convexity is obviously satisfied.

Joint convexity means that the GNEP is first of all player convex, and there is a closed convex set $\mathcal{F} \subseteq \mathcal{H}$, such that $X_\nu(x_{-\nu}) = \{x_\nu \in \mathcal{X}_\nu \mid (x_\nu, x_{-\nu}) \in \mathcal{F}\}$. Under the framework of the problem in this paper, joint convexity is obviously satisfied. Some properties of joint convex GNEP are given below.

Let us now define the pseudo-gradient of GNEP (or NEP), which represents the re-aggregation of the gradient of each player's objective function with respect to its own decision variable, i.e.,

$$\hat{P}(x) = \begin{pmatrix} \nabla_{x_1}\theta_1(x) \\ \vdots \\ \nabla_{x_N}\theta_N(x) \end{pmatrix}.$$

For the jointly convex GNEP, we define a corresponding variational inequality (VI) problem, that is, to find $x^* \in \mathcal{F}$, such that

$$\langle x - x^*, \hat{P}(x^*)\rangle \geq 0 \quad \forall x \in \mathcal{F}.$$

Let us write this variational inequality problem as VI($\mathcal{F}$, $\hat{P}$). According to the literature [4], for joint convex GNEP, the solution of this VI is also the solution of GNEP. The solution of GNEP corresponding to this VI is called the variational equilibrium or normalized equilibrium.

Lemma 5.3 ([26]) *x^* is the variational equilibrium of joint convex GNEP, which is equivalent to* $0 \in \hat{P}(x^*) + N_{\mathcal{F}}(x^*)$.

In this manuscript, we consider the constraints as $\mathcal{F} = \{x \in \mathcal{X} \mid Ax = b\}$, that is, a linear constraints $Ax = b$ and an abstract constraint $x \in \mathcal{X}$. For this purpose, the concept of variational KKT point is introduced.

Definition 5.3 (*Variational KKT point*) The point $(x^*, \lambda^*) \in \mathbb{R}^n \times \mathbb{R}^m$ is called a variational KKT point (or a variational KKT pair) of GNEP (5.1), if it satisfies the KKT-type condition

$$0 \in \hat{P}(x^*) + A^T\lambda^* + N_{\mathcal{X}}(x^*) \qquad Ax^* - b = 0. \tag{5.10}$$

From this definition, it can be intuitively seen that the variational KKT point can be regarded as the KKT point in the case where the multipliers of each player are the same. Regarding the relationship between the variational KKT point and the variational equilibrium, the following results are obtained.

Proposition 5.1 *If (x^*, λ^*) is the variational KKT point of GNEP (5.1), then x^* is also the variational equilibrium for GNEP.*

Note that the converse of Proposition 5.1 does not hold, which can be achieved under the regularity conditions [26].

The above results tell us that a variational KKT point is the variational equilibrium, and a variational equilibrium is the solution of GNEP. Therefore, a solution of the variational KKT point is a candidate for obtaining the solution of GNEP.

Now we write the variational KKT condition in a compact form. Let

$$\mathcal{W} := \mathcal{X}_1 \times \cdots \times \mathcal{X}_N \times \mathbb{R}^m, \quad w := (x_1, \ldots, x_N, \lambda).$$

The pseudo-gradient is consequently expanded and regarded as a function of w, denoted as

$$P(w) := \begin{pmatrix} \hat{P}(x) \\ 0 \end{pmatrix}.$$

Define operator $G : \mathbb{R}^n \times \mathbb{R}^m \to \mathbb{R}^n \times \mathbb{R}^m$ as

$$G(w) := \begin{pmatrix} A_1^T \lambda \\ \vdots \\ A_N^T \lambda \\ b - \sum_{\nu=1}^N A_\nu x_\nu \end{pmatrix}. \tag{5.11}$$

and

$$G_0 := \begin{pmatrix} 0 & \cdots & 0 & A_1^T \\ \vdots & & \vdots & \vdots \\ 0 & \cdots & 0 & A_N^T \\ -A_1 & \cdots & -A_N & 0 \end{pmatrix}, \quad b_0 := \begin{pmatrix} 0 \\ \vdots \\ 0 \\ b \end{pmatrix}.$$

Therefore, we have $G(w) = G_0 w + b_0$. Since G_0 is an antisymmetric matrix, so we have

$$\langle Gx, x \rangle = 0 \quad \forall x \in \mathbb{R}^n.$$

With the above notation, the variational KKT condition can also be expressed in a compact form, see the following lemma.

Lemma 5.4 ([26]) *The point $w^* = (x^*, \lambda^*) \in \mathbb{R}^n \times \mathbb{R}^m$ is a variational KKT point of GNEP (5.1) if and only if $w^* \in \mathcal{W}^*$, where $\mathcal{W}^* = \{w \in \mathbb{R}^n \times \mathbb{R}^m \mid 0 \in P(w) + G(w) + N_{\mathcal{W}}(w)\}$.*

Define a set-valued operator

$$T(w) := P(w) + G(w) + N_{\mathcal{W}}(w). \tag{5.12}$$

Its domain is obviously the non-empty set $\mathcal{W}$, and its zero point is the variational KKT point of GNEP (5.1).

We next prove the maximal monotonicity of the operator T. This conclusion can be proved by proving that each part of the operator T is maximally monotone, and each part satisfies the condition of the addition operation, maintaining maximal monotonicity. Please see the following lemmas, which are generally from [28, 29].

Lemma 5.5 *Let the single-valued operator $A : \mathcal{H} \to \mathcal{H}$ be continuous and monotone, then A is a maximal monotone operator.*

We then illustrate that the operator G defined in (5.11) is maximally monotone.

Lemma 5.6 *Let $\mathcal{W} \subset \mathcal{H}$ be a non-empty closed convex set, $A : \mathcal{W} \to \mathcal{H}$ is a continuous monotone operator, then the operators $N_{\mathcal{W}}(\cdot)$ and $A + N_{\mathcal{W}}(\cdot)$ are both maximally monotone.*

Lemma 5.7 *Let $A, B : \mathcal{H} \to 2^{\mathcal{H}}$ be two maximal monotone operators, if any of the following conditions hold:*

1. $dom B = \mathcal{H}$,
2. $dom A \cap int dom B \neq \emptyset$,

Then $A + B$ is also a maximal monotone operator.

The maximal monotonicity of the operator P also needs to be explained below. Since P is a single-valued mapping, from Lemma 5.5, it is only necessary to assume that it is a monotone operator.

Proposition 5.2 *Suppose operator $P : \mathbb{R}^n \times \mathbb{R}^m \to \mathbb{R}^n \times \mathbb{R}^m$ is monotone, then the set-valued operator T is maximally monotone.*

Till now, we have already obtained that operator T is maximally monotone provided that P is monotone. In the finite-dimensional Euclidean space, the maximal monotonicity of the operator T can lead to the closedness of its image. However, this conclusion is not true in infinite dimensions, and the strong or weak convergence of the sequence needs to be taken into account. Under the condition of infinite dimension, the strong-weak sequential closedness of the operator T can be derived. See the following lemma, and its proof, see [28].

Lemma 5.8 *Let $A : \mathcal{H} \to 2^{\mathcal{H}}$ be a maximal monotone operator, and $\{(x_k, u_k)\}_{k\in\mathbb{N}}$ is a bounded sequence in $gra\ A$, for $(x, u) \in \mathcal{H} \times \mathcal{H}$, if any of the following conditions are true:*

1. $x_k \to x,\ u_k \rightharpoonup u$,
2. $x_k \rightharpoonup x,\ u_k \to u$,

then we have $(x, u) \in gra\ A$.

5.2.3 Splitting-Type Methods

This section introduces some basics about the forward-backward splitting algorithm. We first recall the classic forward-backward splitting algorithm, which is employed for solving the zeros of the set-valued operator $T : \mathcal{H} \to 2^{\mathcal{H}}$, where T can be expressed as $A + B$, with $A : \mathcal{H} \to 2^{\mathcal{H}}$, operator $B : \mathcal{H} \to \mathcal{H}$. Let $x^{k+1} := (I + \beta A)^{-1}(I - \beta B)x^k$. This iteration can also be divided into two steps, denoted as $y^k := x^k - \beta B x^k, x^{k+1} := (I + \beta A)^{-1} y^k$. The explicit calculation of y^k is called the forward step, and the solution of x^{k+1} is called the backward step, which is the so-called forward-backward splitting algorithm.

Under appropriate conditions, the forward-backward splitting algorithm can converge to the zero point of the operator T.

Lemma 5.9 ([29]) (Forward-backward splitting) *Let $T : \mathcal{H} \to 2^{\mathcal{H}}$ be a set-valued map with at least one zero. $T := A + B$, where the mapping $A : \mathcal{H} \to 2^{\mathcal{H}}$ is maximally monotone, the mapping $B : \mathcal{H} \to \mathcal{H}$ is single-valued and α-cocoercive. Let $x^0 \in \mathcal{H}$, and generate the sequence $\{x^k\}_{k\in\mathbb{N}}$ as follows*

$$\begin{cases} y^k = (I - c_k B)(x^k), \\ x^{k+1} = J_{c_k A}(y^k), \end{cases}$$

where the parameter c_k satisfies the existence of $M > m > 0$, such that

$$0 < m \le c_k \le M < 2\alpha \quad \forall k.$$

Then the sequence $\{x^k\}_{k\in\mathbb{N}}$ converges to a zero of T.

Another class of split-type algorithms is proposed by Tseng in [28], a simplified version is that $x^{k+1} = [(I - \beta B)(I + \beta A)^{-1}(I - \beta B) + \beta B]x^k$. Respectively denote $y^k := x^k - \beta B x^k$, $z^k = (I + \beta A)^{-1} y^k$, then $x^{k+1} := z^k + \beta(Bx^k - Bz^k)$. This algorithm is also known as the forward-backward-forward algorithm.

Under some mild conditions, the algorithm can converge to the zero of the operator T.

Lemma 5.10 ([28]) *Let D be a non-empty subset of $\mathcal{H}$, f, g are appropriate convex lower semicontinuous functions on $\mathcal{H}$, $dom\partial f \subset D$, g is G-differentiable on D, $argmin\ (f + g) \neq \emptyset$, and ∇g is $1/\alpha$-Lipschitz continuous on $dom\partial f$ ($\alpha > 0$). Let $x^0 \in \mathcal{H}$, $\beta \in (0, \alpha)$, and generate the sequence $\{x^k\}_{k\in\mathbb{N}}$ as follows*

$$\begin{cases} y^k = x^k - \beta \nabla g(x^k), \\ z^k = Prox_{\beta f}(y^k), \\ x^{k+1} = z^k + \beta(\nabla g(x^k) - \nabla g(z^k)). \end{cases} \tag{5.13}$$

Then the sequence $\{x^k\}_{k\in\mathbb{N}}$ converges to a zero of T.

5.2.4 Assumptions

This subsection introduces four assumptions, which are the basic assumptions that need to be satisfied throughout the paper, as well as the strong monotonicity assumption, the cocoercivity assumption, and the Lipschitz continuity assumption of the pseudo-gradients. They will be used successively in the following convergence proofs.

Assumption 5.1 *For all* $\nu = 1, \ldots, N,\ A_\nu \in \mathbb{R}^{m\times n_\nu},\ b \in \mathbb{R}^m$, *assuming* $\theta_\nu(\cdot, x_{-\nu})$ *is convex and continuously differentiable, and its gradient is continuous about* $x = (x_1, \ldots, x_N)$, *the pseudo-gradient* $\hat{P}$ *is monotone with respect to* $x = (x_1, \ldots, x_N)$. *Let* $\mathcal{X}$ *be a non-empty closed convex set, and the feasible set* $\mathcal{F} = \{x \in \mathcal{X} \mid Ax = b\}$ *is non-empty.*

Assumption 5.2 *Suppose the pseudo-gradient* $\hat{P}$ *is* ρ*-strongly monotone at the solution point* x^*, *that is, for* $\rho > 0$, *we have*

$$\langle \hat{P}(x) - \hat{P}(x^*), x - x^* \rangle \geq \rho \|x - x^*\|^2 \quad \forall x \in \mathbb{R}^n. \tag{5.14}$$

Assumption 5.3 *Suppose the pseudo-gradient* $\hat{P}$ *is* α*-cocoercive, that is, for* $\alpha > 0$, *we have*

$$\langle \hat{P}(x) - \hat{P}(y), x - y \rangle \geq \alpha \|\hat{P}(x) - \hat{P}(y)\|^2 \quad \forall x, y \in \mathbb{R}^n. \tag{5.15}$$

Assumption 5.4 *For all* $\nu = 1, \ldots, N$, *assuming gradient* $\nabla_{x_\nu}\theta_\nu$ *is* L_ν*-Lipschitz continuous. That is, for* $L_\nu > 0,\ \nu = 1, \ldots, N$, *we have*

$$\|\nabla_{x_\nu}\theta_\nu(x) - \nabla_{x_\nu}\theta_\nu(y)\| \leq L_\nu \|x - y\| \quad \forall x, y \in \mathbb{R}^n. \tag{5.16}$$

Strong monotonicity and Assumption 5.4 are employed in [27] for Algorithm 2. Strong monotonicity, Assumptions 5.3, and 5.4 are employed for Algorithm 1 in [26], respectively. Note that the assumption of strong monotonicity in [26, 27] has been weakened in this paper, i.e., which only needs to be satisfied at the solution point.

5.3 Augmented Lagrangian-Type Algorithms

In this section, several types of augmented Lagrangian-type algorithms for solving convex generalized Nash equilibrium problems with separable linearly common equality constraints are presented, which are called regularized linearized augmented Lagrangian algorithm (RLALM), regularized augmented Lagrangian algorithm (RALM), and corrected regularized linearization augmented Lagrangian algorithm (CRLALM), respectively.

Under the assumptions of basic convexity, strong monotonicity, cocoercivity, and Lipschitz continuity of pseudo-gradients, several convergence proofs are given respectively. Specifically, under the assumption of basic convexity, strong monotonicity and Lipschitz continuity of pseudo-gradients, the Fejér monotonicity of the iterative sequence of RLALM can be proved. Similarly, we can prove the Fejér monotonicity of the iterative sequence of RALM only with the basic convexity assumption. Under the coercivity condition, it can be proved that RLALM is equivalent to the forward-backward splitting algorithm, thus proving the convergence. In addition, if an additional correction step is added to RLALM, which is CRLALM, then only the basic convexity assumption and Lipschitz continuity assumption of the gradient (or pseudo-gradient) can be used for convergence. In addition, if an additional correction step is added to RLALM, which is CRLALM, then only the basic convexity assumption and Lipschitz continuity assumption of the gradient (or pseudo-gradient) can be used for convergence. At the end of this section, we take a closer look at GNEP with inequality constraints will also be given. The idea is to convert inequality constraints into equality constraints and then use the algorithms proposed in this section.

For each player, we define the regularized linearized (partial) augmented Lagrangian function as

$$\begin{aligned} L_\nu(x, x^k, \lambda^k) := \nabla_{x_\nu}\theta_\nu(x^k)^T(x_\nu - x_\nu^k) + (\lambda^k)^T\left(\sum_{\nu=1}^N A_\nu x_\nu - b\right) \\ + \frac{\beta}{2}\|\sum_{\nu=1}^N A_\nu x_\nu - b\|^2 + \frac{\gamma}{2}\|x_\nu - x_\nu^k\|^2, \end{aligned} \tag{5.17}$$

where β is penalty parameter, and γ is the regularization parameter. Note that each player will have their own augmented Lagrangian function, and each function is still related to the decision variables of all participants, but no longer has constraints besides their own abstract constraints. In this way, the subproblems of the algorithm will become a convex Nash equilibrium problem (NEP) with convex abstract constraints.

5.3.1 *Regularized Linearized Augmented Lagrangian Algorithm*

The first augmented Lagrangian algorithm proposed in this paper is shown in Algorithm 3, where subproblems (5.18) should be solved at each iteration, we then update the multiplier, where the parameters γ and β are fixed.

Algorithm 3 Regularized linearized augmented Lagrangian algorithm

1: Choose a starting point $(x^0, \lambda^0) \in \mathcal{X} \times \mathbb{R}^m$, parameters $\beta,\ \gamma > 0$.
2: If a suitable termination criterion is satisfied: STOP.
3: Calculate the solution x^{k+1} of the following NEP problem

$$\min_{x_\nu \in \mathcal{X}_\nu} L_\nu(x, x^k, \lambda^k), \quad \nu = 1, 2, 3, \ldots, N. \tag{5.18}$$

4: Compute

$$\lambda^{k+1} = \lambda^k + \beta \left(\sum_{\nu=1}^{N} A_\nu x_\nu^{k+1} - b \right). \tag{5.19}$$

5: Set $k := k + 1$, and go to Step 1.

Note that the subproblems of this algorithm are a convex NEP problem. According to [4], the solution of the NEP is equivalent to the solution of the variational inequality VI($\mathcal{X}$, P_L^k), where P_L^k is defined as

$$P_L^k(x) := \begin{pmatrix} \nabla_{x_1} L_1(x, x^k, \lambda^k) \\ \vdots \\ \nabla_{x_N} L_N(x, x^k, \lambda^k) \end{pmatrix}.$$

Then the subproblem is equivalent to find $x^{k+1} \in \mathcal{X}$, such that

$$\langle x - x^{k+1}, P_L^k(x^{k+1}) \rangle \geq 0 \quad \forall x \in \mathcal{X}. \tag{5.20}$$

The optimality condition of the variational inequality VI($\mathcal{X}$, P_L^k) is

$$\begin{aligned} 0 \in & \begin{pmatrix} \nabla_{x_1} \theta_1(x^k) \\ \vdots \\ \nabla_{x_N} \theta_N(x^k) \end{pmatrix} + \begin{pmatrix} A_1^T \lambda^k \\ \vdots \\ A_N^T \lambda^k \end{pmatrix} + \beta \begin{pmatrix} A_1^T(\sum_{\nu=1}^N A_\nu x_\nu^{k+1} - b) \\ \vdots \\ A_N^T(\sum_{\nu=1}^N A_\nu x_\nu^{k+1} - b) \end{pmatrix} \\ & + \gamma \begin{pmatrix} x_1^{k+1} - x_1^k \\ \vdots \\ x_N^{k+1} - x_N^k \end{pmatrix} + \begin{pmatrix} N_{X_1}(x_1^{k+1}) \\ \vdots \\ N_{X_N}(x_N^{k+1}) \end{pmatrix}. \end{aligned} \tag{5.21}$$

The third term in (5.21) reflects the essential difference between the algorithm and the Jacobi-type ADMM algorithm in Algorithm 1 and the Gauss-Seidel-ADMM-type algorithm in Algorithm 2. First, review the optimality conditions for the subproblems of Algorithms 1 and 2. The optimality conditions for the subproblems of the Jacobi-type ADMM in Algorithm 1 are given below.

$$0 \in \begin{pmatrix} \nabla_{x_1}\theta_1(x^k) \\ \vdots \\ \nabla_{x_N}\theta_N(x^k) \end{pmatrix} + \begin{pmatrix} A_1^T\lambda^k \\ \vdots \\ A_N^T\lambda^k \end{pmatrix} + \beta \begin{pmatrix} A_1^T(A_1x_1^{k+1} + \sum_{\mu\neq 1} A_\mu x_\mu^k - b) \\ \vdots \\ A_N^T(A_Nx_N^{k+1} + \sum_{\mu\neq N} A_\mu x_\mu^k - b) \end{pmatrix}$$
$$+\gamma \begin{pmatrix} x_1^{k+1} - x_1^k \\ \vdots \\ x_N^{k+1} - x_N^k \end{pmatrix} + \begin{pmatrix} N_{X_1}(x_1^{k+1}) \\ \vdots \\ N_{X_N}(x_N^{k+1}) \end{pmatrix}. \tag{5.22}$$

The difference between (5.22) and (5.21) is the third term; the third term of (5.21) is more concise.

The optimality conditions for the subproblems of the Gauss-Seidel-type ADMM in Algorithm 2 are given below.

$$0 \in \begin{pmatrix} \nabla_{x_1}\theta_1(x_1^{k+1}, x_{>1}^k) \\ \vdots \\ \nabla_{x_\nu}\theta_\nu(x_{\leq\nu}^{k+1}, x_{>\nu}^{k+1}) \\ \vdots \\ \nabla_{x_N}\theta_N(x^{k+1}) \end{pmatrix} + \begin{pmatrix} A_1^T\lambda^k \\ \vdots \\ A_\nu^T\lambda^k \\ \vdots \\ A_N^T\lambda^k \end{pmatrix}$$
$$+\beta \begin{pmatrix} A_1^T(A_1x_1^{k+1} + \sum_{\mu>1} A_\mu x_\mu^{k+1} - b) \\ \vdots \\ A_\nu^T(\sum_{\mu\leq\nu} A_\mu x_\mu^{k+1} + \sum_{\mu>\nu} A_\mu x_\mu^{k+1} - b) \\ \vdots \\ A_N^T(\sum_{\mu=1}^N A_\mu x_\mu^{k+1}) \end{pmatrix} \tag{5.23}$$
$$+\gamma \begin{pmatrix} x_1^{k+1} - x_1^k \\ \vdots \\ x_\nu^{k+1} - x_\nu^k \\ \vdots \\ x_N^{k+1} - x_N^k \end{pmatrix} + \begin{pmatrix} N_{X_1}(x_1^{k+1}) \\ \vdots \\ N_{X_\nu}(x_\nu^{k+1}) \\ \vdots \\ N_{X_N}(x_N^{k+1}) \end{pmatrix}.$$

In the (5.23), $x_{\leq\nu}$ represents all decision variable blocks whose subscripts do not exceed ν, and other similar symbols have similar meanings. The difference between (5.23) and (5.21) lies in the first term and the third term. Due to the characteristics of serial calculation of subproblems of the Gauss-Seidel-type ADMM algorithm, the form of its optimality condition is more complicated.

Since the subproblem of Algorithm 3 is to solve a NEP as a whole, the third term in the (5.21) only appears in the iteration point x^{k+1} of the $(k+1)$-th step, which is well symmetry. And combining with (5.19), the second term and the third term on the right-hand side of the above formula can be directly combined, so that the above optimality is simplified.

$$0 \in \begin{pmatrix} \nabla_{x_1}\theta_1(x^k) \\ \vdots \\ \nabla_{x_N}\theta_N(x^k) \end{pmatrix} + \begin{pmatrix} A_1^T\lambda^{k+1} \\ \vdots \\ A_N^T\lambda^{k+1} \end{pmatrix} + \gamma \begin{pmatrix} x_1^{k+1} - x_1^k \\ \vdots \\ x_N^{k+1} - x_N^k \end{pmatrix} + \begin{pmatrix} N_{X_1}(x_1^{k+1}) \\ \vdots \\ N_{X_N}(x_N^{k+1}) \end{pmatrix}. \quad (5.24)$$

Note that (5.19) can also be transformed into a form consistent with the above. Use square brackets $[\cdot]$ to group the terms of (5.19), which becomes

$$0 = [0] + \left[b - \sum_{\nu=1}^{N} A_\nu x_\nu^{k+1} \right] + \left[\frac{1}{\beta}(\lambda^{k+1} - \lambda^k) \right] + [0], \quad (5.25)$$

and considering $N_{\mathbb{R}^m}(\lambda^{k+1}) = 0$, replacing the last term on the right-hand side of the above equation, the optimality condition can be extended to

$$\begin{aligned} 0 \in & \begin{pmatrix} \nabla_{x_1}\theta_1(x^k) \\ \vdots \\ \nabla_{x_N}\theta_N(x^k) \\ 0 \end{pmatrix} + \begin{pmatrix} A_1^T\lambda^{k+1} \\ \vdots \\ A_N^T\lambda^{k+1} \\ b - \sum_{\nu=1}^{N} A_\nu x_\nu^{k+1} \end{pmatrix} \\ & + \begin{pmatrix} \gamma(x_1^{k+1} - x_1^k) \\ \vdots \\ \gamma(x_N^{k+1} - x_N^k) \\ \frac{1}{\beta}(\lambda^{k+1} - \lambda^k) \end{pmatrix} + \begin{pmatrix} N_{X_1}(x_1^{k+1}) \\ \vdots \\ N_{X_N}(x_N^{k+1}) \\ N_{\mathbb{R}^m}(\lambda^{k+1}) \end{pmatrix}. \end{aligned} \quad (5.26)$$

The matrix $Q \in \mathbb{R}^{(n+m)\times(n+m)}$ is introduced as follows.

$$Q = \begin{pmatrix} \gamma I_{n_1} & & & \\ & \ddots & & \\ & & \gamma I_{n_N} & \\ & & & \frac{1}{\beta} I_m \end{pmatrix} = \begin{pmatrix} \gamma I_n & \\ & \frac{1}{\beta} I_m \end{pmatrix}. \quad (5.27)$$

Hence, (5.26) can be reformulated by

$$0 \in P(w^k) + Gw^{k+1} + Q(w^{k+1} - w^k) + N_{\mathcal{W}}(w^{k+1}). \quad (5.28)$$

Based on this formula and different assumptions, two types of proofs are given below. That is to prove the Fejér monotonicity of the iterative point or prove that (5.28) is equivalent to finding the zero of the operator T using a forward-backward splitting algorithm.

Under Assumptions 5.1, 5.2, and 5.4, we will prove the Fejér monotonicity of the iterative sequence of Algorithm 3 , and then use Lemma 5.1 to prove the convergence of Algorithm 3. Now introduce the notation

$$\Delta(w^{k+1}, w^k) = Q(w^{k+1} - w^k) + P(w^k) - P(w^{k+1}). \tag{5.29}$$

By (5.12), $T(w) := P(w) + G(w) + N_{\mathcal{W}}(w)$. Optimality condition (5.28) can be rewritten as

$$0 \in T(w^{k+1}) + \Delta(w^{k+1}, w^k). \tag{5.30}$$

Lemma 5.11 *Let $w^* = (x^*, \lambda^*)$ be the variational KKT point of GNEP (5.1), the pseudo-gradient $\hat{P}(x)$ is the ρ-strongly monotone operator, then for any $w = (x, \lambda) \in \mathcal{X} \times \mathbb{R}^m$, we have*

$$\langle P(w) + Gw, w - w^* \rangle \geq \rho \|x - x^*\|^2. \tag{5.31}$$

Proof According to the characterization of the variational KKT point by Lemma 5.4, we have

$$\langle P(w^*) + Gw^*, w - w^* \rangle \geq 0 \quad \forall w = (x, \lambda) \in \mathcal{X} \times \mathbb{R}^m. \tag{5.32}$$

When the pseudo-gradient $\hat{P}(x)$ is strong monotone, combined with the properties (5.2.2) of G, we have

$$\langle P(w) + Gw - (P(w^*) + Gw^*), w - w^* \rangle \geq \rho \|x - x^*\|^2. \tag{5.33}$$

Adding (5.32) and (5.33) results in (5.31). □

According to the definition of normal cone (5.2.1), the optimality condition can be rewritten in an equality form, see the following lemma.

Lemma 5.12 *The iterative steps (5.18) and (5.19), i.e. (5.26), is equivalent to find $w^{k+1} = (x_1^{k+1}, \ldots, x_N^{k+1}, \lambda^{k+1})$, such that*

$$\begin{aligned} &\langle P(w^{k+1}) + Gw^{k+1} + Q(w^{k+1} - w^k), w - w^{k+1} \rangle \\ &+ \sum_{\nu=1}^{N} \langle \nabla_{x_\nu} \theta_\nu(x^k) - \nabla_{x_\nu} \theta_\nu(x^{k+1}), x_\nu - x_\nu^{k+1} \rangle \geq 0 \quad \forall w \in \mathcal{W}. \end{aligned} \tag{5.34}$$

Based on the above results, the convergence analysis of Algorithm 3 can be given.

Theorem 5.1 *Assume that GNEP (5.1) has variational KKT points. Assumptions 5.1, 5.2, and 5.4 hold. Parameters $\gamma > \varepsilon L > 0,\ \beta > 0$, then the iterative point w^{k+1} generated by Algorithm 3 converges to the variational KKT point $(x_1^*, \ldots, x_N^*, \lambda^*)$ of (5.1).*

Proof Since the parameters $\gamma > 0,\ \beta > 0$, it can be seen that the matrix Q is a symmetric positive definite matrix.

Using Lemma 5.11, we have

$$\langle P(w^{k+1}) + Gw^{k+1}, w^{k+1} - w^* \rangle \geq \rho \|x^{k+1} - x^*\|^2. \tag{5.35}$$

Taking $w = w^*$ into (5.34), using Cauchy-Schwarz inequality and Young's inequality (Lemma 5.2), for parameters ε, ρsatisfying $\varepsilon > \frac{1}{2\rho}$ (consequently $2\rho - \frac{1}{\varepsilon} > 0$), we have

$$
\begin{aligned}
0 \leq & \langle P(w^{k+1}) + Gw^{k+1} + Q(w^{k+1} - w^k), w^* - w^{k+1} \rangle \\
& + \sum_{\nu=1}^{N} \langle \nabla_{x_\nu} \theta_\nu(x^k) - \nabla_{x_\nu} \theta_\nu(x^{k+1}), x_\nu^* - x_\nu^{k+1} \rangle \\
\leq & \langle w^{k+1} - w^k, w^* - w^{k+1} \rangle_Q - \rho \|x^{k+1} - x^*\|^2 \\
& + \sum_{\nu=1}^{N} \langle \nabla_{x_\nu} \theta_\nu(x^k) - \nabla_{x_\nu} \theta_\nu(x^{k+1}), x_\nu^* - x_\nu^{k+1} \rangle \\
\leq & \langle w^{k+1} - w^k, w^* - w^{k+1} \rangle_Q - \rho \|x^{k+1} - x^*\|^2 \\
& + \sum_{\nu=1}^{N} \|\nabla_{x_\nu} \theta_\nu(x^k) - \nabla_{x_\nu} \theta_\nu(x^{k+1})\| \cdot \|x_\nu^* - x_\nu^{k+1}\| \\
\leq & \langle w^{k+1} - w^k, w^* - w^{k+1} \rangle_Q - \rho \|x^{k+1} - x^*\|^2 \\
& + \sum_{\nu=1}^{N} \frac{\varepsilon}{2} \|\nabla_{x_\nu} \theta_\nu(x^k) - \nabla_{x_\nu} \theta_\nu(x^{k+1})\|^2 + \frac{1}{2\varepsilon} \|x_\nu^* - x_\nu^{k+1}\|^2 \\
\leq & \langle w^{k+1} - w^k, w^* - w^{k+1} \rangle_Q - \rho \|x^{k+1} - x^*\|^2 \\
& + \sum_{\nu=1}^{N} \frac{\varepsilon}{2} L_\nu^2 \|x^k - x^{k+1}\|^2 + \frac{1}{2\varepsilon} \|x_\nu^* - x_\nu^{k+1}\|^2 \\
= & - \langle w^{k+1} - w^k, w^{k+1} - w^* \rangle_Q - (\rho - \frac{1}{2\varepsilon}) \|x^{k+1} - x^*\|^2 \\
& + \frac{\varepsilon}{2} L \|x^k - x^{k+1}\|^2.
\end{aligned}
$$

From the identity $2\langle w, v \rangle = \|w\|^2 + \|v\|^2 - \|w - v\|^2$, we have

$$
\begin{aligned}
0 \leq & \frac{1}{2} \|w^k - w^*\|_Q - \frac{1}{2} \|w^{k+1} - w^k\|_Q - \frac{1}{2} \|w^{k+1} - w^*\|_Q \\
& - \left(\rho - \frac{1}{2\varepsilon} \right) \|x^{k+1} - x^*\|^2 + \frac{\varepsilon}{2} L \|x^k - x^{k+1}\|^2.
\end{aligned}
$$

After arranging the above equation, we can get

$$
\begin{aligned}
\|w^{k+1}-w^*\|_Q^2 \leq& \|w^k-w^*\|_Q^2-\|w^{k+1}-w^k\|_Q^2\\
&-\left(2\rho-\frac{1}{\varepsilon}\right)\|x^{k+1}-x^*\|^2+\varepsilon L\|x^k-x^{k+1}\|^2\\
=&\|w^k-w^*\|_Q^2-\gamma\|x^{k+1}-x^k\|^2-\frac{1}{\beta}\|\lambda^{k+1}-\lambda^k\|^2\\
&-\left(2\rho-\frac{1}{\varepsilon}\right)\|x^{k+1}-x^*\|^2+\varepsilon L\|x^k-x^{k+1}\|^2\\
=&\|w^k-w^*\|_Q^2-(\gamma-\varepsilon L)\|x^{k+1}-x^k\|^2-\frac{1}{\beta}\|\lambda^{k+1}-\lambda^k\|^2\\
&-\left(2\rho-\frac{1}{\varepsilon}\right)\|x^{k+1}-x^*\|^2.
\end{aligned}
\tag{5.36}
$$

Note that in the above steps, when using Young's inequality. On the other hand, after selecting the parameter ε, the regular term parameter γ can be sufficiently large, so that $\gamma > \varepsilon L$, then we have $\gamma - \varepsilon L > 0$. In this way, the Fejér monotonicity of the iterative point sequence with respect to the variational KKT point set is obtained. Thus, the sequence $\{w^{k+1}\}$ is bounded, so it has cluster points. Below, we will prove that the cluster point of the sequence is a variational KKT point.

Let $\{w^{k+1}\}_I$ be the subsequence convergent to the cluster point $\bar{w}$ with $I \subset \mathbb{N}$, denote as $w^{k+1} \to_I \bar{w}$. Since $\mathcal{X}_1, \ldots, \mathcal{X}_N$ are closed convex sets, so $\mathcal{W} = \mathcal{X}_1 \times \ldots \times \mathcal{X}_N \times \mathbb{R}^m$ is also closed convex set. So $\bar{w} \in \mathcal{W}$.

Let $\hat{\rho} = 2\rho - \frac{1}{\varepsilon} > 0,\ \hat{\gamma} = \gamma - \varepsilon L > 0$, then Fejér monotonicity can be rewritten as

$$
\hat{\gamma}\|x^{k+1}-x^k\|^2+\frac{1}{\beta}\|\lambda^{k+1}-\lambda^k\|^2+\hat{\rho}\|x^{k+1}-x^*\|^2 \leq \|w^k-w^*\|_Q^2-\|w^{k+1}-w^*\|_Q^2.
\tag{5.37}
$$

In (5.37), summation of $k = 0, 1, \ldots, t$ yields that

$$
\begin{aligned}
&\sum_{k=0}^{t}(\hat{\gamma}\|x^{k+1}-x^k\|^2+\frac{1}{\beta}\|\lambda^{k+1}-\lambda^k\|^2+\hat{\rho}\|x^{k+1}-x^*\|^2)\\
&\leq \sum_{k=0}^{t}(\|w^k-w^*\|_Q^2-\|w^{k+1}-w^*\|_Q^2)\\
&= \|w^0-w^*\|_Q^2-\|w^{t+1}-w^*\|_Q^2\\
&\leq \|w^0-w^*\|_Q^2.
\end{aligned}
\tag{5.38}
$$

Let $t \to \infty$ in the above formula, we can get

$$
\|x^{k+1}-x^k\| \to 0, \quad \|\lambda^{k+1}-\lambda^k\| \to 0, \quad \|x^{k+1}-x^*\| \to 0.
\tag{5.39}
$$

So we have

$$\|w^{k+1} - w^k\| \to 0, \quad x^{k+1} \to x^*. \tag{5.40}$$

Combining $w^{k+1} \to_{\mathcal{I}} \bar{w}$, we get $w^k \to_{\mathcal{I}} \bar{w}$. Then by Assumption 5.4, i.e., the Lipschitz continuity of $\nabla_{x_\nu}\theta_\nu$, we have

$$\|\nabla_{x_\nu}\theta_\nu(x^k) - \nabla_{x_\nu}\theta_\nu(x^{k+1})\| \leq L_\nu\|x^k - x^{k+1}\| \to 0. \tag{5.41}$$

That is

$$\|P(x^k) - P(x^{k+1})\| \to 0, \quad k \to \infty. \tag{5.42}$$

So

$$\|\Delta(w^{k+1}, w^k)\| \leq \|Q\|\|w^k - w^{k+1}\| + \|P(x^k) - P(x^{k+1})\| \to 0, \quad k \to \infty. \tag{5.43}$$

So there is $\Delta(w^{k+1}, w^k) \to_{\mathcal{I}} 0$. (5.30) can be rewritten as $-\Delta(w^{k+1}, w^k) \in T(w^{k+1})$. Using the Lemma 5.8, and $\Delta(w^{k+1}, w^k) \to_{\mathcal{I}} 0,\ w^{k+1} \to_{\mathcal{I}} \bar{w}$, we get $0 \in T(\bar{w})$. So it is proved that $\bar{w}$ is a variational KKT point. By Lemma 5.1, it is known that the iterative sequence $\{w^{k+1}\}$ converges to a certain variational KKT point. □

So far, we have proved the convergence of RLALM under the strong monotonicity and Lipschitz continuity assumptions of pseudo-gradients. Next, we consider "weakening" the assumptions. Note that "weakening" here is not strictly speaking, since the strong monotonicity assumption of pseudo-gradients is only satisfied at the solution points. Below we will illustrate that Algorithm 3 is equivalent to the forward-backward splitting algorithm based on (5.28) under Assumptions 5.1 and 5.3, the convergence analysis will be implied.

Let

$$\overline{Q} := \beta Q = \begin{pmatrix} \beta\gamma I_n & \\ & I_m \end{pmatrix}. \tag{5.44}$$

The parameter $\beta > 0$, so $\overline{Q}$ is also a positive definite matrix. Then the inner product $\langle \cdot, \cdot \rangle_{\bar{Q}}$ and its induced norm $\|\cdot\|_{\overline{Q}}$ can be defined. Convergence is proved under this norm later in this subsection. In finite-dimensional Euclidean space, by the equivalence of any two norms, the corresponding convergence of different norms is also equivalent.

Rewrite (5.28) with $\overline{Q}$:

$$\begin{aligned}
&0 \in P(w^k) + Gw^{k+1} + Q(w^{k+1} - w^k) + N_{\mathcal{W}}(w^{k+1}) \\
&= \frac{1}{\beta}\overline{Q}w^k - P(w^k) \in Gw^{k+1} + \frac{1}{\beta}\overline{Q}w^{k+1} + N_{\mathcal{W}}(w^{k+1}) \\
&= (I - \beta\overline{Q}^{-1}P)(w^k) \in (I + \beta\overline{Q}^{-1}(G + N_{\mathcal{W}}))(w^{k+1}).
\end{aligned}$$

Let $T = P + G + N_{\mathcal{W}} = T_1 + T_2$, where $T_1 = P,\ T_2 = G + N_{\mathcal{W}}$. By Assumption 5.1, Lemmas 5.5, and 5.7, we know that $T_1,\ T_2$ are both maximally monotonic. And denote

$$A := \overline{Q}^{-1} T_2 = \overline{Q}^{-1}(G + N_{\mathcal{W}}), \qquad B := \overline{Q}^{-1} T_1 = \overline{Q}^{-1} P.$$

So

$$A + B = \overline{Q}^{-1}(T_1 + T_2) = \overline{Q}^{-1} T. \tag{5.45}$$

Since $\overline{Q}^{-1}$ is positive definite, it is an invertible matrix. So the zero of the operator $\overline{Q}^{-1} T$ is the same as the zero of T. That is, the zero point of $A + B$ is equivalent to the zero point of T. Below we will prove that Algorithm 3 is equivalent to the forward-backward splitting algorithm that finds the zero point of $\overline{Q}^{-1} T = A + B$.

Although T_1, T_2 are maximal monotone operators, but $A = \overline{Q}^{-1} T_2$, $B = \overline{Q}^{-1} T_1$ is not necessarily a maximal monotone operator. After defining the induced norm $\|\cdot\|_{\overline{Q}}$, it can be proved that A, B are maximally monotone under the induced norm.

Lemma 5.13 *Let $Q \in \mathbb{R}^n$ be a positive definite matrix, the operator $T : \mathbb{R}^n \to 2^{\mathbb{R}^n}$ is a maximal monotone operator about the norm $\|\cdot\|$, then the operator $Q \cdot T$ is a maximal monotone operator about the norm $\|\cdot\|_{Q^{-1}}$*

Proof According to the definition of the maximal monotone operator in view of (5.8), for the operator $Q \cdot T$, and the point $(x, u) \in \mathbb{R}^n \times \mathbb{R}^n$ proves this equivalence.

On the one hand, if $u \in QAx$, that is, $Q^{-1}u \in Ax$, for $\forall v \in QAy$, i.e. $Q^{-1}v \in Ay$, using the maximal monotonicity of A, we have

$$\langle x - y, Q^{-1}u - Q^{-1}v \rangle \geq 0,$$

That is

$$\langle x - y, u - v \rangle_{Q^{-1}} \geq 0.$$

Conversely, for all $v \in QAy$, we have

$$\langle x - y, u - v \rangle_{Q^{-1}} \geq 0,$$

Then for all $v' \in Ay$, there is $Qv' \in QAy$, and

$$\langle x - y, Q^{-1}u - v' \rangle = \langle x - y, u - Qv' \rangle_{Q^{-1}} \geq 0.$$

From the maximal monotonicity of A, we can get $Q^{-1}u \in Ax$ and hence $u \in QAx$. The proof is complete. □

According to Lemma 5.13, the following conclusions can be drawn.

Proposition 5.3 *Let Assumption 5.1 hold, parameters $\beta > 0$, $\gamma > 0$, then under the norm $\|\cdot\|_{\overline{Q}}$, the operators $A = \overline{Q}^{-1} T_2$, $B = \overline{Q}^{-1} T_1$ are both maximally monotonic.*

Based on the above notation, (5.28) can be further rewritten as

$$(I - \beta B)(w^k) \in (I + \beta A)(w^{k+1}),$$

where the operator A is maximally monotonic, then its resolution $J_{\beta A} = (I + \beta A)^{-1}$ is a single-valued mapping. In addition, the operator $T_1 = P$ is obviously a single-valued mapping, so $B = \overline{Q}^{-1} T_1$ is also a single-valued mapping. Then the above formula is equivalent to

$$w^{k+1} = (I + \beta A)^{-1}(I - \beta B)(w^k).$$

This gives the iterative form of the forward-backward splitting algorithm for finding the zeros of $A + B$.

To prove the convergence, we also need to consider the cocoercivity of the operator B. Under Assumption 5.3, i.e.,$\hat{P}$ is α-cocoercive, obviously P is also α-cocoercive. The following results give the effect of the composite matrix $\overline{Q}^{-1}$ on the cocoercivity.

Lemma 5.14 ([26]) *Let the operator $P : \mathbb{R}^n \to 2^{\mathbb{R}^n}$ be α-cocoercive about the inner product $\langle \cdot, \cdot \rangle$, Q is a symmetric positive definite matrix of order n, then the operator $Q^{-1}P$ is $\alpha/\|Q^{-1}\|$-cocoercive about the inner product $\langle \cdot, \cdot \rangle_Q$.*

The inverse of the matrix $\overline{Q}$ can be directly obtained:

$$\overline{Q}^{-1} = \begin{pmatrix} \frac{1}{\beta\gamma} I_n & \\ & I_m \end{pmatrix}.$$

The eigenvalues of $\overline{Q}^{-1}$ are $1/\beta\gamma$ and 1, so

$$\|\overline{Q}^{-1}\| = max \left\{ \frac{1}{\beta\gamma}, 1 \right\},$$

$$\frac{1}{\|\overline{Q}^{-1}\|} = min \{\beta\gamma, 1\}.$$

We want the operator $B = \overline{Q}^{-1} P$ to be α-cocoercive. From the definition of cocoercivity, it can be seen that the operator of δ-cocoercive ($\delta > \alpha$) is also α-cocoercive. So just let $1/\|\overline{Q}^{-1}\| \geq 1$, i.e. $\beta\gamma \geq 1$, then we get the α-cocoercivity of the operator B.

Therefore, applying the convergence Theorem 5.9 of the forward-backward splitting algorithm, the following result can be obtained.

Theorem 5.2 *Assume that GNEP (5.1) has variational KKT points, and that Assumptions 5.1 and 5.3 hold. Parameters $\gamma > 0$, $\beta > 0$, and satisfy $\beta\gamma \geq 1$, $\beta \in (0, 2\alpha)$. Then the iteration point w^{k+1} generated by Algorithm 3 converges to the variational KKT point $(x_1^*, \ldots, x_N^*, \lambda^*)$ of the problem under inner product $\langle \cdot, \cdot \rangle_{\bar{Q}}$ and its induced norm $\| \cdot \|_{\overline{Q}}$.*

5.3.2 Regularized Augmented Lagrangian Algorithm

This subsection will propose the second augmented Lagrangian-type algorithm. Considering that the augmented Lagrangian function is not linearized in the subproblem (5.18), then in addition to Assumption 5.1, the convergence can be proved without any additional conditions. That is, using the following augmented Lagrangian function

$$\begin{aligned}\bar{L}_\nu(x, x^k, \lambda^k) =&\theta_\nu(x) + (\lambda^k)^T\Big(\sum_{\nu=1}^{N} A_\nu x_\nu - b\Big)\\ &+ \frac{\beta}{2}\|\sum_{\nu=1}^{N} A_\nu x_\nu - b\|^2 + \frac{\gamma}{2}\|x_\nu - x_\nu^k\|^2.\end{aligned}$$

The algorithm is modified to the following Algorithm 4.

Algorithm 4 Regularized augmented Lagrangian algorithm

1: Choose a starting point $(x^0, \lambda^0) \in \mathcal{X} \times \mathbb{R}^m$, parameters $\beta,\ \gamma > 0$.
2: If a suitable termination criterion is satisfied: STOP.
3: Calculate the solution x^{k+1} of the following NEP problem

$$\min_{x_\nu \in \mathcal{X}_\nu} \bar{L}_\nu(x, x^k, \lambda^k), \quad \nu = 1, 2, 3, \ldots, N. \tag{5.46}$$

4: Compute

$$\lambda^{k+1} = \lambda^k + \beta\left(\sum_{\nu=1}^{N} A_\nu x_\nu^{k+1} - b\right).$$

5: Set $k := k + 1$, and go to Step 1.

The optimality condition of (5.46) can be rewritten as

$$0 \in P(w^{k+1}) + Gw^{k+1} + Q(w^{k+1} - w^k) + N_{\mathcal{W}}(w^{k+1}).$$

Redefine the residual notation $\Delta(w^{k+1}, w^k) = Q(w^{k+1} - w^k)$. Then the optimality condition is still rewritten as $0 \in T(w^{k+1}) + \Delta(w^{k+1}, w^k)$. Its form is simpler, and convergence can be easily proved by a similar method.

Similar to Lemma 5.11, we have the following result.

Lemma 5.15 *Let $w^* = (x^*, \lambda^*)$ be the variational KKT point of GNEP (5.1), if Assumption 5.1 is established, then for any $w = (x, \lambda) \in \mathcal{X} \times \mathbb{R}^m$, we have*

$$\langle P(w) + Gw, w - w^* \rangle \geq 0. \tag{5.47}$$

Proof According to the characterization of the variational KKT point by Lemma 5.4, we have

$$\langle P(w^*) + Gw^*, w - w^* \rangle \geq 0. \tag{5.48}$$

When Assumption 5.1 is satisfied, by the monotonicity of $\hat{P}(x)$ and the property (5.2.2) of G, we get

$$\langle P(w) + Gw - (P(w^*) + Gw^*), w - w^* \rangle \geq 0. \tag{5.49}$$

Adding (5.48) and (5.49) together yields (5.47). □

Due to the changes made in this lemma and the lack of linearization, compared with the proof of the convergence theorem under the strong monotonic assumption, there will be no more two differences about P and strong monotonic parameter terms in the process. Then we have

$$\begin{aligned} &0 \leq \langle Q(w^{k+1} - w^k), w^* - w^{k+1} \rangle \\ \Leftrightarrow &\|w^{k+1} - w^*\|_Q^2 \leq \|w^k - w^*\|_Q^2 - \|w^{k+1} - w^k\|_Q^2. \end{aligned}$$

That is, the Fejér monotonicity of the variational KKT point set with a simpler form. The following proofs are basically the same. So the convergence theorem of Algorithm 4 can be obtained.

Theorem 5.3 *Assume that GNEP (5.1) has variational KKT points, and that Assumption 5.1 holds, parameters $\gamma > 0$, $\beta > 0$. Then the iterative point w^{k+1} generated by Algorithm 4 converges to the variational KKT point $(x_1^*, \ldots, x_N^*, \lambda^*)$ of (5.1).*

For ALM without linearization in Algorithm 4, the corresponding convergence is easier to prove; however, the solution of algorithmic subproblems tends to be more complicated. The solution method of the subproblems is still similar to the inner iterative solution method, but each inner iteration will recall the pseudo-gradient function once, which will increase a lot of computation. But for small-scale problems, its numerical performance is often better.

5.3.3 *Corrected Regularized Linearized Augmented Lagrangian Algorithm*

This section will focus on the corrected regularized linearized augmented Lagrangian algorithm, which requires weaker assumptions to guarantee the corresponding convergence results. Referring to Tseng's algorithm (Lemma 5.10), which is a forward-backward splitting algorithm, where a correction step is added at each iteration. In this case, the cocoercivity assumption about operator B can be weakened into the

Lipschitz continuity assumption. Inspired by this, the corresponding correction step is added to Algorithms 3, and 5 is obtained.

Algorithm 5 Corrected regularized linearized augmented Lagrangian algorithm

1: Choose a starting point $(x^0, \lambda^0) \in \mathcal{X} \times \mathbb{R}^m$, parameters $\beta,\ \gamma > 0$.
2: If a suitable termination criterion is satisfied: STOP.
3: Calculate the solution y^k of the following NEP problem

$$\min_{x_\nu \in \mathcal{X}_\nu} L_\nu(x, x^k, \lambda^k), \qquad \nu = 1, 2, 3, \ldots, N. \tag{5.50}$$

4: Compute

$$\lambda^{k+1} = \lambda^k + \beta \left(\sum_{\nu=1}^{N} A_\nu y_\nu^k - b \right). \tag{5.51}$$

5: Compute

$$x^{k+1} = y^k + \frac{1}{\gamma} \left(\hat{P}(x^k) - \hat{P}(y^k) \right). \tag{5.52}$$

6: Set $k := k + 1$, and go to Step 1.

A correction step (5.52) has been added to Algorithm 5, which can be calculated explicitly. This section will give the equivalence of Algorithm 5 and Tseng's splitting algorithm under Assumptions 5.1 and 5.4 to prove the convergence. Denote

$$v^{k+1} := \begin{pmatrix} y^k \\ \lambda^{k+1} \end{pmatrix}.$$

The correction step (5.52) should be

$$w^{k+1} = v^{k+1} + \beta \left(B(w^k) - B(v^{k+1}) \right) = v^{k+1} + \beta \left(\bar{Q}^{-1} P(w^k) - \bar{Q}^{-1} P(v^{k+1}) \right).$$

Then the update of the decision variable x should be

$$x^{k+1} = y^k + \beta \cdot \frac{1}{\beta\gamma} (\overline{P}(x^k) - \overline{P}(y^k) = y^k + \frac{1}{\gamma} (\overline{P}(x^k) - \overline{P}(y^k)).$$

This is the correction step in Algorithm 5.

Next, we discuss the Lipschitz continuity of the operator $B = \overline{P}^{-1} T_1 = \overline{P}^{-1} P$. Based on Assumption 5.4, for all $\nu = 1, \ldots, N$, gradient $\nabla_{x_\nu} \theta_\nu$ is L_ν-Lipschitz continuous. Then for the operator P, for $\forall w = (x, \lambda),\ v = (y, \mu) \in \mathbb{R}^{n+m}$, we have

$$\|P(w) - P(v)\| = \left\| \begin{pmatrix} \nabla_{x_1}\theta_1(x) - \nabla_{x_1}\theta_1(y) \\ \vdots \\ \nabla_{x_N}\theta_N(x) - \nabla_{x_N}\theta_N(y) \\ 0 \end{pmatrix} \right\|$$
$$= \sum_{\nu=1}^{N} \|\nabla_{x_\nu}\theta_\nu(x) - \nabla_{x_\nu}\theta_\nu(y)\|$$
$$\leq \left(\sum_{\nu=1}^{N} L_\nu\right) \|x - y\| \leq \left(\sum_{\nu=1}^{N} L_\nu\right) \|w - v\|.$$

Denote $L = \sum_{\nu=1}^{N} L_\nu$, we know that P is L-Lipschitz continuous. Obviously, we know that the operator $B = \overline{Q}^{-1} P$ is also $\|\overline{Q}^{-1}\| L$-Lipschitz continuous. So we just require $\|\overline{Q}^{-1}\| \leq 1$, i.e. $\beta\gamma \geq 1$ to get the L-Lipschitz continuity of the operator B.

Therefore, using Lemma 5.10 of Tseng's splitting algorithm, the following results can be obtained.

Theorem 5.4 *Suppose GNEP (5.1) has variational KKT points, and Assumptions 5.1 and 5.4 hold. Denote* $L = \sum_{\nu=1}^{N} L_\nu$, *parameters* $\gamma > 0$, $\beta > 0$, *and satisfy* $\beta\gamma \geq 1$, $\beta \in (0, L)$. *Then the iteration point* w^{k+1} *generated by Algorithm 5 converges to the variational KKT point* $(x_1^*, \ldots, x_N^*, \lambda^*)$ *of the problem (5.1).*

5.3.4 Treatment of Inequality Constraints

So far, we have considered algorithms for convex GNEP with common linear equality constraints. Now generalize the previous results to linear inequality constraints. By introducing new variables, the inequality constraints are transformed into equality constraints, and the constraints of the new variables are incorporated into the abstract constraints. This allows the use of algorithms for equality constraints.

Börgens and Kanzow [26] give two transformation methods to convert GNEP (5.2) to linear equality-constrained GNEP. One is to introduce a new decision variable s_ν for each player, that is, for $\nu = 1, \ldots, N$, solve

$$\min_{\substack{x_\nu \in X_\nu \\ s_\nu \in \mathbb{R}^m_-}} \theta_\nu(x_\nu, x_{-\nu}) \quad \text{s.t.} \quad \sum_{\nu=1}^{N} A_\nu x_\nu - b - \sum_{\nu=1}^{N} s_\nu = 0. \tag{5.53}$$

Or we just introduce a new decision variable s for the last player, i.e., for $\nu = 1, \ldots, N - 1$, solve

$$\min_{x_\nu \in X_\nu} \theta_\nu(x_\nu, x_{-\nu}) \quad \text{s.t.} \quad \sum_{\nu=1}^{N} A_\nu x_\nu - b - s = 0. \tag{5.54a}$$

And for the N-th player, solve

$$\min_{\substack{x_N \in X_N \\ s \in \mathbb{R}^m_-}} \theta_N(x_N, x_{-N}) \quad \text{s.t.} \quad \sum_{\nu=1}^{N} A_\nu x_\nu - b - s = 0. \tag{5.54b}$$

Regarding the relationship between their solutions and the solution of the original problem, we have the following results.

Proposition 5.4 ([26]) *The following statements are equivalent:*

1. $x^* = (x_1^*, \ldots, x_N^*)$ *is the solution of GNEP (5.2).*
2. *For some* $s_\nu^* \in \mathbb{R}^m_-$, $\nu = 1, \ldots, N$, $(x_1^*, s_1^*, \ldots, x_N^*, s_N^*)$ *is the solution of GNEP (5.53).*
3. *For some* $s^* \in \mathbb{R}^m_-$, $(x_1^*, \ldots, x_N^*, s^*)$ *is the solution of GNEP (5.54).*

Regarding the relationship between their variational KKT points and the variational KKT points of the original problem, we have the following results.

Proposition 5.5 ([26]) *The following statements are equivalent:*

1. $(x^*, \lambda^*) = (x_1^*, \ldots, x_N^*)$ *is the variational KKT point of GNEP (5.2).*
2. *For some* $s_\nu^* \in \mathbb{R}^m_-$, $\nu = 1, \ldots, N$, $((x_1^*, s_1^*, \ldots, x_N^*, s_N^*), \lambda^*)$ *is the variational KKT point of GNEP (5.53).*
3. *For some* $s^* \in \mathbb{R}^m_-$, $((x_1^*, \ldots, x_N^{ast}, s^*), \lambda^*)$ *is the variational KKT point of GNEP (5.54).*

Note that after employing new variables, the gradient of the objective function with respect to the new decision variables no longer maintains strong monotonicity, but its cocoercivity can still be maintained. In addition, regarding the solution of the subproblems at each iteration, the idea of this paper is to solve the old and new decision variables separately.

5.4 Numerical Experiment

In this section, some numerical examples will be given to verify the effectiveness of the proposed algorithms.

5.4.1 Numerical Examples

Example 5.1 The first example was proposed by Facchinei and Fischer [6] and tested by Facchinei and Kanzow [9] and Han et al. [30]. There are two players in this question, each player ν controls a decision variable $x_\nu \in \mathbb{R}$. The objective function and constraints of each player are as follows

$$\begin{aligned}\min_{x_1} \quad & (x_1 - 1)^2 \\ \text{s.t.} \quad & x_1 + x_2 \le 1,\end{aligned}$$

and

$$\begin{aligned}\min_{x_2} \quad & (x_2 - \frac{1}{2})^2 \\ \text{s.t.} \quad & x_1 + x_2 \le 1.\end{aligned}$$

The generalized Nash equilibrium point set of this example is $\{(s, 1-s) \mid s \in [\frac{1}{2}, 1]\}$, and the unique variational equilibrium point is $(\frac{3}{4}, \frac{1}{4})$.

Example 5.2 The second example, taken from [6, 9], also consists of two players. The first player controls a two-dimensional decision variable $x_1 = (x_{11}, x_{12})^T =: (y_1, y_2)^T \in \mathbb{R}^2$, another participant controls a one-dimensional decision variable $x_2 =: y_3 \in \mathbb{R}$. The objective functions of the two participants are

$$\theta_1(x) = y_1^2 + y_1 y_2 + y_2^2 + (y_1 + y_2) y_3 - 25 y_1 - 38 y_2,$$

and

$$\theta_2(x) = y_3^2 + (y_1 + y_2) y_3 - 25 y_3.$$

They all have non-negative constraints $y_1, y_2, y_3 \ge 0$, and common linear constraints

$$\begin{aligned} y_1 + 2y_2 - y_3 &\le 14, \\ 3y_1 + 2y_2 + y_3 &\le 30. \end{aligned}$$

The generalized Nash equilibrium point set of this example is $\{(s, 11-s, 8-s) \mid s \in [0, 2]\}$, and the only variational equilibrium point is $(0, 11, 8)$.

Example 5.3 The third example is a modified version [26] of the duopoly model introduced by Krawczyk and Uryasev [31]. Two players ν control a decision variable $x_\nu \in \mathbb{R}$, representing their production of a given product. The objective function representing their profit is given by

$$\theta_\nu(x) = x_\nu(\rho(x_1 + x_2) + \lambda - d), \qquad \nu = 1, 2.$$

The productive capacity of each participant is limited by individual constraints $x_\nu \in [0, 10]$. Furthermore, both participants have a common resource limit constraint $x_1 + x_2 \le r$. Here we choose parameters $d = 20,\ \lambda = 4,\ \rho = 1,\ r = 9$.

Example 5.4 The fourth example is from Harker [32]. There are two players in this problem, player ν controls a decision variable $x_\nu \in \mathbb{R}$. The objective functions and constraints of each player are as follows

Table 5.1 Parameters of a river basin pollution model

Player ν	$c_{1\nu}$	$c_{2\nu}$	e_ν	$u_{\nu 1}$	$u_{\nu 2}$
1	0.10	0.01	0.50	6.5	4.583
2	0.12	0.05	0.25	5.0	6.250
3	0.15	0.01	0.75	5.5	3.750

$$\begin{aligned} \min_{x_1} \quad & (x_1)^2 + \frac{8}{3}x_1x_2 - 34x_1 \\ \text{s.t.} \quad & 0 \le x_1 \le 10, \\ & x_1 + x_2 \le 15, \end{aligned}$$

and

$$\begin{aligned} \min_{x_2} \quad & (x_2)^2 + \frac{5}{4}x_1x_2 - 24.25x_1 \\ \text{s.t.} \quad & 0 \le x_2 \le 10, \\ & x_1 + x_2 \le 15. \end{aligned}$$

The generalized Nash equilibrium point set for this example is $\{(5, 9)\} \cup \{(s, 15 - s) \mid s \in [9, 10]\}$. The variational equilibrium is $(5, 9)$.

Example 5.5 This example is a river basin pollution model introduced by Krawczyk and Uryasev [31] and tested by Han [30] et al. There are three players in this problem, player ν controls a decision variable $x_\nu \in \mathbb{R}$. The objective function is given by

$$\theta_\nu(x) = x_\nu(c_{1\nu} + c_{2\nu}x_\nu - d_1 + d_2(x_1 + x_2 + x_3)), \qquad \nu = 1, 2, 3.$$

And there are common linear constraints

$$\begin{pmatrix} u_{11}e_1 & u_{21}e_2 & u_{31}e_3 \\ u_{12}e_1 & u_{22}e_2 & u_{32}e_3 \end{pmatrix} \begin{pmatrix} x_1 \\ x_2 \\ x_3 \end{pmatrix} \le \begin{pmatrix} K_1 \\ K_2 \end{pmatrix}.$$

The parameters $c_{1\nu}$, $c_{2\nu}$, e_ν, $u_{\nu 1}$, $u_{\nu 2}$ are shown in Table 5.1, and setting the parameters $K_1 = K_2 = 100, d_1 = 3, d_2 = 0.01$. An approximate solution for this example is (21.1448, 16.0279, 2.7260).

Example 5.6 This example is the oligopoly model described in the literature [33], which was tested by Facchinei and Kanzow [9]. The problem has N players, each player has a decision variable $x_\nu \in \mathbb{R}$, and the objective function is as follows

$$\theta_\nu(x) = f_\nu(x_\nu) - 5000^{1/\eta}x_\nu(x_1 + x_2 + \ldots + x_N)^{-1/\eta}, \quad \nu = 1, 2, \ldots, N,$$

where

Table 5.2 Oligopoly model parameters

Player ν	1	2	3	4	5
c_ν	10	8	6	4	2
K_ν	5	5	5	5	5
δ_ν	1.2	1.1	1.0	0.9	0.8

$$f_\nu(x_\nu) = c_\nu x_\nu + \frac{\delta_\nu}{1+\delta_\nu} K_\nu^{-1/\delta_\nu}(x_\nu)^{\frac{1+\delta_\nu}{\delta_\nu}}, \quad \nu = 1, 2, \ldots, N.$$

All of them have non-negative constraints $x_\nu \geq 0, \ \nu = 1, 2, \ldots, N$, and common linear constraints

$$x_1 + x_2 + \ldots + x_N \leq P.$$

Take the parameters $N = 5, \ \eta = 1.1, \ P = 75$. The remaining parameters $c_\nu, \ K_\nu, \ \delta_\nu$ are shown in Table 5.2. An approximate solution for this example is

$$(10.4038, 13.0359, 15.4074, 17.3815, 18.7713).$$

Example 5.7 A final example is the electricity market model. It is presented in [11] and further studied in [34]. The problem has 2 players, each with a power plant on 2 of the 3 regions, which can be defined on the nodes of the graph (this is a simplified model of the literature [11]). First denote

$$S_1 = 40 - \frac{40}{500}(x_1 + x_4 + x_7 + x_{10}),$$
$$S_2 = 35 - \frac{35}{400}(x_2 + x_5 + x_8 + x_{11}),$$
$$S_3 = 32 - \frac{32}{600}(x_3 + x_6 + x_9 + x_{12}).$$

The objective functions of the two players are as follows

$$\theta_1(x) = (15 - S_1)(x_1 + x_4) + (15 - S_2)(x_2 + x_5) + (15 - S_3)(x_3 + x_6),$$

$$\theta_2(x) = (15 - S_1)(x_7 + x_{10}) + (15 - S_2)(x_8 + x_{11}) + (15 - S_3)(x_9 + x_{12}).$$

They all have non-negative constraints $x_\nu \geq 0, \ \nu = 1, 2, \ldots, 12$, and common linear constraints

$$x_1 + x_2 + x_3 \leq 100,$$
$$x_4 + x_5 + x_6 \leq 50,$$
$$x_7 + x_8 + x_9 \leq 100,$$
$$x_{10} + x_{11} + x_{12} \leq 50,$$

$$S_j - S_i \leq 0, \quad \forall i, j = 1, 2, 3, \quad i \neq j.$$

After testing, an approximate solution for this example is

$$(43.5364, 28.1381, 28.3255, 26.8698, 11.4714, 11.6588,$$
$$43.5364, 28.1381, 28.3255, 26.8698, 11.4714, 11.6588).$$

5.4.2 Numerical Results

We test all examples in Example 5.6 using the regularized linearized augmented Lagrangian algorithm in Algorithm 3 (RLALM), the regularized augmented Lagrangian algorithm in Algorithm 4 (RALM), the corrected regularized linear augmented Lagrangian algorithm in 5 (CRLALM), the Jacobi-type ADMM method in Algorithm 1 (JADMM), and the Gauss-Seidel-type ADMM method in Algorithm 2 (GADMM).

The subproblems for each algorithm are solved by the self-adjusting ratio projected gradient method. The termination condition of the inner loop is that the difference of the iterates $\|x^k - x^{k-1}\|$ is less than 10^{-8}. The termination condition of the outer iteration is taken as

$$\|P(x^k) + A^T \lambda^k\| \leq 10^{-6},$$
$$\| \min(\lambda^k, b - Ax^k)\| \leq 10^{-6}.$$

Regarding the adjustment of parameters, JADMM adopts the same parameter settings in [26] on the one hand, that is, for the α-cocoercive operator P, taking $\beta = \{\alpha, 0.2\}$, $\gamma = 1/\beta^2 + \|M\|$. On the other hand, this paper will manually adjust its parameters, denoted JADMMa, to further obtain better results. For GADMM, we take the adaptive strategy of the parameter γ in [27]. The residual of the iterates is

$$r^k := \sum_{\nu=1}^{N} \|\nabla_{x_\nu} \theta_\nu(x^k) + A_\nu^T \lambda^k + v_\nu^k\|^2 + \|\sum_{\nu=1}^{N} A_\nu x_\nu^k - b\|^2,$$

where $v_\nu^k \in N_{X_\nu}(x^k)$. This residual characterizes how well the iterate satisfies the variational KKT condition. Börgens and Kanzow [27] proposed that the condition that γ is sufficiently large is not a necessary condition for convergence, so they proposed an adaptive method to update γ when the following three conditions are satisfied at the same time:

1. γ is less than the given upper bound.
2. Residuals do not fall sufficiently: $r^{k+1} > \varepsilon' r^k$, where parameter $\varepsilon' \in (0, 1)$.
3. γ has not been updated in the previous κ steps, where κ is a given constant.

Table 5.3 Numerical results of RLALM with different ways of introducing new variables

Example	RLALM		RLALM2	
	Iter.	Time	Iter.	Time
3.1	14	5.4	14	4.9
3.2	22	34.1	22	26.8
3.3	8	3.8	8	2.7
3.4	92	80.8	92	62.4
3.5	23	77.9	31	90.7
3.6	23	44.5	23	48.9
3.7	46	124.6	46	105.0

Table 5.4 Numerical results of several ALM-type algorithms

Example	RLALM		RALM		CRLALM	
	Iter.	Time	Iter.	Time	Iter.	Time
4.1	14	5.4	6	4.1	47	15.2
4.2	22	34.1	9	36.0	93	98.8
4.3	8	3.8	7	3.0	29	11.7
4.4	92	80.8	5	11.5	326	88.8
4.5	23	77.9	10	66.8	97	92.4
4.6	23	47.5	7	29.5	346	149.8
4.7	46	124.6	13	131.1	114	194.8

γ can be updated by accumulation or multiplication, in other words, $\gamma^{k+1} := \gamma^k + \tau$ or $\gamma^{k+1} := \tau \cdot \gamma^k$. Here, we use the accumulation principle to update γ.

First, we test and compare the two ways of converting inequality constraints into equality constraints.

In Table 5.3, the column RLALM represents the transformation of inequality constraints into equality constraints by introducing N block new variables, then using the regularized linearized augmented Lagrangian algorithm for equality constraints proposed in this paper to solve the problem. The column RLALM2 indicates that only one new variable is introduced, and then the regularized linearized augmented Lagrangian algorithm is used to solve it. It can be seen that the numerical performance of the two is relatively close in terms of the number of iterations and time. All the following algorithms will uniformly adopt the method of introducing N blocks of new variables (Table 5.4).

Several augmented Lagrangian-type algorithms proposed in this paper are compared. It can be found from Algorithm 5.4 that for RALM, generally when the regular parameter γ is relatively small, the performance is better. Compared with RLALM, the number of iterations of RLALM is significantly smaller, and the iteration time of some examples is also significantly shorter. The parameters of CRLALM have to

Table 5.5 Numerical results for JADMM and RLALM

Example	JADMM		JADMMa		RLALM	
	Iter.	Time	Iter.	Time	Iter.	Time
4.1	53	21.3	22	8.5	14	5.4
4.2	87	99.1	34	31.0	22	34.1
4.3	112	39.6	23	9.5	8	3.8
4.4	536	171.0	202	81.6	92	80.8
4.5	4511	13868.4	361	499.7	23	101.5
4.6	–	–	156	175.0	23	61.0
4.7	788	874.4	190	261.9	46	124.6

Table 5.6 Numerical results for GADMM and RLALM

Example	GADMM		LGADMM		RLALM	
	Iter.	Time	Iter.	Time	Iter.	Time
4.1	22	10.5	35	18.8	14	5.4
4.2	36	32.5	366	512.3	22	34.1
4.3	36	29.5	197	93.7	8	3.8
4.4	100	72.0	364	191.2	92	80.8
4.5	500	772.9	711	990.2	23	101.5
4.6	60	86.9	187	221.8	23	61.0
4.7	91	428.5	160	580.6	46	124.6

be re-adjusted to have better results. Its numerical performance is obviously inferior to that of RLALM.

Next, we examine the comparison between JADMM in [26] and RLALM in this paper (Table 5.5).

It can be seen from Algorithm 5.5 that the iteration of JADMM using the original parameter settings in [26] is too many, and JADMMa with manual parameter adjustment can greatly speed up the convergence. And RLALM is also significantly faster than the tuned JADMM. Note that since Example 5.6 is difficult to track the cocoercivity parameters of the pseudo-gradient, the numerical results of JADMM are not given.

Finally, we will examine the numerical performance comparison between the Gauss-Seidel-type ADMM method of the literature [27] and the RLALM (Table 5.6).

Both GADMM and LGADMM in Algorithm 5.6 use the adaptive strategy in the literature [27] to adjust the parameters γ. LGADMM represents the linearization of the objective function part of the GADMM algorithm. We examine the effect of this operation on numerical performance. The parameter requirements of LGADMM are different from those of GADMM, and it needs a larger initial γ and growth span than GADMM in order to have better convergence performance. But its numerical performance is not as good as GADMM. In addition, the numerical performance of RLALM is obviously better than that of GADMM.

5.5 Conclusions

In this paper, we propose several augmented Lagrange-type algorithms for solving generalized Nash equilibrium problems with linear common equality or inequality constraints. Under different assumptions, there were different technical methods for convergence, i.e., the Fejér monotonicity of the corresponding iterative sequences or the equivalence with forward-backward splitting theoretically. In the future, we may consider adapting the parameters to give full play to the efficiency of the algorithms.

Acknowledgements This work was supported by National Natural Science Foundation of China grants 11971238.

References

1. Nikaido, H., Isoda, K.: Note on noncooperative convex games. Pac. J. Math. **5**, 807–815 (1955)
2. Bensoussan, A.: Points de Nash dans le cas de fonctionnelles quadratiqueset jeux differentiels lineaires a N personnes. SIAM J. Control Optim. **12**, 460–499 (1974)
3. Rosen, J.B.: Existence and uniqueness of equilibrium points for concave n-person games. Econometrica **33**, 520–534 (1965)
4. Facchinei, F., Kanzow, C.: Generalized Nash equilibrium problems. Ann. Oper. Res. **175**(1), 177–211 (2010)
5. Fischer, A., Herrich, M., Schönefeld, K.: Generalized Nash equilibrium problems—recent advances and challenges. Ann. Oper. Res. **34**(3), 521–558 (2014)
6. Facchinei, F., Fischer, A., Piccialli, V.: Generalized Nash equilibrium problems and Newton methods. Math. Program. **117**, 163–194 (2009)
7. Santos, P.J.S., Santos, P.S.M., Scheimberg, S.: A newton-type method for quasi-equilibrium problems and applications. Optimization **71**(1), 7–32 (2022)
8. Fukushima, M.: Restricted generalized Nash equilibria and controlled penalty algorithm. CMS **8**(3), 201–218 (2011)
9. Facchinei, F., Kanzow, C.: Penalty methods for the solution of generalized Nash equilibrium problems. SIAM J. Optim. **5**, 2228–2253 (2010)
10. Facchinei, F., Lampariello, L.: Partial penalization for the solution of generalized Nash equilibrium problems. J. Global Optim. **50**(1), 39–57 (2011)
11. Pang, J., Fukushima, M.: Quasi-variational inequalities, generalized Nash equilibria, and multi-leader-follower games. Math. Program. **2**, 21–56 (2005)
12. Kanzow, C.: On the multiplier-penalty-approach for quasi-variational inequalities. Math. Program. **160**(1–2), 33–63 (2016)
13. Kanzow, C., Steck, D.: Augmented Lagrangian methods for the solution of generalized Nash equilibrium problems. SIAM J. Optim. **26**, 2034–2058 (2016)
14. Facchinei, F., Piccialli, V., Sciandrone, M.: Decomposition algorithms for generalized potential games. Comput. Optim. Appl. **50**, 237–262 (2011)

15. Lei, J., Shanbhag, U.V., Pang, J., Sen, S.: On synchronous, asynchronous, and randomized best-response schemes for stochastic Nash games. Math. Oper. Res. **45**, 157–190 (2020)
16. Shanbhag, U.V., Pang, J., Sen, S.: Inexact best-response schemes for stochastic Nash games: linear convergence and iteration complexity analysis. In: Proceedings of the 55th Conference on Decision and Control, pp. 3591–3596. IEEE (2016)
17. Boţ, R.I., Csetnek, E.R., Nguyen, D.-K.: A proximal minimization algorithm for structured nonconvex and nonsmooth problems. SIAM J. Optim. **29**, 1300–1329 (2019)
18. Boţ, R.I., Nguyen, D.-K.: The proximal alternating direction method of multipliers in the nonconvex setting: convergence analysis and rates. Math. Oper. Res. **54**, 682–712 (2020)
19. Börgens, E., Kanzow, C.: Regularized Jacobi-type ADMM-methods for a class of separable convex optimization problems in Hilbert spaces. Comput. Optim. Appl. **73**, 755–790 (2019)
20. Boyd, S., Parikh, N., Chu, E., Peleato, B., Eckstein, J.: Distributed optimization and statistical learning via the alternating direction method of multipliers. Found. Trends Mach. Learn. **3**, 1–122 (2011)
21. Eckstein, J., Yao, W.: Understanding the convergence of the alternating direction method of multipliers: theoretical and computational perspectives. Pac. J. Optim. **11**, 619–644 (2015)
22. He, B., Hou, L., Yuan, X.: On full Jacobian decomposition of the augmented Lagrangian method for separable convex programming. SIAM J. Optim. **25**, 2274–2312 (2015)
23. He, B., Tao, M., Yuan, X.: Alternating direction method with Gaussian back substitution for separable convex programming. SIAM J. Optim. **22**, 313–340 (2012)
24. He, B., Xu, H.-K., Yuan, X.: On the proximal Jacobian decomposition of ALM for multipleblock separable convex minimization problems and its relationship to ADMM. J. Sci. Comput. **66**, 1204–1217 (2016)
25. He, B.: PPA-like contraction methods for convex optimization: a framework using variational inequality approach. J. Oper. Res. Soc. China **3**, 391–420 (2015)
26. Börgens, E., Kanzow, C.: A distributed regularized Jacobi-type ADMM-method for generalized Nash equilibrium problems in Hilbert spaces. Numer. Funct. Anal. Optim. **39**, 1316–1349 (2018)
27. Börgens, E., Kanzow, C.: ADMM-type methods for generalized Nash equilibrium problems in Hilbert spaces. SIAM J. Optim. **31**, 377–403 (2021)
28. Bauschke, H.H., Combettes, P.L.: Convex Analysis and Monotone Operator Theory in Hilbert Spaces. Springer (2010)
29. Facchinei, F., Pang, J.: Finite-Dimensional Variational Inequalities and Complementarity Problems. Springer (2003)
30. Han, D., Zhang, H., Qian, G., Xu, L.: An improved two-step method for solving generalized Nash equilibrium problems. Eur. J. Oper. Res. **216**, 613–623 (2012)
31. Krawczyk, J.B., Uryasev, S.: Relaxation algorithms to find Nash equilibria with economic applications. Environ. Model. Assess. **5**(1), 63–73 (2000)
32. Harker, P.T.: Generalized Nash games and quasi-variational inequalities. Eur. J. Oper. Res. **54**, 81–94 (1991)
33. Outrta, J.V., Kocvara, M., Zowe, J.: Nonsmooth approach to optimization problems with equilibrium constraints. J. Environ. Manag. **103**(4), 74–82 (1997)
34. Nabetani, K., Tseng, P., Fukushima, M.: Parametrized variational inequality approaches to generalized Nash equilibrium problems with shared constraints. Comput. Optim. Appl. **48**(3), 423–452 (2011)

GPSR Compliance

The European Union's (EU) General Product Safety Regulation (GPSR) is a set of rules that requires consumer products to be safe and our obligations to ensure this.

If you have any concerns about our products, you can contact us on ProductSafety@springernature.com

In case Publisher is established outside the EU, the EU authorized representative is:

Springer Nature Customer Service Center GmbH
Europaplatz 3
69115 Heidelberg, Germany

Batch number: 10725798

Printed by Printforce, the Netherlands

Olaf Irretier, Marco Jost (Hrsg.)

Handbuch HärtereiPraxis

Verfahren | Anwendungen | Innovationen

Vulkan Verlag

Bibliografische Information der Deutschen Nationalbibliothek
Die Deutsche Nationalbibliothek verzeichnet diese Publikation in der Deutschen Nationalbibliografie; detaillierte bibliografische Daten sind im Internet über www.dnb.de abrufbar.

Handbuch HärtereiPraxis
Verfahren | Anwendungen | Innovationen
Olaf Irretier, Marco Jost (Hrsg.)
5. Ausgabe 2026

ISBN: 978-3-8027-3196-9 (Print)
ISBN: 978-3-8027-3197-6 (eBook)

Friedrich-Ebert-Straße 55, 45127 Essen, Deutschland
E-Mail: bestellung@vulkan-verlag.de
Telefon: +49 201 820 02-0
Internet: www.vulkan-verlag.de

Projektmanagement: Marie-Therese Hanschmann, Vulkan-Verlag GmbH, Essen
Lektorat: Marie-Therese Hanschmann, Annamaria Weinert, Vulkan-Verlag GmbH, Essen
Herstellung: Melanie Zöller, Vulkan-Verlag GmbH, Essen
Umschlaggestaltung: Melanie Zöller, Vulkan-Verlag GmbH, Essen
Titelbild: © SMS group
Satz: Brigitte Schmidt, Schmidt Media Design, München
Druckerei: mediaprint solutions GmbH, 33100 Paderborn

Vorwort

Die Frage, ob wir uns auf dem richtigen Weg befinden, sollte kein Tabu sein

Als ich im Jahr 2000 in die Welt der Wärmebehandlung einstieg, wurde mir als jungem Techniker schnell eines beigebracht: Bestimmte Themen sind im internationalen Geschäft tabu – insbesondere Religion, Politik oder gesellschaftlich kontroverse Fragen.

Heute, über 25 Jahre später, hat sich mein Blick auf die Dinge gewandelt. Ich durfte viele Länder bereisen, Erfahrungen sammeln und unterschiedliche Perspektiven kennenlernen. Und ich habe festgestellt: Die strikte Trennung zwischen Fachthemen und gesellschaftspolitischen Fragen lässt sich heute kaum noch aufrechterhalten.

Ich erinnere mich an ein Gespräch mit chinesischen Kollegen, in dem ich – vielleicht etwas unbedacht – das Thema der Uiguren ansprach oder an einen Termin mit einem österreichischen Interessenten, dem ich erläuterte, dass vermeintlich „kostenlose" erneuerbare Energie oft durch deutsche Verbraucher mitfinanziert wird – etwa durch den Export von Überschussstrom tagsüber und den teuren Rückimport am Abend über Pumpspeicherkraftwerke.

Diese Beispiele zeigen: Politische und wirtschaftliche Rahmenbedingungen beeinflussen unsere tägliche Arbeit stärker denn je. Auf Messen, Konferenzen oder in Kundengesprächen – politische Themen stehen immer häufiger im Raum, selbst bei ersten Begegnungen. Und das ist verständlich: Unsere Branche ist eng mit der Industriepolitik, Energieversorgung und Standortattraktivität verknüpft.

Was mir derzeit Sorgen bereitet: Während die wirtschaftliche Dynamik international – insbesondere außerhalb Europas – wieder anzieht, bleibt der Aufschwung hierzulande aus. Das einst wirtschaftlich starke Deutschland verliert an Zugkraft – und dient anderen Ländern inzwischen eher als abschreckendes Beispiel.

Die Belastungen für Unternehmen sind vielfältig: steigende Bürokratie, hohe Sozialabgaben und weltweit einzigartige Energiekosten. Ich werde oft von internationalen Partnern darauf angesprochen – und ich kann zahlreiche Beispiele nennen, wo gut gemeinte Subventionen aus dem Ruder gelaufen sind.

Die Frage, ob wir uns auf dem richtigen Weg befinden, sollte kein Tabu sein. Im Gegenteil: Gerade in unserer Branche, die auf Innovationskraft und Wettbewerbsfähigkeit angewiesen ist, müssen wir offen und kritisch über Rahmenbedingungen, politische Weichenstellungen und Zukunftsperspektiven sprechen dürfen.

Ich hoffe, dass der eine oder andere Beitrag hier für Sie von Interesse ist – und vielleicht sogar zu neuen Ideen oder innovativen Ansätzen in Ihrem Arbeitsalltag inspiriert.

Mit kollegialen Grüßen

Gerald Hiller
ECM GmbH

Beratung

- Werkstoffe und Wärmebehandlungsverfahren
- Modernisierung vorhandener oder Implementierung neuer Wärmebehandlungsprozesse
- Auslegung der Industrieöfen und der zugehörigen Anlagentechnik
- Überprüfung und Erstellung von Lastenheften

Anlagentechnik

- Individuell zugeschnittene Industrieöfen
- Kühlwassertechnik und Reinigungsanlagen
- Ersatzteile und Chargiermittel aus hitzebeständigen Edelstählen
- CFC-Gestelle

Inhouse-Schulungen

- Lebendig, individuell, fachlich kompetent
- Einsteiger- oder Intensivseminare
- individuell an das Arbeitsumfeld angepasst

Intensivseminare

- in Düsseldorf, Dortmund, Nürnberg, Frankfurt und München
- Werkstoffkunde und Wärmebehandlung von Stahl und Aluminium
- Anlagentechnik und Energieeffizienz in der Wärmebehandlung
- Vergüten von Stahl
- Nitrieren, Einsatzhärten, Borieren und Randschichthärten

Infos zur allen Intensivseminaren auf:
www.ibw-irretier.de

IBW Dr. Irretier GmbH
Mühsol 44 • 47533 Kleve
+49 (2821) 715 39 48 • info@ibw-irretier.de

IBW
Industrieberatung für Wärmebehandlungstechnik

Vorwort

Die aktuelle Krise wird irreparable Schäden in der deutschen Industrie hinterlassen

Die deutsche Wirtschaft kommt nicht aus der Krise – im Gegenteil: Die Hiobsbotschaften mehren sich, tausende Stellen fallen weg und die Wachstumsaussichten sind ernüchternd. Fast sämtliche Branchen sind davon betroffen.

Alle Blicke richten sich auf die Regierung in Berlin. Sie hatte versprochen, die Wirtschaft würde schon im Sommer 2025 die Wende spüren. Das ist nicht passiert. Man kann den Politikern auch nicht vorwerfen, sie tun nichts. Eingeladen wurde zum Autogipfel, Stahlgipfel, einem Innovationsgipfel usw. Doch man spürt davon leider keine Belebung der Wirtschaft. Was Deutschland braucht, sind keine großen Absichtserklärungen mit tollen Fotos, sondern: niedrigere Energie- und Lohnkosten, weniger Steuern und Bürokratie und so wenig politische Einmischung (Stichwort: Verbrennerverbot) wie möglich – und das so schnell wie möglich!

Auch im automobilen Mittelstand herrscht eine Dauerkrise. Früher war die automobile Welt klar aufgeteilt. Die Deutschen erfanden das Automobil, die Amerikaner führten die Massenproduktion ein, die Kunden kauften. Die großen Automobilhersteller kamen aus den USA sowie aus Deutschland und weiteren Ländern Europas. Dann begann in den 1960er-Jahren die japanische globale Offensive, gefolgt von den Koreanern. Eine chinesische Industrie gab es nicht – und auch keinen Markt. Das änderte sich in den 1980er-Jahren. VW baute in China Werke und versorgte die Menschen dort erstmals mit Autos. Über Jahrzehnte hinweg war der chinesische Markt fest in der Hand ausländischer Marken. Der Absatz wuchs und wuchs. Dann begann der Aufstieg chinesischer Autokonzerne. Gleichzeitig expandierten chinesische Hersteller und Zulieferer nach Europa, vornehmlich im Bereich der Elektromobilität. Das Verhältnis zwischen europäischen und chinesischen Firmen verändert sich drastisch und im Eiltempo. Der globale Automarkt polarisiert sich zunehmend, wobei mehrere entkoppelte regionale Ökosysteme entstehen. Inzwischen ist China das Land, in dem die meisten Autos verkauft und auch produziert werden, für alle namhaften Automobilhersteller außerhalb der USA der größte und wichtigste Markt. Für Europas Volumenhersteller, dazu zählen auch die deutschen Unternehmen VW, BMW und Mercedes, ist es kaum (nicht) möglich, den chinesischen Markt – trotz des harten Wettbewerbs – links liegen zu lassen.

Jetzt stehen die deutschen Automobilhersteller, deren Zulieferer und eingebundene Dienstleistungsunternehmen, zu denen auch die Lohnwärmebehandler zählen, seit Jahren unter Druck. Neue Formen der Mobilität, ein volatiler Markt, Kostendruck von außen und innen – es ist Crunch-Time im Mittelstand. Aktuell sehen 60 % der Unternehmen eine fehlende Wachstumsperspektive am Markt. Globale Trends wie Elektromobilität, neue Marktteilnehmer und geopolitische Unsicherheiten treiben Marktdynamiken wie nie zuvor an und setzen den deutschen Mittelstand unter extremen Handlungszwang. Zudem erschweren Vorgaben und Beschränkungen der EU-Politik technische Möglichkeiten der Unternehmen.

Inzwischen arbeiten alle an Gegenmaßnahmen. Restrukturierungen, Sanierungen, Kostenprogramme und Prozessoptimierungen wurden überall in Angriff genommen und vielfach auch schon umgesetzt. Dies hat zur Folge, dass allein innerhalb von zwölf Monaten in der deutschen Automobilindustrie mehr als 50.000 Arbeitsplätze unwiederbringlich abgebaut wurden. Bis zum Jahr 2030 werden hier höchstwahrscheinlich leider noch mehr als 100.000 weitere Arbeitsplätze abgebaut. Für 2025 wird erwartet, dass die Insolvenzen im Automobilzulieferbereich um 30 % zulegen werden. Diese Entwicklung wird sich wahrscheinlich auch im Jahr 2026 fortsetzen. Bei Zulieferern, die rein im Verbrennerbereich aktiv sind, fehlen die Perspektiven – eine Zäsur für die deutsche Automobilindustrie.

Mittelständische Zulieferer stehen nun an einem Wendepunkt: Fokussierung oder Diversifizierung? Das Ergebnis einer aktuellen Studie ist dabei eindeutig. Fast vier von fünf Unternehmen betrachten Diversifizierung als strategisch wertvoll. Viele Unternehmen erhoffen sich Wachstumspotenzial außerhalb des Automotive-Umfelds. Jeder zweite Umfrageteilnehmer ist der Ansicht, dass Diversifizierung die Resilienz gegenüber geopolitischen und regulatorischen Risiken erhöht.

Aufgrund der augenblicklichen politischen Situation an der NATO-Ostgrenze und des schon zu lange andauernden Ukrainekrieges zeigen Automotive-Zulieferer und auch Lohnwärmebehandlungsbetriebe verstärkt Interesse an der Rüstungsindustrie. Auch die Luft- und Raumfahrtindustrie ist für dem Mittelstand attraktiv. Allerdings ist die Erschließung des Rüstungsmarktes leichter gesagt als getan. Es gibt hohe Eintrittsbarrieren, neue Netzwerke und Vertriebskanäle müssen aufgebaut werden und politische wie regulatorische Rahmenbedingungen beachtet werden. All dies braucht Zeit. Hier stellt sich jetzt die Frage: Hat gerade der familiengeführte Mittelstand die Zeit und monetären Rücklagen für solch grundlegende Veränderungen bzw. Anpassungen?

Auch wenn die Stimmung in Deutschland als eher negativ zu bewerten ist, gibt es auch immer noch Lichtblicke. Die Firma Bosch ist weiterhin die Nummer eins unter den weltgrößten Zulieferern. Unter den Top-100-Unternehmen der Zulieferbranche kommen 20 Unternehmen aus Deutschland, drei Unternehmen sind sogar unter den Top 10. Darauf kann man stolz sein, sich aber nicht ausruhen. Das tut auch keiner, aber mit politischer Unterstützung für eine komplette Technologieoffenheit, günstigeren Energiepreisen, weniger Bürokratiehürden und mehr Flexibilität hinsichtlich der Umsetzung von Maßnahmen kann die deutsche Industrie auch weiterhin positiv nach vorne blicken.

Die Lohnwärmebehandler sind ein gutes Beispiel. Man trifft sich zu Tagungen, Seminaren und Fortbildungen. Es werden Wärmebehandlungsanlagen, Organisationsabläufe, Energiereduzierungsmaßnahmen und Prozessparameter optimiert bzw. weiterentwickelt. Interessante Ergebnisse werden in vielen Berichten leichtverständlich dargestellt und werden auch in diesem Buch attraktiv präsentiert. Persönlich freue ich mich, für dieses Fachbuch das Vorwort geschrieben zu haben und wünsche Ihnen jetzt eine interessante und informative Lektüre.

Uwe Schmelzing
Aalberts Surface Technologies GmbH

Inhaltsverzeichnis

1. HärtereiPraxis

1.1 Ein mehr als hundertjähriges Duell: Verbrennungsmotor versus Elektroantrieb

Uwe Schmelzing

1.1.1 Einführung

Viele Menschen sind der Überzeugung, dass Elektromobilität nach moderner Technik klingt und eine Entwicklung des 21. Jahrhunderts ist. Doch den ersten Elektromotor für ein Schiff gab es schon in den 1830er-Jahren. Es ist daher ein eher wiederentdeckter Antrieb, der mit heutigen technischen Möglichkeiten auf ein neues Level gehoben wurde. Tatsächlich waren bereits die ersten Autos, die auf europäischen und US-amerikanischen Straßen fuhren, elektrisch betrieben. Das erste in Deutschland gefertigte E-Auto gab es schon zum Ende des 19. Jahrhunderts und es löste einen richtigen Boom aus. Das Duell zwischen dem Verbrennungsmotor und dem Elektroantrieb ist damit mehr als ein Jahrhundert alt und eine faszinierende Geschichte technologischen Fortschritts, wirtschaftlicher Interessen und gesellschaftlicher Veränderung. Während die frühen Entwicklungen oft belächelt und übersehen wurden, prägen sie bis heute die Debatte und Zukunft der Mobilität.

1.1.2 Geschichtliche Entwicklung der Elektromobilität

Zu Beginn des 20. Jahrhunderts spielte die Elektromobilität in Deutschland und Europa eine führende Rolle, denn die Anzahl der elektrisch betriebenen Fahrzeuge war fast doppelt so hoch wie die mit Verbrennungsmotoren. Auch in den USA präsentierten Erfinder wie William Morrison, Henry G. Morris und Thomas Parker funktionstüchtige Elektrofahrzeuge. 1897 begann in den USA die kommerzielle Produktion und bis 1912 erlebte der E-Auto-Markt einen wahren Boom. Rund 34.000 Elektrofahrzeuge mit Reichweiten von bis zu 100 km und mehr wurden registriert und machten damit 38 % der Fahrzeuge auf amerikanischen Straßen aus. Im Vergleich dazu hatten nur 22 % der Autos einen Benzinmotor. Der größte Konkurrent der damaligen E-Autos waren Dampffahrzeuge, die mit 40 % den Großteil des Straßenverkehrs dominierten.

Die Anfänge der Elektromobilität gehen auf Michael Faraday zurück. Er legte 1821 den Grundstein für viele Entwicklungen im Bereich der Elektromotoren. Er zeigte erstmals, wie Bewegung durch Elektromagnetismus entsteht. Die Erfindung des Gleichstrom-Motors durch Thomas Davenport machte es möglich, dass im Jahr 1881 in Paris das erste elektrisch betriebene Dreirad vorgestellt wurde – und von da an nahm alles seinen Lauf.

Seinen Anfang nahm die Geschichte der E-Autos mit Gustave Trouve. Der französische Erfinder gilt als entscheidender Wegbereiter. Sein Trouve Tricycle war 1881 das erste richtige Elektrofahrzeug der Welt. Streng genommen handelte es sich dabei allerdings um ein Dreirad und nicht um ein Automobil mit vier Rädern. Das Trouve Tricycle wurde mit neuen modernen Blei-Akkus betrieben, erreichte eine Geschwindigkeit von 12 km/h und hatte eine Reichweite von 14 bis 16 km. Schon früher tüftelte man an Fahrzeugen mit E-Antrieb, wie etwa der Schotte Robert Anderson. Er stellte 1835 (manche Quellen sagen auch 1839) einen Prototypen vor, der eine mit Rohöl betriebene Einwegbatterie nutzte.

1888 entwickelte der Coburger Maschinenfabrikant Andreas Flocken den „Flocken Motorwagen" – das erste Elektroauto Deutschlands. Das Modell war mit einem neuen Elektro-

Bild 1: Lohner-Porsche „Semper Vivus“ (BEV) im Jahr 1900

Bild 2: Ferdinand Porsche im Jahr 1900

motor mit einer Leistung von 0,7 kW ausgestattet, der die Kraft über Lederriemen auf die Hinterachse übertrug.

Im Jahr 1898 konstruierte Ferdinand Porsche als Mitarbeiter bei den Vereinigten Elektrizitätswerken von Bela Egger & Co in Wien für die Firma Lohner den „Egger-Lohner C.2 Phaeton“, eine Art Kutsche mit Elektroantrieb, der auch als „P1“ bekannt wurde. In alle wichtigen Bauteile seiner Konstruktion hatte er das Kürzel „P1“ geschlagen, das dem Elektrowagen seinen inoffiziellen Namen gab. Es erreichte eine Spitzengeschwindigkeit von 25 km/h und eine Reichweite von 80 km – beeindruckende Werte für diese Zeit. Ab 1899 arbeitete Ferdinand Porsche als technischer Leiter bei der Wiener k.u.k. Kutschen- und Hofwagenfabrik Ludwig Lohner & Co. Hier begann seine Karriere in der Automobilindustrie. Dort entwickelte er den elektrischen Radnabenmotor. Im Jahre 1900 erfolgte die Präsentation des ersten Lohner-Porsches „Semper Vivus“ (immer lebendig) (**Bild 1**) auf der Weltausstellung in Paris und gewann eine Goldmedaille. Mit 2 x 2,5 PS erreichte er 37 km/h Höchstgeschwindigkeit und eine Reichweite von ca. 50 km. Lohners Grund, gemeinsam mit Ferdinand Porsche (**Bild 2**) ein Fahrzeug mit einem Elektromotor zu entwickeln, klingt heute so aktuell wie damals: „Die enorme Lärmbelästigung, dazu wird die Luft von den in großer Anzahl auftretenden Benzinmotoren durch die beginnende Massenmotorisierung erbarmungslos verdorben“.

Im Jahr 1902 entwarf Porsche das erste funktionsfähige Hybridautomobil der Welt, den Lohner-Porsche „Mixte“. Das Prinzip dahinter scheint so simpel wie genial: Der Daimler-Verbrennungsmotor erzeugte in Verbindung mit einem Generator Strom und versorgte so über Batterien die Elektromotoren.

Diese Fahrzeuge standen jedoch vor großen Herausforderungen: Hohe Anschaffungskosten, eine unzureichende Ladeinfrastruktur und die damals geringe Batteriekapazität – ein kompletter Ladevorgang dauerte mehrere Tage – machten die Elektromobilität im Alltag und als Massenprodukt schwer umsetzbar.

Der Rekordwagen „La Jamais Contente – die nie Zufriedene“ von dem belgischen Automobilrennfahrer und Konstrukteur von Elektroautos Camille Jenatzky (**Bild 3**) setzte 1899 dennoch ein Zeichen, indem er als erstes Fahrzeug eine Geschwindigkeit von über 100 km/h erreichte – ein Meilenstein, der die Leistungsfähigkeit elektrischer Fahrzeuge demonstrierte. Angetrieben wurde das Fahrzeug mit zwei 25 kW-Elektromotoren, die von Batterien mit einer Spannung von 200 Volt und 125 Ampere gespeist wurden. Dadurch leistete „La Jamais Contente“ etwa 68 PS und hatte ein Eigengewicht von 1.450 kg. Ermittelt wurde der Rekord auf einer 1 km langen Strecke mit fliegendem Start. Jenatzky benötigte dafür 34 Sekunden.

Bild 3: Camille Jenatzky mit dem Rekordwagen „La Jamais Contente“ im Jahr 1899

Die Entwicklung des Verbrennungsmotors ab 1886 veränderte die Automobilwelt nachhaltig. Carl Benz stellte mit seinem Benz-Patent-Motorwagen das erste Benzinauto vor. Der Erfolg des Verbrennungsmotors war auf seine Leistung und die aufkommende Verfügbarkeit von Benzin zurückzuführen. Zudem waren die Preise für Benzin im Vergleich zu Strom damals deutlich geringer.

In den Goldenen Zwanzigern wurden die Elektrofahrzeuge jedoch an den Rand der Bedeutungslosigkeit verdrängt. Dass Verbrenner am Ende doch beliebter wurden als E-Autos lag ausgerechnet an einer elektrischen Erfindung. Im Jahr 1911 wurde der elektrische Anlasser für Benzinautos vorgestellt. Damit entfiel das lästige und körperlich anstrengende Kurbeln, ohne welches man einen Verbrenner bis dato nicht starten konnte. Doch das war nicht der alleinige Grund, warum immer mehr Menschen sich gegen ein elektrisches Auto entschieden. Benziner wurden immer günstiger und hatten zu der Zeit einfach die deutlich höhere Reichweite. Dazu wurde auch Öl immer billiger und überall schossen Tankstellen aus dem Boden.

Auch wenn die Verkaufszahlen von E-Autos ab 1912 rückläufig waren und Verbrenner seit den 1920er-Jahren immer mehr die Straßen dominierten – so richtig weg waren elektrische Fahrzeuge nie. Allerdings konzentrierte man sich hauptsächlich auf den Einsatz kleiner Nutzfahrzeuge: Ob als Flurfahrzeuge in Lagerhäusern und Hallen oder als Kleinsttransporter in Markthallen – der eine oder andere Einsatzort fand sich für die Elektroautos immer.

1.1.3 Wiederbelebung der Elektromobilität

Nach Jahrzehnten im Schatten der Verbrennertechnologien versuchte man in den 1970er-Jahren die Elektrotechnologie für den privaten Autofahrer wieder neu zu beleben. Die Projekte blieben aber weitgehend erfolglos und wurden eingestellt. Erst in den 1990er-Jahren wurden E-Autos dann auch langsam wieder für die Straße interessant. Inzwischen hatte sich die Technik verbessert. Die Einführung der Lithium-Ionen-Batterien war ein technischer Durchbruch, der Elektrofahrzeugen eine realistische Chance auf Marktfähigkeit ermöglichte. Zudem herrschten nun ein größeres Umweltbewusstsein und strengere, politisch diktierte Umweltauflagen, die die Automobilindustrie zwangen, ihre CO_2-Emissionen zu reduzieren. Große Erfolge konnten die ersten Modelle mit Elektro-Antrieb allerdings nicht feiern. Zwischen 1992 und 1996 versuchte VW mit dem Golf Citystromer ein Elektroauto zu etablieren. Doch nach der Produktion von 120 Exemplaren, die nur in den Flotten von Energieversorgern eingesetzt wurden, scheiterte das Projekt aufgrund geringer Nachfrage. Am besten angenommen wurde noch der 1997 erstmals vorgestellte Toyota Prius mit seinem Hybrid-Antrieb. Trotz dieser Rückschläge intensivierten Autohersteller weltweit

Bild 4: Tesla Roadster im Jahr 2008

ihre Bemühungen, Elektroautos in größeren Stückzahlen zu produzieren. Ein Auslöser für diese Richtungsänderung dürfte u. a. die durch den Golfkrieg ausgelöste Ölkrise in den 1990er-Jahren gewesen sein.

Interessant wurden Elektrofahrzeuge erst wieder mit dem Tesla Roadster (**Bild 4**), der 2006 auf den Markt kam. Er war mit einer Technik ausgestattet, die es ermöglichte, auch längere Strecken zu fahren.

Dem US-Hersteller Tesla kommt wohl der Verdienst zu, fast im Alleingang dem E-Auto wieder zu einem coolen Image verholfen zu haben. Seit der Einführung des Tesla Models S im Jahr 2012 können Elektrofahrzeuge problemlos in fast allen technischen Belangen mit Benzinern und Dieselfahrzeugen konkurrieren. Das Tesla Model 3 war das erste E-Auto, das so richtig den Massenmarkt erobern konnte und mit dem Model Y lieferte Tesla einen regelrechten Verkaufshit ab. Die Modelle 3 und Y beherrschten über Jahre unangefochten den Weltmarkt der Elektrofahrzeuge und gehören aktuell immer noch zu den am meisten verkauften E-Fahrzeugen weltweit.

Parallel zu Tesla trieb auch BMW die Elektromobilität voran. Mit Gründung der Submarke BMW i präsentierte der deutsche Hersteller im Jahr 2011 nicht nur den ersten Plug-in-Hybrid Sportwagen BMW i8, sondern auch das Konzept des BMW i3 als erstes rein elektrisches Stadtauto. Der BMW i3 ging 2013 in Serie und wurde in Leipzig gefertigt. Der i3 zeichnete sich durch seine innovative Bauweise mit einem leichten Carbonfaser-Chassis und einem modernen, nachhaltigen Design aus. Bis 2022 wurden rund 250.000 Fahrzeuge verkauft, was den BMW i3 zu einem wichtigen Meilenstein in der Geschichte der Elektroautos machte.

Der zweite Boom der Elektroautos, angeführt von Pionieren wie Tesla und BMW, hat die Automobilindustrie nachhaltig verändert. Tesla etablierte Elektroautos als leistungsfähige und attraktive Alternative zu konventionellen Fahrzeugen, während BMW mit dem i3 und dem i8 zeigte, dass Elektromobilität auch in unterschiedlichen Segmenten erfolgreich umgesetzt werden kann.

Heute verfolgt jeder Automobilhersteller seine eigene Elektromobilitätsstrategie. Aufgrund gesetzlicher Vorgaben haben so gut wie alle traditionellen Hersteller Elektroautos in ihrem Verkaufsportfolio. Dazu kommen Start-ups und kleine Manufakturen, die sich ausschließlich der Elektromobilität verschrieben haben. Auch immer mehr chinesische Hersteller drängen in den europäischen Markt. Elektrofahrzeuge sind heute leistungsfähig, alltagstauglich und aus dem modernen Stadtbild nicht mehr wegzudenken. Die Zukunft der Mobilität wird immer mehr elektrisch und durch kontinuierliche technologische Fortschritte und eine wachsende Infrastruktur weiter vorangetrieben.

1.1.4 Zukunft der Elektromobilität

Die Zukunft der Elektromobilität wird stark von technologischen Innovationen geprägt sein. Neue Batterietechnologien, wie Feststoffbatterien, versprechen höhere Energiedichten und kürzere Ladezeiten. Auch die Integration von Künstlicher Intelligenz (KI) in Fahrzeugen wird voranschreiten, was zu autonomen Fahrfunktionen und verbesserten Fahrassistenzsystemen führen könnte.

Die politischen und wirtschaftlichen Rahmenbedingungen werden maßgeblich die Entwicklung der Elektromobilität beeinflussen. Regierungen weltweit setzen auf Förderprogramme und gesetzliche Vorgaben, um den Übergang zu nachhaltiger Mobilität zu beschleunigen. Auch die deutsche Regierung hat mit Förderprogrammen versucht, die E-Mobilität in Deutschland für die Transformation hin zu CO_2-Neutralität zu etablieren. Von 2016 bis Dezember 2023 wurden ca. 10 Mrd. Euro dafür investiert. Aber nach dem plötzlichen Ende der Subventionen brach der Verkauf von E-Autos deutlich ein. Im Jahr 2024 wurde gegenüber 2023 ein Rückgang von 27,4 % bei den BEV-Modellen (BEV = battery electric vehicle / Batterie-Elektrofahrzeug) verzeichnet. Der Anteil der Neuzulassungen rein batteriebetriebener PKW belief sich 2024 auf 13,5 %. Das politische Ziel, dass in Deutschland im Jahr 2030 ca. 15 Mio. E-Autos auf den Straßen fahren, erscheint momentan utopisch. Aktuell sind ca. 1,65 Mio. BEV-Modelle zugelassen; das entspricht einem Anteil an den Gesamtzulassungen von ca. 3 %. Rechnerisch müssten von nun an in jedem Jahr rund 1,7 Mio. E-Fahrzeuge neu zugelassen werden, und dies bei insgesamt zuletzt rund 3 Mio. Gesamtneuzulassungen.

Neben den fehlenden Anreizen von Subventionen ist der Preis für Elektrofahrzeuge eine weitere Verkaufsbremse. Es gibt augenblicklich in Deutschland kein Fahrzeug unter 30.000,- Euro zu erwerben, außer einigen asiatischen Modellen. Neuere kleinere deutsche Modelle mit einem Preis zwischen 20.000,- und 25.000,- Euro sollen erst ab 2027 zur Verfügung stehen. Im Schnitt sind Elektroautos in der Anschaffung sehr viel teurer als ein Auto derselben Marke bzw. derselben Modellreihe mit Benzin- oder Dieselmotor. Weitere große Herausforderungen bleiben der Ausbau der Ladeinfrastruktur und der Strompreis in Deutschland. Es wird notwendig sein, ein flächendeckendes Netz von Schnellladestationen zu schaffen, um die Akzeptanz von Elektroautos weiter zu erhöhen. Schaut man auf die Gesamtkosten, schneiden E-Autos grundsätzlich weder besser noch schlechter ab als Autos mit Verbrennungsmotor. Hier kommt es auf das jeweilige Modell, aber auch auf die

Audi Q5 quattro	Merkmale	VW ID.4
	**Design**	
Dieselmotor 2.0 l Hubraum 204 PS	**Motorisierung**	Vollelektrisch (BEV) Batteriekapazität 77 kWh Elektromotor 150 kW/204 PS
ca. 1.080 km	**Maximale Reichweite** bei vergleichbarer moderater Fahrweise Stadtverkehr und Autobahn	**ca. 400 km**
ca. 5 Minuten	**Tank- bzw. Ladezyklus** ⇦ fast leeren Dieseltank volltanken – 68 Liter Batterie aufladen von ca. 15 % auf 80 % ⇨	**ca. 30 Minuten** mit SLP > 50 kW **ca. 420 Minuten** mit Wallbox 11 kW
ca. 6,4 Liter	**Verbrauch pro 100 km** nach knapp 10.000 km bei vergleichbarer Fahrweise und Fahrtzyklen	**ca. 17,8 kWh**
Diesel 1,35 €/l = 8,64 € Diesel 1,50 €/l = 9,60 € **Diesel 1,65 €/l = 10,56 €** Diesel 1,80 €/l = 11,52 €	**Verbrauchskosten pro 100 km**	**Strom 0,35 €/kWh = 6,23 €** Strom 0,50 €/kWh = 8,90 € Strom 0,60 €/kWh = 10,68 € **Strom 0,79 €/kWh = 14,06 €**

Bild 5: Vergleich Reichweite, Tank- bzw. Ladezeiten und Verbrauchskosten von Audi Q5 und VW ID.4

Entwicklung der Kraftstoffpreise und der Ladekosten an, die sich individuell sehr unterscheiden können (**Bild 5**). Wer ausschließlich öffentlich lädt, hat deutlich höhere Kosten als beim Laden an der eigenen Wallbox. Kostenseitig besonders attraktiv wird das Elektroauto, wenn der genutzte Strom von der eigenen PV-Anlage stammt – eine Win-Win-Situation für Umwelt und Verbraucher. Aber nur ein Bruchteil der fahrenden Bevölkerung hat die Möglichkeit einer eigenen PV-Anlage.

Nach ein paar Jahren mit einem positiven Entwicklungstrend ist aktuell eine spürbare Zurückhaltung zu verzeichnen, sogar in solchem Ausmaß, dass sehr viele Automobilhersteller, deren Zulieferer und Dienstleister aufgrund milliardenschwerer Investitionen in die Transformation bei jetzt fehlenden bzw. ungenügenden Umsätzen in wirtschaftliche Schieflage geraten sind. Restrukturierungsmaßnahmen in der gesamten Automobilbranche sind die Folge.

Man muss aber festhalten: Der PKW-Gesamtmarkt schwächelt wegen stark gestiegener Lebenshaltungskosten und aktueller politischer Unsicherheit. Besonders die Einfuhrzoll-Kapriolen des amerikanischen Präsidenten ab April 2025 werden noch lange spürbar bleiben. In Deutschland hat sich das Verhältnis zwischen privat und gewerblich zugelassenen E-Autos mit dem Wegfall der Kaufprämie gedreht. Gab es 2023 mehr private Zulassungen als gewerbliche, lagen 2024 die gewerblichen Käufe trotz deutlich gesunkenem Zulassungs-Niveau vorne (**Bild 6**). Eine Absatzprognose für das Jahr 2025 ist kaum möglich, weil eine neue Bundesregierung sich erst auf Rahmenbedingungen einigen muss. Im 1. Quartal 2025 lag der Anteil der BEVs bei den privaten Zulassungen lediglich bei ca. 3 %.

Der Flottenmarkt gilt als Treiber für die Transformation zur E-Mobilität, doch die Wende stockt. Viele Unternehmen drehen die Entwicklung zurück und nehmen BEVs aus ihren Fuhrparks. Hierzu wurden in Deutschland 300 Unternehmen befragt. Insgesamt waren es mehr als 8.000 Firmen in 28 Ländern. Die Skepsis gegenüber E-Autos zeigt sich in vielen befragten Ländern, ganz erheblich in Deutschland. Firmen in Deutschland zögern, die E-Wende weiter voranzutreiben. Sie gehen davon aus, dass in drei Jahren der BEV-Anteil lediglich bei 27 % liegen wird. Benziner und Diesel stellen weiterhin das Gros. Laut den deutschen Studienergebnissen haben 28 % der befragten Unternehmen reine E-Autos in ihrer PKW-Flotte, 5 % weniger als ein Jahr zuvor. Noch schlechter ist es um die Wende bei leichten Nutzfahrzeugen bestellt. Hier drehen die Unternehmen noch deutlicher das Rad zurück. Der alternative Antrieb in der Transporter-Flotte liegt durchschnittlich bei nur noch

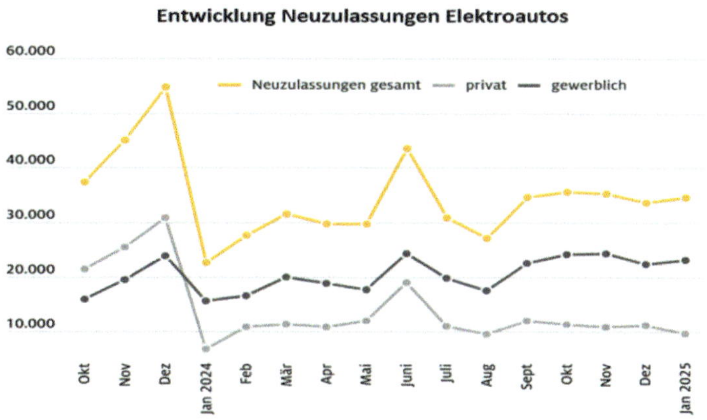

Bild 6: Entwicklung Neuzulassungen Elektroautos privat vs. gewerblich im Jahr 2024 (Quelle: Kraftfahrzeug-Bundesamt)

14 %. Bei den Autovermietern kommt noch ein entscheidender Punkt hinzu: Im vergangenen Jahr sind die Preise für Gebrauchtwagen massiv gefallen – für gebrauchte Elektroautos noch stärker als für Verbrenner. Dies ist gerade für Vermieter, die Fahrzeuge i. d. R. nur sechs bis neun Monate in ihrer Flotte haben, ein wesentlicher Faktor. Sie werden jetzt die Fahrzeuge nicht mehr los oder nur wenn sie bereit sind, hohe Verluste in Kauf zu nehmen. Der Markt wird sich erst dann wieder normalisieren, wenn die private Nachfrage wieder anspringt. Derzeit sind dafür aber keine Impulse erkennbar. Diesen Trend bekommt Tesla aktuell besonders zu spüren. Das Flottengeschäft brach ähnlich ein, wie der Absatz an Privatkunden. Nun bringt das politische Engagement von CEO Elon Musk weitere Probleme. Deshalb strich bereits 2024 Drogist Rossmann die Modelle aus der Flotte. Weitere Unternehmen erwägen, dem Beispiel zu folgen.

Auch die zunehmende Bedeutung chinesischer Hersteller wie BYD, Geely, NIO usw. auf dem chinesischen Markt tangiert die deutsche Automobilindustrie in großem Maße, besonders im Hinblick auf die Verkaufszahlen der deutschen Hersteller in China. Hier wurden die Erwartungen, besonders bei den E-Fahrzeugen, deutlich verfehlt. Ob es sich hierbei um einen aktuell wahrzunehmenden konjunkturellen Einfluss handelt (auch Chinas Wirtschaft schwächelt nicht unwesentlich) oder ob es ein länger anhaltender Zustand bleibt, ist abzuwarten. Dabei ist die Abhängigkeit der deutschen Automobilhersteller vom chinesischen Markt signifikant: Etwa 39 % der weltweit verkauften Volkswagen inklusive Audi (Verbrenner & E-Autos) und rund ein Drittel aller Fahrzeuge von Mercedes-Benz und BMW gehen nach China oder werden bereits in China hergestellt.

Über Jahrzehnte war der Premiummarkt in China fest in der Hand von „BBA“: Benz, BMW, Audi. Doch mit Elektromobilität, Fahrzeugdigitalisierung und einem neuen Selbstbewusstsein der Chinesen wurde der Markt neu verteilt. Chinesische Kunden kaufen sich heute lieber ein einheimisches Fabrikat. Deren Anteil liegt schon bei 65 %. Für die deutschen Konzerne ist der Absatzverlust auf dem weltgrößten Automarkt schmerzhaft. Aber sie glauben an ein Comeback mit besseren, digitaleren Produkten. Es gibt derzeit in China mit allein ca. 130 chinesischen Marken zu viele Anbieter. Diese Automarkt-Übersättigung führt zu einer ausgedehnten Konsolidierung in der chinesischen Branche. Es wird erwartet das nur wenige Giganten (ca. 30) übrigbleiben. Nach den Konsolidierungsjahren 2025 und 2026, wenn der Rauch verzogen ist, wird sich zeigen, wie „BBA“ wirklich dasteht und wie hoch der Wert von „Made in Germany“ auf dem Leitmarkt noch ist. Auch Tesla gehört in China inzwischen zum alten Eisen.

Viele chinesische Modelle zeigen die hohe Innovationskraft und das Engagement von China, die globale Führungsposition in der Elektromobilität zu übernehmen. Deutlich geringere Preise, bessere Batterietechnologie, ein ansprechendes Design und eine zurzeit überlegene Digitalisierung in den Fahrzeugen sprechen für sich. Hier haben sowohl die deutschen als auch viele europäische Automobilhersteller zu lange auf hochpreisige Modelle und nicht mehr marktgerechte Prioritäten gesetzt. Gleichheit mit großen Volumina ist die neue Stellschraube. Das erlaubt Quantensprünge bei der Senkung der Produktionskosten. Die Modelle Y und 3 machen 97 % aller Tesla-Verkäufe aus, stehen auf einer gemeinsamen Plattform und werden in nur wenigen Außenfarben und Ausstattungen angeboten. Aus Stahl- und Lack-Vielfalt Individualität zu erzeugen, scheint ausgedient zu haben. Vielfalt entsteht heute durch innovative Software. Elon Musk mit Tesla und Wang Chuanfu mit BYD sind die Ersten, die diese Erkenntnis in der Autoindustrie umgesetzt und damit das „New Manufacturing“ definiert haben. Zudem unternimmt das Unternehmen BYD, mittlerweile größter E-Auto-Hersteller weltweit, noch weitere Wege, um sich in Europa kurzfristig und nachhaltig zu etablieren. Man geht Kooperationen mit Auto-Vermietungs-Unternehmen ein und errichtet aktuell in Ungarn ein Produktionswerk für die Herstellung von bis zu 300.000 Fahrzeugen. So umgeht man auch mögliche europäische Einfuhrzölle und redu-

Bild 7: Umfrage über digitale Innovationskraft, China vs. Deutschland, im April 2025

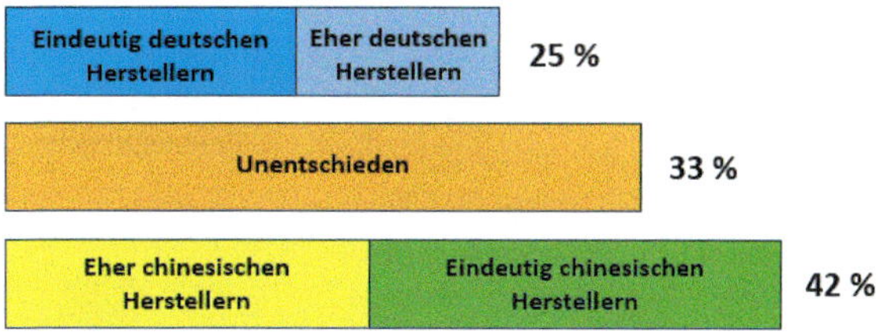

ziert Transportkosten und Lieferzeiten. Den Mut darf man zumindest bewundern. Im Heimatland stark, aber nicht profitabel, im Ausland bisher ohne echten Durchbuch.

Es ist auch schon sehr erstaunlich – eigentlich erschreckend – dass bei einer Umfrage unter deutschen Konsumenten der chinesischen Innovationskraft mehr Vertrauen geschenkt wird als der deutschen. Auf die Frage „Trauen Sie eher chinesischen oder eher deutschen Automobilherstellern mehr digitale Innovation zu (z. B. autonomes Fahren, Konnektivität, Infotainment)?" wurde wie in **Bild 7** dargestellt geantwortet.

Im globalen Wettkampf um das Elektroauto stehen sich drei radikal unterschiedliche Strategien gegenüber: China setzt auf eine entschlossene und ambitionierte industrielle Strategie. Die USA setzen auf Wirtschaftsförderung, zudem neuerdings auf hohe Einfuhrzölle und Europa reguliert, was das Zeug hält. In Europa ohne jegliche Flexibilität einfach nur Bußgelder zu verhängen und Deadlines zu setzen, ist sehr gefährlich. Die Geschwindigkeit des Hochlaufs der automobilen Elektrifizierung ist halb so schnell wie das, was man bräuchte, um die Ziele zu erreichen, bei denen keine Bußgelder drohen. Wenn die europäische Politik schnelles Wachstum für Elektroautos will, muss sie etwas dafür tun. Wahrscheinlich braucht Europa ein Mischmodell mit kurzfristigem Schutz der Industrie und Kooperation. Europa kann auch nicht auf eine Kooperation mit China verzichten. China hat nun einmal eine Fahrzeuggeneration Vorsprung in der Elektromobilität und die Kontrolle über die gesamte Wertschöpfungskette. Ein Beispiel sind die Seltenen Erden, die für den Bau moderner Elektromotoren unabdingbar sind. Dauermagnete können ohne diese Materialien nicht produziert werden. China beherrscht deren globale Fertigung zu 90 %, ebenso die Weiterverarbeitung aller 17 Seltenen Erden. Jetzt gibt es zudem neue Ausfuhrbeschränkungen Chinas bei den Seltenen Erden, die zu einer ernsten Versorgungskrise beim Bau von Elektromotoren außerhalb Chinas führen könnten. China setzt damit gezielte Nadelstiche mit einer großen Hebelwirkung. Eine schnelle Abhilfe ist nicht in Sicht.

Trotz des Wettbewerbsdrucks durch die Elektromobilität ist eine vollständige Elektrifizierung der weltweiten Fahrzeugflotten noch lange nicht absehbar. Die Infrastruktur für Elektrofahrzeuge ist in vielen Ländern unzureichend und in ländlichen Regionen bleiben Verbrenner eine notwendige Alternative. Zudem sind in vielen Ländern die Stromquellen noch auf Jahrzehnte fossil, sodass Elektroautos zwar die städtische Luftqualität verbessern, aber wenig zur Reduktion der globalen CO_2-Emissionen beitragen.

Politische Maßnahmen sollen einerseits den Übergang zur Elektromobilität beschleunigen, andererseits sehen sich Autofahrer mit hohen Kosten konfrontiert. Parkmöglichkeiten werden gerade in den Großstädten bewusst zurückgebaut, zugleich steigen die Preise für Parkplätze in den deutschen Innenstädten dramatisch. Neue Regelungen sollen den Verkehr in den Städten reduzieren, obwohl auch der Politik bekannt ist, dass sich das Verkehrsaufkommen in den nächsten Jahrzehnten – sowohl privat als auch gewerblich – noch deutlich steigern wird (**Bild 8**).

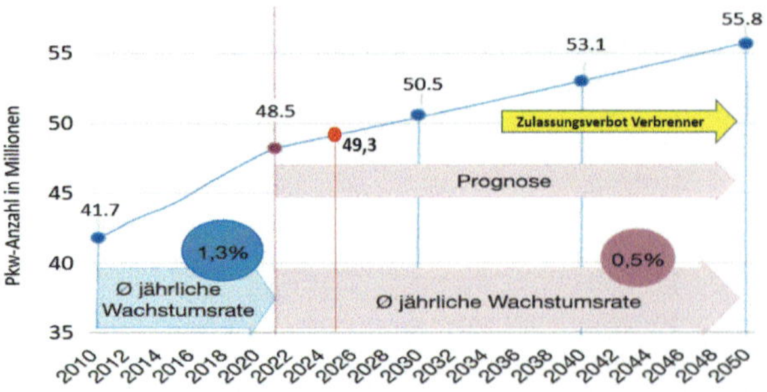

Bild 8: Entwicklung der PKW-Dichte in Deutschland bis ins Jahr 2050 (Quelle: Kraftfahrt-Bundesamt)

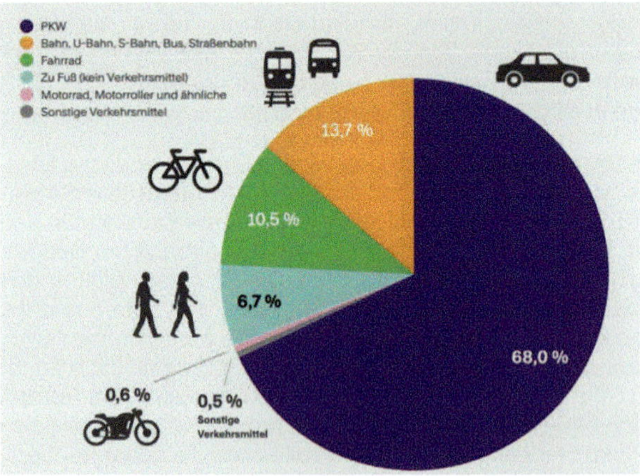

Bild 9: Angewandte Mobilität der Erwerbstätigen in Deutschland im Jahr 2023 (Quelle: Kraftfahrzeug-Bundesamt)

Gleichzeitig wächst auch der Widerstand gegen das Zulassungsverbot von Neuwagen mit einem Verbrennungsmotor ab 2035 in Europa. In Deutschland und anderen Ländern wird die politische Debatte von Stimmen begleitet, die ein stufenweises Vorgehen und alternative Kraftstoffe wie den synthetischen Diesel HVO 100 oder weitere sogenannte E-Fuels fordern.

Auch das Angebot an Carsharing-Fahrzeugen in Deutschland ist immer noch gering und gerade für Regionen außerhalb der Großstädte nicht wirtschaftlich und sehr unflexibel. Die arbeitende deutsche Bevölkerung ist nach wie vor stark auf den privaten PKW angewiesen. Bei einer mittlerweile durchschnittlichen Pendelstrecke von 10 bis 50 km zum Arbeitsplatz bleibt das eigene Fahrzeug oft die praktischste und kostengünstigste Lösung (**Bild 9**).

Die Meinungen zum Elektroauto gehen weit auseinander. Während die einen behaupten, Elektroautos würden sich gegenüber dem PKW mit Verbrennungsmotor nie durchsetzen, erklären viele E-Auto-Erfahrene, dass sie nie mehr ein Auto mit Verbrennungsmotor fahren würden. Das sind die Extrempositionen. Was müsste passieren, damit die Elektromobilität in Deutschland sich auch bei skeptischen Menschen durchsetzt? Ist ein Kipp-Punkt denkbar, wo – ähnlich wie beim Smartphone – Kundinnen und Kunden in Scharen zu den elektrischen Angeboten der Automobilhersteller greifen? Stirbt der Verbrenner als „old fashioned" und technologisch rückständig irgendwann aus, wie es beim „unsmarten" Mobiltelefon oder bei analogen Fotoapparaten passiert ist? Es gibt durchaus Trends und gute Gründe, die dem Thema Elektroauto in den nächsten Jahren Schwung verleihen werden (könnten):

1. Fortschrittliche Batterien führen zu höheren Reichweiten und kürzeren Ladezeiten.
2. Öffentliche Ladeinfrastruktur entwickelt sich und Ladestrom wird günstiger.
3. Anschaffungspreise sinken und Wiederverkaufspreise werden attraktiver.

1.1.5 Fazit

Es gibt ihn, den mehr als 100-jährigen Wettbewerb der unterschiedlichen Antriebssysteme in der Automobilindustrie – aktuell ohne erkennbaren Sieger und Verlierer. Das war zu Beginn der Massenmotorisierung Anfang des letzten Jahrhunderts anders. Hier hatten sich die Verbrennungsmotoren, aufgrund der zur Verfügung stehenden technischen Möglichkeiten, kommerziell durchgesetzt. Dabei ist die Geschichte der Elektroautos eine faszinierende Reise durch die Zeit, die zeigt, wie technologische Innovationen und gesellschaftliche Veränderungen Hand in Hand gehen. Von den bescheidenen Anfängen im 19. Jahrhundert über die Höhen und Tiefen des 20. Jahrhunderts bis hin zur modernen Ära, in der Elektroautos eine zentrale Rolle in der Mobilitätswende spielen, hat sich viel getan. Heute steht man an einem Wendepunkt, an dem Elektromobilität nicht nur eine umweltfreundliche Alternative, sondern auch ein Symbol für Fortschritt und Nachhaltigkeit darstellt. Die Entwicklung der Elektroautos ist ein beeindruckendes Beispiel dafür, wie visionäre Ideen und kontinuierliche Weiterentwicklung die Welt verändern können. Dennoch sind noch immer wirtschaftliche und technische Hürden zu überwinden, um eine flächendeckende Elektrifizierung der Mobilität zu ermöglichen.

Während Elektroautos lokal emissionsfrei fahren und damit die unmittelbare Umgebung nicht mit Luftschadstoffen belasten und dazu auch noch geräuscharm unterwegs sind, bleiben Fahrzeuge mit einem Verbrennungsmotor für viele Regionen und Situationen unverzichtbar. In Anbetracht globaler Anforderungen an die Mobilität wird eine gemischte Flotte aus Elektro- und Verbrennerfahrzeugen wohl noch über Jahrzehnte Bestand haben. Der Weg zur Klimaneutralität in der Mobilität bleibt herausfordernd.

Dabei muss Deutschland als traditionelles „Autoland" den Fokus darauf richten, dass es auch in Zukunft noch zu den weltweit führenden Automobilherstellern mit Produktionsstandorten im eigenen Land zählt. Hier sind die Politik mit Bereitstellung der notwendigen Rahmenbedingungen und die Automobilhersteller mit maximaler Flexibilität, um auf die unterschiedlichsten Forderungen und Wünsche des globalen Marktes zu reagieren, mehr als bisher gefordert. Es muss ein schnelles Umdenken und zeitnahes Handeln stattfinden.

Die EU sollte vermeiden, die deutsche und die EU-Automobilindustrie durch das Verbrennerverbot ab 2035 sowie unflexible und sich schnell verschärfende Flottengrenzwerte faktisch zur einseitigen Fokussierung auf die Entwicklung von Elektrofahrzeugen zu zwingen – bei gleichzeitigem Weiterverkauf von ineffizienten Verbrennermotoren, die nicht mehr weiterentwickelt werden. In Nachhinein würden europäische Hersteller dann erneut mit dem Vorwurf konfrontiert werden, sie hätten eine wichtige globale Entwicklung verschlafen – diesmal jedoch den Trend zu effizienten und kostengünstigeren Hybriden und Range-Extender-Fahrzeugen, die mit alternativen Kraftstoffen betankt werden können. Die Politik müsste dann ehrlicherweise eingestehen, dass sie durch eine zu einseitige Regulierung die falschen Weichen für die europäische Autoindustrie gestellt hat.

Anfang März 2025 hat die EU einen Aktionsplan zur Stärkung der Wettbewerbsfähigkeit der europäischen Automobilindustrie veröffentlicht. Die dort aufgeführten kurzfristigen Entlastungsmaßnahmen, um drohende Strafzahlungen wegen der verschärften Flottengrenzwerte ab 2025 zu vermeiden, reichen allerdings langfristig nicht zur Überwindung der Krise aus. Stattdessen ist ein größerer Wurf erforderlich.

Warum nicht wie der weltgrößte Wettbewerber China handeln? Die chinesische Automobilindustrie verfolgt einen technologieoffenen Ansatz und ermöglicht bis 2060 die Zulassung von Hybriden und effizienten Verbrennern, die mit alternativen Kraftstoffen betrieben werden. Auch den europäischen Fahrzeugherstellern sollte die Möglichkeit gegeben werden, effiziente Verbrennungsmotoren und Hybride zu entwickeln und über 2035 hinaus auf ihrem loyalen Heimatmarkt anzubieten.

Europa muss sich zudem vor den chinesischen Wettbewerbern auch nicht verstecken. Chinas Hersteller, so viel gelobt und so sehr gefürchtet, haben gerade genug mit sich selbst zu tun. Sie lernen gerade ihre eigenen Grenzen kennen. Der chinesische Staat hat die Hersteller lange Zeit unterstützt. Nun scheint Peking eine Auslese für geboten zu halten. Der perfekte Darwinismus im chinesischen Automarkt.

Quellen

[1] www.fahrtraum.at

[2] Kraftfahrtbundesamt

[3] Statista, [https://de.statista.com/]

[4] demostalgiker.at

[5] www.wirtschaftsdienst.eu

[6] Statistisches Bundesamt, 2022

[7] ADAC e.V., 01/2025

[8] Automobilwoche

[9] Umweltbundesamt

[10] Handelsblatt

[11] WirtschaftsWoche, Mai 2024

[12] EnBW Energie Baden-Württemberg AG

[13] Civey GmbH

1.2 Hitzebeständige Stähle: Einsatz in Wärmebehandlung und Industrieofenbau

Julian Irretier, Marco Jost, Dr. Olaf Irretier, Dr. Giuseppe Polignano, Michael Schlerkmann

Wärmebehandlung, Härterei und der hierfür erforderliche Industrieofenbau wären ohne den Einsatz hitzebeständiger Stähle in der Herstellung des Schleuder- und Formgusses nicht denkbar. Glüh- und Härtetemperaturen von über 1.000 °C und die immer höheren Abforderungen in eine automatisierte Industrieofentechnik verlangen Legierungszusammensetzungen und angepasste Gussverfahren, die in diesen thermischen Grenzbereichen die Anforderungen erfüllen. Dabei gilt es, das Expertenwissen der Werkstoffkunde/Legierungslehre mit dem Expertenwissen der Urformung, sprich der Gießereitechnik, in statischer und dynamischer Auslegung der Komponenten zu vereinen. Der vorliegende Beitrag beschäftigt sich in einer Übersicht mit den Herausforderungen der Auswahl hitzebeständiger Stähle als Gussbauteile und deren Einsatz in Wärmebehandlung und Industrieofenbau.

Es gibt viele verschiedene Arten von hitzebeständigen Stählen mit unterschiedlichen Legierungselementen und Eigenschaften, die je nach den spezifischen Anforderungen in der Wärmebehandlung und im Industrieofenbau zur Anwendung kommen. Je nach Anforderung an die Hochtemperatur-, Ermüdungs- oder Kriechbeständigkeit, an die Oxidations-, Korrosions- und Temperaturwechselbeständigkeit sowie an die Aufkohlungsbeständigkeit und Schweißbarkeit können eine Vielzahl unterschiedlicher Legierungen erschmolzen und vergossen werden. So werden je nach Einsatzbedingungen neben austenitischen, ferritischen und martensitischen Güten austenitische Super-Legierungen, Nickel- oder Kobaltbasislegierungen sowie Duplex-Stähle gegossen.

1.2.1 Konstruktive Auswahl und Dimensionierung von Gussbauteilen

Mit der Auswahl der für den Anwendungsfall geeigneten Gusslegierung ist die Art und Weise der Ausführung des Gussteils eng verbunden. Die Konstruktion und die Dimensionierung der herzustellenden Komponenten unter Berücksichtigung der maximal auftretenden Spannungen werden seit vielen Jahren sehr genau über FEM-Berechnungsprogramme durchgeführt. Geeignete Prozesssimulationen kommen für die optimale Auslegung von Gusskomponenten zum Einsatz.

Ausgehend von einer ersten Fertigungszeichnung, die i. d. R. vom Endkunden auf Basis der Ausführung der Wärmebehandlungsanlage erstellt wird, beginnen die eigentliche Dimensionierung des Bauteils und die Auswahl der geeigneten Legierung. Hierzu wird zunächst eine Beladung oder mechanische Belastung des Bauteils simuliert. Die Einsatzbedingungen und Belastungsangaben sind i. d. R. vom Kunden in einer Spezifikation zusammengefasst. Diese ist die Grundlage für die Auslegung.

Über geeignete 3D-Programme für die FEM-Berechnung wird die Beladungssituation simuliert. Im Falle des Grundrostes (**Bild 1**) können in diesem Beispiel verschiedene zylindrische Bauteile mit unterschiedlichen Gewichten und Durchmessern geladen werden.

Über die FEM-Berechnung werden im gesamten Bauteil die auftretenden Spannungen für jeweils einen Belastungsfall ermittelt und mögliche Spannungsspitzen erkannt (**Bild 2**). Alle in der Spezifikation definierten Belastungsfälle werden simuliert und der „worst case" als Referenz für die Auslegung herangezogen. Anschließend wird auf Basis der Temperatursituation die Legierung ausgewählt, die bei der maximalen Ofentemperatur noch ausreichende mechanische Festigkeitswerte besitzt, um auf Dauer der Belastung standzuhalten.

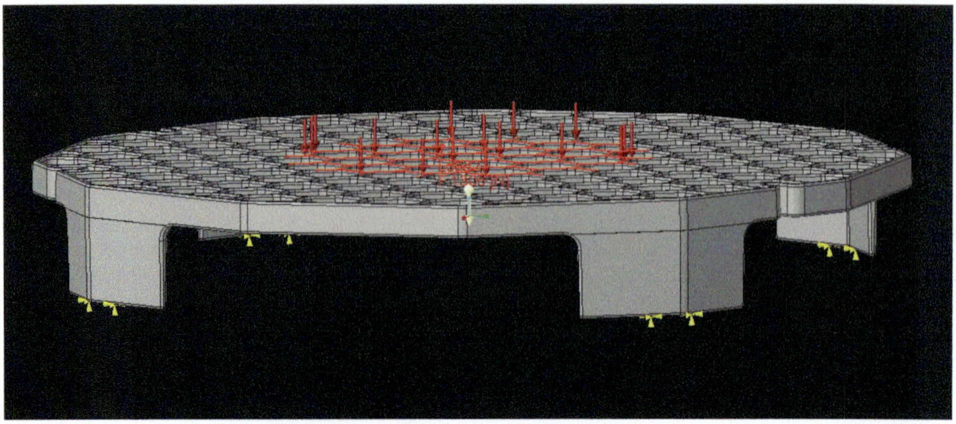

Bild 1: Belastungssimulation eines Grundrostes durch maximale Beladung bei zentraler Positionierung (Quelle: FAI-FTC)

Nach der ersten Festlegung einer Legierung beginnt der eigentliche Prozess der Dimensionierung. Durch Variation von z. B. Wandstärken und Wabengeometrie können die auftretenden Spannungen variiert werden, was zusätzlich eine andere Materialauswahl zur Folge haben kann. Aufgabe der Ingenieure ist es, die optimale Lösung zu finden, welche die einwandfreie und dauerhafte Funktion des Bauteils garantiert, die aber auch Kostenüberlegungen mit einbezieht.

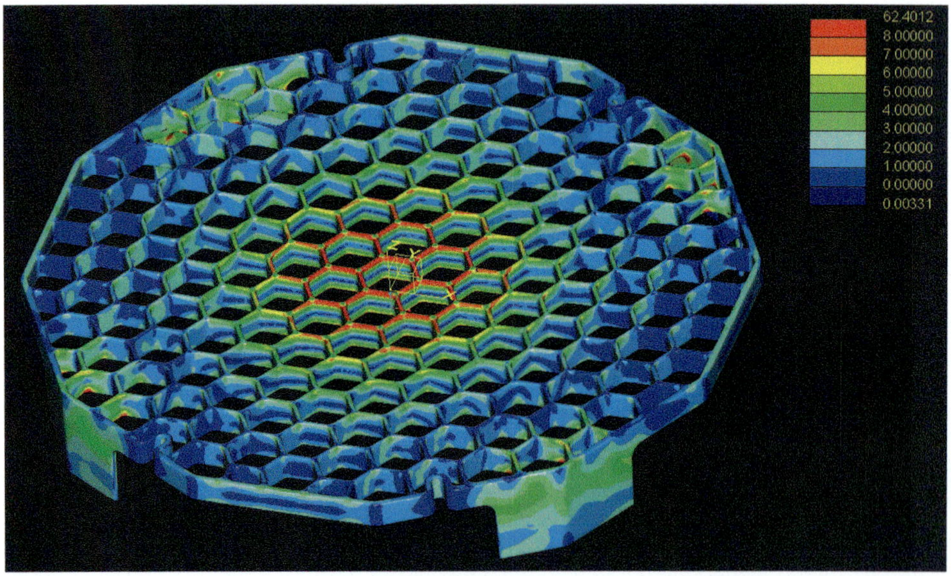

Bild 2: FEM-Berechnung der maximal auftretenden Spannungen in einem Grundrost (Quelle: FAI-FTC)

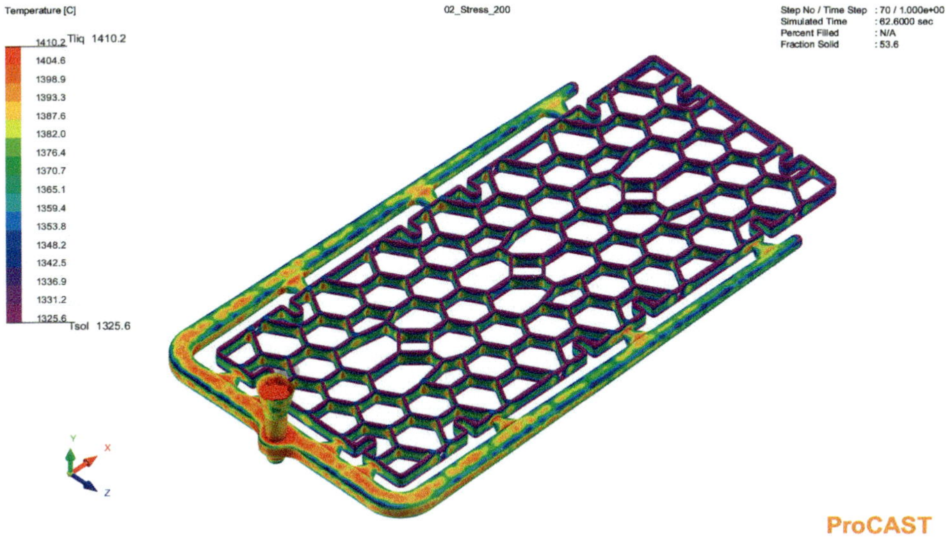

Bild 3: Simulation eines Abkühlprozesses eines Grundrostes nach dem Abguss (Quelle: FAI-FTC)

Nach abschließender Festlegung der Ausführung des Gussteils wird die entsprechende Fertigungszeichnung erstellt. Danach beginnt die Phase der Optimierung des Gießprozesses. Auch hierzu werden Simulationen herangezogen, die eine geeignete Dimensionierung des Gießsystems ermöglichen. Eine Prototypenfertigung zur finalen Kontrolle der Gussqualität ist allerdings unumgänglich. **Bild 3** zeigt z. B. eine Simulation des Abkühlprozesses eines Grundrostes nach dem Abguss.

1.2.2 Werkstofftechnische Auswahl von Gussbauteilen

Hitzebeständige Edelstähle in Gussbauteilen sind spezielle Legierungen, die eine lange Beständigkeit gegenüber hohen Temperaturen aufweisen, wie sie beispielsweise in der Wärmebehandlung und Härtereitechnik vorkommen. Sie zeichnen sich dabei durch ihre Fähigkeit aus, ihre mechanischen Eigenschaften auch bei hohen Temperaturen beizubehalten. Sie haben eine gute Oxidationsbeständigkeit und können auch unter oxidierenden Bedingungen eingesetzt werden, ohne dass sie schnell an Festigkeit verlieren.

1.2.3 Hitzebeständige hochlegierte Stähle

1.2.3.1 Definitionen und Legierungsklassifizierungen

Hitzebeständige Gusslegierungen werden in erster Linie für Anwendungen verwendet, bei denen die Betriebstemperaturen 650 °C überschreiten und bis zu 1.315 °C erreichen können. Die Festigkeit bei erhöhten Temperaturen steht bei der Auswahl der Legierung oft an erster Stelle. Die Beständigkeit der Legierung gegen Umwelteinflüsse, das Vorhandensein aggressiver Korrosionsmittel und die Einwirkung zyklischer Spannungen und Temperaturen sind jedoch einige der anderen Faktoren, die ebenfalls berücksichtigt werden müssen.

Tabelle 1: Bruchspannungen von Walz- und Gusswerkstoffen

Form	Legierung	10.000-Stunden-Bruchspannung [MPa]	
		870 °C	**980 °C**
Walzwerkstoff	AISI 309	10,3	3,4
Gusswerkstoff	1.4837	1.4837	7,0
Walzwerkstoff	AISI 310	10,3	3,4
Gusswerkstoff	1.4848	23,0	9,6
Walzwerkstoff	AISI 330	11,7	4,1
Gusswerkstoff	1.4806	27,0	12,5

Einige der hitzebeständigen Gusslegierungen sind in ihrer Zusammensetzung mit den nichtrostenden Knetstählen verwandt. Der Hauptunterschied zwischen diesen Werkstoffen ist ihr Kohlenstoffgehalt. Bis auf wenige Ausnahmen liegt der Kohlenstoffgehalt in den hitzebeständigen Gusslegierungen zwischen 0,3 und 0,6 %, während er bei den korrosionsbeständigen Knet- und Gussstählen normalerweise zwischen 0,01 und 0,25 % liegt. Dieser Unterschied im Kohlenstoffgehalt führt zu erheblichen Veränderungen der Eigenschaften, z. B. der höheren Bruchfestigkeit der Gusslegierungen, die in **Tabelle 1** mit denen der Knetlegierungen verglichen werden. Es ist daher wichtig, diese Unterscheidungen zu kennen, da jede Gruppe von Legierungen ihre eigene geeignete Anwendung und die entsprechenden Spezifikationen hat.

Für den Bedarf der Industrie wurde eine breite Palette von hochlegierten hitzebeständigen Gussstählen entwickelt. Die Zusammensetzungen einiger dieser Legierungen sind in **Tabelle 2** aufgeführt. Die Standardsorten liegen in einem Bereich von 0 bis 62 % Nickel mit 8 bis 35 % Chrom und dem Rest hauptsächlich aus Eisen plus mit bis zu 2,5 % Silizium und 2 % Mangan. Proprietär-Legierungen und andere, die inzwischen allgemein bekannt sind, sind höher legiert. Einzelne oder mehrere Zusätze der Elemente Aluminium, Kobalt, Molybdän, Niob, Titan, Wolfram und Zirkonium werden hinzugefügt, um bestimmte Eigenschaften wie Hochtemperaturfestigkeit, Aufkohlungsbeständigkeit und Beständigkeit gegen Temperaturwechsel zu verbessern.

1.2.3.2 Eisen-Chrom-Nickel-Legierungen

Die Legierungen dieser Gruppe enthalten 18 bis 32 % Chrom und 8 bis 22 % Nickel, wobei der Chromanteil immer höher ist als der Nickelanteil. Zu ihnen gehören die Sorten 1.4825, 1.4826, 1.4837, 1.4832 und 1.4848. Während diese Legierungen als austenitisch gelten, enthalten die niedrigeren Nickelzusammensetzungen etwas Ferrit. Die Umwandlung von Ferrit in die spröde Sigma-Phase ist bei dieser Gruppe ein Problem (selbst bei den höheren Nickelsorten), vor allem wenn das Gleichgewicht der Zusammensetzung zu Ferrit tendiert.

1.2.3.3 Eisen-Nickel-Chrom-Legierungen

Die Standardsorten dieser Gruppe, 1.4855, 1.4805, 1.4806, 1.4865, 1.4849, 1.4857, 1.4852, 1.4838, 1.4836, enthalten 15 bis 28 % Chrom und 23 bis 41 % Nickel. Der Nickelanteil übersteigt stets den Chromanteil. Diese Legierungen weisen ein stabiles austenitisches Gefüge, eine gute Hochtemperaturfestigkeit, eine erhöhte Beständigkeit gegen Temperaturwechsel

Tabelle 2: Beispiele und chemische Zusammensetzung einiger hitzebeständiger Stähle

Werkstoff	Chemische Zusammensetzung [%]										
	C	Si	Mn	Cr	Ni	Nb	W	Co	Ti	Al	Add.
1.4825	0,3	1,5	1	18	9						
1.4826	0,4	1,5	1	22	10						
1.4832	0,3	1,5	1	20	14						
1.4837	0,4	1,5	1	25	12						
1.4848	0,4	1,5	1	25	20						
1.4855	0,4	1,5	1	24	24	1					
1.4805	0,4	1,5	1	21	25						
1.4806	0,4	1,5	1	17	35						
1.4865	0,4	1,5	1	19	38						
1.4849	0,4	1,5	1	19	38	1,5					
1.4859	0,1	1,5	1	20	32	1					
1.4857	0,4	1,5	1	25	35						
1.4852	0,4	1,5	1	25	35	1,5					
1.4838	0,4	1,5	1	25	35	1,5			Add.		
1.4836	0,4	1,5	1	25	35		5				
1.4839	0,4	1,5	1	35	45				Add.		
1.4869	0,5	1,5	1	25	35		5	15			
2.4879	0,4	1,5	1	28	48		5				
2.4881	0,45	1,5	1	28	48		5	3			
2.4815	0,4	1,5	1	15	62						
M-PM 4	0,3	0,3	0,3	33	50		16			Add.	
H-PM 5	0,4	1,5	1	18	52	1					
L-PM 6	0,1	0,3	0,3	16	75					5	Zr, Y

und thermische Spannungen sowie eine hohe Beständigkeit gegen oxidierende und reduzierende Umgebungen auf.

1.2.3.4 Nickel-Eisen-Chrom-Legierungen

Eine Standardsorte, 2.4815, fällt in diese Gruppe, die 58 bis 68 % Nickel und 10 bis 19 % Chrom enthält. Gewöhnlich als hochlegierte Stähle bezeichnet, sind diese Werkstoffe korrekter als „Nickelbasislegierungen" zu bezeichnen. Sie weisen zwar eine mäßige Hochtemperaturbruchfestigkeit auf, ihre Kriechfestigkeit ist jedoch gering. Diese Sorten weisen die höchste Aufkohlungsbeständigkeit der Standardlegierungen und die höchste Temperaturwechselbeständigkeit auf. Diese Werkstoffe haben einen hohen elektrischen Widerstand, der für Heizelemente verwendet wird. Andere Anwendungen sind Herde, Muffeln, Abschreckvorrichtungen und Behälter für geschmolzenes Blei.

1.2.4 Auswahl der Legierung

Bei der Auswahl von Gussstücken aus hitzebeständigen Legierungen oder von geschweißten Baugruppen ist eine Reihe von Faktoren zu berücksichtigen:

- Erwartete normale Betriebstemperatur sowie die wahrscheinlich auftretenden Mindest- und Höchsttemperaturen
- In Extremfällen können die Liquidus- und Solidustemperaturen erforderlich sein
- Häufigkeit und Geschwindigkeit der Temperaturwechsel
- Stärke der thermischen Gradienten im Bauteil
- Wärmeausdehnungseigenschaften der Legierung
- Angewandte Lasten und Art der Belastung
- Auslegungslebensdauer
- Hochtemperatur-Zugeigenschaften und Langzeit-Kriech- und -Bruchfestigkeit
- Betriebsumgebung, z. B. oxidierend, reduzierend, sulfidierend, aufkohlend usw., oder Vorhandensein von korrodierenden Stoffen wie Natrium, Chloride usw. oder flüssigen Metallen
- Schweißbarkeit im gegossenen Zustand und nach der Beanspruchung
- Bearbeitbarkeit.

Aufbauend auf diesen Daten wählt der Konstrukteur die am besten geeignete Legierung aus, welche ausreichende Temperatur- und Korrosionsfestigkeit für den Einsatzzweck besitzt, eine kontrollierte Verformung aufweist und mit den strukturellen Einschränkungen kompatibel ist, deren Kriechverhalten für die Belastung des Bauteils geeignet ist und die eine ausreichende Zugfestigkeit besitzt.

Datenblätter und technische Broschüren der Legierungen sind bei den Herstellern erhältlich. Die Materialkosten sollten nicht davon abhalten, auf eine höher legierte Sorte umzusteigen. Der aufgerüstete Werkstoff kann eine höhere Festigkeit aufweisen, sodass das Bauteil auf kleinere Wandstärken umgestaltet werden kann. Erfahrene Metallurgen und Gießereiingenieure sollten bei der Auswahl der Legierung und der Konstruktion des Bauteils zu Rate gezogen werden.

1.2.5 Verfestigung bei erhöhter Temperatur

Die Beständigkeit von hitzebeständigen Gusslegierungen gegen Kriechverformung ist auf die Verfestigung durch dispergierte Partikel und Mischkristallverfestigung zurückzuführen. Bei den ACI-Standardlegierungen wird diese Verfestigung durch die im Ausgangszustand vorhandenen primären Chromkarbidphasen bewirkt. Der Effekt wird durch die Ausscheidung von wesentlich feineren Sekundärkarbiden verstärkt, die bei Temperaturen von über 650 °C auftreten. Diese Karbide verhindern die plastische Verformung bei hohen Temperaturen und erhöhen somit die Festigkeit.

1.2.6 Oxidation

Hitzebeständige Legierungen müssen mehr als 12 % Chrom enthalten, um eine wirksame Beständigkeit gegen Oxidablagerungen bei Temperaturen von über 675 °C zu gewährleisten. Ein höherer Chromgehalt erhöht die Oberflächenstabilität durch die Bildung dichter Oxidschichten, die das Eindringen von Oxid verhindern und die weitere Oxidation verringern. Die Oxidationsbeständigkeit wird auch durch eine Erhöhung des Nickelgehalts verbessert, was den Einsatz bei höheren Betriebstemperaturen ermöglicht. Silizium ist ebenfalls wichtig für die Bildung von Schutzoxiden. Ohne diesen Schutz würde die fortschreitende Verzunderung die Querschnittsfläche eines Bauteils verkleinern und die Tragfähigkeit verringern.

Zyklische Temperaturen oder abwechselnd oxidierende und reduzierende Bedingungen können zu Oxidabplatzungen oder zur chemischen Reduktion der Schutzoxide führen. Mit Niob modifizierte Legierungen sind bei Temperaturen von über 1.065 °C empfindlicher gegen Abplatzungen als die Grundzusammensetzung, während Legierungszusätze von Wolfram oder Molybdän weniger nachteilige Auswirkungen haben.

1.2.7 Schleuder- und Formguss für die Wärmebehandlung

Für Wärmebehandlung und Härterei werden für den Betrieb in Industrieöfen eine Vielzahl an Komponenten und Betriebsmitteln wie Mantelstrahlrohre für Gasbrenner oder Ofenrollen, aber auch unterschiedlichste Chargiermittel wie z. B. Roste und Körbe sowie verzahnte Antriebs- und Umlenkrollen, untere und obere Stützrollen, Gussglieder oder auch komplette Gussgliederbänder sowie diverse Ausführungen von Ventilatoren benötigt.

Bild 4: Herstellung von nahtlosen Rohren im Schleudergussverfahren (Quelle: FAI-FTC)

Diese Komponenten und Betriebsmittel werden überwiegend im Gussverfahren wie dem Schleuder- oder Formgießen hergestellt. Schleuderguss ist ein Verfahren zur Herstellung von Metallteilen, bei dem flüssiges Metall in eine rotierende Kokille gegossen wird. Durch die Zentrifugalkraft wird das Metall gleichmäßig an die Innenwand der Kokille gedrückt und erstarrt dort. Dadurch entstehen Teile mit hoher Dichte und geringen Poren, was zu einer verbesserten Festigkeit und Haltbarkeit führt. **Schleuderguss** ist daher für die Herstellung von Mantelstrahlrohren, Ofenrollen und anderen komplexen Metallteilen prädestiniert (**Bild 4**).

Beim **Formguss** wird flüssiges Metall in eine Form gegossen und anschließend abgekühlt, um das gewünschte Teil zu formen. Die Form kann aus Sand, Keramik oder Metall bestehen und wird vor dem Gießen entsprechend des gewünschten Teils hergestellt. Formguss wird häufig für die Massenproduktion von Teilen verwendet, da es kostengünstig und effizient ist. Er eignet sich daher sehr gut für die Herstellung komplexer Formen und detaillierter Oberflächen wie z. B. Roste, Körbe und Ventilatoren für die Wärmebehandlung (**Bild 5**).

1.2.8 Fertigung für Gusskomponenten

Schlussendlich ist auch die weiterführende Fertigungsweise von Gusskomponenten ein wesentlicher Herstellungsschritt, der über die Eignung und die Lebensdauer von Gusskomponenten in der Wärmebehandlung entscheidend ist. Mittels der im Speziellen bei FAI-FTC zur Verfügung stehenden Anlagentechnik können heute nahtlose Rohre mit Durchmessern von 77 bis 1.390 mm bei Wandstärken von 8 bis 100 mm und einer maximalen Länge von bis zu 5.700 mm im Schleuderguss hergestellt werden. Das maximale Gewicht eines Rohres kann bis zu 3.000 kg betragen.

Bild 5: Formguss (Quelle: FAI-FTC)

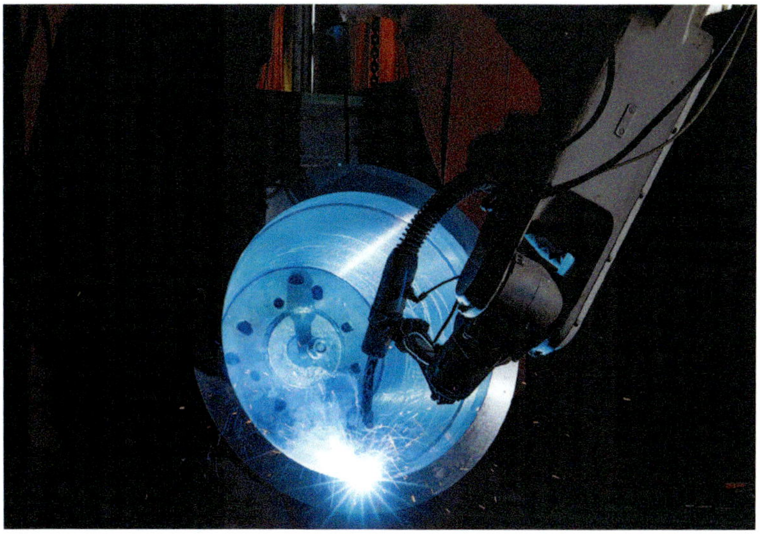

Bild 6: Schweißroboter beim Fügen von Gusskomponenten (Quelle: FAI-FTC)

Die **mechanische Fertigung** basiert in modernen Gießereien auf hochmodernen CNC-Drehmaschinen, sodass die komplette Fertigung einbaufertiger Ofenkomponenten im Herstellungswerk gewährleistet ist.

Im Bereich „**Schweißen und Zusammenbau**" werden Schweißprozesse von Gusskomponenten (TIG, MIG und Elektrodenschweißen) nach den gültigen Normen und Spezifikationen durchgeführt. Neben dem manuellen Schweißen werden hier seit vielen Jahren bereits Schweißroboter eingesetzt (**Bild 6**). Dadurch ist die hohe und reproduzierbare Schweißnahtqualität der Produkte, die beispielsweise die absolute Gasdichtheit von Mantelrohren für Gasbrenner ermöglicht, gewährleistet. Unverzichtbar ist die hohe Schweißnahtgüte auch bei mechanisch hochbelasteten Ofenrollen. Die hohen Anforderungen an die Auswuchtgüte von Komponenten wie Ofenrollen wird durch eine dynamische Auswuchtmaschine sichergestellt.

Gemäß dem Motto „**alles aus einer Hand**" (über Beratung, Werkstoffauswahl, Konstruktion, Formgebung, mechanische Fertigung mit Schweißen und Montage) können ein hoher Qualitätsstandard und die lückenlose Qualitätskontrolle aller Arbeitsschritte durch die eigenen Mitarbeiter gewährleistet werden.

1.2.9 Anwendungsbeispiele für Gussbauteile in der Wärmebehandlung und im Industrieofenbau

1.2.9.1 Mantelstrahlrohre für Gasbrenner

Mantelstrahlrohr-Gasbrenner werden in Industrieöfen für Wärmebehandlung und das Härten unter Schutzgasatmosphäre verwendet, wie z. B. bei thermochemischen Härteverfahren wie Einsatzhärten oder auch Nitrieren. Innerhalb des Mantelrohres wird durch den Brenner eine mehr oder weniger gleichmäßige Flamme erzeugt. Diese Flamme ist sehr stabil und hat eine hohe Temperatur, sodass sich Mantelstrahlrohre für verschiedene Anwendungen, wie z. B. in Industrieöfen, eignen.

Durch die Trennung der Brennerflamme vom reaktiven Behandlungsraum im Industrieofen können Schutzgase geregelt eingesetzt werden, die Oxidationen der Oberfläche der Bauteile oder auch eine Entkohlung verhindern. Mantelstrahlrohre bestehen aus nahtlos im Schleudergussverfahren hergestellten Rohren, Formguss-Verschlusskappen und Brennerflanschen. Eine perfekte Schweißnahtvorbereitung und hohe Schweißnahtgüte ist zwingend notwendig, um eine Gasdichtheit der Mantelrohre zu gewährleisten.

Vorteile von Mantelstrahlrohren, die im Schleudergussverfahren hergestellt werden, liegen in einer hohen Effizienz bei geringem Energieverbrauch. Sie ermöglichen eine präzise Steuerung der Flammenintensität und können daher für unterschiedliche Prozesse optimiert werden. Zudem sind sie relativ wartungsarm und langlebig. Insgesamt sind Mantelstrahlrohre für Gasbrenner eine effektive Lösung für die Erzeugung von Wärme in verschiedenen industriellen Anwendungen.

Mantelstrahlrohre aus Schleuderguss zeichnen sich zudem durch eine hohe Temperaturbeständigkeit, eine hohe Oxidations- und Korrosionsbeständigkeit sowie eine gute Thermoschockbeständigkeit aus. Aus Gründen der erforderlichen Gasdichtheit werden hohe Anforderungen an die Schweißnahtgüte gestellt. Die Schweißnähte der fertigen Mantelstrahlrohre werden mittels Farbeindringprüfung auf Rissfreiheit geprüft und die Dichtheit durch ein Abdrücken mit Druckluft kontrolliert.

1.2.9.2 Ofenrollen

Ein wesentliches Einsatzgebiet von Schleudergussrohren ist die Fertigung von Ofenrollen, wie sie z. B. in Rollenherdofenanlagen zum Einsatz kommen. Bei der Auslegung und der Festlegung des Designs der Rolle wird neben der individuellen Belastung der jeweiligen Ofenrolle die Auswahl der Legierung als auch die Umgebungs- bzw. Atmosphärenbedingungen sowie die Einsatztemperaturen berücksichtigt. Die dynamische und thermische Analyse der Belastung sind dabei für die Auslegung der Rollen entscheidend (**Bild 7**). Langjährige Erfahrung in Auslegung und Konstruktion von Ofenrollen sind für eine optimale Her-

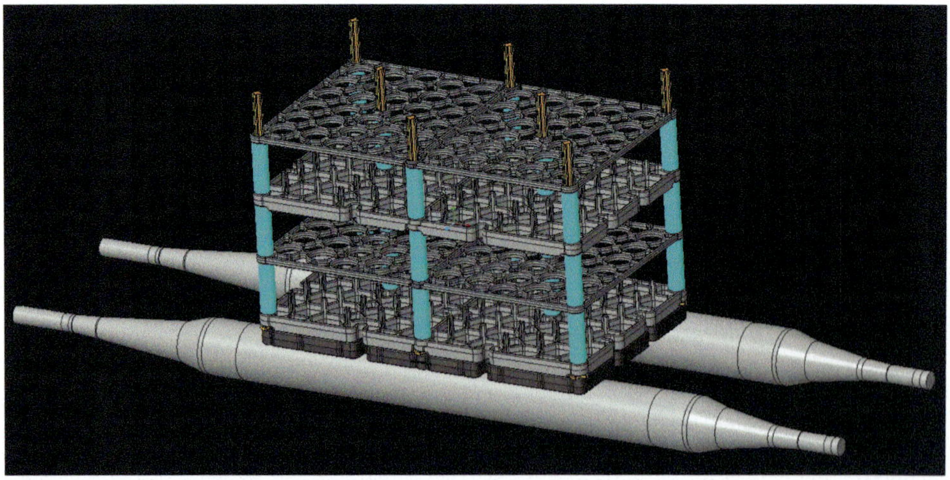

Bild 7: Analyse der dynamischen und thermischen Belastung der Ofenrollen (Quelle: FAI-FTC)

stellung unerlässlich. Anpassungen von Ofenrollen an neue Produktionsbedingungen in bestehenden Ofenanlagen sind Beispiele für die vielfältigen Anforderungen der Anwender.

Zur Festlegung des finalen Designs der Ofenrollen können FEM-Analysen (Mesh Development) durchgeführt werden. Über speziell entwickelte Berechnungsprogramme erfolgt eine Bewertung der voraussichtlichen Lebensdauer der Ofenrollen.

1.2.9.3 Chargierhilfsmittel

Bei Chargierhilfsmitteln für Wärmebehandlung und Härterei werden i. d. R. hitzebeständige Werkstoffe verwendet, die den hohen Temperaturen und chemischen Belastungen standhalten können. Die häufigsten Werkstoffe sind Edelstähle, aufgrund ihrer Korrosionsbeständigkeit und Hitzebeständigkeit (Tabelle 2), und Nickelbasislegierungen. Die Wahl des geeigneten Werkstoffs hängt von verschiedenen Faktoren ab, wie z. B. der spezifischen Anwendung, den Betriebsbedingungen und den Anforderungen an die Haltbarkeit und Lebensdauer des Chargiergestells.

Neben der Konzepterstellung und der individuellen Beratung der Anwender wird je nach Anforderung eine optimale Werkstoffauswahl getroffen und die Konstruktion an die Belastung und die Einsatzbedingungen angepasst – sowohl bei der Optimierung vorhandener Chargiermittel als auch bei Neukonstruktionen. Bei der Konstruktion von Chargiermitteln kommen, wie bei der Auslegung von Ofenrollen, FEM-Belastungsberechnungen zum Einsatz. **Bild 8** zeigt ein typisches Chargiergestell, welches für spezielle Bauteile unter Berücksichtigung der Einsatzbedingungen ausgelegt wurde.

1.2.9.4 Gussgliederbandsysteme für Durchlaufanlagen

Gussgliederbandöfen werden zur Behandlung von Schüttgütern eingesetzt. Dabei erfolgt der kontinuierliche Teiletransport auf einem Gussgliederband, welches im Behandlungsraum der Anlage durch eine verzahnte Antriebsrolle angetrieben wird. Nachdem das Schüttgut durch den Ofenraum transportiert wurde, in dem die gewünschte Behandlung unter einer Schutzgasatmosphäre abläuft, fallen die Bauteile zur Abschreckung über einen

Bild 8: Individuell ausgelegtes Chargiergestell (Quelle: FAI-FTC)

Bild 9: Gussgliederband mit verzahnter Antriebsrolle (Quelle: FAI-FTC)

Fallschacht ins Abschreckbecken (Öl, Polymere, Salzbad). Gussgliederbandöfen werden typischerweise mit Bandbreiten von bis zu 1.600 mm und Ofenlängen von bis zu 12.000 mm hergestellt. Es werden Durchsatzleistungen bis über 2.500 kg/h erreicht. Von den führenden Herstellern wird die Belieferung kompletter Transportsysteme, bestehend aus dem Gussgliederband sowie den Antriebs-, Trag- und Umlenkrollen, erwartet, wobei die einzelnen Komponenten optimal aufeinander abgestimmt werden müssen. **Bild 9** zeigt den Aufbau einer typischen verzahnten Antriebsrolle, die das Gussgliederband in der Anlage antreibt.

1.2.10 Zusammenfassung

Der vorliegende Fachbeitrag zeigt die Aspekte auf, die mit der Herstellung von Gusskomponenten für die Wärmebehandlung und Härtereitechnik als auch dem Industrieofenbau verbunden sind. Die Herstellung qualitativ hochwertiger rostfreier und hitzebeständiger Spezialstähle für den Einsatz in der Stahlindustrie und der Wärmebehandlung basiert dabei auf einem hohen Maß an werkstoff- und verfahrenstechnischem Fachwissen. Europaweit führende Hersteller von Gusskomponenten als auch Experten in Planung und Beratung entsprechender Prozesse gewährleisten den Einsatz und Betrieb dieser komplexen Komponenten wie Ofenrollen, Mantelstrahlrohre, Rohre, konische oder zylindrische Trommelmäntel, Buchsen oder diverse Typen von Unterstützungen.

1.3 DualQuench: Flexible Abschrecktechnik für Durchstoßanlagen

Dr. Klaus Buchner

Klimawandel, Künstliche Intelligenz und Globalisierung sind nur einige Schlagworte, die das tägliche Handeln prägen. Auch der Bereich der Thermoprozesstechnik muss sich den Herausforderungen der Zeit stellen. Im Spannungsfeld zwischen Energiekosten und CO_2-Fußabdruck gilt es, mit einer flexiblen Anlagentechnik zielgerichtet auf schwankende Auslastungen und verändernde Anforderungen zu reagieren. Die flexible Abschrecktechnik ist dabei ein wesentlicher Aspekt.

Angesichts der aktuellen wirtschaftlichen Situation steht Europa vor einem signifikanten Umbruch. Globale Umwälzungen und politische Spannungen bringen mannigfaltige Herausforderungen mit sich – zusätzlich gilt es, den demographischen Wandel zu meistern. Dies wirkt sich auch auf den Alltag der Thermoprozesstechnik aus. Um flexibel auf veränderte Rahmenbedingungen reagieren zu können, ist eine angepasste Anlagentechnik erforderlich, wobei die Abschrecktechnik einen wesentlichen Bestandteil der Prozesskette darstellt.

1.3.1 Wärmebehandlungsprozesse und Abschreckkonzepte

Wärmebehandlungsprozesse von Metallen sind grundsätzlich durch einen definierten Zeit-Temperatur-Zyklus gekennzeichnet, der die Erwärmungs-, Halte- und Abkühlphase umfasst. Das Gefüge und die damit verbundenen Eigenschaften eines wärmebehandelten Teils hängen von der Abkühlungsgeschwindigkeit ab, die wiederum von Faktoren wie Bauteilmasse, Abschreckmedium, Strömungsgeschwindigkeit des Abschreckmediums etc. bestimmt wird. Zudem ist das Abschrecken auch mit der Ausbildung von Wärme- und Umwandlungsspannungen verbunden, die zu Änderungen in Größe und Form führen [1]. Hierbei ist jedoch festzuhalten, dass der Bauteilverzug von vielfältigen Parametern in der gesamten Fertigungskette beeinflusst wird [2]. Das Hauptziel des Abschreckens besteht somit darin, das gewünschte Gefüge sowie die mechanischen Kennwerte zu erreichen und gleichzeitig Eigenspannungen und Verzug zu minimieren. In der betrieblichen Praxis sind die gebräuchlichsten Abschreckmittel Flüssigkeiten wie Wasser (fallweise mit Zusatz von Salzen), wässrige Polymerlösungen und Härteöle. Darüber hinaus kommen auch Inertgase und Salzschmelzen zum Einsatz, während Metallschmelzen und Wirbelschichten seltener verwendet werden. **Tabelle 1** gibt einen kurzen Überblick bezüglich der Vor- und Nachteile der unterschiedlichen Abschreckmedien.

Tabelle 1: Vor- und Nachteile unterschiedlicher Abschreckmedien (Quelle: Aichelin Holding GmbH)

Abschreckmedium	Aggregatzustand	Intensität	Gleichmäßigkeit	Handling/ Investment
Verdampfende Flüssigkeiten	Flüssig und gasförmig	+	-	+
Nicht-verdampfende Flüssigkeiten	Flüssig (Schmelzpunkt > Raumtemperatur)	○	+	○
Gase	Gasförmig	-	+	-

1.3.2 Verdampfende Abschreckmedien

Beim Abschrecken in flüssigen Medien mit Siedetemperaturen, die deutlich unter der Härtetemperatur des Bauteils liegen, treten drei Phasen der Wärmeabfuhr mit jeweils unterschiedlichen Wärmeübergangscharakteristika auf. Diese werden als Dampfhautphase (A), Koch-/Siedephase (B) und Konvektionsphase (C) bezeichnet. Während der Dampfhautphase ist die Oberflächentemperatur des Werkstücks ausreichend hoch, um die Abschreckflüssigkeit zu verdampfen und einen stabilen Film um das Teil zu bilden. Der Dampffilm wirkt isolierend, sodass die Abkühlgeschwindigkeit während des Filmsiedens relativ gering ist (Leidenfrost-Phänomen). Beim Unterschreiten der Leidenfrost-Temperatur bricht der Dampffilm zusammen und die Koch-/Siedephase beginnt. In diesem Stadium verdampft die mit der heißen Oberfläche in Kontakt stehende Flüssigkeit und die Dampfblasen verlassen sofort die Oberfläche. Dadurch entsteht eine ausgeprägte lokale Turbulenz, die zu einer hohen Wärmeübertragungsrate führt – die Abkühlrate erreicht ihr Maximum. Kurz vor Erreichen der Siedetemperatur beginnt die dritte Phase des Abschreckens – die Konvektionsphase. Die Abkühlgeschwindigkeit ist gering und wird hauptsächlich durch die Konvektionsgeschwindigkeit und die Viskosität des flüssigen Abschreckmediums bestimmt.

Eine schematische Darstellung der Temperatur-Zeit-Abkühlungskurve, die diese drei Stufen veranschaulicht, ist **Bild 1** zu entnehmen.

Aufgrund der komplexen Geometrie des realen Bauteils treten diese drei Wärmeübertragungsphasen zeitlich versetzt an unterschiedlichen Positionen auf. Infolge der ungleichmäßigen Wärmeübergangsgeschwindigkeit an der Bauteiloberfläche entstehen zusätzliche Temperaturgradienten im Bauteil, die beim Abschrecken in verdampfenden Medien einen wesentlichen Verzugsfaktor darstellen. Darüber hinaus ist der Einfluss der Benetzungskinematik auf die metallurgischen Eigenschaften zu berücksichtigen.

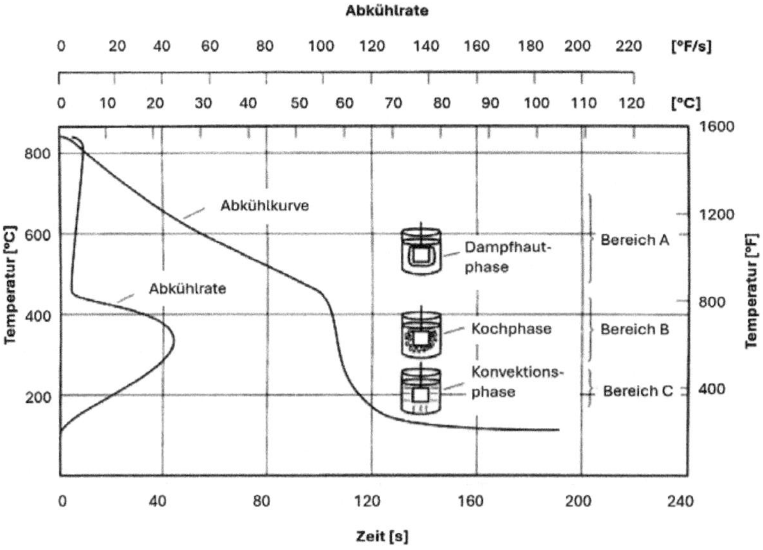

Bild 1: Abkühlkurve und Abkühlrate, Wasser 95 °C (Quelle: [3])

1.3.3 Nicht-verdampfende Abschreckmedien

Neben der Gasabschreckung/Hochdruckgasabschreckung sind Salzschmelzen die wichtigsten Vertreter der nicht verdampfenden Abschreckmedien. Es handelt sich dabei um Nitrit-/Nitrat-Abschrecksalze, die für Anwendungen zwischen 160 °C und 550 °C geeignet sind und sich durch eine Siedetemperatur oberhalb der Härtetemperatur des abgeschreckten Bauteils auszeichnen.

Das Abschrecken beruht also ausschließlich auf konvektiver Wärmeübertragung, ohne die drei verschiedenen Abkühlphasen, die zuvor für verdampfende Flüssigkeiten erwähnt wurden. Gegenüber dem Abschrecken mit Öl oder Wasser ergeben sich somit folgende Vorteile [4]:

- Geringerer Verzug und geringere Eigenspannungen bei vergleichbarer Härte bei martensitischer Härtung, signifikant reduzierter Verzug bei hoher Duktilität, Zähigkeit und Festigkeit mit einem gewissen Härteverlust beim Bainitisieren, annähernd homogener Wärmeübergang trotz unterschiedlicher Abschrecktemperaturen und Strömungsverhältnisse.
- Die Abschreckintensitäten (Abschreckraten) können in begrenztem Umfang durch Wasserzugabe variiert werden.
- Effizientes Waschen und Recycling ermöglichen einen ressourcen- und umweltschonenden Betrieb.

1.3.4 Hochdruck-Gasabschreckung

Da beim Abschrecken in Gasen das Leidenfrostphänomen nicht auftritt, sind eine höhere Reproduzierbarkeit der Härteergebnisse und eine geringere Streuung der Maß- und Formänderungen gegeben. Eine Nachreinigung der wärmebehandelten Bauteile ist verfahrensbedingt nicht erforderlich. Üblicherweise beschränkt man sich bei der Betrachtung der Einflussgrößen des Hochdruck-Gasabschreckens auf die Anlagentechnik, die Gasart und den Druckbereich, um die zu erwarteten Härtewerte und Gefügecharakteristika am spezifischen Bauteil zu erzielen.

1.3.5 Hochdruckgasabschreckung aus Sicht der Nachhaltigkeit

Es zeigt sich, dass durch den Einsatz von Helium und eine Erhöhung des Gasdruckes auf 20 bar auch Einsatzstähle und niedriglegierte Stähle erfolgreich gehärtet werden können. **Bild 2** zeigt beispielhaft gemessene Wärmeübergangskoeffizienten für verschiedene Gase und Abschreckdrücke.

Weiter sei auf die Besonderheiten von Wasserstoff-Stickstoff-Gemischen hinsichtlich der Abschreckwirkung hingewiesen. Dabei zeigt sich, dass Gemische einen höheren Wärmeübergangskoeffizienten als reiner Wasserstoff aufweisen, wobei dies im Wesentlichen von den Strömungsverhältnissen abhängt [6]. Die bessere Abschreckwirkung von Wasserstoff in Verbindung mit höheren Drücken im Vergleich zu Stickstoff ermöglicht eine martensitische Härtung von Bauteilen mit größeren Querschnitten. In **Tabelle 2** sind drei Kaltarbeitsstähle hinsichtlich des durchhärtbaren Querschnitts für unterschiedliche Medien bzw. Abschreckdrücke aufgeführt [7]. Wenngleich die Literaturangaben zum CO_2-Fußabdruck von Helium stark schwanken – einerseits bedingt durch die Art der Gewinnung, andererseits durch die gewählten Bilanzgrenzen – so muss an dieser Stelle festgehalten werden, dass Helium direkt mit der Erdgasgewinnung gekoppelt ist und 85 % der globale Jahresproduktion von nur zwei Ländern abgedeckt werden [8]. Im Sinne der Nachhaltigkeit und

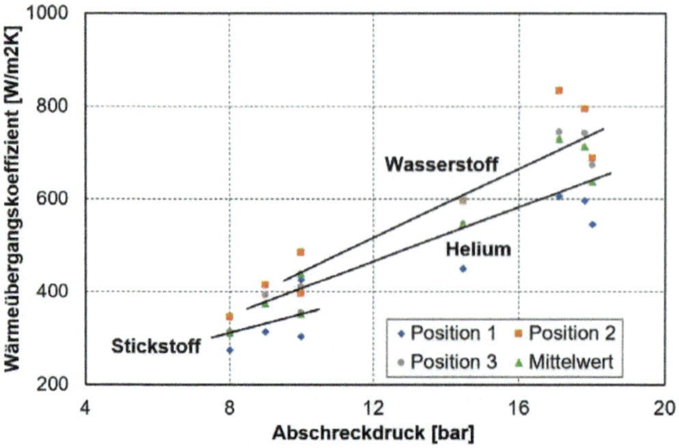

Bild 2: Einfluss von Abschreckdruck und Gasart auf den mittleren Wärmeübergangskoeffizienten (Quelle: [5])

Tabelle 2: Durchhärtbare Querschnitte verschiedener Kaltarbeitsstähle [7]

Werkstoff	6 bar N_2	6 bar H_2	20 bar H_2
1.2080	70 bis 80	110	200
1.2842	< 15	30	50
1.2067	-	5	15

der Versorgungssicherheit stellt die alternative Nutzung von Wasserstoff als Abschreckmedium eine weitere Option dar, zumal grünem Wasserstoff als möglichem zukünftigen Energieträger eine wesentliche Rolle zukommen kann. Unabhängig davon ist eine Gasrückgewinnungsanlage aufgrund der Medienkosten für Helium wirtschaftlich sinnvoll, wenngleich dies eine höhere Anlagenkomplexität bedingt.

1.3.6 Aufkohlung [9]

Das Einsatzhärten ist seit Jahrzehnten der dominierende Wärmebehandlungsprozess für hochbeanspruchte Bauteile im Maschinenbau. Abgesehen von Sonderanwendungen haben sich gegenwärtig zwei Verfahrensvarianten in der Industrie etabliert – die Gasaufkohlung und die Niederdruckaufkohlung. Beim Vergleich der beider Verfahren ist zwischen dem Aufkohlungs- und dem Abschreckprozess zu unterschieden. Sowohl Prozessgasanlagen als auch Vakuumanlagen können mit Gas- oder mit Ölabschreckung ausgeführt werden.

1.3.7 Niederdruckaufkohlung

Die Aufkohlung erfolgt durch Zersetzung der Prozessgase (Acetylen, seltener Propan, im Druckbereich von wenigen mbar) direkt an der metallischen Bauteiloberfläche, die katalytisch wirkt. Aufgrund dieser Zersetzungsreaktion diffundiert ein sehr großer Kohlenstoffmassestrom in die Bauteiloberfläche ein. Dies führt zu einem raschen Anstieg des Rand-C-Gehaltes, aber auch zu einer raschen Übersättigung, sofern die Prozessgaszufuhr nicht unterbrochen wird. Aus diesem Grund hat sich eine pulsierende Prozessgaszufuhr durchgesetzt, bei der Aufkohlungs- und Diffusionsphasen einander abwechseln. Durch eine geeignete Wahl von Begasungsmenge, Puls- und Pausenzeiten kann die Karbidbildung reduziert und in relativ kurzer Zeit eine ausreichend dicke, aufgekohlte Randschicht erreicht werden. Die Aufkohlung erfolgt dabei rezeptgesteuert und liefert nach entsprechender Optimierung gute und reproduzierbare Ergebnisse. Konzepte zur Prozessregelung wurden zwar entwickelt, haben sich aber bis dato nicht am Markt etabliert.

1.3.8 Gasaufkohlung

Als Trägergas werden in der Prozessgastechnik meist Endogas oder Stickstoff/Methanol eingesetzt, wobei im ersten Fall ein Endogasgenerator Erdgas (seltener Propan) und Luft in einem Katalysatorbett bei ca. 1.000–1.050 °C vollständig zu Kohlenmonoxid, Wasserstoff und Stickstoff umsetzt. Im Gegensatz zur Niederdruckaufkohlung ist bei der Prozessgasaufkohlung eine Prozessregelung mittels Sauerstoff- oder λ-Sonde Stand der Technik; alternative Regelsysteme basieren auf der Kohlenmonoxid-Kohlendioxid-Messung. Die Regelung des Kohlenstoffpegels erfolgt durch Zugabe von Zusatzgas (Erdgas oder Propan, zur Erhöhung des Kohlenstoffpegels) oder Luft (zur Absenkung des Kohlenstoffpegels). Der Vollständigkeit halber sei noch auf die Direktbegasung mit Kohlenwasserstoffen und Luft hingewiesen. Aufgrund der mangelnden Gleichgewichtseinstellung und der dadurch erschwerten Regelbarkeit bzw. der eingeschränkten Einsatzbereiche ist sie heute als Sonderanwendung anzusehen.

1.3.9 Anlagentechnik bei der atmosphärische Gasaufkohlung

Während Thermoprozessanlagen auf Basis der Niederdruckaufkohlung modular aufgebaut sind und somit einen Batch-Charakter aufweisen, ermöglicht die Gasaufkohlung auch kontinuierliche Anlagenlösungen. Speziell bei hohen Durchsatzleistungen wie in der Automobilindustrie bieten Durchstoß- oder Ringherdofenanlagen in Verbindung mit modernster Anlagentechnik hinsichtlich Beheizung und thermischer Isolierung eine effiziente und wirtschaftliche Wärmebehandlung. An dieser Stelle sei erwähnt, dass verzugsempfindliche Bauteile, die in einer Fixtur abgeschreckt werden müssen, in atmosphärischen Gasaufkohlungsanlagen in einem Arbeitsgang erwärmt, aufgekohlt und direkt gehärtet werden können. Insbesondere unter dem Aspekt der Energiekosten und des CO_2-Fußabdrucks ist das Direkthärten der Einfachhärtung mit Wiedererwärmung vorzuziehen.

1.3.10 Flexible Abschrecktechnik für Durchstoßanlagen

Wie einleitend ausgeführt, stellt eine flexible Abschrecktechnik einen wesentlichen Aspekt dar, um zielgerichtet auf schwankende Auslastungen und verändernde Anforderungen reagieren zu können. Anhand realisierter Anlagenausführungen werden nachfolgend Beispiele für Durchstoßanlagen mit kombinierter Hochdruckgasabschreckkammer und Ölbad sowie Ölbad und Einzelentnahme mit Härtepresse diskutiert.

Das Funktionsprinzip eines Durchstoßofens beruht auf dem kontinuierlichen Transport der auf Grundrosten chargierten Bauteile durch die Thermoprozessanlage. Der Begriff „Durchstoß" beschreibt den Transportmechanismus, bei dem die Grundroste mit konstanter Taktzeit schrittweise um eine Position weiterbefördert werden. Moderne mehrbahnige Anlagen ermöglichen unterschiedliche Taktzeiten für die zwei oder drei parallel angeordneten Stoßbahnen, wodurch unterschiedliche Prozesszeiten innerhalb des Ofens realisiert werden können. Wird der Grundrost um mehr als eine Position weiterbewegt, ergibt sich eine zusätzliche Flexibilität. So können verschiedene Bauteile/Bauteilfamilien mit unterschied-

Bild 3: Durchstoßanlage mit Hochdruckgasabschreckung und Ölbad (Quelle: Aichelin Holding GmbH)

lichen Prozesszeiten gleichzeitig vollautomatisch wärmebehandelt werden. Speziell die Kombination mit einem Hochregallager ermöglicht eine optimierte Automatisierung, sodass die erforderliche Flexibilität und damit Kosten- und Energieeffizienz gewährleistet sind.

1.3.11 DualQuench: Hochdruckgasabschreckung und Ölbadabschreckung

Einsatzhärten ist ein wesentliches Wärmebehandlungsverfahren für Bauteile unterschiedlichster Geometrien und Anforderungen. Dies bedingt für den jeweiligen Anwendungsfall angepasste Abschrecktechnologien.

Aufgrund des Produktspektrums wurde eine mehrbahnige Gasaufkohlungsanlage mit einer kombinierten Hochdruckgas- und Ölbadabschreckung ausgeführt. Über eine Zwischenkammer wird die Charge wahlweise in die Hochdruckgasabschreckkammer, welche für das Abschrecken mit Stickstoff von bis zu 20 bar konzipiert ist, oder in das Ölbad transportiert (**Bild 3**).

In **Bild 4a** und **Bild 4b** ist der Einfluss des Abschreckmediums auf die Zahnflanken-Winkelabweichung einer verzugskritischen Antriebswelle dargestellt. Bild 4a zeigt die Flankenlinie der Verzahnung nach Ölabschreckung. Im Gegensatz dazu führt die Hochdruckgasabschreckung zu einer reduzierten Balligkeit und einer gleichmäßig konkaven Verzahnungsgeometrie (Bild 4b). Andererseits gewährleistet das Ölbadhärten bei nicht verzugskriti-

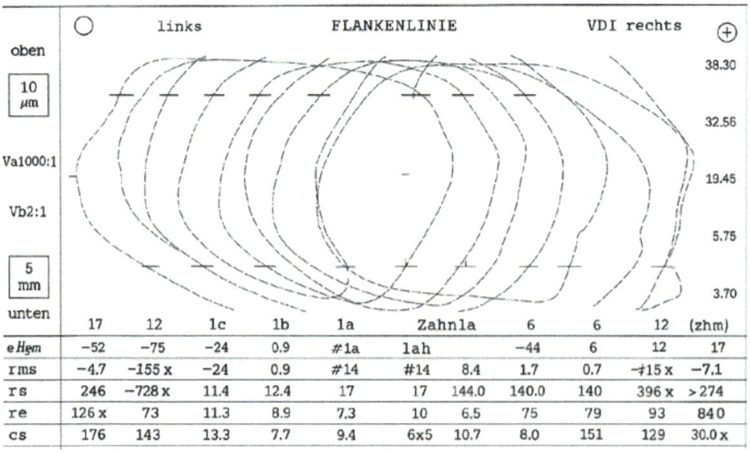

	17	12	1c	1b	1a	Zahnla		6	6	12	(zhm)
eHgm	−52	−75	−24	0.9	#1a	1ah		−44	6	12	17
rms	−4.7	−155 x	−24	0.9	#14	#14	8.4	1.7	0.7	−‡15 x	−7.1
rs	246	−728 x	11.4	12.4	17	17	144.0	140.0	140	396 x	>274
re	126 x	73	11.3	8.9	7.3	10	6.5	75	79	93	840
cs	176	143	13.3	7.7	9.4	6x5	10.7	8.0	151	129	30.0 x

Bild 4a: Flankenlinien einer Antriebswelle nach Ölabschrecken (Quelle: Aichelin Holding GmbH)

1

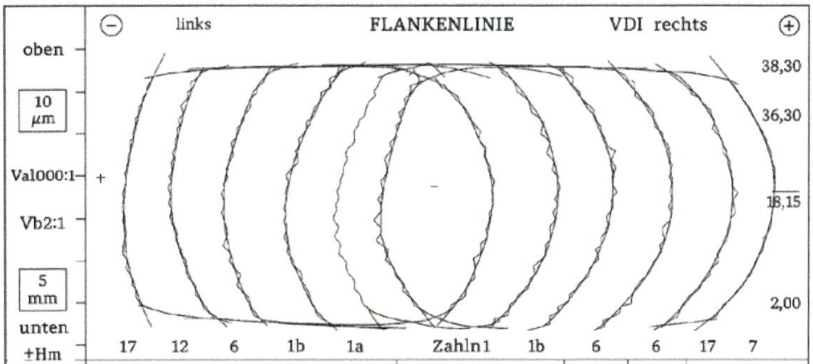

±Hm	17	12	6	1b	1a	Zahln1	1b	6	6	17	7
µm	28		v	8,9		2,0		2,0	5,5	5,5	
FHβ	−3,7	−3,4	8,9	2,3	2,3	+15	17	16,1	9,5	3,8	5,5
fβ	9,6	6,9	5,2	3,4	5,7	17	16,1	16,1	9,5	9,5	9,6
Fβ	4,7	3,8	8,5	5,7	4,7	14,7	x10,5 100,0	17,0	8,9	9,5	10,6
Cβ	5,1	7,0	8,9	8,5	8,5	8,5	35/15	13,4	11,7	12,0	18,4

Bild 4b: Flankenlinien einer Antriebswelle nach Hochdruckgasabschrecken (Quelle: Aichelin Holding GmbH)

schen Bauteilen aufgrund der höheren Abschreckintensität auch bei mäßiger Härtbarkeit des Werkstoffes eine ausreichende Zahnfußfestigkeit.

1.3.12 DualQuench: Ölbadabschreckung und Einzelteilentnahme mit Härtepresse

Beim Einsatzhärten von Tellerrädern ist zu beachten, dass der auftretende Bauteilverzug ein kostenintensives Verzahnungsschleifen erforderlich macht. Daher ist eine systematische Verzugsbeherrschung in der gesamten Fertigungskette vom Stahlwerk bis zum einsatzgehärteten Tellerrad wünschenswert [10]. Üblicherweise werden verzugsempfindliche Getriebeteile wie Tellerräder, Synchronringe etc. daher in Fixturen abgeschreckt. Häufig erfolgt der Aufkohlungsprozess getrennt von der Wiedererwärmung und dem Fixturhärten, wobei für den zweiten Prozessschritt aufgrund der Einzelteilentnahme atmosphärische Thermoprozessanlagen eingesetzt werden. Obwohl durch die Wiedererwärmung eine Kornfeinung eintritt, ist es energetisch – und somit ökonomisch und ökologisch – nahelie-

Bild 5: Durchstoßofen mit Einzelteilentnahme (Quelle: Aichelin Holding GmbH)

gend, unmittelbar nach der Aufkohlungsphase und der Temperaturabsenkung abzuschrecken (Direkthärten), zumal Feinkornstähle dies auch aus metallurgischer Sicht zulassen.

Bei der anlagentechnischen Umsetzung sorgt eine separate Einzelentnahmekammer mit Zwischentür dafür, dass die Prozessgasatmosphäre in der Thermoprozessanlage auch dann stabil bleibt, wenn die Klapptür geöffnet wird und der Roboter das Bauteil entnimmt und in der Fixturhärtemaschine ablegt (**Bild 5**). Alternativ dazu kann die Charge im Härteölbehälter abgeschreckt werden. Mit dieser Anlagenkonfiguration können sowohl Chargen mit Tellerrädern als auch Chargen mit Kegelrädern gleichzeitig aufgekohlt und getrennt in Fixturhärtemaschinen und Ölbädern abgeschreckt werden, bevor sie gemeinsam der Nachreinigung, dem Anlassofen und der Entladung zugeführt werden.

1.3.13 Fazit

Die Wärmebehandlungsbranche steht vor zahlreichen Herausforderungen, die es zu bewältigen gilt – die Notwendigkeit, ressourcen- und kosteneffizienter zu arbeiten, ist dringlicher denn je. Durch den Einsatz moderner Anlagentechnik kann eine höhere Flexibilität in den Produktionsprozessen erreicht werden. Die Kombination unterschiedlicher Abschrecktechniken ermöglicht es Unternehmen nicht nur, ihre Produktionskapazitäten zu optimieren, sondern auch nachhaltige Lösungen zu entwickeln. Diese Innovationen tragen dazu bei, den ökologischen Fußabdruck der Branche zu verringern und gleichzeitig den Anforderungen des globalen Marktes gerecht zu werden.

Am Beispiel energieeffizienter Durchstoßanlagen kann gezeigt werden, dass durch die Kombination verschiedener Abschrecktechnologien zusätzliche Potenziale für die atmosphärische Wärmebehandlung erschlossen werden können. Die Integration mit vollautomatisierten Hochregallagersystemen und Anlagen mit unterschiedlichen Takt-/Prozesszeiten ermöglicht die erforderliche Flexibilität für eine effiziente Produktion.

Literatur

[1] Hoferer, M.; Lübben, T.; Majorek, A.; Thoden, B.; Zoch, H.-W.: Grundlagen der Maß- und Formänderungsentstehung bei der Wärmebehandlung. In: HTM Journal of Heat Treatment and Materials, 51 (1996) 1, S. 56–68

[2] Hoffmann, F.; Keßler, O.; Lübben, T.; Mayr, P.: „Distortion Engineering" – Verzugsbeherrschung in der Fertigung. In: HTM Journal of Heat Treatment and Materials, 57 (2002) 3, S. 213–217

[3] Bates, C. E; Totten G. E.; Brennan, R. L.: Quenching of Steel. In: Heat Treating Volume 4 of ASM Handbook, OH: ASM International, 1991

[4] Foreman, R. W.: Salt Bath Quenching. In: Totten, G. E. (Hrsg.): Conference Proceedings Quenching and Distortion Control, Illinois, USA, 1992

[5] Hoffmann, F.; Gondesen, B.; Lohrmann, M.; Lübben, T.; Mayr, P.: Möglichkeiten und Grenzen des Gasabschreckens. In: HTM Journal of Heat Treatment and Materials, 53 (1998) 2, S. 81–86

[6] Laumen, C.; Holm, T.; Lübben, T.; Hoffmann, F.; Mayr, P: Hochdruck-Gasabschrecken mit Wasserstoff. In: HTM Journal of Heat Treatment and Materials, 53 (1998) 2, S. 72–80

[7] Altena, H.: Hochdruck-Wasserstoffabschreckung. In. HTM Journal of Heat Treatment and Materials, 50 (1995) 1, S. 27–30

[8] U.S. Geological Survey: Mineral commodity summaries 2024. U.S. Geological Survey

[9] Altena, H.; Dopler T.; Korlath N.: Einsatzhärten von Großserien-Getriebebauteilen für die Automobilindustrien. In: Irretier, O.; Jost, M. (Hrsg): Handbuch HärtereiPraxi, Vulkan-Verlag GmbH, Essen, 2017, S. 134–135

[10] Surm, H.; Hermann M.; Sattelberger K.; Küper, A.; Hoffmann, F.: Identifikation verzugsrelevanter Einflussgrößen und Wechselwirkungen in der Fertigungskette „Einsatzgehärtetes Tellerrad". In: HTM Journal of Heat Treatment and Materials, 65 (2010) 2, S. 85–95

1.4 Eigenschaften wärmebehandelter korrosionsbeständiger Stähle

Dr. Thomas Waldenmaier

Dieser Beitrag beleuchtet die Eigenschaften und die Notwendigkeit der Wärmebehandlung korrosionsbeständiger Stähle, insbesondere durch Nitrieren, Nitrocarburieren und Carburieren. Es wird die Motivation für diese Behandlungen dargelegt, wobei der Fokus auf der Verbesserung der Korrosions- und Verschleißbeständigkeit sowie auf der Festigkeit liegt. Ein besonderer Schwerpunkt liegt auf dem Vergleich von Nieder- und Hochtemperaturverfahren und deren spezifischen Prozessführungen und Schichtcharakterisierungen. Detaillierte Untersuchungen zu Schichtaufbau, Härteprofilen, Schweißeignung, Schwingverschleiß und Eigenspannungen werden für Niedertemperaturverfahren präsentiert. Die Hochtemperaturverfahren werden hinsichtlich ihrer Stickstofflöslichkeit, Härtetiefenprofile und Schwingfestigkeit bewertet. Der Beitrag schließt mit einem vergleichenden Überblick beider Verfahrenstypen und diskutiert deren Vor- und Nachteile für austenitische und ferritisch/martensitische Stähle.

1.4.1 Einleitung

Korrosionsbeständige Stähle finden aufgrund ihrer hervorragenden Beständigkeit gegenüber korrosiven Medien breite Anwendung in der Industrie. In vielen Applikationen, insbesondere in der Automobilindustrie, wie beispielsweise bei Komponenten für Hochdruck-Kraftstoffpumpen, sind jedoch zusätzlich hohe Anforderungen an die mechanische Belastbarkeit, wie Verschleißbeständigkeit und Festigkeit, sowie spezifische magnetische Eigenschaften zu erfüllen. Unbehandelte, korrosionsbeständige Stähle können diesen kombinierten Anforderungen oft nicht gerecht werden, was zu Problemen wie Lochfraß, interkristalliner Korrosion oder mangelnder Verschleißbeständigkeit führt. Die Wärmebehandlung durch Nitrieren, Nitrocarburieren und Carburieren bietet eine etablierte Methode, um diese Eigenschaften gezielt zu verbessern.

Das Problem bei der Wärmebehandlung korrosionsbeständiger Stähle liegt in der Gefahr der Chromnitrid- oder -carbidbildung, die zu einem Verlust des im Stahl gelösten Chroms und damit zu einer Abnahme der Korrosionsbeständigkeit führen kann. Die Herausforderung besteht darin, Wärmebehandlungsprozesse einzusetzen, die die gewünschten mechanischen Eigenschaften verbessern, ohne die inhärente Korrosionsbeständigkeit des Materials zu beeinträchtigen. Dies ist besonders kritisch bei Betriebsbedingungen, die korrosive Medien, schlagenden und gleitenden Verschleiß sowie Adhäsionsverschleiß umfassen, und wenn dünne Verschleißschutzschichten auf ferritischen oder austenitischen Stählen aufgrund mangelnder Stützwirkung des Grundwerkstoffs zu einem „Eierschaleneffekt" neigen, der abrasiven Verschleiß fördert.

Die vorliegende Arbeit, basierend auf einer Präsentation auf der „HärtereiPraxis 2025" von Dr. Thomas Waldenmaier (Bosch), stellt eine systematische Betrachtung von Wärmebehandlungsprozessen für korrosionsbeständige Stähle vor. Ziel ist es, die Eigenschaften dieser Stähle so zu modifizieren, dass sie den komplexen Anforderungen moderner Anwendungen gerecht werden. Dabei werden sowohl Nieder- als auch Hochtemperaturverfahren betrachtet und deren Effekte auf Schichtcharakteristik, Härte, Korrosionsbeständigkeit, Schweißeignung, Schwingverschleiß und Eigenspannungen detailliert analysiert. Die gewonnenen Erkenntnisse sollen die Prozessentwicklung in der Industrie unterstützen und die Anwendung wärmebehandelter korrosionsbeständiger Stähle in anspruchsvollen Umgebungen ermöglichen.

1.4.2 Grundlegendes zu Wärmebehandlungsverfahren korrosionsbeständiger Stähle

1.4.2.1 Geeignete Stähle und Mechanismen der Korrosionsbeständigkeit

Alle nichtrostenden Stähle, die nominell über 13 Gew.-% Chrom enthalten, sind grundsätzlich für Nitrier-, Nitrocarburier- und Carburierprozesse geeignet. Dazu gehören Ferrite, Martensite, Maraging-Stähle, Duplex- und Austenitstähle. Typische Beispiele sind 1.4511 (X3CrNb17), 1.4005 (X12CrS13), 1.4568 (X7CrNiAl17-7), 1.4462 (X2CrNiMoN22-5-3) und 1.4301 (X5CrNi18-10).

Die Korrosionsbeständigkeit dieser Stähle wird maßgeblich durch den im Stahl gelösten Chromanteil bestimmt. Ein hoher Chromgehalt führt zu einer hohen Affinität zur Stickstoff- und Kohlenstoffaufnahme während der Wärmebehandlung. Allerdings führt die Bildung von Carbiden und Nitriden – insbesondere Chromnitriden – zu einem Verlust an gelöstem Chrom im Stahl. Dies beeinträchtigt die Passivierfähigkeit und damit die Korrosionsbeständigkeit. Daher ist es entscheidend, dass nach der Wärmebehandlung noch mindestens 12 Gew.-% Chrom in der Lösung verbleiben, um die Korrosionsbeständigkeit zu gewährleisten.

1.4.2.2 Unterscheidung thermochemischer Behandlungen und Beschichtungen

Bei der Modifikation der Oberfläche von korrosionsbeständigen Stählen wird zwischen Beschichtungen (z. B. PVD, CVD) und thermochemischen Behandlungen unterschieden. Thermochemische Behandlungen wie Nitrieren, Nitrocarburieren und Carburieren sind keine klassischen Schichtausbildungen im Sinne einer separaten Schicht auf dem Grundwerkstoff, sondern vielmehr Eindiffusionszonen von Stickstoff und/oder Kohlenstoff in die Bauteiloberfläche. Im Gegensatz zu Beschichtungen treten hier i. d. R. keine Schichthaftungsprobleme auf, da es sich um intrinsische Schichten handelt.

Die thermochemischen Wärmebehandlungen lassen sich in Niedertemperatur- und Hochtemperaturverfahren unterteilen:

Niedertemperaturverfahren: Diese Verfahren arbeiten typischerweise mit N+C-Gehalten von unter 13 Gew.-% in der Diffusionszone. Sie sind bei zu langer Prozessdauer anfällig für die Bildung von Chromnitriden in Abhängigkeit von der Behandlungstemperatur. Darüber hinaus können Oberflächenverunreinigungen die Ergebnisse negativ beeinflussen.

Hochtemperaturverfahren: Hier liegen die N+C-Gehalte typischerweise unter 0,6 Gew.-%. Bei diesen Verfahren können Verunreinigungen auf der Werkstoffoberfläche i. d. R. vernachlässigt werden und die Gefahr der Chromnitridbildung ist gering, da Chromnitrid bei den üblichen Temperaturen der Hochtemperaturverfahren thermodynamisch nicht stabil ist.

1.4.2.3 Prozessparameterbereiche zur Vermeidung von Chromausscheidungen

Die Vermeidung von Chromausscheidungen (insbesondere Chromnitriden wie CrN und Cr_2N) ist entscheidend für den Erhalt der Korrosionsbeständigkeit. Die Bildung dieser Ausscheidungen hängt stark von der Temperatur und der Dauer des Wärmebehandlungsprozesses ab.

Niedertemperaturprozesse: Für Niedertemperaturprozesse ist ein Bereich zwischen 350 °C und 480 °C optimal. Hier ist die Prozessdauer entscheidend, da bei längeren Behandlungszeiten die Kinetik der Chromnitridbildung zu CrN einsetzt und die Korrosionsbeständigkeit negativ beeinflusst wird.

Hochtemperaturprozesse: Für Hochtemperaturprozesse ist ein Temperaturbereich je nach Legierung zwischen 1.000 °C und 1.300 °C geeignet. Innerhalb dieses Bereichs findet zeitunabhängig keine Bildung von Chromnitriden statt. Voraussetzung für eine Vermeidung von Chromnitriden während des gesamten Verfahrens ist aber auch eine ausreichend schnelle Abkühlung oder Abschreckung, um die Ausscheidungskinetik zu unterdrücken.

1.4.3 Prozessführungen

Die Prozessführung der Nitrier-, Nitrocarburier- und Carburierverfahren unterscheidet sich grundlegend zwischen den Nieder- und Hochtemperaturverfahren in Bezug auf Temperatur, Zeit und Gaszusammensetzung bzw. Druck.

1.4.3.1 Vergleich Nieder- mit Hochtemperaturverfahren

Ein systematischer Vergleich der Prozessparameter zeigt die fundamentalen Unterschiede:

Niedertemperaturprozess (NT-Prozess):

- Temperatur: Niedrigere Temperaturen, typischerweise zwischen 350 °C und 480 °C.
- Zeit: Die Prozesszeiten sind deutlich länger, da die Diffusionsgeschwindigkeiten bei niedrigeren Temperaturen geringer sind. Für Austenitstähle sind typischerweise 12 bis 48 h erforderlich, während Verfahren für Ferrite und Martensite deutlich kürzere Behandlungszeiten aufweisen.
- Gaszusammensetzung: Die Atmosphäre ist entscheidend und wird durch eine spezifische Gaszusammensetzung kontrolliert, um die Stickstoff- und/oder Kohlenstoffaufnahme zu steuern.

Hochtemperaturprozess (HT-Prozess):

- Temperatur: Deutlich höhere Temperaturen, typischerweise über 1.000 °C, oft bis 1.200 °C.
- Zeit: Die Zeit bei Maximaltemperatur ist oft deutlich kürzer als bei NT-Prozessen, da die Diffusionsgeschwindigkeit höher ist. Übliche Prozessdauern sind 30 min bis mehrere Stunden.
- Druck (N_2): Der Prozess wird unter kontrolliertem Stickstoffdruck durchgeführt und je nach gewünschter Stickstoffaufnahme abhängig von der Legierung gewählt.
- Abschrecken/Tiefkühlen/Anlassen: Nach der Hochtemperaturbehandlung ist ein schnelles Abkühlen erforderlich, um die Ausscheidung von Chromnitriden zu unterdrücken. Bei ferritischen und martensitischen Stählen kann nach dem Abschrecken ein Tiefkühlen zur Minimierung des Restaustenits und zum Erreichen maximaler Härte durchgeführt werden. Anschließend folgt ein Anlassen zur Einstellung der gewünschten mechanischen Eigenschaften.

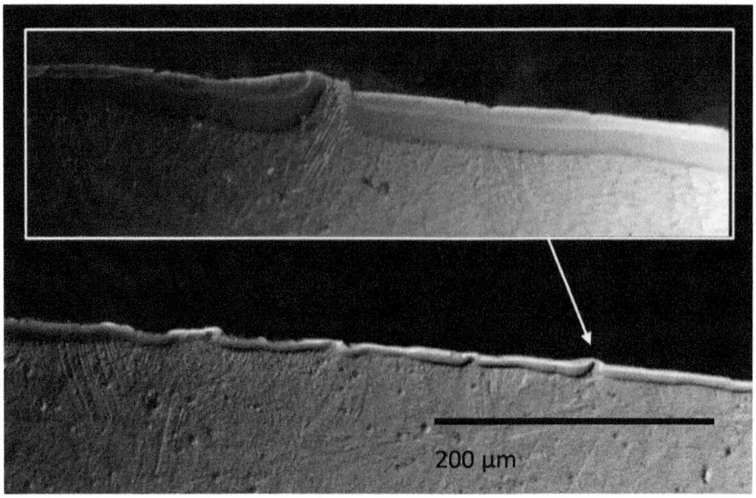

Bild 1: Ungleichmäßig niedertemperaturnitrocarburierter Austenit durch Oberflächenverunreinigung (Quelle: Fa. Expanite)

1.4.3.2 Niedertemperaturverfahren: Depassivierung

Ein entscheidender Schritt bei Niedertemperaturverfahren ist das Depassivieren der Edelstahloberfläche. Ohne diesen Prozessschritt, der oft firmenspezifisches Wissen erfordert (Prozess-Know-how), ist keine erfolgreiche thermochemische Behandlung möglich. Das Depassivieren beinhaltet das Entfernen oder Umgehen der diffusionshemmenden Chromoxidschicht auf der Stahloberfläche. Dies kann entweder vor oder während der eigentlichen thermochemischen Behandlung erfolgen. Eine ungleichmäßige Depassivierung, beispielsweise durch Oberflächenverunreinigungen, kann zu ungleichmäßig nitrierten/nitrocarburierten Schichten führen, s. **Bild 1**.

1.4.4 Schichtcharakterisierung

Die Charakterisierung der erzeugten Schichten ist essenziell, um die Wirksamkeit der Wärmebehandlungsverfahren zu beurteilen.

1.4.4.1 Schichtcharakterisierung: Allgemeine Grundlagen

Sowohl Nieder- als auch Hochtemperaturnitrieren sind thermochemische Verfahren, die auf der Eindiffusion von Stickstoff und/oder Kohlenstoff in die Bauteiloberfläche basieren. Ziel ist die Ausbildung einer Diffusionszone ohne die Bildung einer spröden Verbindungsschicht. Idealerweise werden dabei keine Chromnitride gebildet, um die Korrosionsbeständigkeit zu erhalten.

Niedertemperaturnitriert/-nitrocarburiert (z. B. X5CrNi18-10): Typischerweise bildet sich eine Schicht mit „expanded Austenite“ oder auch eine sogenannte „S-Phase“ mit Stickstoffanteilen von bis zu 13 Gew.-% unterhalb der Oberfläche. Diese Schicht zeichnet sich durch eine hohe Härte und hohe Druckeigenspannungen aus.

Hochtemperaturnitriert (z. B. X12CrS13): Bei Ferriten und Martensiten entsteht ein Stickstoffmartensit mit Stickstoffanteilen von bis zu 0,6 Gew.-%. Hierbei kann die Härte vergleichbar zum Einsatzhärten gesteigert werden. Austenite erfahren durch eine Mischkristallverfestigung ebenfalls eine Härtesteigerung, wenngleich die Härtezunahme vergleichsweise gering ausfällt.

1.4.4.2 Niedertemperaturverfahren - Schichtcharakterisierung

Die Niedertemperaturverfahren bilden eine Randschicht aus, die je nach Verfahren (Nitrieren, Carburieren, Nitrocarburieren) als Einfach- oder Doppelschicht aufgebaut ist, s. **Bild 2**.

Härte und Härtetiefenprofile

Die Randhärte der mit Niedertemperaturverfahren behandelten Stähle liegt zwischen 900 und 1.500 HV0.05, wie anhand von Härteeindrücken im Randbereich eines nitrocarburierten X5CrNi18-10 (1.4301) gezeigt wurde. Die Schichtdicken können dabei zwischen 5 µm und 45 µm variieren.

Vergleichende Härtetiefenprofile von nitrocarburierten, nitrierten und carburierten X5CrNi18-10 (1.4301) zeigen deutliche Unterschiede, s. **Bild 3**:

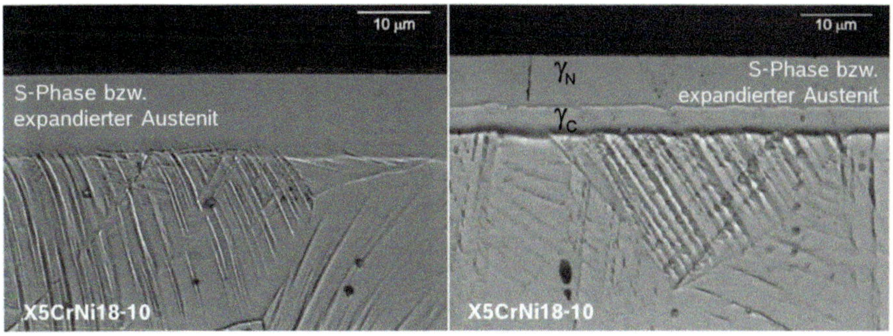

Bild 2: Links: niedertemperaturnitrierter Austenit; rechts: niedertemperaturnitrocarburierter Austenit mit einer Doppelschicht aus einer Schicht mit überwiegender Stickstoffanreicherung (oberflächennah) und darunterliegend einer Schicht mit überwiegender Kohlenstoffanreicherung

Bild 3: Vergleichende Härteprofile

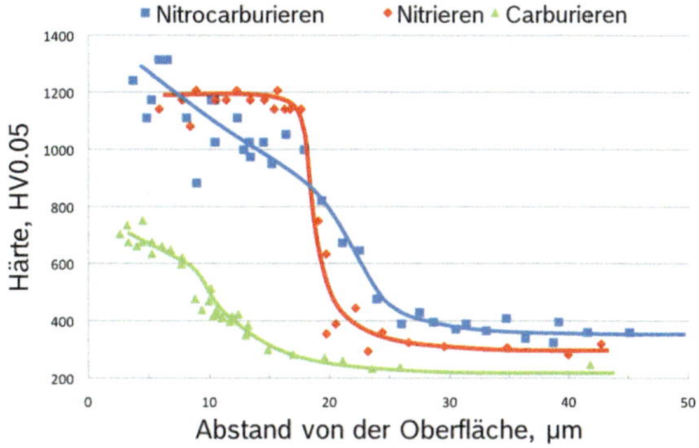

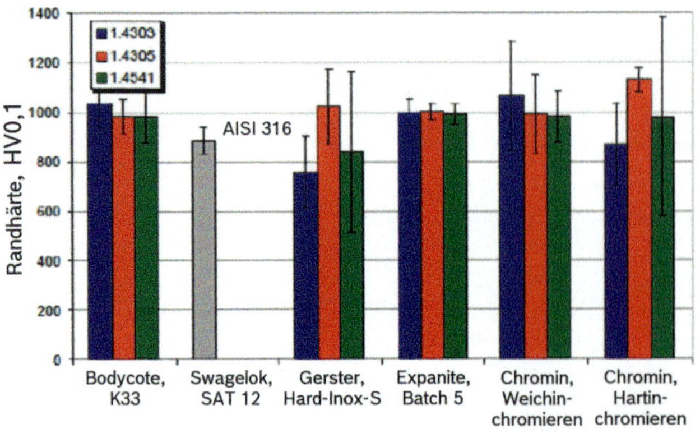

Bild 4: Vergleich erzielbarer Oberflächenhärten

- Nitrocarburieren: Erreicht die höchste Randhärte und zeigt einen moderaten Härteabfall in die Tiefe.
- Nitrieren: Führt zu einem starken Härteabfall.
- Carburieren: Zeigt eine moderate Randhärte und einen moderaten Härteabfall.

Die Oberflächenhärte (HV0.1) variiert nur geringfügig mit der Stahlzusammensetzung der verschiedenen Austenitstähle (z. B. 1.4303, 1.4305, 1.4541, AISI 316), s. **Bild 4**. Dies deutet darauf hin, dass der Prozess die Oberflächeneigenschaften weitgehend unabhängig von Legierungsunterschieden erhöht.

Korrosionsbeständigkeit

Die Korrosionsbeständigkeit der mit Niedertemperaturverfahren behandelten Proben ist entscheidend. Ein Salzsprühnebeltest an AISI 316 nach 400 Stunden zeigt, dass bei Proben, die mittels Niedertemperaturverfahren behandelt wurden, keine signifikante Änderung im Vergleich zum unbehandelten Grundwerkstoff festgestellt wurde. Dies gilt ebenfalls für Proben aus X4CrNi18-12 (1.4303), die mit anderen Niedertemperaturverfahren wärmebehandelt wurden und 30 Tage in E24+ ausgelagert wurden, was den Erhalt der Korrosionsbeständigkeit bei korrekter Durchführung der Verfahren unterstreicht, s. **Bild 5**.

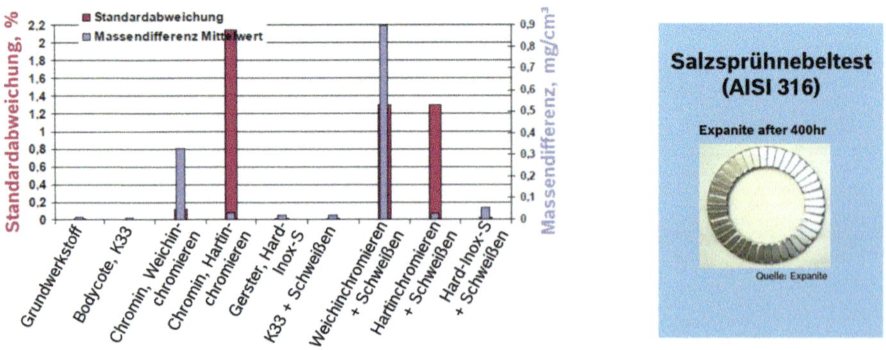

Bild 5: Korrosionsbeständigkeit verschieden wärmebehandelter und beschichteter Proben aus X4CrNi18-12 (1.4303)

Tabelle 1: Übersicht der für die Untersuchung zur Schweißeignung eingestellten Schichtdicken und Stickstoffanteile

Gewichtsanteil N	5–6 Gew.-% (reduced)	8–10 Gew.-% (standard)	13 Gew.-% (high)**B
Schichtdicke			
5–10 µm	Batch 1	Batch 2	Batch 3
15–10 µm	Batch 4	Batch 5*	Batch 6
25–30 µm	Batch 7	Batch 8	Batch 9

Schweißeignung

Die Schweißeignung von niedertemperaturnitrierten Stählen ist ein kritischer Aspekt. Untersuchungen der N-/C-Menge in der Randschicht in Abhängigkeit von der Wärmebehandlung zeigen, dass die Profile der Stickstoff- und Kohlenstoffgehalte durch Niedertemperaturnitrieren eingestellt werden können, s. **Tabelle 1**.

In **Bild 6** sind die resultierenden Stickstoff- und Kohlenstoffmengen dargestellt; die Zahlenwerte kommen durch das Integrieren der Fläche unter den Profilkurven zustande. Die Profile selbst wurden mit einem Glimmentladungsspektroskop ermittelt.

Eine Übersicht der Schweißergebnisse ist aus **Bild 7** zu entnehmen.

Zusammenfassend zeigt die Bewertung der Schweißeignung im Querschliff:

- Es gibt keinen signifikanten Unterschied zwischen einer Schweißgeschwindigkeit von 1 und 4 m/min.
- Der Gesamtgehalt an Stickstoff kann als Maß für die Schweißeignung herangezogen werden.
- Eine systematische Untersuchung des N-Gehalts ermöglicht die Bewertung der Schweißeignung.
- Proben mit einem zu hohen Stickstoffgehalt (Batches 6 bis 9) sind hinsichtlich der Schweißeignung als kritisch einzustufen.

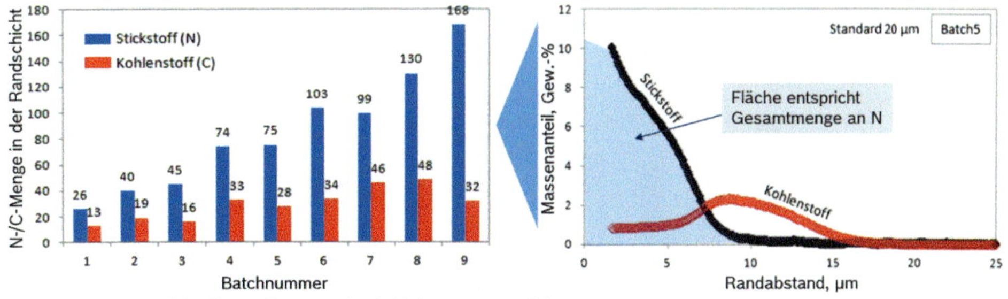

Eingestellte Stickstoff-/Kohlenstoffmengen durch Niedertemperaturnitrieren

Bild 6: Links: eingestellte Stickstoff- und Kohlenstoffmengen; rechts: exemplarisches Ergebnis zum Ermitteln der integralen Fläche unter den Profilkurven mittels Glimmentladungsspektroskopie

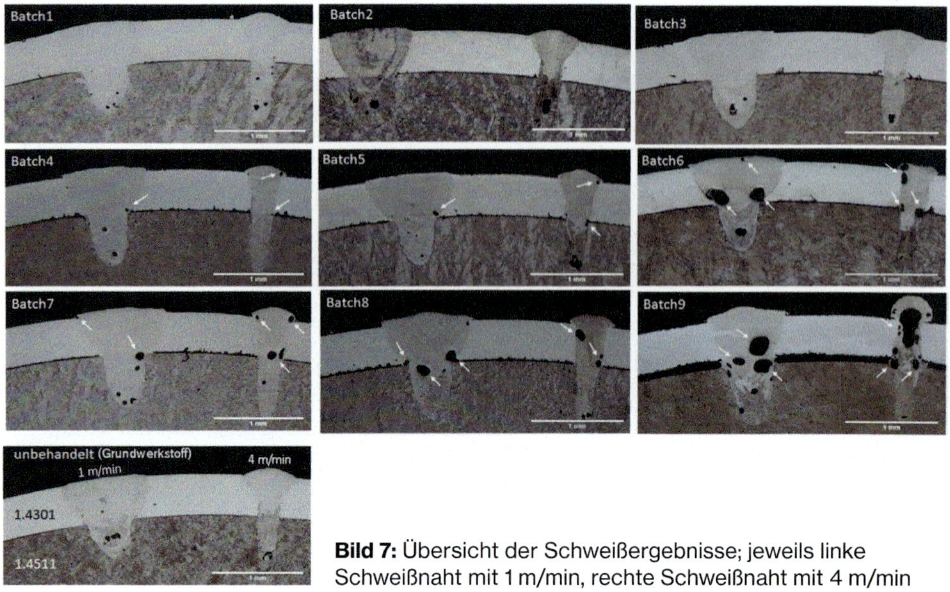

Bild 7: Übersicht der Schweißergebnisse; jeweils linke Schweißnaht mit 1 m/min, rechte Schweißnaht mit 4 m/min

Schwingverschleiß

Neben der Schweißeignung wurden mit den Proben der verschiedenen Batches ebenfalls Schwingverschleißversuche unter Kraftstoff (Winterdiesel) mit einer Si3N4-Kugel (Ø 12,7 mm, keine Adhäsion) durchgeführt. Die Ergebnisse an X5CrNi18-10 (1.4301) mit einer Oberfläche von *Rz* = 4 und der geringsten Schichtdicke (Batch 1 mit 9 µm) zeigen, dass im Rahmen der Messgenauigkeit keine eindeutige Differenzierung der Schichten möglich ist. Dies deutet darauf hin, dass sich unabhängig vom Stickstoffanteil oder der Schichtdicke der Verschleiß unter diesen spezifischen Bedingungen nicht signifikant anders darstellt, solange die Verschleißtiefe die Schichtdicke nicht überschreitet, s. **Bild 8**.

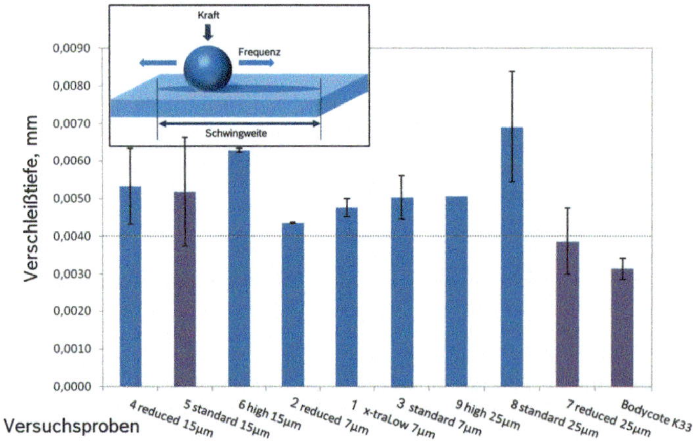

Bild 8: Ergebnisse zur Verschleißuntersuchung

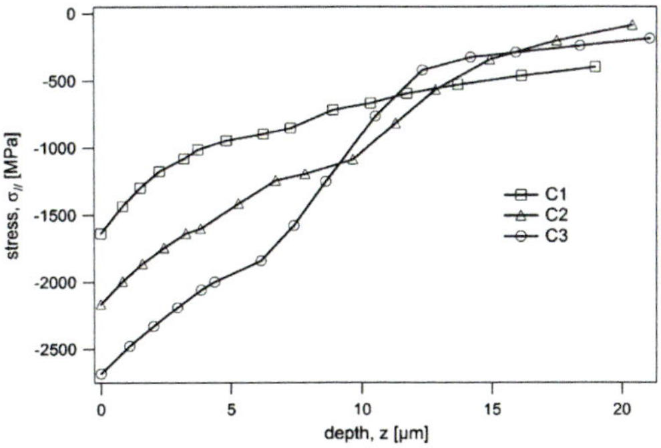

Versuch	Dauer, h	CO/H_2-Anteil, %	Temperatur, °C
C1	2	10/90	520
C2	5	20/80	515
C3	2,5	90/10	520

Element	Cr	Ni	Mo	Mn	Si	Fe
Ma.-%	17,9	13,4	2,4	1,7	0,7	63,9

Bild 9: Exemplarische Eigenspannungsverläufe niedertemperaturnitrocarburierter Proben aus AISI 316 [1]

Bild 10: Ergebnisse eines Umlaufbiegeversuchs mit Proben aus AISI 316L (1.4404) [2]

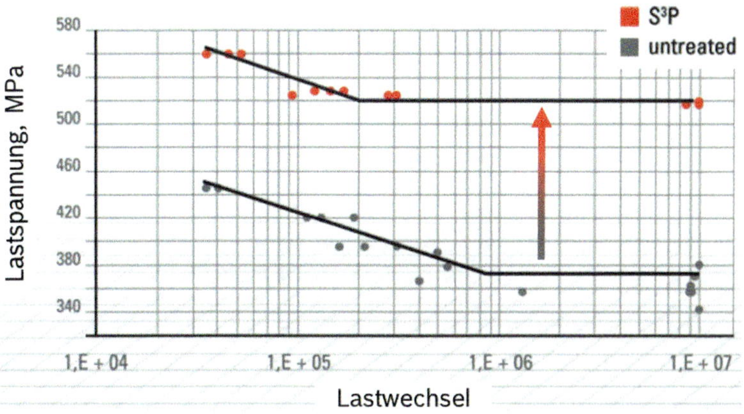

Eigenspannungen und Ermüdungsbeständigkeit

Niedertemperaturverfahren erzeugen hohe Druckeigenspannungen im Bereich von –1.600 MPa bis –2.800 MPa; beispielhaft für Eigenspannungsverläufe sind Ergebnisse in **Bild 9** für den Werkstoff AISI 316 dargestellt [1].

Diese hohen Druckeigenspannungen sind dafür bekannt, die Ermüdungsbeständigkeit von Bauteilen signifikant zu verbessern. Entsprechend konnte eine ca. 40 % erhöhte Ermüdungsfestigkeit bei Proben aus AISI 316L (1.4404), hergestellt in einem Umlaufbiegeversuch, nachgewiesen werden, was den positiven Einfluss von Druckeigenspannungen auf die Ermüdungsfestigkeit belegt, s. **Bild 10** [2].

1.4.4.3 Hochtemperaturverfahren - Schichtcharakterisierung Austenit

Stickstofflöslichkeit

Die Stickstofflöslichkeit in X5CrNi18-10 bei Hochtemperaturverfahren ist druck- und temperaturabhängig. Bei einem Prozessdruck von 1 bar N_2 und 1.150 °C sind maximal 0,6 Gew.-%

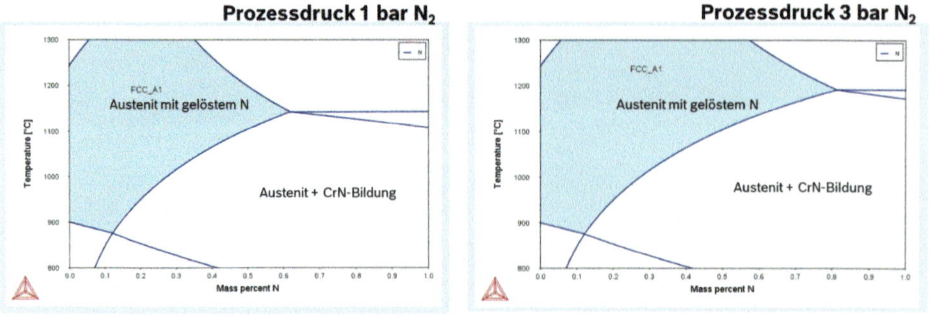

Max. 0,6 Gew.-% N löslich (bei 1150 °C) Max. 0,8 Gew.-% N löslich (bei 1180 °C)

Bild 11: Phasendiagramme für X5CrNi18-10 (1.4301), berechnet mit Thermo-Calc

Stickstoff löslich. Bei einem erhöhten Prozessdruck von 3 bar N_2 und 1.180 °C steigt die maximale Stickstofflöslichkeit auf 0,8 Gew.-% an. Die Phasendiagramme visualisieren diese Abhängigkeiten, s. **Bild 11**.

Härtetiefenprofil

Berechnete Verläufe des Stickstoffanteils für 1.180 °C und 1 bar N_2 zeigen, wie der Stickstoffgehalt mit zunehmender Behandlungsdauer tiefer in den Werkstoff eindringt. Abhängig von der gewünschten Diffusionstiefe kann eine geeignete Behandlungsdauer festgelegt werden, s. **Bild 12**.

Die Härtetiefenprofile einer teilaufgestickten Probe (2,82 mm Dicke, 1.4301) bei 1.180 °C, 1 bar und 6 h Prozessdauer zeigen, dass die Randhärte gegenüber dem Kern der Probe von ca. 180 HV0.05 auf ca. 300 HV0.05 erhöht werden konnte, s. **Bild 13**. Dieser Wert entspricht der üblicherweise maximal erreichbaren Randhärte für einen hochtemperaturnitrierten Austeniten.

Fließkurve

Die Fließkurve von X5CrNi18-10 (1.4301) nach Hochtemperaturnitrieren zeigt eine signifikante Erhöhung der Festigkeit im Vergleich zum geglühten Zustand. Die Zugversuche nach

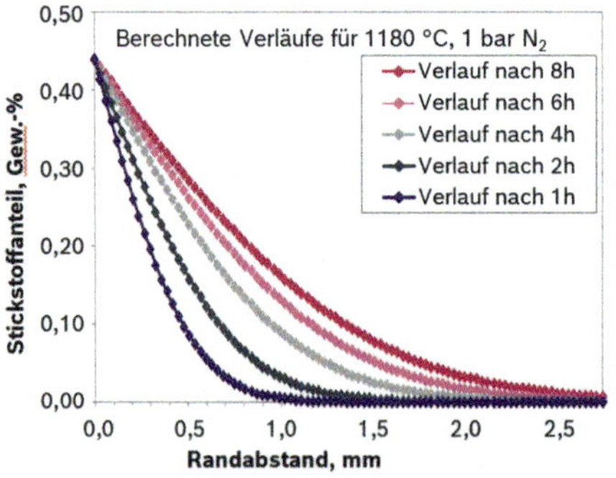

Bild 12: Berechnete Profile von Stickstoffanteilen bei hochtemperaturnitriertem 1.4301 bei 1.180 °C und 1 bar Stickstoffdruck

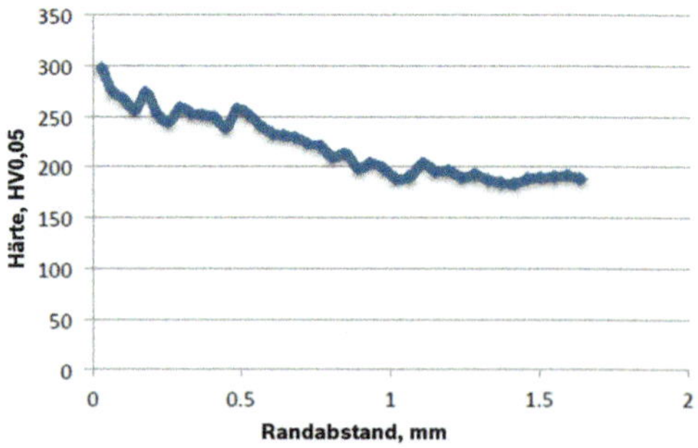

Bild 13: Härteprofil an einer teilaufgestickten Probe bei 1.180 °C und 1 bar Stickstoff für 6 h, mit 2,82 mm Dicke aus X5CrNi18-10 (1.4301)

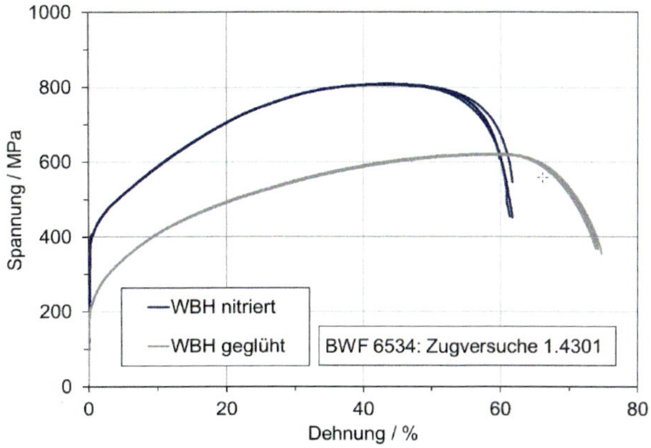

Bild 14: Fließkurven durchgestickter Flachzugproben mit einer Dicke von 0,8 mm nach einem Hochtemperaturnitrieren bei 1.180 °C, 1 bar Stickstoff und 6 h

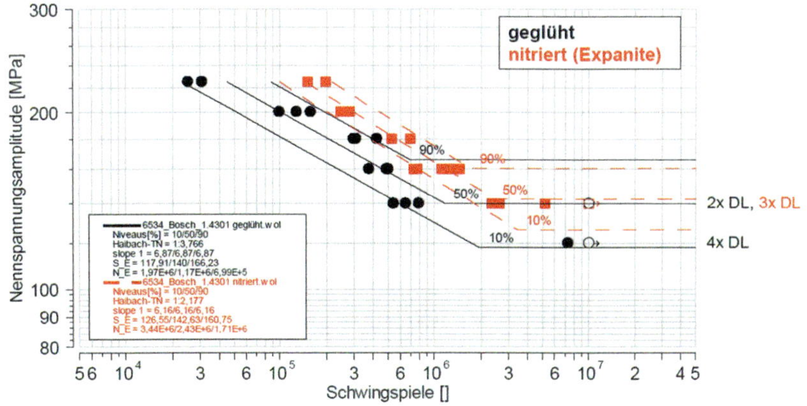

Bild 15: Ergebnisse einer vergleichenden Schwingfestigkeitsuntersuchung zwischen hochtemperaturnitrierten Proben und der Referenz (geglühter Ausgangszustand)

BWF 6534 belegen, dass die nitrierte Probe eine höhere Streckgrenze und Zugfestigkeit aufweist, während die Bruchdehnung entsprechend abnimmt, s. **Bild 14**.

Schwingfestigkeit

Trotz der verbesserten mechanischen Eigenschaften, nachgewiesen durch die Zugversuche, zeigt die Schwingfestigkeit keine nennenswerte Verbesserung. Dies kann auf die Grobkornbildung zurückgeführt werden, die bei zu hohen Temperaturen auftreten kann und die Vorteile der Verfestigung durch Stickstoff teilweise kompensiert, s. **Bild 15**.

1.4.4.4 Schichtcharakterisierung: Martensit (Hochtemperaturverfahren)

Bei Hochtemperaturverfahren an martensitischen Stählen, wie X12CrS13 (1.4005), zeigt sich die Bildung einer martensitischen Randschicht, ähnlich dem Einsatzhärten. Mikroskopische Aufnahmen zeigen eine aufgestickte Randschicht mit hellen Feldern (Ferrit) und

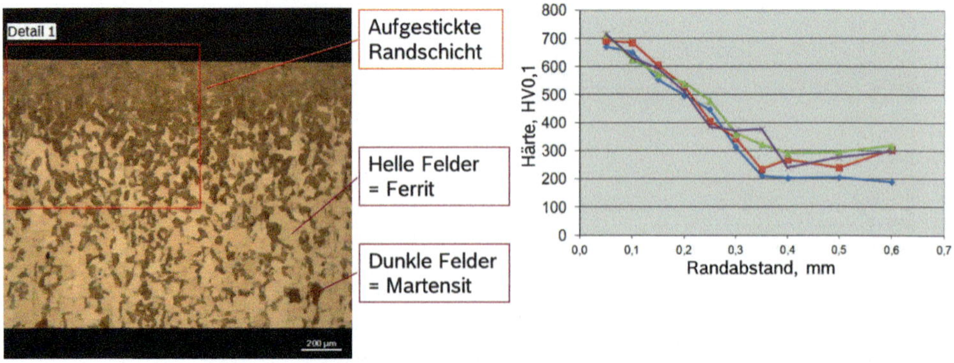

Bild 16: Hochtemperaturnitrierter X12CrS13 (1.4005); links: lichtmikroskopische Gefügeaufnahme; rechts: Härteprofil

dunklen Feldern (Martensit). Das Härtetiefenprofil dieser Stähle zeigt eine deutlich erhöhte Härte an der Oberfläche (ca. 700 HV0.1), die mit zunehmendem Randabstand abnimmt, s. **Bild 16**.

1.4.5 Vergleich und Diskussion der Verfahren

Die Wahl zwischen Niedertemperatur- und Hochtemperaturverfahren hängt von den spezifischen Anforderungen an den Stahl und das Bauteil ab. Eine zusammenfassende Übersicht soll die spezifischen Vor- und Nachteile verdeutlichen.

1.4.5.1 Niedertemperaturverfahren für austenitische Stähle

Vorteile:

- Erhalt der Korrosionsbeständigkeit: Durch die Vermeidung von Chromnitridbildung bei korrekter Prozessführung
- Vernachlässigbare Form-, Farb- und Maßänderung: Geringere thermische Belastung
- Erhöhung der Verschleißbeständigkeit und Ermüdungsbeständigkeit: Durch die Bildung von expandiertem Austenit und hohe Druckeigenspannungen
- Kein Abplatzen der Randschicht: Intrinsische Schichtbildung
- Erhalt der Kaltverfestigung: Für bereits kaltverfestigte Bauteile von Vorteil.

Nachteile/Risiken:

- Probleme beim Schweißen: Verminderte Schweißeignung durch erhöhten N/C-Gehalt in der Randschicht
- Inhomogenitäten in der Schichtdicke: Bei nicht ausreichender Depassivierung vor oder während der Wärmebehandlung
- Relativ teuer durch lange Prozesszeiten
- Bildung unerwünschter Phasen: Bei falscher Prozessführung (z. B. zu lange Prozessdauer bei unzureichender Temperaturkontrolle) kann es zu Chromnitridbildung kom-

men, die zur Versprödung der Randschicht und Verlust der Korrosionsbeständigkeit führt.

1.4.5.2 Hochtemperaturverfahren für austenitische Stähle

Vorteile:

- Inhomogenitäten in der Schichtdicke: Durch die hohen Behandlungstemperaturen wird die Chromoxidschicht der korrosionsbeständigen Stähle automatisch abgebaut und andere Passivierungen sind kaum bekannt.
- Bildung CrN: Thermodynamisch auf Behandlungstemperatur nicht möglich; bei ausreichend schneller Abkühlung kann die Chromnitridkinetik unterdrückt werden.
- Schichtdicke: 50 bis 1.000 µm; deutlich höhere Schichtdicken als bei Niedertemperaturverfahren sind möglich.
- Schweißeignung: Ist nur geringfügig vermindert.

Nachteile/Risiken:

- Härte: < 300 HV1, deutlich geringer
- Kornwachstum: Kornvergröberung durch zu hoch gewählte Behandlungstemperatur
- Kaltverfestigung: Wird abgebaut.

1.4.5.3 Hochtemperaturverfahren für ferritisch/martensitische Stähle

Vorteile:

- Inhomogenitäten in der Schichtdicke: Geringes Risiko, wie bei den Austeniten.
- Bildung CrN: Wie bei Austeniten; für die martensitische Härtung ist eine ausreichend schnelle Abkühlung ohnehin unerlässlich.
- Schichtdicke: Wie bei den Austeniten zwischen 50 und 1.000 µm.
- Schweißeignung: Ist nur geringfügig vermindert.
- Härte: Randhärten bis ca. 700 HV1 erreichbar.

Nachteile/Risiken:

- Kornwachstum: Kornvergröberung durch zu hoch gewählte Behandlungstemperatur.
- Kaltverfestigung: Wird abgebaut.

1.4.5.4 Allgemeine Diskussion

Die Ergebnisse zeigen, dass Niedertemperaturverfahren eine vielversprechende Methode darstellen, um die Oberflächeneigenschaften korrosionsbeständiger Stähle zu verbessern, insbesondere im Hinblick auf Härte, Verschleiß- und Ermüdungsbeständigkeit, ohne die Korrosionsbeständigkeit zu opfern. Die Steuerung der Chromnitridbildung durch präzise Prozessparameter (Temperatur, Zeit) ist hierbei von höchster Relevanz. Die Notwendigkeit der Depassivierung als kritischer Prozessschritt verdeutlicht das erforderliche technologische Know-how.

Hochtemperaturverfahren bieten größere Schichtdicken und können ebenfalls eine hohe Festigkeit erzeugen (wie durch Fließkurven gezeigt), sind aber anfällig für Kornwachstum und eine ggf. vorhandene vorhergehende Kaltverfestigung wird abgebaut. Die Schweißeignung ist bei den Hochtemperaturverfahren nur gering vermindert.

1.4.5.5 Ausblick auf zukünftige Forschung

Zukünftige Forschungsarbeiten sollten sich auf die weitere Optimierung der Prozessparameter konzentrieren, um die Balance zwischen mechanischen Eigenschaften und Korrosionsbeständigkeit noch besser einzustellen. Dies könnte die Entwicklung neuer Analysemethoden zur Echtzeitüberwachung der Chromnitridbildung umfassen. Weiterhin sind detailliertere Untersuchungen zur Langzeitstabilität der modifizierten Schichten unter realen Betriebsbedingungen, insbesondere im Hinblick auf Schwingverschleiß unter verschiedenen Belastungen und Medien, erforderlich. Die Verbesserung der Schweißeignung für höher stickstofflegierte Oberflächenbereiche stellt ebenfalls einen wichtigen Forschungsbereich dar.

1.4.6 Fazit

Die Wärmebehandlung von korrosionsbeständigen Stählen durch Nitrieren, Nitrocarburieren und Carburieren ist eine essenzielle Technologie zur Verbesserung der Oberflächeneigenschaften und zur Erfüllung anspruchsvoller industrieller Anforderungen. Die vorliegende Arbeit hat gezeigt, dass sowohl Niedertemperatur- als auch Hochtemperaturverfahren spezifische Vorteile bieten, die je nach Anwendung sorgfältig abgewogen werden müssen. Niedertemperaturverfahren zeichnen sich durch den möglichen Erhalt der Korrosionsbeständigkeit, signifikante Härtesteigerung, verbesserte Verschleiß- und Ermüdungsbeständigkeit sowie minimale Form- und Maßänderungen aus. Hochtemperaturverfahren ermöglichen größere Schichtdicken und eine erhöhte Festigkeit, gehen aber oft mit Kornwachstum und einem Abbau der Kaltverfestigung einher. Ein kritisches Element bei allen Verfahren ist die präzise Prozessführung zur Vermeidung von Chromnitridbildung. Die hier präsentierten Erkenntnisse tragen dazu bei, das Verständnis dieser komplexen Prozesse zu vertiefen und die Entwicklung maßgeschneiderter Lösungen für korrosionsbeständige Stähle voranzutreiben, um deren Einsatzspektrum in der modernen Technik weiter zu erweitern.

Literatur

[1] Christiansen, T. L., Somers, M. A. J.: Stress and Composition of Carbon Stabilized Expanded Austenite on Stainless Steel. In: Metallurgical And Materials Transactions A, 40a (2009) 8, S. 1.791–1.798

[2] Gerritsen, S., et al.: S3P – Innovative Surface Treatment to Increase the Wear Resistance of Stainless Steel Components. In: HTM J. Heat Treatm. Mat., 77 (2022) 6, S. 425–436

2. Nachhaltigkeit und Energieeffizienz

2.1 Energieeffizienz: Gesetzliche Pflichten und wirtschaftliche Anreize

Jörg Scheyhing

Der effiziente Umgang mit Energie sollte aus betriebswirtschaftlicher Sicht selbstverständlich sein. Dennoch vertraut der Gesetzgeber hierbei nicht allein auf den Eigenantrieb der Unternehmen. Er verpflichtet Betriebe dazu, gesetzliche Mindestanforderungen, Berichtspflichten und ökologische Gegenleistungen zu erfüllen. Immerhin wird die Umsetzung von Maßnahmen im Gegenzug häufig durch Förderprogramme unterstützt. Dieser Beitrag soll einen aktuellen Überblick ermöglichen.

2.1.1 Energiemanagementsysteme: Kür oder Pflicht?

Energiemanagementsysteme nach ISO 50001 (**Bild 1**) oder EMAS-Systeme bieten einen systematischen und wiederkehrenden Ansatz für den Umgang mit Energie im eigenen Unternehmen. Die mittlerweile recht weite Verbreitung insbesondere der ISO 50001 ist in Deutschland der Tatsache geschuldet, dass der Gesetzgeber bereits ab den 2000er-Jahren begonnen hatte, Entlastungen im Abgaben- und Umlagenbereich auf den Energiebezug an die Gegenleistung des Betriebs in Form eines Energiemanagementsystems zu knüpfen. Exemplarisch sei hier die frühere Entlastung von der EEG-Umlage oder der steuerliche Spitzenausgleich für stromkostenintensive Unternehmen erwähnt. Inzwischen reichen die gesetzlichen Regelungen jedoch weit über dieses Prinzip von „Zuckerbrot und Peitsche" hinaus. So verpflichtet das Ende 2023 in Kraft getretene Energieeffizienzgesetz (EnEfG) Betriebe mit einem gemittelten jährlichen Endenergieverbrauch von mindestens 7,5 Mio. kWh dazu, bis spätestens zum 18. Juli 2025 ein Energiemanagementsystem nach ISO 50001 oder ein EMAS-System einzuführen. Dies ist eine einseitige gesetzliche Verpflichtung, aus der keine Gegenleistungen in Form von Entlastungen für das betrachtete

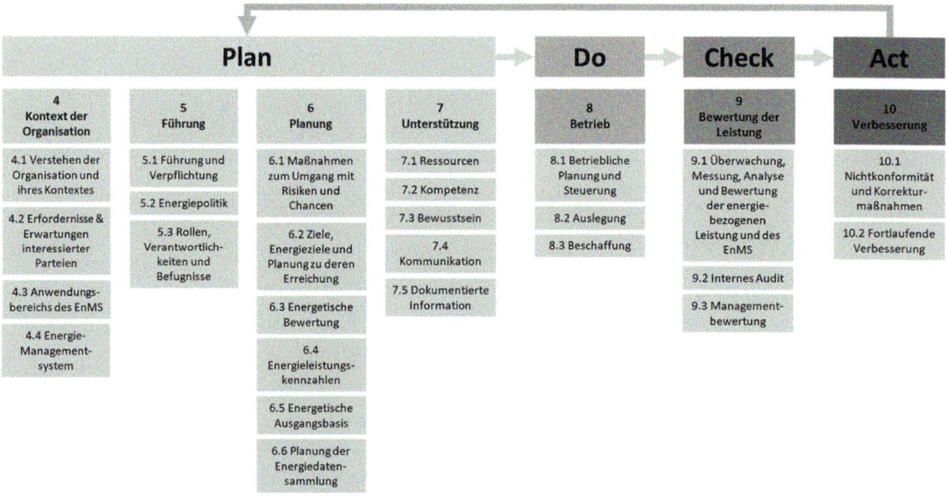

Bild 1: Normkapitel der ISO 50001

Unternehmen resultieren. Sollte ein Unternehmen diese Vorgabe nicht beachten, drohen Bußgelder von bis zu 100.000 €, die das zuständige Bundesamt für Wirtschaft und Ausfuhrkontrolle (BAFA) im Rahmen von Stichprobenverfahren aussprechen kann.

Aber auch für kleinere Energieverbraucher kann sich ein solches Energiemanagementsystem als sinnvolle freiwillige Investition erweisen. So besteht für Unternehmen bestimmter Branchen die Möglichkeit, sich Umlagen auf den Strombezugspreis gegen den Nachweis des Betriebs eines Energiemanagementsystems sowie von ökologischen Gegenleistungen auf Antrag reduzieren zu lassen. Dies gilt beispielsweise im Rahmen des Energiefinanzierungsgesetzes (EnFG) für die KWKG- und Offshorenetz-Umlage, beim Brennstoffemissionshandelsgesetz (BEHG) für die CO_2-Abgabe auf im Betrieb eingesetzte Energieerzeugnisse sowie gegebenenfalls für die sogenannte Strompreiskompensation bestimmter begünstigter Anlagen.

Natürlich gibt es für viele Betriebe auch ohne gesetzliche Verpflichtung gute Gründe für den Betrieb eines Energiemanagementsystems. So gibt es zunehmende Anforderungen von der Kundenseite und die Chance auf einen möglichst kosteneffizienten Umgang mit Energie im Betrieb wird selbstredend gesteigert.

Die weitere Entwicklung der Gesetzgebung sollte jedoch aufmerksam verfolgt werden: So wurde im Koalitionsvertrag der neuen Bundesregierung die Absicht aufgenommen, die Verpflichtungen über das Energiemanagement erst für deutlich größere Energieverbraucher vorzusehen. Es könnte hier also zukünftig zur Reduzierung der Anforderungen auf das europäische Mindestmaß kommen.

2.1.2 Effizienzmaßnahmen: Von der Berichts- zur Umsetzungspflicht

Im Zuge verstärkter Bemühungen zur Steigerung der Energieeffizienz bzw. des Klimaschutzes hat der Gesetzgeber in den vergangenen Jahren sukzessive neue Berichts- bzw. Umsetzungspflichten in verschiedenen energierelevanten Gesetzen eingeführt.

Einen Vorgeschmack auf folgende Regelungen bot bereits im Jahr 2022 die Energiesicherungsmittelfristmaßnahmenverordnung (EnSimiMaV), die im Zuge der Energiepreiskrise von größeren Verbrauchern (ab 10 Mio. kWh/a Energiebedarf) die nachweisliche Umsetzung ausreichender wirtschaftlicher Maßnahmen forderte. Diese Verordnung hatte aber nur eine begrenzte zeitliche Gültigkeit und ist mittlerweile wieder außer Kraft getreten. Bemerkenswert an ihr war jedoch der Ansatz, dass Unternehmen sogar zu Investitionen in Effizienzmaßnahmen gezwungen werden konnten.

Umfangreiche zusätzliche Verpflichtungen sind über das im November 2023 in Kraft getretene Energieeffizienzgesetz (EnEfG) gefolgt. Neben den bereits oben ausgeführten Pflichten zum Betrieb eines geeigneten Energiemanagementsystems für größere Energieverbraucher sind bereits Unternehmen mit einem jährlichen Energieeinsatz von mindestens 2,5 Mio. kWh dazu verpflichtet, Umsetzungspläne zu erstellen, diese von einer prüfungsberechtigten Stelle überprüfen zu lassen und letztlich geeignet zu veröffentlichen (innerhalb eines Zeitraums von 36 Monaten).

Weitere Transparenz- und Umsetzungspflichten bestehen für Unternehmen, die Sonderregelungen zur Reduzierung von Energienebenkosten in Anspruch nehmen wollen. In diesem Fall wird von der Erbringung ökologischer Gegenleistungen (**Bild 2**) durch die im Rahmen der beihilferechtlich relevanten (Entlastungs-)Sachverhalte gesprochen. Entsprechende Regelungen sind im Kontext der besonderen Ausgleichsregelung zum Energiefinanzierungsgesetz (EnFG) zur Reduzierung der KWKG- und Offshorenetzumlage, aber auch der Reduzierung der CO_2-Abgabe im Rahmen der Carbone-Leakage-Verordnung (BECV)

Art	Beihilfen/gesetzliche Pflichten	Investitionshöhe	Def. Wirtschaftlichkeit	Stichtag für das EnMS
Ökologische Gegenleistung	BesAR (EnFG)	Investitionssumme (Antragsjahr) ≥ 50% Begrenzungsbetrag (Antragsjahr + 1)	Pos. Kapitalwert nach **max. 60%** der Nutzungsdauer	30.06. im Antragsjahr
Ökologische Gegenleistung	BECV	Investitionssumme (Abrechnungsjahr) ≥ 50% Beihilfesumme (Abrechnungsjahr - 1)	Pos. Kapitalwert nach **max. 60%** der Nutzungsdauer	Gültiges System zum Zeitpunkt der Antragstellung
Ökologische Gegenleistung	SPK	Investitionssumme (Abrechnungsjahr) ≥ 100% Beihilfesumme (Abrechnungsjahr)	Amortisationsdauer ≤ **3 Jahren** oder nach § 11 Abs. 1 BECV	Gültiges System zum Zeitpunkt der Antragstellung
Gesetzliche Verpflichtung	EnSimiMaV	- (alle wirtschaftlichen Maßnahmen)	Pos. Kapitalwert nach max. **20%** der Nutzungsdauer	-
Gesetzliche Verpflichtung	EnEfG	- (keine Umsetzungs-verpflichtungen)	Pos. Kapitalwert nach max. **50%** der Nutzungsdauer	18.07.2025 bzw. 20 Monate nach Erlangung Status

Bild 2: Ökologische Gegenleistungen und ihre Wirtschaftlichkeit

sowie der Strompreiskompensation zu beachten. Im Grundsatz muss stets der Nachweis erbracht werden, dass wirtschaftliche Maßnahmen in ausreichendem Umfang erbracht bzw. umgesetzt werden. Was wirtschaftlich ist, bestimmt sich auf Basis der DIN 17463 (sogenannte Valeri-Norm über die Wirtschaftlichkeit von Energieeffizienzmaßnahmen), die je nach Sachverhalt auf unterschiedliche Zeiträume bezogen auf die Nutzungsdauer der betrachteten Anlage anzuwenden ist. Aufgrund der Komplexität kann betroffenen Betrieben geraten werden, ein geeignetes Maßnahmenmanagement aufzubauen und laufend fortzuführen.

2.1.3 Abwärmepotenziale

Die Grundidee des Gesetzgebers leuchtet zunächst ein: Durch Berichtspflichten für Unternehmen über bestehende Abwärmepotenziale und durch auf kommunaler Ebene ermittelte Wärmepotenziale im Fernwärmenetz (kommunale Wärmeplanung) sollen Quellen und Senken mit der Zielsetzung Wärmewende zusammengeführt werden.

Unternehmen mit einem Energieverbrauch von mindestens 2,5 Mio. kWh/a sind über die Regelungen des Energieeffizienzgesetzes (EnEfG) dazu verpflichtet, zu prüfen, ob Abwärmepotenziale im Sinn des EnEfG vorliegen. Wenn dies zutreffen sollte, sind diese jährlich über ein hierfür vorgesehenes Online-Meldeportal bei der Bundesstelle für Energieeffizienz zu melden. Durch Präzisierungen in den Ausführungsbestimmungen (Merkblatt „Abwärmeportal" der Bundesstelle für Energieeffizienz) bestehen hier glücklicherweise einige Erleichterungen, die dazu führen können, dass betroffene Unternehmen ggf. doch keine Meldung abzugeben haben (**Bild 3**).

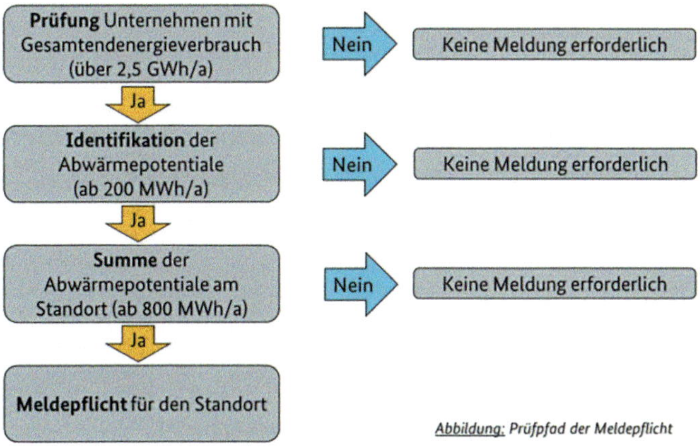

Bild 3: Entscheidungspfad Abwärmemeldung (Quelle: BAFA)

2.1.4 Das Energiedienstleistungsgesetz (EDL-G)

Bereits im Jahr 2015 trat das Energiedienstleistungsgesetz (EDL-G) in Kraft, das von sogenannten Nicht-KMU (also Unternehmen, die größer als die KMU-Definition mit 250 Mitarbeitern bzw. 50 Mio. € Jahresumsatz sind) alle vier Jahre die Durchführung eines Energieaudits nach DIN EN 16247-1 verlangt (**Bild 4**). Unternehmen, die eine ISO 50001 oder ein EMAS-System betreiben, sind von dieser Auditpflicht befreit.

Inhaltlich besteht ein EDL-G-Audit aus den drei Stufen der Erfassung der eingesetzten Energieträger: der Analyse des Energieverbrauchs auf Verbraucherebene sowie der Identifikation und der Bewertung von Effizienzmaßnahmen.

Die vormalige Ampelkoalition hatte bereits für 2024 eine Novelle des EDL-G geplant, die bis dato noch nicht umgesetzt wurde. Demnach war geplant, die Pflicht zur regelmäßigen

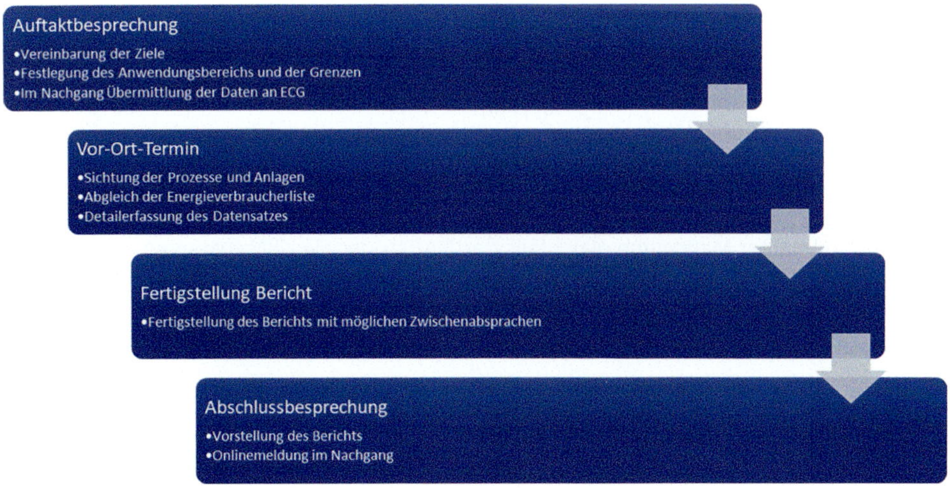

Bild 4: Ablauf eines Energieaudits nach DIN EN 16247-1

Durchführung von Energieaudits nicht mehr am Nicht-KMU-Status, sondern an einem Energieeinsatz von mindestens 2,77 Mio. kWh/a festzumachen. Derzeit kann nicht abgesehen werden, ob und gegebenenfalls wann diese Gesetzesänderung greifen wird.

2.1.5 Förderprogramme zur Steigerung der Energieeffizienz

Eine Vielzahl von staatlichen Stellen auf unterschiedlichen Verwaltungsebenen bietet Fördermöglichkeiten für Vorhaben der Steigerung der Energieeffizienz an. Auf Bundesebene ist das Programm „Bundesförderung für Energie- und Ressourceneffizienz in der Wirtschaft“ das mächtigste Instrument, das sich speziell an Unternehmen richtet. Soll das Vorhaben aus eigenen Mitteln finanziert werden, ist der Antrag auf Förderung beim Bundesamt für Wirtschaft und Ausfuhrkontrolle (BAFA) zu stellen, andernfalls über die Hausbank bei der Kreditanstalt für Wiederaufbau (KfW). Hinsichtlich des Förderrahmens hat der Antragsteller die Wahl aus de-minimis-Zuschüssen (bis zu 300.000 € Zuschuss binnen dreier Steuerjahre) bezogen auf die Investitionskosten oder sogenannter AGVO-Förderung zur Bezuschussung der Investitionsmehrkosten zur Steigerung der Energieeffizienz. Das Programm besteht heute aus sechs Modulen. Über Modul 1 wird die Anschaffung effizienter Querschnittstechnologien gefördert, beispielsweise energieeffizienter Antriebe und Pumpen, Ventilatoren, Drucklufterzeuger sowie Wärmetauscher. Über Modul 2 werden Vorhaben der Prozesswärmebereitstellung aus erneuerbaren Energiequellen (u. a. Solarkollektoren, Umweltwärme, Geothermie) bezuschusst. Das Modul 3 sieht eine Bezuschussung von Investitionen in Messtechnik und zugehörige Software für den Fall des Betriebs eines Energiemanagementsystems vor. Über das Modul 4 lassen sich grundsätzlich sämtliche Vorhaben zur Effizienzsteigerung von Produktionsprozessen fördern. Hierbei wird nach der Basisförderung für bestimmte Standardanlagen (E-Stapler, Spritzgießmaschinen, Lackierkabinen, Laserschneidanlagen usw.) und der Premiumförderung, über die technologieoffen sämtliche Vorhaben, die im Umfang der Basisförderung nicht erwähnt werden, bezuschusst werden können. Aufgrund des spezifischen Charakters der Vorhaben im Bereich der Premiumförderung setzt diese die Erstellung eines geregelten Einsparkonzepts voraus. Modul 5 umfasst die Förderung der Erstellung von Transformationsplänen für die Transformation des eigenen Unternehmens bis hin zur Klimaneutralität. Das Modul 6 wendet sich an kleine Unternehmen, die Prozesse elektrifizieren möchten.

Die Fördersätze in den Modulen 1 bis 6 variieren nach Unternehmensgröße, Energie- bzw. Klimaeffizienz und betragen zwischen 20 und 60 % bezogen auf den geplanten Invest bzw. die Investitionsmehrkosten. Die Förderung beschränkt sich je nach Modul teilweise auf kleine bzw. mittlere Unternehmen; je größer die Unternehmen sind, umso niedriger fallen tendenziell die Förderquoten aus.

Aber natürlich darf für Investitionen im Gebäudebereich die „Bundesförderung für effiziente Gebäude“ (BEG) nicht unerwähnt bleiben. Auch für den Nichtwohngebäudebereich werden in diesem Programm zahlreiche Effizienzinvestitionen bezuschusst. Für dieses Programm ist je nach geplanter Maßnahme gleichfalls das BAFA oder die KfW zuständig.

2.2 Dekarbonisierung in Härterei und Industrieofenbau – technische und wirtschaftlich Aspekte

Dr. Klaus Buchner

Unter dem Gesichtspunkt veränderter ökologischer und ökonomischer Rahmenbedingungen gewinnt die Thematik CO_2-Neutralität zunehmend an Bedeutung – auch der Sektor der Wärmebehandlung muss sich diesbezüglich seiner Verantwortung stellen. Hierbei gilt es, sowohl den Wärmebehandlungsprozess und das Anlagenkonzept kritisch zu hinterfragen als auch konkrete Antworten im Hinblick auf die Anlagentechnik zu geben.

2.2.1 Einleitung

Unabhängig von den rechtlichen Verpflichtungen zur ressourcen- und umweltschonenden Produktion stellen die aktuellen Energiepreise eine zunehmende wirtschaftliche Herausforderung für europäische Unternehmen dar. Nur durch konsequente Prozess- und Anlagenoptimierung kann die Wärmebehandlungsbranche im internationalen Wettbewerb erfolgreich bestehen.

Nachfolgend sollen anhand konkreter Beispiele die Aspekte der Wärmebehandlungskonzepte, der Optimierung der Energieeffizienz im Bereich der Beheizung und der Abwärmeintegration beleuchtet werden. Wesentlich dabei ist die Bereitschaft, Bestehendes kritisch zu hinterfragen und neue Wege zu gehen.

2.2.2 Anlagen- und Wärmebehandlungskonzepte [1], [2]

Waren bis dato primär wirtschaftliche Aspekte für den erforderlichen Wärmebehandlungsprozess eines Bauteils in Abhängigkeit von dessen Anforderungen und dem eingesetzten Werkstoff ausschlaggebend, so führen Weiterentwicklungen im Bereich der Werkstofftechnik sowie steigende Energiekosten zu veränderten Prozessrouten in der Fertigung.

Am Beispiel des Direkthärtens anstelle des Einfachhärtens in Kombination mit Aufkohlungsprozessen zeigt sich, welch signifikantes Einsparungspotenzial hinsichtlich Energie und CO_2-Emissionen gegeben ist. Aufgrund der Maß- und Formänderungen infolge der Wärmebehandlung müssen verzugsempfindliche Teile häufig in Fixturen abgeschreckt werden. Oftmals werden die Wärmebehandlungsteile in Mehrzweck-Kammeröfen oder kleinen, kontinuierlichen Anlagen aufgekohlt, langsam unter Schutzgas abgekühlt, anschließend in einem Drehherdofen wiedererwärmt und in einer Härtepresse abgeschreckt. Alternativ dazu ermöglichen Durchstoß- und bevorzugt Ringherdofenanlagen die Aufkohlung und nachfolgende Direkthärtung in der Härtepresse in einem Behandlungsschritt – die Wiedererwärmung entfällt. Ein exemplarischer Vergleich für die Wärmebehandlung von 500.000 Tellerräder/Jahr aus der CHD-Gruppe 1 mm (CHD: core hardness depth) zeigt, dass ein Ringherdofenkonzept im Vergleich zu einer Kammerofenvariante (bestehend aus vier Einkammeröfen, einem Drehherdofen mit Härtepresse und zwei Anlassöfen sowie zwei Schutzgaserzeugern) ein CO_2-Einsparungspotenzial von ca. 25 % aufweist. Unter Berücksichtigung der reduzierten Gesamtdurchlaufzeit sowie des optimierten Personalaufwandes ergeben sich somit reduzierte Wärmebehandlungskosten in der Größenordnung von 20 bis 25 %.

Dem Leitgedanken der Energie- und CO_2-Einsparung durch Optimierung der Prozessführung folgend, sei an dieser Stelle noch auf den Aspekt des Hochtemperaturaufkohlens ver-

Tabelle 1: Einsparungspotenzial bei erhöhter Prozesstemperatur beim Gasaufkohlen (Durchstoßofenanlage, 20MnCr5, CHD-Gruppe 1 mm (CHD: core hardness depth)) [4]

Prozesstemperatur [°C]	930	950	980	1.000
Kohlenstoffpegel [%C]	1,10	1,20	1,30	1,35
Prozesszeiten [Minuten]	385	330	280	260
Durchsatzleistung [%]	100	123	151	167
Energieverbrauch [%]	100	110	125	135
Relativer Energieverbrauch [%]	100	89	83	81

wiesen. Die Anhebung der Prozesstemperatur beim Einsatzhärten führt zu einer Zeit- und Kostenreduktion [3]. Hinsichtlich einer prozesssicheren Anwendung gilt es, neben der Anlagen- und Regeltechnik auch die metallurgischen Aspekte bezüglich Kornwachstum, Randoxidation (atmosphärische Wärmebehandlung), Effusionseffekte (Niederdruckwärmebehandlung) und Vorwärmebehandlung (Glühprozess) sowie den Einfluss auf Maß- und Formänderung zu berücksichtigen. **Tabelle 1** stellt Prozesszeit/Durchsatzleistung und Energieverbrauch einer Durchstoßofenanlage in Abhängigkeit von der Aufkohlungstemperatur einander gegenüber. Hierbei zeigt sich, dass der zusätzliche Energiebedarf geringer als die erzielbare Steigerung der Durchsatzleistung ist, sodass der relative Energiebedarf pro kg Wärmebehandlungsgut mit steigender Prozesstemperatur abnimmt.

Vergleichbare Überlegungen sind auch im Bereich des Vergütens in der Schmiedeindustrie gegeben, wobei **Bild 1** einen Überblick über das Energieeinsparungspotenzial gibt.

Hinsichtlich des Vergütens aus der Schmiedehitze ist zu berücksichtigen, dass kornfeinende Umwandlungsschritte entfallen und höhere bzw. inhomogenere Bauteiltemperaturen vor dem Abschrecken vorliegen können. Mögliche Auswirkungen auf die Rissfreiheit und

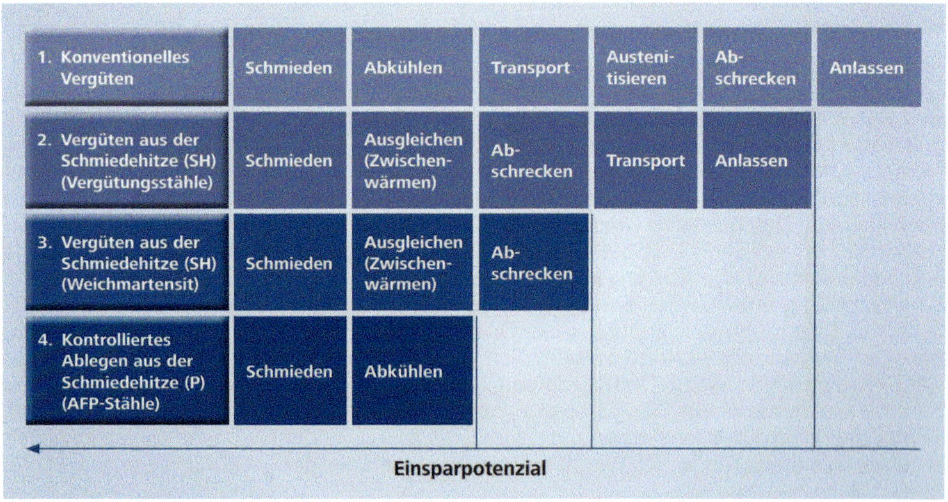

Bild 1: Prozessbedingte Einsparpotenziale in der Wärmebehandlung in der Schmiedeindustrie [5]

die Festigkeits-bzw. Zähigkeitseigenschaften – niobmikrolegierte Stähle können den Abfall der Kerbschlagarbeit oftmals nur bedingt kompensieren – sind daher kritisch zu betrachten. Im Rahmen einer industriellen Umsetzung das Vergüten aus der Schmiedehitze wurde durch die Installation eines elektrisch beheizten Temperaturausgleichsofens mit integriertem Polymerabschreckbad die homogene Temperaturverteilung im Wärmebehandlungsgut sichergestellt. Dabei ist ein zusätzliches automatisiertes Einzelteilbeschickungssystem mit berührungsloser Temperaturmessung hervorzuheben. Durch dieses Anlagenkonzept kann eine überwachte Bauteiltemperatur mit optimierter Austenitisierungstemperatur und damit eine reproduzierbare Gefügeausbildung gewährleistet werden; gleichzeitig kann durch den Entfall der Wiedererwärmung eine Reduktion des Energiebedarfs um ca. 75 % realisiert werden.

2.2.3 Beheizungskonzepte

Dem Bestreben der Dekarbonisierung in Härterei und Industrieofenbau folgend sind hinsichtlich der Anlagenbeheizung zwei Stoßrichtungen der Beheizung denkbar: der Einsatz regenerativer Brennstoffe (z. B. Wasserstoff, Ammoniak etc.) oder erneuerbarer elektrischer Energie.

Vorab gilt es festzuhalten, dass, dem Prinzip der Gasbeheizung geschuldet, Abgasverluste unvermeidbar sind, aber durch Abwärmenutzung, im Bereich der Thermoprozesstechnik üblicherweise zur regenerativen oder rekuperativen Verbrennungsluftvorwärmung, signifikant reduziert werden können. Aktuelle Optimierungen im Bereich der Rekuperatorausführung (komplexe Geometrien / 3D-Druck [6], Spaltstrom-Wärmetauscher [7] oder kombinierte Metallschaumausführungen [8]) zeigen, dass in diesem Bereich gegenüber älteren Bestandsanlagen ein weiteres Einsparpotenzial in der Größenordnung von 10 bis 15 % vorliegt, sodass Brennerwirkungsgrade von bis zu 85 % realisiert werden können.

Mittlerweile bieten mehrere Brennerhersteller Wasserstoffbrenner an und arbeiten zusätzlich an Konzepten, um diese auch mit Erdgas-Wasserstoff-Gemischen betreiben zu können. Parallel dazu wird zunehmend auch die Verbrennung von Ammoniak bzw. Ammoniak-Spaltgas untersucht. Ammoniak, basierend auf Stickstoff und Wasserstoff aus erneuerbaren Energien, kann als regenerativer Brennstoff angesehen werden. Im Vergleich zu Wasserstoff zeichnet sich Ammoniak durch einfachere Transport- und Speichereigenschaften (Verflüssigung bei -33 °C unter Normaldruck bzw. 20 °C bei 9 bar) aus. Wenngleich die Reaktionsmechanismen bekannt sind und erste positive Erfahrungen gesammelt wurden [9], ist hier teilweise noch Forschungsbedarf gegeben.

Auch im Hinblick auf den Einsatz von Sauerstoffbrennern – eine weitere Maßnahme zur Reduktion von Abgasverlusten (**Bild 2**) – ist anzumerken, dass mit zunehmender Verbrennungstemperatur bzw. Temperaturspitzen die Bildung von thermischen Stickoxiden ansteigt.

Somit ist, über den Aspekt der Dekarbonisierung hinausgehend, auch die Thematik der Reduktion von Stickoxiden zu berücksichtigen. Hierbei ist zwischen primären Maßnahmen (gestufte Verbrennung bzw.

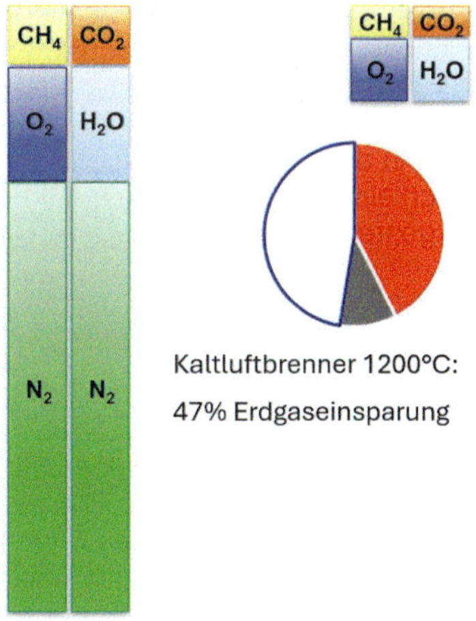

Bild 2: Einsparpotenzial Sauerstoffbrenner (Oxy-Fuel)

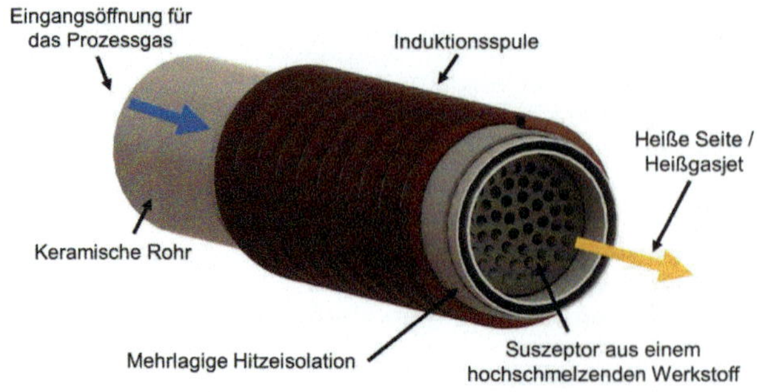

Bild 3: Heißgas-Jet [11]

flammlose Verbrennung, gekennzeichnet durch eine hohe Abgasrezirkulation, welche lokale Temperaturspitzen verhindert) und sekundären Maßnahmen (selektive katalytische Reduktion (SCR) bzw. selektive nicht katalytische Reduktion (SNCR)) zu unterscheiden. Im Bereich der Thermoprozesstechnik werden sowohl Brenner mit flammloser Verbrennung als auch Systeme mit Harnstoffeinspeisung (SCR) [10] eingesetzt.

Dem gegenüber stehen elektrische Beheizungssysteme, deren CO_2-Fußabdruck direkt mit dem jeweiligen geographischen Strommix gekoppelt ist. Somit sind keine verallgemeinerten Aussagen hinsichtlich der CO_2-Emissionen zulässig. Darüber hinaus ist anzumerken, dass anlagenspezifische konstruktive Gegebenheiten einer Thermoprozessanlage platzlimitierend hinsichtlich der zu installierenden elektrischen Heizleistung sein können.

Alternative elektrische Beheizungssysteme abseits breiter industrieller Anwendungen: Wenngleich die Wärmebehandlungsbranche dem ESG-Gedanken (Environmental, Social and Governance / Umwelt, Soziales und Unternehmensführung) verpflichtet ist, so müssen die getroffenen Maßnahmen auch wirtschaftlich vertretbar sein. Unter diesem Gesichtspunkt sind alternative elektrische Beheizungssysteme als Sonderlösung anzusehen, sollen aber dennoch nicht unerwähnt bleiben. Plasmabrenner können sehr vielseitig eingesetzt werden; insbesondere lässt sich nahezu jedes Gas (Luft, N_2, Ar, CO_2, O_2, ...) mittels der eingekoppelten elektrischen Energie in ein hochenergetisches, reaktives Plasma überführen, sodass sauerstofffreie Prozesse realisiert werden können. Hinsichtlich des Betriebskonzeptes kann zwischen gleichstrom-, induktions- und mikrowellenbasierten Systemen unterschieden werden. Weitgehen im Forschungsstadium sind induktivbasierte Heißgasbrenner. Hierbei durchströmt Prozessgas, beispielsweise Luft oder ein inertes Gas, ein feuerfestes Rohr in dem ein hochschmelzender, oxidationsbeständiger metallischer Suszeptor induktiv durch eine außenliegende Induktionsspule erhitzt wird und einen Heißgas-Jet, wie in **Bild 3** dargestellt, erzeugt.

2.2.4 Abwärmeintegration

Wenngleich das Bestreben der Abwärmenutzung in der Wärmebehandlung seit Jahrzehnten gegeben ist, so ist es oftmals die in den letzten Jahren rechtliche Notwendigkeiten der Energieerfassung und Zieldefinition, die hier eine zusätzliche, notwendige Transparenz liefert. Im Rahmen von detaillierten Analysen können die Energieströme und Temperaturniveaus zeitabhängig erfasst werden; gegebenenfalls zeigt sich dabei ein Missverhältnis zwischen Wärmequelle und Wärmesenke. An dieser Stelle sollen ausschließlich die Herausforderungen im Bereich der Thermoprozesstechnik skizziert werden, wobei Wärmetauschersysteme als technisch etabliert angesehen werden können – sie werden daher an dieser Stelle nicht weiter betrachtet.

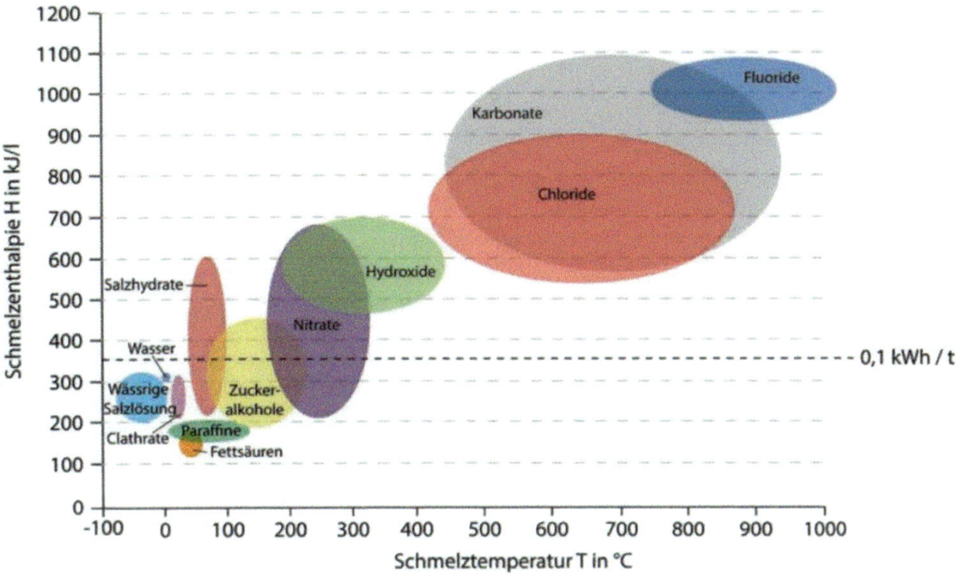

Bild 4: Klassifizierung von Latentwärmespeichern [13]

Zur Überbrückung der zeitlichen Diskrepanz ist auf die Möglichkeit der Abwärmespeicherung hinzuweisen [12], die von einfach zu realisierenden Thermoölpufferspeichern bis hin zu komplexen Latentwärmespeichern und thermochemischen Wärmespeichern reicht. Speziell letztere Systeme haben, aufgrund der Einsatztemperatur und der Kapazitäten, ein hohes Entwicklungspotenzial, stellen aber oftmals eine Sonderanwendung dar. **Bild 4** gibt einen Überblick über Latentwärmespeicher hinsichtlich Speicherkapazität und Temperaturbereich. Der Vorteil von latenten gegenüber sensiblen Wärmespeichern ist, dass Wärme auf einem konstanten Temperaturniveau gespeichert und entnommen werden kann und dadurch größere Speicherkapazitäten erzielt werden können.

Hinsichtlich der Herausforderung des Temperaturniveaus ist die Verfügbarkeit von Wärmepumpen für hohe Heizleistungen und hohe Vorlauftemperaturen in den letzten Jahren stetig gewachsen, sodass hier eine weitere Möglichkeit für dezentrale Anwendungen gegeben ist. Aufgrund des Temperaturniveaus bzw. des Temperaturhubs können Wärmepumpen in drei Kategorien eingeteilt werden. Das Temperaturniveau in konventionellen Wärmepumpen (WP) erreicht Senkentemperaturen von bis zu etwa 80 °C. Dabei wird allgemein von industriellen Hochtemperatur-Wärmepumpen (HTWP) gesprochen. Über 100 °C wird fallweise der Begriff Höchsttemperatur-Wärmepumpe (VHTWP) verwendet. Physikalisch bedingt sinkt die Effizienz der Wärmepumpe mit zunehmendem Temperaturhub ab (Carnot-Beziehung). Die Leistungszahl der Wärmepumpe (COP) beschreibt hierbei das Verhältnis von nutzbarer Wärme zu eingebrachter elektrischer Energie. **Bild 5** stellt die experimentellen COP-Werte für ausgewählte industrielle Wärmepumpen, welche sich im industriellen Betrieb erfolgreich bewährt haben, in Abhängigkeit von dem jeweiligen Temperaturhub dar. Die COP-Werte liegen zwischen 1,6 und 5,8 bei einem Temperaturhub von 130 bis 25 K. Hierbei ist ersichtlich, dass aktuell COP-Werte in der Größenordnung von 45 % des theoretischen maximalen COP-Wertes erzielt werden.

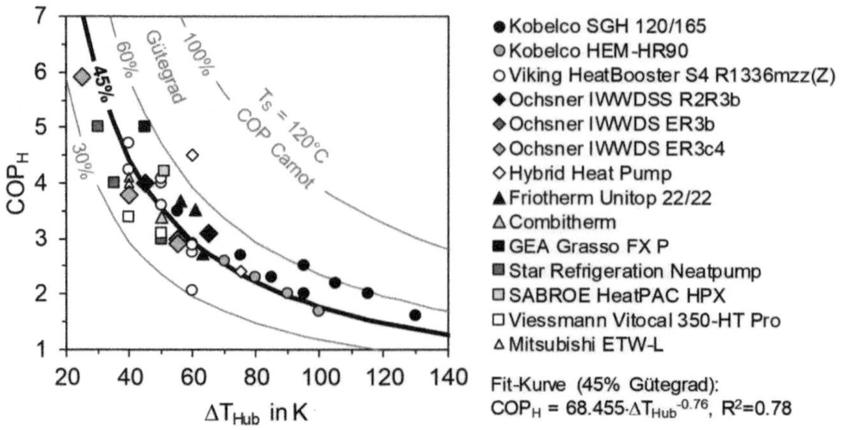

Bild 5: COP-Werte ausgewählter industrieller Wärmepumpen in Abhängigkeit vom Temperaturhub [14]

2.2.5 Fazit

Der Weg zur erfolgreichen Dekarbonisierung erfordert eine enge Zusammenarbeit zwischen Industrie, Forschung und Politik, um innovative Lösungen zu entwickeln und umzusetzen. Dabei zeigt sich, dass es einer Vielzahl sich ergänzender Maßnahmen bedarf, die sowohl auf Neuanlagen als auch auf Bestandsanlagen im Bereich der thermischen bzw. thermochemischen Wärmebehandlung übertragbar sind. Wesentlich ist es, bestehende Systeme und Prozesse kritisch zu hinterfragen sowie bisher, primär aus ökologischen Gründen, nicht umgesetzte Entwicklungen unter den veränderten Rahmenbedingungen neu zu bewerten.

Literatur

[1] Buchner, K.: Reduktion des CO_2-Fußabdrucks in der Wärmebehandlung. In: Prozesswärme, 1 (2023), S. 42–45

[2] Buchner, K.: Energieeffizienz neu gedacht – Vergüten aus der Schmiedehitze. In: HTM J. Heat. Treatm. Mat., 79 (2024) 4, S. A24–A25

[3] Grosch, J.; Liedtke, D.; Kallhardt, K.; Tacke, D.; Hoffmann, R.; Luiten, C. H.; Eysell, F. W.: Gasaufkohlen bei Temperaturen oberhalb 950 °C in konventionellen Öfen und in Vakuumöfen (Podiumsdiskussion beim 36. HK, 1980). In: HTM J. Heat. Treatm. Mat., 36 (1981) 5, S. 262–269

[4] Altena, H.: Prozessgasaufkohlung von Getriebeteilen – Möglichkeiten der Prozesszeitverkürzung und Kostenreduktion. In: Gaswärme International, 5 (2014), S. 59–65

[5] Ade, H.: Massivumformung in Deutschland – Eine energieintensive energieeffiziente Branche. Industrieverband Massiumumformung e.V., Hagen, 2017

[6] Fuchs, M.; Luo, X.; Bender, W.; Schwarz, P.; Kabelac, S.: Kompakte Hochleistungs-Rekuperatoren mit 3D-Druck-Bauteilen für höchste Anforderungen. In: Echterhof, T.; Pfeifer, H. (Hrsg.): Proceedings from the 3. Aachener Ofenbau- und Thermoprozess Kolloquium. 07. bis 08. Oktober 2021, Aachen, S. 365–371

[7] Wünning, J. G.: Spaltstrombrenner im industriellen Einsatz. In: Gaswärme International, 1 (2013), S. 35–38

[8] Rakette, R.; Mäder, D.; Neumann, B.: Hochgeschwindigkeitsbrenner mit Rippenrohr-Rekuperator für höchste Energieeffizienz. In: Gaswärme International, 1 (2017), S. 41–44

[9] Biebl, M.; Leicher, J.; Giese, A.; Wieland, C.: A comprehensive study of non-premixed combustion of ammonia and its blends: Flame stability and emission reduction. In: Fuel, 386 (2025), Artikel 134.501

[10] Uhlig, J.; Wolf, M.: Wirkungsgradsteigerung und NO_x-Reduzierung im Anlagenbestand. In: Echterhof, T.; Pfeifer, H. (Hrsg.): Proceedings from the 3. Aachener Ofenbau- und Thermoprozess Kolloquium. 07. bis 08. Oktober 2021, Aachen, S. 283–289

[11] Wolf, G.; Keßler, A.: Perspektiven zur CO_2-freien Schmelztechnik. In: Giesserei, 4 (2023), S. 78–83

[12] OWI Science for Fuels gGmbH: Potentiale von thermischen Speichern in der Thermoprozessindustrie. Forschungsbericht 251/133 Forschungsgemeinschaft Industrieofenbau e.V., 2022

[13] Sterner, M.; Stadler, I.: Energiespeicher – Bedarf, Technologien, Integration. Springer, Berlin; Heidelberg, 2017

[14] Arpagaus, C.: Hochtemperatur-Wärmepumpen: Marktübersicht, Stand der Technik und Anwendungspotenziale. VDE Verlag, Offenbach, 2018

2.3 CO_2-freie Prozesswärme durch den Einsatz elektrischer Beheizung

Andreas Kraus

Das Thema Energieeffizienz hat für viele Unternehmen nicht zuletzt angesichts der stark gestiegenen Energiepreise enorm an Bedeutung gewonnen. Doch welche Möglichkeiten gibt es, als Anbieter im Maschinen- und Anlagenbau zu marktgerechten Preisen zu produzieren und gleichzeitig die Produktion nachhaltiger auszurichten, um den eigenen CO_2-Fußabdruck zu reduzieren?

Die Vorgaben sind klar: Der CO_2-Ausstoß muss – je nach Branche – bis 2030 um 65 % gesenkt werden. Bis 2045 gilt „Klimaneutralität" als Ziel, so will es das Bundes-Klimaschutzgesetz. Hiervon betroffen sind vor allem Unternehmen in energieintensiven Branchen wie beispielsweise die Hersteller von Stahl, Metall, Glas, Papier oder Zement (**Bild 1**). In diesen Sektoren war die Optimierung der Prozesswärme lange Zeit kein Thema. Management und Controlling hatten den Blick eher auf die Prozess- als auf die Gesamtkosten gerichtet. Tatsächlich hatte der Faktor Energie bis vor dem Ukrainekrieg keine signifikanten Auswirkungen auf die Stückkosten. Das ist nun anders: Durch die Energiekostenexplosion erhält der Faktor Energie eine zentrale Bedeutung und bestimmt die Ausrichtung eines Produktes wesentlich mit (**Bild 2**).

Elektrische Energie zur Wärmeerzeugung ist in vielen industriellen Anwendungen Stand der Technik. Die Wärmeerzeugung kann mit einem Heizelement oder Strahler erfolgen, der je nach Bauart die elektrische Energie in Wärme umwandelt. In der Vergangenheit wurden elektrische Lasten mithilfe von mechanischen Schützen geschaltet. Heutzutage kommen hier Halbleiterrelais (**Bild 3**) oder Leistungssteller (**Bild 4**) zum Einsatz. Viele Energiemanager in den Unternehmen fragen sich vor dem Hintergrund stark gestiegener Energiekosten: Lassen sich Schütze so einfach gegen die elektronischen Schalter austauschen? Und was muss beim Einsatz berücksichtigt werden? Mechanisches Schalten von elektrischen Las-

Bild 1: Bis 2045 soll Klimaneutralität erreicht werden, daher muss der CO_2-Ausstoß um bis zu 65 % gesenkt werden.

Bild 2: Durch die Energiekostenexplosion erhält der Faktor Energie eine zentrale Bedeutung und bestimmt die Ausrichtung eines Produktes wesentlich mit (exempl. Darstellung eines Industrieofens).

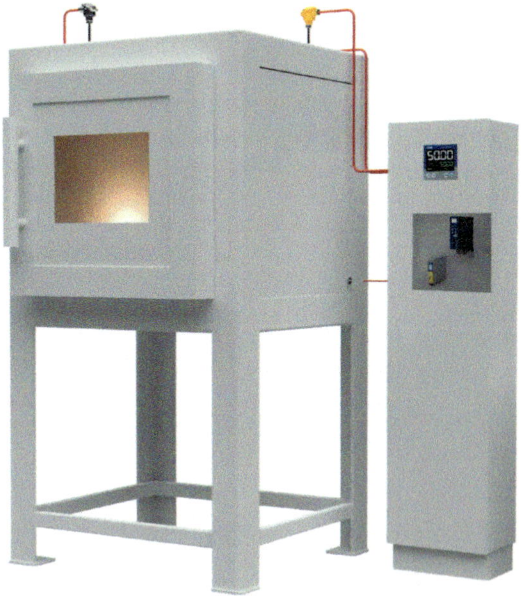

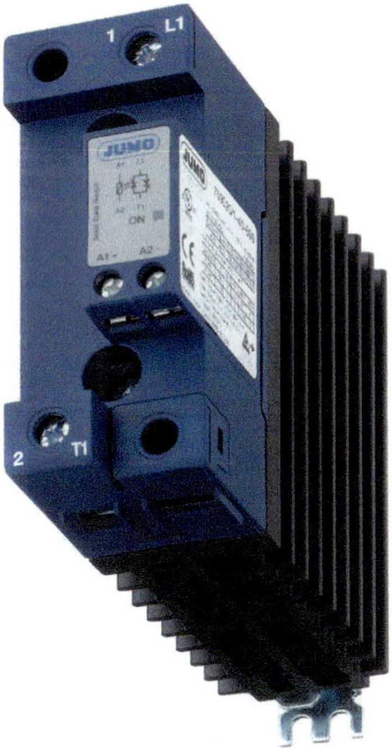

◀ **Bild 3:** Vorteile des Halbleiterrelais liegen in der schnelleren Schaltung und dem geringen Verschleiß aufgrund der fehlenden Mechanik.

▶ **Bild 4:** Leistungssteller können mittels adaptiver Regelung auf Netzspannungsschwankungen reagieren, ehe es zu einer Temperaturänderung kommt.

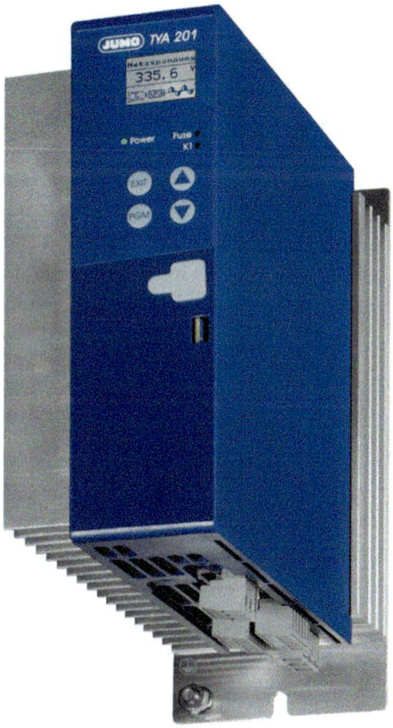

ten ist meist nur noch bei Anwendungen zu finden, die eine geringe Schalthäufigkeit benötigen oder sehr hohe Leistungen schalten, wie z. B. im Kraftwerksbau. Die Vorteile der elektronischen Schalter und Leistungssteller liegen in der schnelleren Schaltung und der längeren Lebensdauer aufgrund der fehlenden Mechanik. Dadurch ist auch der Verschleiß geringer. In der Leistungselektronik kommen häufig antiparallel geschaltete Thyristoren zum Einsatz.

Für rein ohmsche Lasten mit geringen Anforderungen sind elektronische Schalter, auch Halbleiterrelais genannt, meist ausreichend. Diese Geräte werden wie Schütze angesteuert. Dabei bestimmt die übergeordnete Steuerung oder Regelung die Ein- und Ausschaltfolge des Stellgliedes. Eine Rückmeldung bezüglich des Schaltzustandes und der Last ist ohne zusätzlichen Hardwareaufwand meist nicht möglich.

Möchte hingegen der Anwender mehr Informationen über den Schaltzustand und die angeschlossene Last erhalten oder hat die Last einen stark temperaturabhängigen Widerstand oder eine induktive Komponente, dann empfiehlt sich der Einsatz eines Leistungsstellers. Diese können mittels adaptiver Regelung auf Netzspannungsschwankungen reagieren, ehe es zu einer Temperaturänderung kommt. Das Ergebnis sind eine konstante Temperatur und eine gleichbleibende Qualität.

2.3.1 Unterschiedliche Gegebenheiten erfordern angepasste Regelungen

Hier steht der Anwender vor zwei Fragen: Welche Eigenschaften muss der Leistungssteller für meine Anwendung haben? Worauf kann verzichtet werden? Geklärt werden muss folglich die Lastart; daraus ergibt sich die Regelungsart und meist auch die Betriebsart des Gerätes. Optional wird noch eine Teillastbruchüberwachung in Betracht gezogen, um einen Ausfall einer Last zu detektieren.

Hat die Last einen konstanten Widerstand und einen geringen Temperaturkoeffizienten, so reicht eine U^2-Regelung. Hier wird die Ausgangsspannung nicht linear zum Sollwert erhöht, sondern gemäß einer Wurzelfunktion. Dadurch ergibt sich ein linearer Leistungsanstieg am Ausgang, da die Leistung gemäß $P = U^2/R$ bei konstantem Widerstand quadratisch von der Spannung abhängt.

Das Gleiche gilt für die I^2-Regelung. Bei der I^2-Regelung wird der Ausgangsstrom nicht linear zum Sollwert erhöht, sondern proportional zur Quadratwurzel des Sollwerts angepasst. Hintergrund ist die Beziehung $P = I^2 * R$, bei konstantem Widerstand: Um eine lineare Leistungsabgabe zu erreichen, muss der Strom gemäß einer Wurzelfunktion zum Sollwert steigen. Beide Regelungsarten funktionieren gut bei konstantem Lastwiderstand. Ist dies nicht der Fall, so benötigt man eine *P*-Regelung. Hier werden Laststrom und Lastspannung zur Berechnung der Ausgangsleistung herangezogen. Ein weiterer Punkt ist die Betriebsart: Hier wird zwischen Impulsgruppenbetrieb und Phasenanschnitt unterschieden. Doch wie unterscheiden sich diese Betriebsarten?

Bei Impulsgruppenbetrieb werden ganze Wechselspannungszüge, beginnend im Nullpunkt, ein- und ausgeschaltet. Das Verhältnis zwischen Ein- und Ausschaltzeit bestimmt die prozentuale Leistungsabgabe. Diese Betriebsart eignet sich für Heizelemente mit verhältnismäßig trägem Ansprechverhalten. Der Vorteil dieser Betriebsart ist, dass dadurch nur sehr geringe Netzstörungen entstehen und der Blindleistungsanteil sehr geringgehalten wird, was wiederum geringe Betriebskosten zur Folge hat.

Bei der Phasenanschnittsteuerung wird der Stromfluss durch einen Thyristor gesteuert. Nach dem Nulldurchgang der Wechselspannung leitet der Thyristor den Strom so lange nicht, bis er einen Zündimpuls erhält; ab diesem Zeitpunkt wird die Last bis zum nächsten

Nulldurchgang mit Energie versorgt. Je später der Thyristor gezündet wird, desto geringer ist die mittlere Leistung.

Somit lassen sich Ausgangsspannung und Ausgangsstrom regeln. Diese Betriebsart eignet sich für Heizelemente mit sehr kurzen Reaktionszeiten und Strahlern. Durch den nicht synchronen Verlauf von Wechselspannung und Wechselstrom entsteht bei dieser Betriebsart eine Verzerrungsblindleistung. Zudem entstehen durch die steilen Einschaltflanken harmonische Oberwellen, die sich auf die Netzspannung überlagern und dadurch andere elektronische Verbraucher am gleichen Netz stören können. Bei großen Strömen und einem großen Phasenwinkel ist eine externe Beschaltung mit Netzfilter und Netzdrossel nötig, um den Gesamtoberschwingungsgehalt auf den in der Norm EN 50160 geforderten Wert, kleiner 8% THD_U, zu beschränken.

2.3.2 Was sind die typischen Lastarten für einen Leistungssteller?

Da sind einmal die Infrarot-Strahler; diese gibt es als Quarz- und Keramikstrahler. Letztere sind meist mittel- bis langwellig und lassen sich wie ein ohmscher Verbraucher mittels Impulsgruppenbetrieb und unterlagerter U^2-Regelung ansteuern. Bei den Quarzstrahlern unterscheidet man kurz-, mittel und langwellige Ausführungen. Hier muss auf das Verhältnis Warmwiderstand/Kaltwiderstand geachtet werden. Kurzwellige Strahler können im Kaltzustand maximal das Vierzigfache des Nennstromes ziehen. Daher ist es unbedingt wichtig, diese Verbraucher mittels Softstart im Phasenanschnitt anzufahren. Eine Softstartzeit von zwei bis drei Sekunden kann bereits ausreichen. Ein Betrieb in Phasenanschnitt garantiert eine flickerfreie Strahlerleistung; zudem schont diese Betriebsart den Glühwedel und trägt zu einer verlängerten Standzeit der Strahler bei. Bereits mittelwellige Strahler haben ein trägeres Glühverhalten und können je nach Anwendung im Impulsgruppenbetrieb angesteuert werden. Werden mehrere Strahler parallel betrieben, so macht es Sinn, einen Strahlerausfall mittels Teilastbruchalarm zu detektieren.

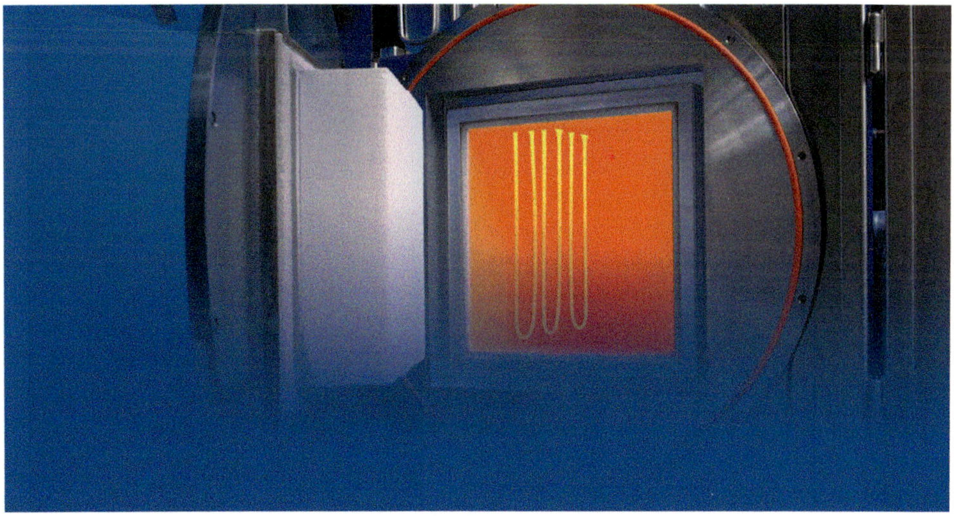

Bild 5: In Hochtemperaturöfen kommen häufig Elemente aus Molybdändisilizid ($MoSi_2$) zum Einsatz.

Die meisten elektrischen Öfen werden mit normalen metallischen Heizern betrieben. Sind diese Heizelemente für 3-Phasen- Wechselstrom ausgelegt, so können diese mittels Sparschaltung betrieben werden. Hierbei werden nur zwei Phasen der Last geschaltet, die dritte Phase ist direkt mit der Last verbunden. Sparschaltung ist gegeben, da ein Thyristorelement eingespart wird. Das bedeutet: ein Drittel mehr Platz im Schaltschrank, ein Drittel weniger Verlustleistung und ca. ein Drittel geringere Investitionskosten. Sicherheitstechnisch muss hierbei geprüft werden, dass im Fehlerfall kein Kontakt zwischen Heizelement und Gehäuse besteht, da dies zu einem unkontrollierten Stromfluss im Heizelement führt.

In Hochtemperaturöfen kommen häufig Elemente aus Molybdändisilizid ($MoSi_2$) zum Einsatz (**Bild 5**). Diese sind bei Raumtemperatur sehr niederohmig und stellen beim Anfahren einen Kurzschluss dar. Um den Kurzschlussstrom so gering wie möglich zu halten, muss der Steller über eine Strombegrenzung verfügen. Eine weitere Eigenschaft von $MoSi_2$ ist, dass die Oberflächentemperatur des Elements proportional zum Widerstand steigt. Wird nun der Widerstand überwacht, so kann die Maximalleistung begrenzt und das Element geschont werden. Zur Regelung von $MoSi_2$ wird die I^2-Regelung eingesetzt.

Weitere Heizelemente sind Siliciumcarbid-Elemente. Sie erreichen Temperaturen von bis zu 1.650 °C. Das Typische ist die Widerstandskennlinie in Form einer „Badewanne“. Das Element hat bis zu einer Temperatur von ca. 800 °C einen negativen Temperaturkoeffizienten, danach wechselt dieser in einen positiven bis hin zur Maximaltemperatur. Ein weiterer Faktor ist das Alterungsverhalten: Je nach umgebender Atmosphäre und Arbeitstemperatur erhöht sich der Widerstand im Laufe der Betriebszeit. Um hier die Ausgangsleistung konstant zu halten, ist es unbedingt notwendig, dass der Steller über eine P-Regelung verfügt.

Leistungssteller erzeugen bauartbedingt eine Verlustleistung, die ca. das 1,3-Fache des Laststroms beträgt. Für ausreichende Kühlung muss gesorgt werden.

2.4 Hallenkonditionierung leicht gemacht: Nachhaltig. Leistungsstark. Wirtschaftlich.

Dr.-Ing. H.-J. Rönnecke

Die klimatischen Verhältnisse in Mitteleuropa weisen verstärkt daraufhin, dass sommerliche Temperaturen um die 30 °C verstärkt zu hohen Belastungen der Mitarbeiter, vor allem in Produktionsbetrieben, führen. Wer möchte schon gerne unter diesen Bedingungen die erforderliche Arbeitsleistung erbringen? Verminderte Leistungsfähigkeit und zunehmendes Unbehagen am Arbeitsplatz sind die Folgen.

Somit stehen moderne Industriehallen zunehmend vor der Herausforderung, ein ganzjährig angenehmes und sicheres Raumklima zu gewährleisten – bei gleichzeitig steigenden Anforderungen an Nachhaltigkeit und Energieeffizienz. Mitarbeiter und Mitarbeiterinnen, aber auch Maschinen und Produkte benötigen stabile, saubere und temperierte Hallenluftverhältnisse, ohne dass der Energieverbrauch explodiert oder hohe Investitionskosten anfallen. Die Lösung: Sustainable Hall Conditioning (SHC) – ein modulares, wirtschaftliches und ökologisch durchdachtes System zur Hallenkonditionierung (**Bild 1**).

2.4.1 Luftqualität und Temperatur: zentrale Erfolgsfaktoren

In der Praxis stellen sich wiederkehrend immer dieselben Fragen:

- Wie lassen sich in Produktionshallen angenehme Temperaturen für Mitarbeitende und Maschinen erreichen – kosteneffizient und nachhaltig?
- Wie schützt man Beschäftigte vor überhitzten Arbeitsplätzen, insbesondere im Sommer?
- Wie lässt sich die Frischluftzufuhr erhöhen, ohne Heiz- und Kühlkosten explodieren zu lassen?
- Wie bleibt die Hallenluft dauerhaft sauber – trotz Produktionsabwärme, Staub und Ölnebel?

Bild 1: Hallenluftbelastung (Quelle: INFRANORM® Technologie GmbH)

2.4.2 Sustainable Hall Conditioning (SHC): Die passende Antwort mit System

Das SHC-System begegnet diesen Herausforderungen mit einem durchdachten und in der Praxis bewährten Ansatz. Herzstück ist eine zweistufige adiabate Kühlung, die die Luft reinigt und temperiert (**Bild 2**, **Bild 3**). Gegenüber konventionellen Klimaanlagen wird dabei bis zu 90 % weniger CO_2 emittiert und durch die Verwendung von Stadtwasser kommen keine klimaschädlichen Kältemittel zum Einsatz.

Die erste Stufe nutzt kaltes Wasser, das durch einen indirekten adiabaten Kühlprozess erzeugt wird, zur Kühlung. In der zweiten Stufe erfolgt eine zusätzliche Temperatursenkung mithilfe patentierter Kühlwaben mit antibakterieller Beschichtung. Diese reduzieren den Wasserverbrauch um bis zu 30 %, senken die Feuchtelast und ermöglichen einen hygienischen, wartungsarmen Betrieb unter Einhaltung aller hygienetechnischen Anforderungen.

2.4.3 Hallen- oder Zonenkonditionierung: Effizient genau dort, wo es darauf ankommt

SHC ist kein starrer Klimakasten, sondern ein flexibles, modulares Baukastensystem. Die Luft wird gezielt dort eingebracht, wo Menschen arbeiten oder Maschinen Wärme abgeben. Textilschläuche sorgen für zugfreie Frischluftzufuhr. Ablufteinheiten führen die aufsteigende Warmluft effizient ab. Optional reinigen Ölnebel- und Rauchfilter die Luft und erhöhen die Arbeitssicherheit. Die Kombination aus Abwärmenutzung, intelligenter Steuerung und adaptiver Luftführung erlaubt selbst im Winter einen energieeffizienten Betrieb – inklusive direkter Abwärmenutzung.

2.4.4 Variantenvielfalt: PreCool, Winterbetrieb, automatische Regelung

Besonders spannend ist die Möglichkeit, bestehende Lüftungsanlagen nachzurüsten. Mit PreCool-Modulen lässt sich SHC platzsparend und effizient auch vor vorhandene Systeme

Bild 2: Kühlmodul
(Quelle: INFRANORM® Technologie GmbH)

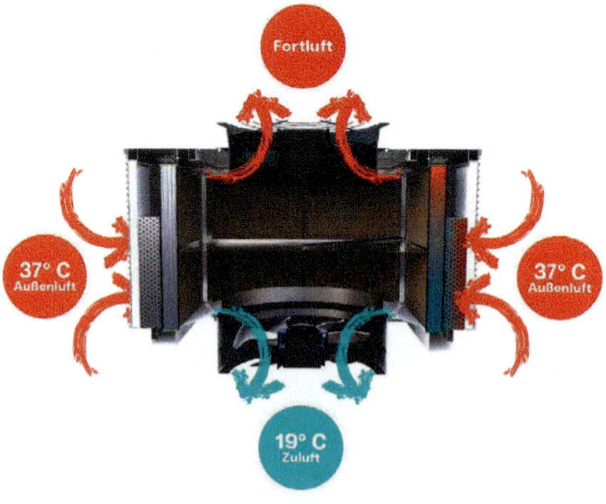

Bild 3: Prinzip zweistufige Adiabatik
(Quelle: INFRANORM® Technologie GmbH)

(z. B. Raumlufttechnische Anlagen – RTL-Anlagen) schalten – entweder einstufig oder zweistufig.

Für jede Witterung gibt es eine passende Betriebsstrategie:

- Energieeffizientes Lüften bei kühlen Außentemperaturen
- Dynamische Vorkühlung an warmen Tagen
- Leistungsstarke Kühlung im Hochsommer – alles automatisch geregelt.

2.4.5 Wirtschaftlichkeit, Nachhaltigkeit und Hygiene

Die Vorteile des Systems sind:

- Bis zu 95 % geringere Betriebskosten gegenüber konventionellen Kühlsystemen
- Bis zu 95 % weniger Stromverbrauch
- 100 % Frischluftqualität durch aktive Zuluftkühlung
- Erfüllung aller Hygieneanforderungen nach VDI 6022 durch zertifizierte Materialien und tägliche Selbsttrocknung
- Wirtschaftlicher Betrieb durch lange Standzeiten der Komponenten (3–5 Jahre) und geringe Wartungskosten.

2.4.6 Simulation schafft Transparenz

Vor der Umsetzung wird das System digital simuliert – stundengenau, auf Basis realer Standortdaten, Hallengeometrie und interner Wärmelasten. So entsteht eine maßgeschneiderte Lösung, die auf Effizienz und Wirkung optimiert ist.

2.4.7 SHC als nachhaltiger Standard

SHC steht für eine neue Generation der Hallenkonditionierung. Es vereint moderne Kühltechnologie, ökologische Verantwortung und wirtschaftlichen Betrieb – alles aus einer Hand. Ob Neuplanung oder Nachrüstung: Sustainable Hall Conditioning (SHC) ist die nachhaltige Antwort auf die Herausforderungen des Klimawandels im industriellen Umfeld.

2.5 Reduzierung der CO_2-Emissionen durch faserverbundkeramische Beschichtungen

Björn Kintscher, Dr. Mathias Kunz

Die Reduzierung der CO_2-Emissionen ist ein klares Ziel aller Industriezweige, auch der Stahlindustrie. Effizientere Produktionswege, wie das CSP®-Verfahren (Compact Strip Production) – die Kombination aus Gieß- und Walzprozess – sind ein großer Schritt, um Energie und damit Emissionen einzusparen. Doch auch hier gibt es Potenzial, noch effizienter zu werden. Eine Schutzbeschichtung für das Isolationsmaterial der Tunnelofenrollen aus Faserverbundkeramik (OCMC) hilft, den Energieverbrauch unter den rauen Bedingungen im CSP®-Tunnelofen zu senken.

Der Eisen- und Stahlsektor verursacht etwa 7–9 % der gesamten weltweiten CO_2-Emissionen und gehört damit zu den Branchen, die am meisten dazu beitragen [1]. Während die Gesamtemissionen auf einem hohen Niveau liegen, gibt es bereits Lösungen, um diese zu senken. Es gibt Lösungen im großen Maßstab, wie z. B. die Ersetzung bestehender Stahlherstellungsprozesse durch wasserstoffbasierte Direktreduktionsprozesse. Aber auch in kleinerem Maßstab gibt es wichtige Bereiche, in denen der Verbrauch fossiler Brennstoffe und damit die CO_2-Emissionen erheblich gesenkt werden können. Der Bereich, der in diesem Beitrag behandelt wird, ist der Heizprozess innerhalb einer kombinierten Gieß- und Walzanlage.

Die konventionelle Herstellung von Stahlbändern erfolgt in einem getrennten Gieß- und Walzverfahren. Der geschmolzene flüssige Stahl wird mit einer Stranggussanlage zu Brammen gegossen, die in einem Brammenlager gelagert werden. Die kalte Bramme wird vor

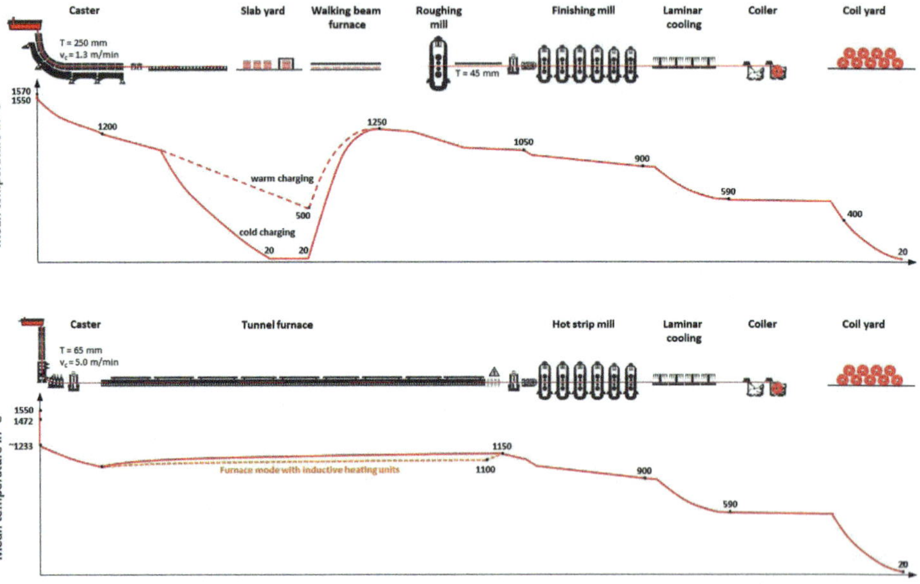

Bild 1: Temperaturprofil der konventionellen Prozessroute gegenüber der CSP®-Prozessroute

dem Warmwalzen in einem Hubbalkenofen auf 1.250 °C erwärmt. Die Warmbreitbandstraße walzt die warme Bramme zu einem Band, das zu einem warmgewalzten Coil aufgewickelt wird.

Die CSP®-Anlagen von SMS haben einen klaren Vorteil in Bezug auf die Energieeffizienz. Hier sind der Gieß- und der Walzprozess in einer Anlage kombiniert. Eine dünne Bramme – typischerweise 50 bis 120 mm dick – wird gegossen und direkt zum Warmwalzwerk für den Walzprozess transportiert. Die Wärme des Schmelzprozesses wird genutzt, sodass das energieintensive Wiederaufheizen der Bramme von 20 °C auf 1.250 °C entfällt. Dieser Unterschied macht diesen Anlagentyp so effizient. **Bild 1** zeigt den Vergleich der Anlagenkonzepte im Hinblick auf das typische Temperaturprofil entlang der Anlage.

Der Tunnelofen einer CSP®-Anlage spielt eine Schlüsselrolle innerhalb des CSP®-Prozesses und hat vier Hauptaufgaben: Transport der Dünnbramme von der Gießmaschine zum Walzwerk, Pufferung der Brammen beim Wechsel der Arbeitswalzen im Walzwerk, Aufheizen auf das gewünschte Temperaturniveau (falls erforderlich) und Homogenisierung der Brammentemperatur. Die Temperatur der Ofenatmosphäre liegt normalerweise zwischen 1.150 °C und 1.200 °C.

Der Tunnelofen ist i. d. R. eine lange, schmale, isolierte Kammer, die mit Erdgasbrennern in den Seitenwänden beheizt wird (**Bild 2**). Alternativ kann der Ofen auch mit elektrischen Heizelementen ausgestattet werden, um die Emissionen auf null zu reduzieren. Ein solcher elektrischer Tunnelofen wird als Teil einer CSP® Nexus-Anlage für Stegra [2] realisiert. Die in Nord-Schweden geplante Anlage für Stegra wird die erste emissionsfreie Großproduktionsanlage der Welt sein. Es wird erwartet, dass hier ab 2026 grüner Stahl produziert wird. Die allgemeine Forderung der Stahlindustrie nach einer Verringerung des CO_2-Ausstoßes und einer Steigerung der Effizienz wird zum wichtigsten Antriebsfaktor für die Entwicklung

Bild 2: Tunnelofen einer CSP®-Anlage

Bild 3a: Herkömmliche Ofenrolle mit feuerfester Betonisolierung

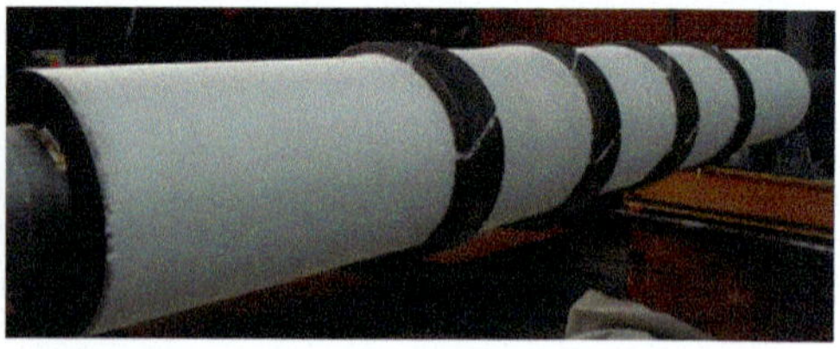

Bild 3b: Ofenrolle mit Faserisolierung

und Verbesserung der Anlagentechnik. Ein Beispiel für eine solche Verbesserung sind die Ofenrollen, die die Aufgabe haben, die Bramme im Tunnelofen zu stützen und zu transportieren. Diese wassergekühlten Rollen bestehen aus einer rohrförmigen Kohlenstoffstahlwelle, hitzebeständigen hochlegierten Laufrädern und einer isolierenden Feuerfestauskleidung zwischen den Laufrädern. Durch die notwendige Wasserkühlung werden nicht nur die Rollen gekühlt, sondern es wird auch Energie aus dem Ofenraum in das Kühlwasser transportiert. Somit spielt die Isolierung der Rollen gegen den Ofenraum eine große Rolle für die Effizienz des gesamten Ofens. I. d. R. wird eine Betonauskleidung als feuerfestes Material für die Rollen verwendet, um den rauen Bedingungen im Inneren des Tunnelofens standzuhalten. Die durchschnittliche Lebensdauer einer solchen Rolle beträgt etwa drei Jahre. Als energiesparendere Alternative können die Rollen mit einer Faserisolierung ausgestattet werden. Beide Arten von Isoliermaterial sind in **Bild 3a** und **Bild 3b** dargestellt.

Die Verwendung einer Faserisolierung führt zu einer Nettoenergieeinsparung von mehr als 30 % im Vergleich zu einer herkömmlichen Betonisolierung (**Bild 4**).

Die Realität zeigt jedoch, dass die ungeschützte Faserisolierung einen gewissen Einfluss auf die Lebensdauer der Rolle hat. Es ist unvermeidlich, dass das Eisen der Bramme mit dem überschüssigen Sauerstoff in der Tunnelofenatmosphäre reagiert und Eisenoxid bildet, das auch als Zunder bekannt ist. Zunderteile zerstören mechanisch das weiche Isolati-

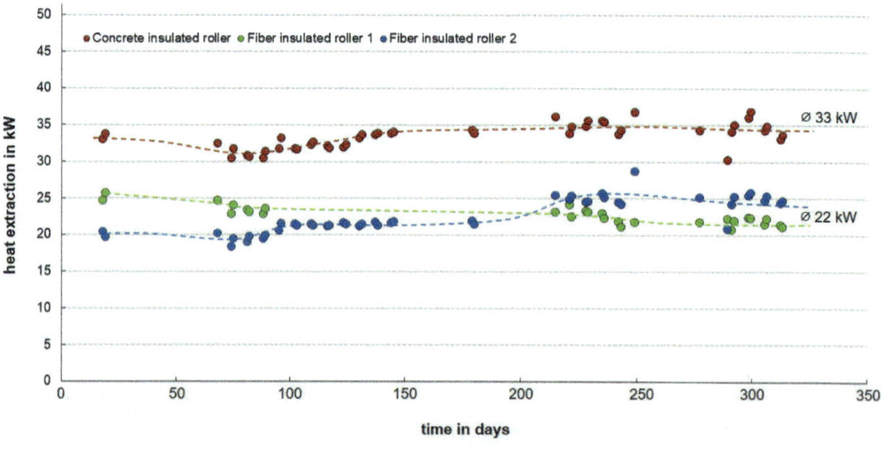

Bild 4: Wärmeentzug einer betonisolierten Rolle gegenüber einer faserisolierten Rolle; Netto-Energieeinsparung von > 30 %

onsmaterial, was dazu führt, dass nach einer Lebensdauer von 12 bis 18 Monaten der energetische Vorteil gegenüber Rollen mit Betonauskleidung nicht mehr gegeben ist.

Um das weiche Faserisoliermaterial der Ofenrolle zu schützen, wurden deshalb in einer nordamerikanischen CSP®-Anlage Langzeitversuche mit einer schützenden, faserverbundkeramischen Ummantelung durchgeführt.

Die in den Versuchen verwendete Ummantelung basiert auf einer Weiterentwicklung der Oxid-Faserverbundkeramik WHIPOX® durch die WPX Faserkeramik GmbH. Oxid-Faserverbundkeramiken (Oxide Ceramic Matrix Composites, OCMC) bilden eine innovative Werkstoffklasse, die ihre einzigartigen Vorteile im Vergleich zu monolithischen Keramiken sowie im Vergleich zu Hochtemperaturmetallen in so unterschiedlichen Anwendungen wie Luft- und Raumfahrt, industrieller Wärmebehandlung und Energieerzeugung unter Beweis gestellt hat [3]. OCMC bestehen aus oxidkeramischen Faserbündeln oder Geweben, die in eine oxidkeramische Matrix eingebettet sind [4]. Im Gegensatz zu monolithischen Keramiken sind sie äußerst stabil gegenüber mechanischen und thermomechanischen Schocks, was den Einsatz von OCMC in sehr rauen Umgebungen bei Temperaturen von bis zu 1.250 °C langfristig und 1.500 °C kurzfristig ermöglicht. Solche Schocks führen bei monolithischer Keramik zu Rissen und katastrophalem Versagen, was deren Einsatz in den oben genannten Anwendungen ausschließt. Im Gegensatz zu Hochtemperaturmetallen, z. B. Molybdän- oder Nickellegierungen, sind viele OCMCs unter thermomechanischer Belastung bei Temperaturen über 1.000 °C kriechfest [5]. Weitere Vorteile von OCMCs sind ihre geringe thermische und elektrische Leitfähigkeit, ihr chemisch inertes Verhalten in oxidierenden Atmosphären und ihre Stabilität gegenüber korrosiven Flüssigkeiten und Atmosphären.

2

OCMCs, die aus Mullit- (Al_2O_3/SiO_2) oder Aluminiumoxid- (Al_2O_3) Fasern bzw. Matrices bestehen, sind besonders für T > 1.000 °C geeignet. Solche keramischen Fasern sind von mehreren Anbietern weltweit kommerziell erhältlich und entsprechende proprietäre Matrices wurden sowohl von akademischen als auch von industriellen Spezialisten auf dem Gebiet der OCMC-Technologie, wie der WPX Faserkeramik GmbH, entwickelt.

Für einen wirksamen, langfristigen Schutz der thermischen Isolierung der wassergekühlten Transportwalzen bot sich die Verwendung einer solchen OCMC-Ummantelung auf Mullitbasis an, da sie a) mechanisch stabil gegen Zunderteile mit hoher Geschwindigkeit ist und somit die weiche thermische Isolierung vor einer solchen Zerstörung schützt (**Bild 5**), b) in einer chemisch aggressiven Hochtemperaturatmosphäre stabil ist, c) eine zusätzliche thermische Isolierung bietet und d) stabil gegen thermische Schocks aufgrund des schnellen

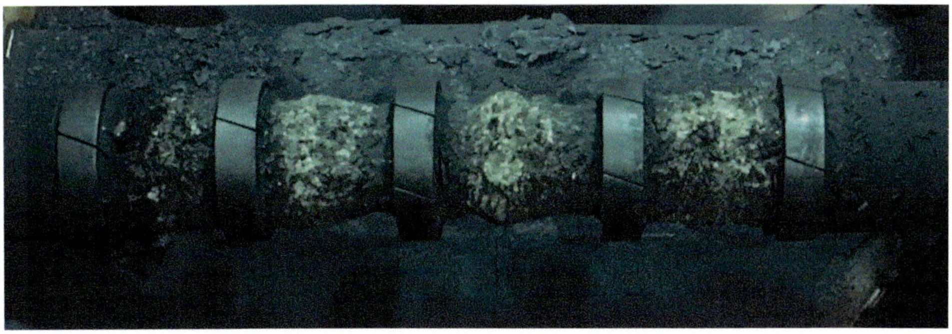

Bild 5: Ungeschützte Faserisolierung nach zwölf Monaten Zundereinwirkung

Bild 6: Provisorische In-situ-Applikation von OCMC-Prepregs

Austauschs in Betriebssituationen ist, in denen komplette Transportwalzen aus dem Ofen genommen werden müssen und die OCMC-Oberfläche eines 1.200 °C heißen Walzensystems innerhalb von Minuten der Raumtemperatur ausgesetzt wird.

Eine große Herausforderung war, dass OCMC-Systeme auf Mullit- oder Aluminiumoxidbasis eine Sinterung bei ca. 1.200 °C am Ende des Herstellungsprozesses erfordern, wodurch auch alle anderen Komponenten des Systems dieser Sintertemperatur ausgesetzt würden. Hier wurde deshalb die Einbausituation mit 1.200 °C für eine In-situ-Sinterung genutzt, indem die Walze mit einer zunächst ungesinterten OCMC-Ummantelung in den Tunnelofen eingesetzt wurde (**Bild 6**).

Das Konzept der OCMC-Ummantelung wurde bei einem ersten Pilotversuch erfolgreich unter Beweis gestellt, da die beiden ausgewählten Rollen den Betriebsbedingungen im Ofen viel besser standhielten als ungeschützte Rollen und über zwei Jahre ohne größere Schäden überdauerten, wodurch die Degradation der Faserisolierung des Feuerfestmaterials erheblich verringert und der Kühlungsbedarf effektiv reduziert wurde. Die Prepregs nutzten jedoch nicht die Kosten- und Performance-Vorteile der WPX-OCMC-Wickeltechnologie, bei der Endlosfaser-Rovings mit einer keramischen Matrix infiltriert und auf einen Kern gewickelt werden, wobei die hohe Zugfestigkeit der Keramikfaser zur Positionierung des OCMC auf dem Wickelkern genutzt wird.

Im zweiten Pilotversuch wurde dieser Vorteil genutzt, indem das grüne OCMC am WPX-Standort direkt auf die Abschnitte der Stahlwalze gewickelt wurde, an denen die Ummantelung die Faserisolierung der Feuerfestmasse schützen soll. Mithilfe einer patentierten, proprietären WPX-Mullit-Matrix wurde so ein OCMC-System geschaffen, das auch im grünen Zustand mechanisch robust für die Handhabung und den Überseetransport der Walzen zur CSP®-Anlage ist.

Vier solcher Walzen wurden dann unter Betriebsbedingungen in den heißen Tunnelofen der CSP®-Anlage eingesetzt. Sie haben sowohl den extremen Temperaturschock beim Einsetzen als auch den mittlerweile dreijährigen kontinuierlichen „Beschuss mit Zunderprojekti-

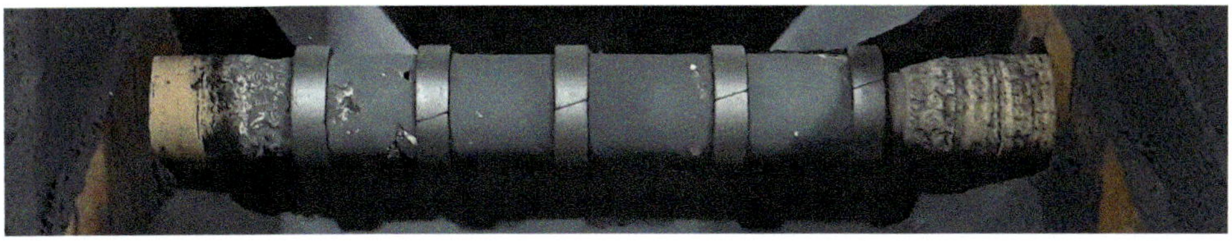

Bild 7: OCMC-Ummantelung einer Faserisolierung nach 18 Monaten Zundereinwirkung

len" in einer korrosiven Umgebung bei einer Temperatur von 1.200 °C überstanden, wobei angesichts des derzeitigen Zustands der Walzen mit mehreren weiteren Jahren intakter Leistung zu rechnen ist (**Bild 7**).

Die Versuche haben gezeigt, dass eine OCMC-Ummantelung der Feuerfestisolierung von wassergekühlten Stahlwalzen a) die Degradation der Feuerfestisolierung erheblich verzögern oder sogar stoppen kann, wodurch die Wärmeverluste im Wasserkühlsystem reduziert werden und somit der CO_2-Fußabdruck des Tunnelofens verringert wird, b) die Wartungsintervalle zwischen dem Austausch von beschädigten Stahlwalzenisolierungen verlängert werden, c) die Lebensdauer der Stahlwalzen verlängert wird, indem die thermische Belastung und die metallurgische Degradation des Stahls verringert werden, und d) die Produkt- und Prozessqualität verbessert wird, indem der Betrieb des Tunnelofens ohne Einschränkungen aufgrund begrenzter Kühlkapazität ermöglicht wird.

Die Wirtschaftlichkeit und der verringerte CO_2-Fußabdruck dieser Anwendung sind insbesondere für bestehende und künftige Stahlwerke in der Europäischen Union attraktiv, wo relativ hohe Energiekosten und zunehmende regulatorische Beschränkungen für die Stahlproduktion auf technische und politische Bemühungen zur Einführung von Green Steel treffen.

Literatur

[1] W. S. Association: Climate change and the production of iron and steel. World Steel Association, Belgien, 2021

[2] https://www.sms-group.com/en-de/press-and-media/press-releases/ press-release-detail/sms-group-supplies-process-technology-for-theworlds- first-large-scale-green-steel-production-plant

[3] Kunz, M.: Effizienz- und Qualitätsverbesserung in der Wärmebehandlung durch Werkstückträger aus Oxid-Faserkeramik. In: Irretier, O.; Jost, M. (Hrsg.): Handbuch HärtereiPraxis, Vulkan-Verlag GmbH, Essen, 2017

[4] Krenkel, W. (Hrsg.): Ceramic Matrix Composites, Fiber Reinforced Ceramics and their Applications. Wiley-VCH GmbH, Weinheim, 2008

[5] Schmücker, M.; Mechnich, P.: All-Oxide Ceramic Matrix Composites with Porous Matrices. In: Krenkel, W. (Hrsg.): Ceramic Matrix Composites, Fiber Reinforced Ceramics and their Applications. Wiley-VCH GmbH, Weinheim, 2008, S. 205–229

2.6 Recycling von Nitrierchemikalien – Ein Riesenschritt zur Erhöhung der Nachhaltigkeit

Dr. Joachim Boßlet

Was ist Nitrieren und Nitrocarburieren? Nitrieren und Nitrocarburieren sind etablierte Verfahren zur Verbesserung der Verschleiß- und Korrosionsbeständigkeit, die in der Wärmebehandlung heute nicht mehr weg zu denken sind. Prinzipiell lassen sich diese Prozesse entweder in Gasatmosphären, im Unterdruck mithilfe von Plasma oder in Salzschmelzen durchführen. Hierbei diffundieren Stickstoffatome (Nitrieren) oder Stickstoff- und Kohlenstoffatome (Nitrocarburieren) in Bauteile aus Stahl oder Gusseisen. Durch die Reaktion dieser Elemente mit dem Eisen und, falls vorhanden, mit entsprechenden Legierungselementen bildet sich in relativ kurzer Zeit eine harte Randschicht. Diese ist in der Praxis üblicherweise bis zu 20 µm dick und zeichnet sich durch ihren quasi keramischen Charakter aus. Im Gegensatz zum Einsatzhärten liegen die Behandlungstemperaturen mit 500–600 °C unterhalb der Umwandlungstemperatur des Eisens, sodass keine Gefügeveränderungen im Kern stattfinden. Somit sind diese Wärmebehandlungsverfahren nahezu verzugsfrei.

2.6.1 Nitrocarburieren in Salzschmelzen

Salzschmelzen bieten den großen Vorteil, dass

- wegen ihrer sehr hohen Wärmeleitfähigkeit,
- außergewöhnlicher Temperaturgleichmäßigkeit sowie
- dem mit Abstand größten Stickstoffangebot

Schichten mit sehr hohen Stickstoffgehalt in kürzester Zeit erzeugt werden. Nicht unerwähnt bleiben sollte die exzellente Robustheit dieser Verfahren.

Die HEF©-Gruppe besitzt fast 100 Jahre Erfahrung auf dem Gebiet des Nitrocarburierens in Salzschmelzen. Die Verfahren werden unter der geschützten Bezeichnung CLIN™ (Controlled Liquid Ionic Nitriding) vermarktet. Die Warenzeichen ARCOR©, TENIFER©, TUFFTRIDE© oder MELONITE© sind weltweit bekannt. Nicht nur in der Automobilindustrie, sondern auch im Maschinen- oder Werkzeugbau, Elektronik-, Öl- und Hydraulikindustrie sowie in der Luftfahrt und der Wehrtechnik werden CLIN™-behandelte Bauteile mit großem Erfolg eingesetzt (**Bild 1**).

Die Kunden schätzen insbesondere

- den hervorragenden Korrosionsschutz,
- die erzielbaren Oberflächenhärten,
- den ausgezeichneten Verschleißschutz,
- die sehr guten tribologischen Eigenschaften sowie
- die hohe Beständigkeit gegen Materialermüdung.

Angesichts verschärfter EU-Richtlinien bezüglich Hartverchromen rückt die Suche nach umweltfreundlichen Alternativen immer mehr in den Vordergrund. Die CLIN™-Nitro-

Bild 1: CLIN^{TM}-behandelte Bauteile

carburierungsverfahren sind praktikable und validierte Alternativen aus technischer, wirtschaftlicher und ökologischer Sicht.

2.6.2 Was passiert mit den Abfällen?

Da beim Nitrieren/Nitrocarburiern chemische Reaktionen stattfinden, werden entsprechend Chemikalien verbraucht. Bei den Gasverfahren handelt es sich um Ammoniakmischungen, die am Ende des Prozesses mithilfe von Propan direkt an den Anlagen mehr oder weniger vollständig verbrannt werden. Diese Entgiftung ist energieintensiv und belastet die Umwelt. Bei den Plasmaverfahren werden Stickstoff-/Wasserstoffmischungen bzw. beim Nitrocarburieren zusätzlich noch Methan – jedoch in deutlich geringeren Mengen – eingesetzt. Allerdings gelangen die zum Teil giftigen Reaktionsprodukte ohne weitere Bearbeitung in die Luft. Bei beiden Verfahren wird der Prozess in geschlossenen Öfen durchgeführt.

Im Gegensatz hierzu kommen beim Nitrocarburieren in Salzschmelzen halboffene Linien zum Einsatz. Die Öfen sind heutzutage aus Energieeinsparungsgründen mit Deckeln verschließbar (**Bild 2**). Ähnlich wie in der Galvanik oder beim Phosphatieren findet jeder Prozessschritt von der Vorreinigung über das Nitrocarburieren bis zum abschließenden Waschen der Bauteile in separaten Stationen statt. Bislang war es üblich, anfallende Schlämme und die mit den Bauteilen ausgetragenen Salzreste entweder nasschemisch zu entgiften oder in geeignete Deponien zu entsorgen.

Bild 2: CLIN™-Anlage

2.6.3 Recycling von Nitriersalzen

Um in einem Umfeld mit knapper werdenden Ressourcen und auch vor dem Hintergrund des weiterwachsenden Umweltbewusstseins für die Zukunft gut gewappnet zu sein, hat die HEF©-Gruppe vor drei Jahren am französischen Hauptstandort eine selbstentwickelte Pilotanlage zum Recyceln von verbrauchten Nitrierchemikalien und Schlämmen gestartet. Die Ergebnisse waren so vielversprechend, dass man nach Erhalt der ganzen bürokratischen Genehmigungen das sogenannte Green Center am Standort Andrézieux nahe Lyon in Betrieb genommen hat. Mittlerweile werden die Altsalze von vier französischen Dienstleistungszentren und auch an den beiden deutschen Standorten in wiederverwendbare Nitriersalze recycelt.

Das patentierte ECO-CLIN™ Verfahren besteht aus mehreren Schritten (**Bild 3**). Zuerst wird sinnvollerweise das Abwasser vor Ort in speziellen Vakuumverdampfer (ECO-CLIN™ PREP) aufkonzentriert. Das verdunstete Wasser hat VE-Qualität und wird wieder den Waschkaskaden zugeführt. Dies trägt zu einer erheblichen Einsparung der Ressource Wasser bei. Des Weiteren werden die Mengen an zu transportierendem Abfall deutlich reduziert. Im Green Center wird dieser chemisch aufbereitet (ECO-CLIN™ CHEM) und anschließend final getrocknet (ECO-CLIN™ DRY). Mit dieser Technologie lassen sich die Abfälle auf nahezu null reduzieren.

Bis 2026 soll ein weltweites Netzwerk von Green Centern in China, Indien, Deutschland, USA, Mexiko und Brasilien aufgebaut werden, damit lokal die Abfälle wiederverwertet werden können. Diese Investitionen vermeiden globale Transporte von Abfällen und tragen deutlich zur Verringerung des CO_2-Fußabdrucks der CLIN™-Verfahren bei. Die HEF©-

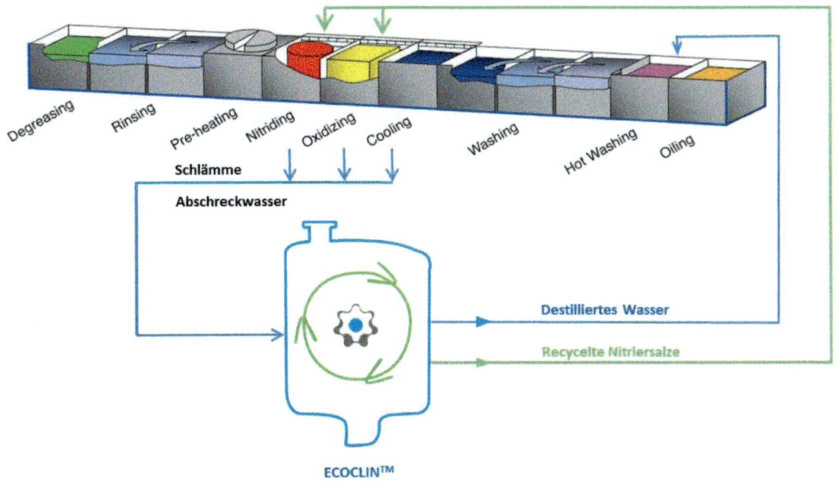

Bild 3: Schema Recycling

Gruppe ist eines der ersten Unternehmen weltweit, das seine Nitrierchemikalien recycelt und direkt wieder in den Produktionszyklus integriert.

2.6.4 Zukunft der CLIN™-Technologien

Die bei den CLIN™-Verfahren verwendeten Chemikalien enthalten keine Stoffe, die entweder als CMR (krebserzeugend, erbschädigend oder reproduktionstoxisch) bzw. als PBT (persistent bioakkumulierend toxisch) oder als vPvB (sehr persistent und sehr bioakkumulierend) eingestuft sind. Daher besteht nicht die latente Gefahr, dass diese irgendwann wie Chrom(VI)-Verbindungen auf der REACH-Verbotsliste erscheinen. Zusammen mit der Wiederverwertung der verwendeten Nitriersalzen wurden erfolgreich auch in ökologischer Sicht die Weichen für die Zukunft gestellt.

3. Automatisierung und Digitalisierung

3.1 Automatisierte Planungsoptimierung der Wärmebehandlung

Dr. Sebastian Goderbauer, Dr. Frank Hippenstiel, Jonas Metzger, Michael Müller

Die klimaneutrale Edelstahlproduktion stellt aufgrund der umfangreichen Wärmebehandlungen eine große Herausforderung dar. Die gapzero GmbH hat gemeinsam mit der BGH Edelstahl Siegen GmbH ein Programm zur automatisierten Planungsoptimierung der Wärmebehandlung entwickelt. Durch präzisere Prozesse und erhöhte Nutzungszeiten der Wärmebehandlungsöfen wird der CO_2-Fußabdruck reduziert. Aufgrund der positiven Erfahrungen bieten sich weitere Einsatzgebiete der Optimierungstechnik in der Fertigung an.

Nachhaltiges Produzieren steht bei der Boschgotthardshütte (BGH) in Siegen, die seit 1466 im Siegerland produziert, seit jeher im Vordergrund. Weit vor der Diskussion über den Klimawandel und der daraus resultierenden Notwendigkeit, CO_2 einzusparen, hatte die BGH bereits in den 1980er-Jahren in einem Forschungsvorhaben mit Universitäten das direkte Abschrecken von Stabstahl aus der Schmiedehitze vorangetrieben. Mittlerweile hat sich dieser Prozess, nicht zuletzt aufgrund des enormen CO_2-Einsparpotenzials, etabliert. Im letzten Schritt wurde eine Induktionsanlage hinter der Schmiedelinie eingerichtet, die Stabstahl aus Werkstoffen mit Lösungsglühtemperaturen oberhalb der Schmiedeendtemperatur vor dem Abschrecken nacherwärmt.

Voraussetzung für die optimale Fertigung von Stabstahl in dem verketteten Prozess von der Erwärmung der Rohblöcke bis hin zum fertig geschmiedeten, wärmebehandelten Stabstahl ist eine entsprechende Programmplanung. Hierbei sind einige Herausforderungen zu berücksichtigten, die sich aus den spezifischen Gegebenheiten bei der BGH, aber auch aus werkstofftechnischen, qualitativen Gesichtspunkten ergeben:

1. Werkstofftechnische Gründe wie z. B. das Management von Karbidausscheidungen
2. Erhöhung des Durchsatzes in dem Bereich Warmumformung und Wärmebehandlung
3. Optimierung der Herstellkosten und der Produktqualität (mechansich-technologische Eigenschaften)
4. Reduzierung des Umlaufbestands, im Wesentlichen aufgrund von Platzgründen.

Bedingt durch das Herstellprogramm werden neben dem direkten Abschrecken aus der Schmiedehitze eine Vielzahl von Möglichkeiten der Wärmebehandlung durchgeführt. In Gänze stehen allein im Werk Weidenau der BGH insgesamt 33 Wärmebehandlungsöfen und drei Abschreckbecken zur Verfügung, die 24/7 genutzt werden. **Bild 1** zeigt den Bereich der Wärmebehandlung 3. Hier kann Stabstahl bis zu einer Länge von 18.000 mm vergütet werden.

3.1.1 Bisheriger Planungsprozess

Die übergeordnete Produktionsplanung erfolgt mittels ERP-System. In den einzelnen Abteilungen erfolgt auf der Basis der Gesamtplanung eine Feinplanung der einzelnen Produktionsaggregate. Die Programmplanung für die Schmiedelinie in Weidenau inklusive Wärmebehandlung erfolgt wöchentlich für die jeweilige Folgewoche, wobei die Reihenfolgeplanung bisher mit geringer digitaler Unterstützung durch ein Gantt-Diagramm erstellt wurde. Wesentlicher Aspekt bei dieser Planung ist die Expertise der Mitarbeiter und ein möglichst

Bild 1: Im Bereich der Wärmebehandlung 3 kann Stabstahl bis zu einer Länge von 18.000 mm vergütet werden (Quelle: BGH Edelstahl Siegen).

störungsfreier Betrieb aller geplanten Anlagen und Öfen, um unter der Woche möglichst wenige Änderungen vornehmen zu müssen.

Es wurden immer wieder Projekte gestartet, um die Abläufe in der Arbeitsvorbereitung mit konventionellen EDV-Lösungen zu automatisieren, was aber bisher immer gescheitert war. Grundsätzlich stand bei diesen Bemühungen stets die Reduzierung des Planungsaufwands im Vordergrund. Bedingt durch die Notwendigkeit, CO_2-Emissionen zu reduzieren und gleichzeitig die Energiekosten zu senken, bekam dieser Wunsch nach einem automatisierten Planungswerkzeug wieder eine neue Priorität. BGH hatte die Problemstellung erneut für eine Ausschreibung formuliert. Gapzero stellte mit ihrer mathematischen Optimierung das schlüssigste Konzept vor und wurde deshalb mit der Erarbeitung der Lösung betraut.

3.1.2 Wärmebehandlungsplaner

Gapzero ist eine Ausgründung der RWTH Aachen und darauf spezialisiert, komplexe Planungsaufgaben softwaretechnisch und automatisiert zu lösen. Dazu wandelt gapzero hochkombinatorische Planungssituationen mit ökologischer und/oder ökonomischer Relevanz in mathematische Zwillinge um und entwickelt Algorithmen, die auf Knopfdruck bestmögliche Handlungsempfehlungen ermitteln. Das kann z. B. ein Produktionsplan sein, der knappe Ressourcen besser nutzt als es ein Mensch aufgrund der Vielzahl von Möglichkeiten und Abhängigkeiten je könnte. Die spezialisierten Algorithmen zur Planungsoptimierung bettet gapzero in eine Software ein, die insbesondere Planungslösungen interaktiv darstellt und auch manuelle Eingriffsmöglichkeiten durch menschliche Experten erlaubt.

Die Zusammenarbeit mit gapzero begann mit einer ersten mathematischen Analyse der Planungssituation, dem Aufzeigen von Optimierungspotenzial sowie der Konzeptionierung einer Lösung und Abstimmung eines entsprechenden Entwicklungsfahrplans. Die Entwicklung des Wärmebehandlungsplaners (WB-Planer) durch gapzero startete im Mai 2022. Das Projektteam auf Seiten der BGH besteht aus den an der Planung beteiligten Personen rund um die Betriebsleiter Schmiede und Wärmebehandlung sowie einem Verantwortlichen aus dem Bereich IT-Management/Kommunikationstechnik. Der erste Projektfokus lag darauf, eine passgenaue Software zur Planungsunterstützung und algorithmischen Optimierung der Belegung der Wärmebehandlungsöfen zu entwickeln und in Anwendung zu bringen. Die Belegungsplanung von Öfen und Abschreckbecken ist ein Teilaspekt der Wärmebehandlungsplanung, der einerseits in der betrieblichen Praxis die größte Volatilität aufweist und deswegen durchgängig Umplanungen hervorruft. Andererseits ist die Ofenbelegungsplanung ein großer Stellhebel zur Optimierung von Durchlaufzeiten und Energieverbrauch.

Der WB-Planer ist eine browserbasierte Software, die vollständig auf Servern der BGH gehostet ist. Die Software ist an die bestehende LDAP-Benutzerauthentifizierung angebunden, sodass sich Mitarbeitende wie gewohnt anmelden können. Für den Austausch aller planungsrelevanten Informationen ist eine automatische, bidirektionale Schnittstelle zum Fertigungsleitsystem implementiert. Ein detailliertes Rechtesystem ermöglicht es, Nutzende mit unterschiedlichen Ansichts- und Bearbeitungsrechten auszustatten.

Planerische Unterstützung bei der Anlagenbelegungsplanung bietet der WB-Planer auf zwei verknüpfte Weisen: Auf der einen Seite hebt der WB-Planer das Planen per Hand auf den aktuellen Stand der Technik. Die bisher verwendete statische Planungsansicht mittels einer Excel-Visualisierung wird durch eine interaktive Visualisierung mit Drag-&-Drop-Funktion und intelligenten Planungshinweisen abgelöst. Zusätzlich ist der WB-Planer mit einer Planungs-KI ausgestattet, die auf Knopfdruck die ganze Planungsarbeit übernimmt und in Sekunden einen optimierten Fertigungsplan vorschlägt.

3.1.3 Intelligente Unterstützung bei der manuellen Planung

Das Herzstück des WB-Planers ist die Planungsansicht (**Bild 2**). In einem Gantt-Diagramm sind die Belegungen der Wärmebehandlungsanlagen zeitlich aufgetragen. Auf der vertikalen Achse sind die Öfen und Becken, gruppiert nach den Hallen, aufgeführt. Abschreckbecken sind blau hinterlegt. Die obersten Zeilen gehören zu dem der Wärmebehandlung vorgelagerten Prozess der Schmiede inklusive Schmiedemaschine und nachgeschalteten Induktions- sowie Tauchanlage. Der Schmiedetermin bildet den zeitlichen Startpunkt der anschließenden Wärmebehandlungen. Somit können die Schmiedetermine im WB-Planer

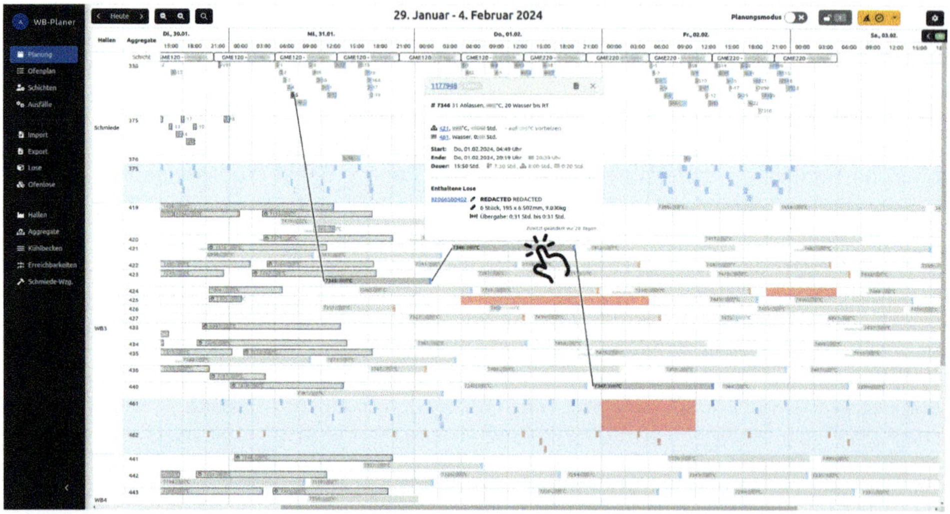

Bild 2: Die Planungsansicht zeigt ein Gantt-Diagramm mit den Belegungen der Wärmebehandlungsanlagen (Quelle: BGH Edelstahl Siegen).

mitgeplant werden. Die dabei planungsrelevanten Arbeitsschichten der Schmiede, inklusive Schmiedewerkzeug und angestrebter Tonnage sind in der Planungsansicht angegeben. Zeiträume, in denen eine Anlage nicht zur Verfügung steht, werden in Rot dargestellt.

Die Planungsansicht ist interaktiv. Für jeden Fertigungsprozess können per Mausklick Detailinformationen aufgerufen werden. Zu den Wärmebehandlungsschritten sind so Details der Behandlung und des behandelten Materials aufgeführt (vgl. Bild 2). Prozessverbindende Linien zeigen den Fluss des Materials vom Schmiedetermin über die Wärmebehandlungen mit eventuellen Abschreckungen in den Becken an.

Durch Aktivierung des Planungsmodus können Um- sowie Neuplanungen vorgenommen werden. Per intuitivem Drag-&-Drop können grün dargestellte Prozesselemente mit der Maus verschoben werden. Zusätzlich öffnet sich ganz rechts eine Spalte, in der alle per Datenschnittstelle automatisch importierten Prozessschritte auftauchen, die noch nicht eingeplant wurden. Auch diese können per Drag-&-Drop jeweils einer Zeit und einer Anlage zugeordnet werden. Bei der Ein- bzw. Umplanung erhalten die Nutzenden visuelle Leitplanken (**Bild 3**). So werden waagerecht Anlagen ausgegraut, die den festgehaltenen Prozessschritt nicht bearbeiten können. In der Vertikalen wird das Zeitintervall weiß hervorgehoben, in dem der Prozessschritt stattzufinden hat, um zeitliche Übergaberestriktionen zum Vorgängerprozess zu erfüllen. Im dargestellten Fall (Bild 3) soll z. B. das Härten innerhalb von 24 Stunden nach dem Schmieden beginnen.

Nutzenden des WB-Planers steht es frei, diese visuellen Leitplanken zu ignorieren. Die Software erkennt das und kommuniziert dies in Form eines kleinen roten Ausrufezeichens in der Ansicht. Ein solcher Planungshinweis erscheint immer dann, wenn eine einzuhaltende Restriktion nicht erfüllt wird. Die bestehende Planung wird somit durchgängig und automatisch auf missachtete Planungsregeln überprüft. Durch den Klick auf ein rotes Dreieck erhalten Nutzende Details darüber, welche Planungsregel im jeweiligen Fall gebrochen wird. In **Bild 4** beispielsweise werden Nutzende darauf hingewiesen, dass Material von zwei Ofenlosen gleichzeitig im selben Wasserbecken abzuschrecken ist und dies zu einer Überbelegung

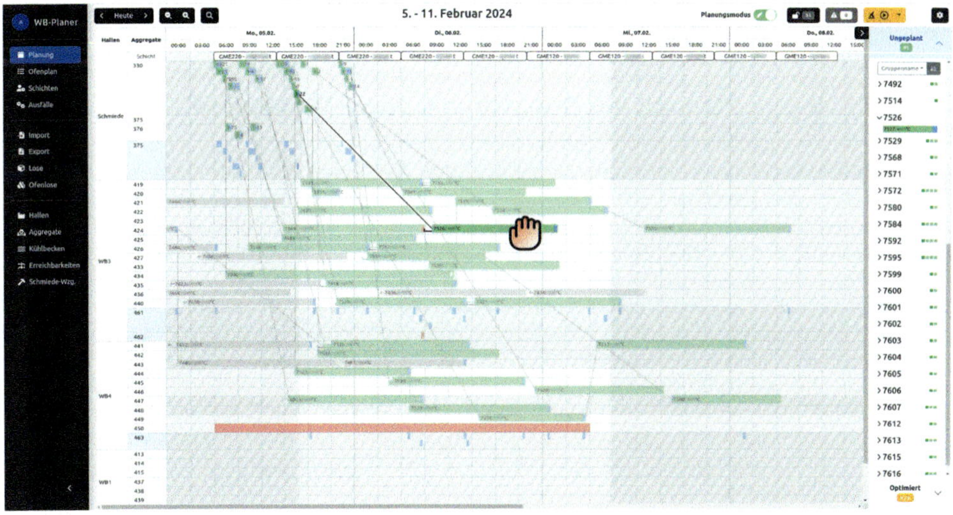

Bild 3: Bei Ein- bzw. Umplanungen weisen visuelle Leitplanken auf missachtete Planungsregeln hin (Quelle: BGH Edelstahl Siegen).

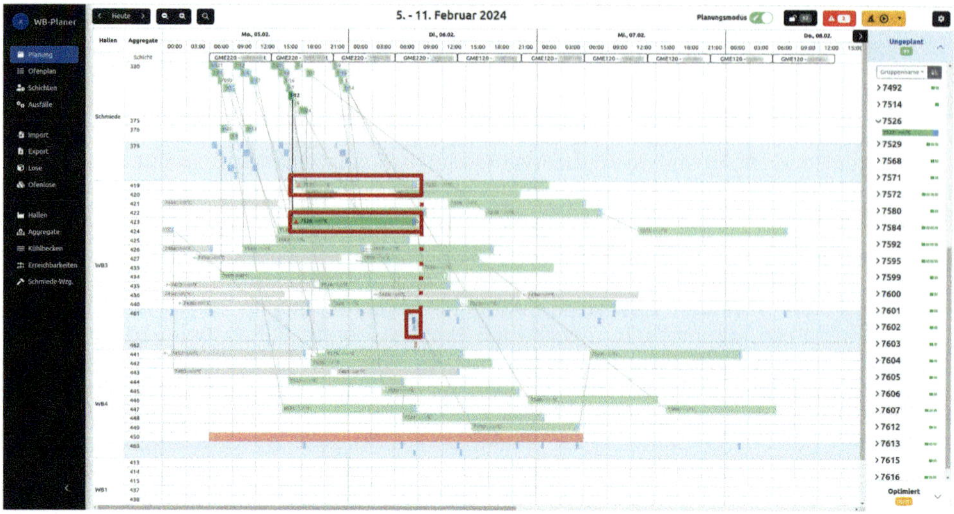

Bild 4: Durch Klick auf ein rotes Dreieck erhalten Nutzende Details darüber, welche Planungsregel im jeweiligen Fall gebrochen wird (Quelle: BGH Edelstahl Siegen).

des Beckens führt. Die Planungsdiskrepanz kann durch Verschieben eines der Ofenlose aufgelöst werden. Sobald die Planung zufriedenstellend ist, können Nutzende diese festsetzen (dargestellt durch graue Prozessbalken) und so an die Fertigung kommunizieren.

Die Planungsansicht wird laufend aktualisiert. Da durch die Datenschnittstelle automatisch die Rückmeldungen aus der Fertigung aufgenommen werden, bildet die Planungsansicht sowohl die Historie von real abgeschlossenen Prozessen (schwarz umrandete, graue Balken) als auch den aktuellen Ist-Zustand von momentan durchgeführten Prozessen sowie zusätzlich einen Blick in die geplante Zukunft ab.

3.1.4 Optimierte Planung auf Knopfdruck: durch Planungs-KI

Anstelle der manuellen Planung können sich Nutzende einer spezialisierten Planungs-KI bedienen. Diese definiert automatisch und in kurzer Zeit einen Plan, der nicht nur sämtliche Restriktionen und Vorgaben erfüllt, sondern auch bezüglich der nutzerseits ausgewählten Bewertungsziele optimiert. Dazu wird das Cockpit zur Planungsoptimierung aufgerufen. Darin ist anhand des Schichtplans in der Schmiede der Zeithorizont auszuwählen, der beplant werden soll. Außerdem ist die Menge an Aufträgen in Form von Schmiedeprozessen und Ofenlosen auszuwählen, die eingeplant werden sollen. Schon hier findet ein ausführlicher Datenabgleich der auf diese Weise definierten Planungsaufgabe statt. Falls z. B. Aufträge ein im Planungszeitraum nicht eingesetztes Schmiedewerkzeug fordern oder planungsrelevante Daten an den Fertigungsaufträgen fehlen bzw. widersprüchlich sind, wird anhand von Planungshinweisen darauf hingewiesen. Abschließend können Nutzende das Optimierungsziel festlegen. Zur Auswahl stehen entweder die Priorisierung von schnellstmöglicher Fertigstellung oder der Eco-Modus zur ökologischen und ökonomischen Optimierung der Prozesse.

Während der Optimierungsrechnung gibt der WB-Planer Informationen über Kennzahlen des bisher besten gefundenen Plans aus (**Bild 5**). Der optimierte Plan wird in der Planungs-

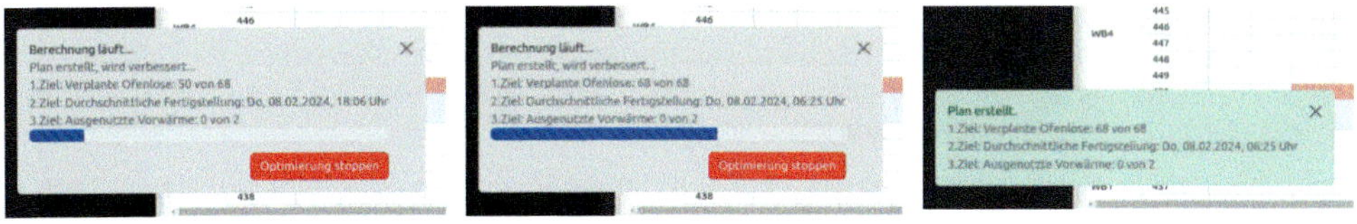

Bild 5: Stand der Planungsoptimierung nach 5, 20 und 30 Sekunden (Quelle: BGH Edelstahl Siegen)

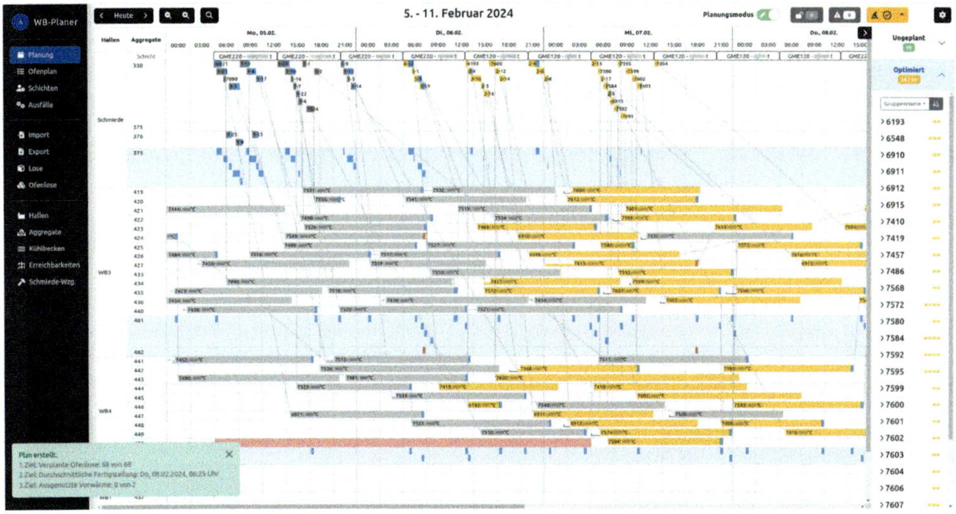

Bild 6: Die Planungsansicht zeigt den optimierten Plan (Quelle: BGH Edelstahl Siegen).

ansicht präsentiert (**Bild 6**). So werden in Sekundenschnelle Fertigungspläne für mehrere Tage ermöglicht, die eine optimale Ofenauslastung bei maximaler Tonnage umsetzen. Eine optimale Ofenauslastung bedeutet, Ofenstillstände zu reduzieren und – im Falle des Eco-Modus – Restwärme nach einer abgeschlossenen Ofenreise für die nächste zu nutzen. Es ist sichergestellt, dass alle Restriktionen, Kapazitäten und Vorgaben an die Planung, Wärmebehandlungen und Anlagen eingehalten sind. Es ist ausgeschlossen, dass eine geforderte Restriktion von dem Planungsalgorithmus missachtet wird.

Es bleibt dem Benutzenden dennoch unbenommen, den optimierten Plan mithilfe der oben vorgestellten Drag-&-Drop-Funktionalitäten manuell zu übersteuern. Dabei ist es wichtig, dass der menschliche Planungsverantwortliche die Oberhand über die Planung behält. Durch die visuelle Darstellung der Planung kann sich der Benutzende interaktiv von der Qualität der optimierten Planung überzeugen.

3.1.5 Zugrunde liegende Mathematik und KI

Dass eine Aufgabe, für die ein Mensch mehrere Stunden aufzuwenden hat, nun auf Knopfdruck in Sekundenschnelle erledigt wird, ist passgenau eingesetzter Mathematik und dar-

3

auf basierenden Algorithmen zuzuschreiben. Das Feld der Künstlichen Intelligenz (KI) ist vielfältig und sehr umfassend. Entscheidungstragende, die KI erfolgreich einsetzen wollen, sollten KI-Technologien und vor allem deren Einsatzmöglichkeiten und Lösungsfähigkeiten in ihren Grundzügen zu unterscheiden wissen.

Der am weitesten verbreitete Teilbereich der KI ist die Prognose-KI mit ihrem bekanntesten Vertreter, dem Maschinellen Lernen (englisch: Machine Learning). Maschinelles Lernen ist in der Lage, große Mengen historischer Daten zu verarbeiten, Muster zu erkennen und so aus der Vergangenheit zu lernen, um Vorhersagen und Aussagen über die Zukunft zu treffen. Mit dieser Technik kann z. B. ein Navigationsdienst wie Google Maps Stauprognosen erstellen oder eine Anlagensoftware abschätzen, wann welches Bauteil ausfallen wird und damit eine Wartung nötig ist (Predictive Maintenance).

Für eine automatische Planungsoptimierung der Wärmebehandlung ist die Fähigkeit der Vorhersage nicht zielführend. Eine automatische Nachahmung der menschlichen Planer würde zwar den Planungsaufwand verringern, aber eine verbesserte Planqualität z. B. im Sinne der Ressourceneffizienz ist damit nicht erreichbar. Ganz im Gegenteil: Anhand der Historie nachgeahmter Pläne würden dieselben Schwächen oder gar Planungsfehler enthalten sein.

Der WB-Planer basiert auf einer Planungs-KI, genauer gesagt der Technologie der Mathematischen Optimierung (englisch: Mathematical Optimization, s. auch Operations Research oder Decision Intelligence). Mathematische Optimierung lernt nicht aus historischen Plänen, wie bisher geplant wurde. Vielmehr wird der Mathematischen Optimierung im Vorfeld vollständig beigebracht, wie geplant werden darf, um für die Zukunft bestmögliche Handlungsempfehlungen zu geben. In enger Zusammenarbeit mit den Experten der BGH durchdrang das Team von gapzero alle planungsrelevanten Details und erarbeitet eine mathematische Beschreibung, eine Art mathematischen Zwilling der Planungssituation, die die mathematische Optimierung versteht. Sie bildet das gesamte Spektrum der zulässigen Planungsmöglichkeiten ab, beschrieben und abgegrenzt durch Restriktionen, Regeln und Vorgaben, aber auch Freiheiten der Planungsaufgabe. Dies sind simple Vorgaben wie „Ein Ofen mit einer Länge von 15 Metern kann nur Material bis zu dieser Länge behandeln" oder auch verknüpfende wie zeitliche Restriktionen zwischen zwei Wärmebehandlungsschritten. Anhand einer weiteren Beschreibung des Optimierungsziels, z. B., dass das Ausnutzen von Ofenwärme bevorzugt wird, erhält die mathematische Optimierung die Fähigkeit, Pläne qualitativ bewerten und so miteinander vergleichen zu können. Es entsteht ein hochkombinatorisches Konstrukt aus ineinander verflochtener und abhängiger Teilentscheidungen, das die mathematische Optimierung beherrschen kann.

Mathematische Optimierung ist erprobt und besonders bei führenden globalen Unternehmen verbreitet. So setzen 85 % der 500 umsatzstärksten US-Unternehmen diese Planungs-KI in ihrem Betrieb ein. Digitale Herausforderer wie Flaschenpost, Uber oder HelloFresh revolutionieren ganze Branchen mithilfe Mathematischer Optimierung.

3.1.6 Erste Anwendungserfahrungen

Der Status quo ante war eine Softwareumgebung, in der z. B. Zeitpunkte Ziffer für Ziffer eingegeben werden mussten und ein statisches Gantt-Diagramm in einem separaten Programm generiert werden musste. Für die Planung eines Produktionstages benötigten Nutzende einen knappen Arbeitstag. Die Inbetriebnahme der manuellen Planungsoberfläche im Januar 2023 hat bereits deutliche Verbesserungen gebracht. So können nun einzelne Wärmebehandlungslose in der manuellen Planung einfach und zeitsparend mithilfe von Drag-&-Drop umgeplant werden. Ebenso ist die Ansicht der Planung übersichtlicher und für

Mitarbeitende in der Wärmebehandlung digital einsehbar. Erforderliche Informationen zu den Wärmebehandlungslosen sind nun per Mausklick abrufbar. Damit wird ein potenzieller Übertragungsfehler von relevanten Prozessdaten nunmehr ausgeschlossen.

Mit der Inbetriebnahme der automatischen Planungsoptimierung im August 2023 konnte der Aufwand noch zusätzlich deutlich reduziert werden. Das Programm für einen kompletten Planungstag kann nun innerhalb von maximal 1,5 Stunden komplettiert werden. Dies gibt den Nutzenden Freiraum für wichtigere Tätigkeiten. Des Weiteren konnte die Planung wesentlich präzisiert werden, sodass die Kapazitäten in der Wärmebehandlung optimal ausgenutzt werden können und „Staus" von Material, das umgehend in Wärmebehandlungsöfen übernommen werden muss, vermieden werden.

3.1.7 Ausblick: weitere Optimierungsunterstützung

Die Planungssoftware wird derzeit für eine weitere Fertigungslinie im Werk Eintracht angepasst. Aufgrund von überwiegenden Direktübernahmen aus dem Stahlwerk müssen Schmiede und Wärmebehandlung auf Änderungen im Betriebsablauf reagieren können. Das Konzept von gapzero sieht vor, die Planungsansicht und -algorithmen des WB-Planers so anzupassen, dass mit den vorhandenen Unsicherheiten bestmöglich umgegangen werden kann. Abweichungen im Betriebsablauf optimal zu meistern – das ist auch die Überschrift der Weiterentwicklung des WB-Planers für den Standort Siegen-Weidenau. Die im ersten Ausbauschritt für die Vorplanung eingesetzte Planungsansicht und die eingesetzten Optimierungsalgorithmen werden zu einer Live-Plantafel erweitert. Für den Fall, dass eine Ofenkurve nicht wie geplant gefahren werden kann, sodass sich der Abschluss der Wärmebehandlung verzögert oder ein Aggregat spontan ausfällt, sorgt eine automatische Re-Optimierung für die beste Handlungsempfehlung. In diesem Zuge wird auch eine hiermit verknüpfte, neue Ansicht für das Personal an den Öfen entwickelt. Ähnlich wie bei einer Abflugtafel am Flughafen wird dargestellt, welche Arbeit an welchem Ofen als nächstes ansteht. In derselben Ansicht können Rückmeldungen einfacher und detaillierter erfasst werden, als es im bisher angewendeten Prozess der Fall war. Diese neuen Informationen können direkt bei der Re-Optimierung genutzt werden.

Darüber hinaus liegen auf Seiten von gapzero Konzepte und erste Projekterfahrungen vor, um die vorgeschaltete Planungsarbeit der Bildung von Wärmebehandlungslosen automatisch und vor allem ressourcenoptimiert zu ermöglichen. Die Zusammenstellung von Ofenlosen ist ein wichtiger Hebel, um Ofenauslastung und somit den spezifischen Energieeinsatz zu verbessern.

Literatur

[1] Beer, S.; Müller, M.; Hippenstiel, F.; Janz, W.: Optimierung der Wärmebehandlung durch eine fortschrittliche Induktionsanlage. In: elektrowärme international, 02 (2017), S. 51–55

[2] Hippenstiel, F.; Janz, W., Müller, M.: Direct Quenching After Forging and Inductive Re-heating – A New Efficient Process Technology. In: BHM, 163 (2018) 9, S. 384–389

[3] Gurobi Optimization, LLC: State of Mathematical Optimization Report. 2021

[4] Goderbauer, S.: Vortrag auf der 25-Jahresfeier der Effizienz-Agentur NRW zum Thema Ressourceneffizienz 4.0. 24. Oktober 2023

3

3.2 Heating Up Efficiency – Digitalisierung, die einfach funktioniert!

Björn Schuh

In der Industrie gewinnt die Kombination von Digitalisierung, Prozessautomatisierung und Künstlicher Intelligenz zunehmend an Bedeutung. In diesem Artikel wird dargestellt, wie modulare Automatisierungsökosysteme Produktionsprozesse in der Wärmebehandlung effizienter, sicherer und zukunftsfähig machen (**Infokasten 1**, **Infokasten 2**). Wie das funktioniert, wird exemplarisch anhand des Demig-Produktangebotes vorgestellt.

Wärmebehandlungsbetriebe stehen heute in einem komplexen Spannungsfeld: Einerseits steigen die Anforderungen an normgerechte Wärmebehandlung und deren Dokumentation, energieeffiziente Prozesse und IT-Sicherheit; andererseits nehmen die Betriebskosten, insbesondere für Energie, drastisch zu. Gleichzeitig wird es immer schwieriger, geeignetes Fachpersonal zu finden. Damit geht nicht nur Kapazität verloren, sondern auch wertvolles Know-how. In diesem herausfordernden Umfeld gewinnen digitale Ökosysteme in der industriellen Produktion zunehmend an Bedeutung. Sie ermöglichen einen ganzheitlichen Lösungsansatz, der den zentralen Herausforderungen der Branche strukturiert begegnet.

Infokasten 1

Stärken des Automatisierungsökosystems:
Modulare Systemarchitektur: Skalierbare Systeme ermöglichen durchgängige Kontrolle und einfache Erweiterung.
Zuverlässige Prozesssteuerung: Thermische Prozesse werden präzise gesteuert und sicher überwacht – für gleichbleibende Qualität und hohe Effizienz.
Intelligente Softwarelösungen: Automatisierung und smarte Tools optimieren Abläufe, sparen Ressourcen und integrieren sich nahtlos in bestehende IT-Strukturen.
Standardkonforme Dokumentation: Relevante Daten werden automatisch, vollständig und normgerecht dokumentiert – ideal für Qualitätssicherung und Audits ohne Aufwand.
Schnittstellenoffenheit & Kompatibilität: Hohe Systemkompatibilität ermöglicht die Integration verschiedener Anlagen und reduziert Investitionsrisiken.
Cybersicherheit & Systempflege: Regelmäßige Updates und schnelle Wiederherstellung sichern den Betrieb und schützen vor Ausfällen.
Benutzerzentrierte Bedienung: Intuitive Oberflächen und mobile Zugänge erleichtern die Nutzung – weltweit und mehrsprachig.
Wissensaufbau & Support: Schulungen und Beratung stärken das Know-how im Betrieb und machen Mitarbeitende zu Prozessprofis.

Infokasten 2

Warum regelmäßige Software-Updates unverzichtbar sind:
Sicherheit: Schließen Schwachstellen, schützen vor Cyberangriffen.
Stabilität: Beheben Fehler, verhindern Abstürze und reduzieren Ausfallzeiten.
Effizienz: Optimieren Leistung und Ressourcennutzung – für mehr Produktivität.
Kompatibilität: Sichern Anschlussfähigkeit an neue Geräte und Systeme.
Innovation: Bringen neue Funktionen und erweitern den Nutzwert.
Compliance: Erfüllen gesetzliche Vorgaben und Datenschutzstandards.
Kostenersparnis: Verhindern teure Ausfälle und aufwendige Systemwechsel.

Doch was genau ist ein industrielles IT- und Prozess-Ökosystem?

Ein industrietaugliches digitales Prozess-Ökosystem besteht aus einer modular aufgebauten, eng verzahnten Kombination von Hardware, Software, Dateninfrastruktur und Dienstleistungen. Es integriert Maschinen, Sensorik, Steuerungssysteme, Prozessleittechnik, Analyse-Tools und Mensch-Maschine-Schnittstellen (HMI) in ein zusammenhängendes Gesamtsystem. Dabei geht es nicht um eine Aneinanderreihung von Einzelkomponenten, sondern um ein abgestimmtes System, das nahtlos ineinandergreift, Synergien schafft und Prozesse über alle Ebenen hinweg steuert und optimiert. Klingt kompliziert, ist es aber nicht, wenn die Interaktion von der Konzipierung an mitgedacht wird.

Gerade in sicherheitskritischen Bereichen wie der Wärmebehandlung bringen solche Plattformlösungen messbare Vorteile: von der Effizienzsteigerung über verbesserte Datenverfügbarkeit bis hin zur erhöhten IT- und Betriebssicherheit. Ein gut konzipiertes Ökosystem ist überdies skalierbar, offen für neue Technologien und trägt entscheidend dazu bei, die Komplexität moderner Produktionsumgebungen beherrschbar zu machen.

3.2.1 Effizient, normkonform und zukunftssicher: die Vorteile eines integrierten Automatisierungsansatzes

Moderne Wärmebehandlungsprozesse erfordern ein hohes Maß an Präzision, Dokumentationssicherheit und Effizienz. Das Demig-Automatisierungs-Ökosystem (**Bild 1**) liefert hierzu einen umfassenden Lösungsansatz, der alle Ebenen der Prozessautomatisierung und der Produktion sinnvoll miteinander verknüpft. Ein zentrales Element ist die ganzheitliche Prozesskontrolle: Von Einzelreglern bis hin zum ERP-System ist die Dokumentation automatisiert und durchgängig verfügbar, und zwar ohne Eingriff des Personals. Beispiele hierfür sind die permanente Datenerfassung und -speicherung durch den Prozessregler und die Übergabe an das Prozessleitsystem als automatisiertes Back-up. So lassen sich thermische Abläufe exakt steuern, überwachen und betriebssicher speichern sowie rechtssicher dokumentieren – durchgängig, transparent und nachvollziehbar. Gleichzeitig trägt das System maßgeblich zur Effizienzsteigerung bei. Manuelle Arbeitsschritte werden durch intelligente Automatisierungslösungen ersetzt, wodurch sich Fehlermöglichkeiten minimieren, Produktionszeiten verkürzen und der Ressourcenverbrauch deutlich senken lässt.

Als Basis hierzu dient die Definition der Systemvariablen bzw. „Symbole“, die durchgängig über alle Ebenen der Automatisierungspyramide zugeordnet bleibt und redundant gesichert wird. Dazu dient eine Software namens „Winconfig“, mit der die Systemkonfiguration,

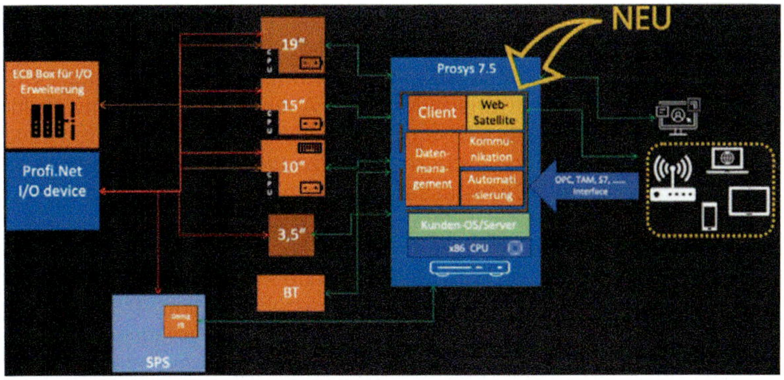

Bild 1: Smarte Prozesse brauchen smarte Tools: das Demig-Ökosystem auf einen Blick (Quelle: Demig Prozessautomatisierung GmbH)

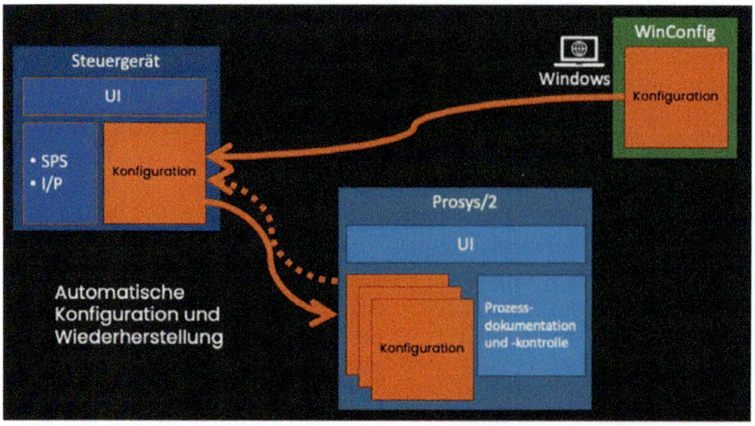

Bild 2: Die Systemkonfiguration ist das Herzstück des Demig-Ökosystems (Quelle: Demig Prozessautomatisierung GmbH)

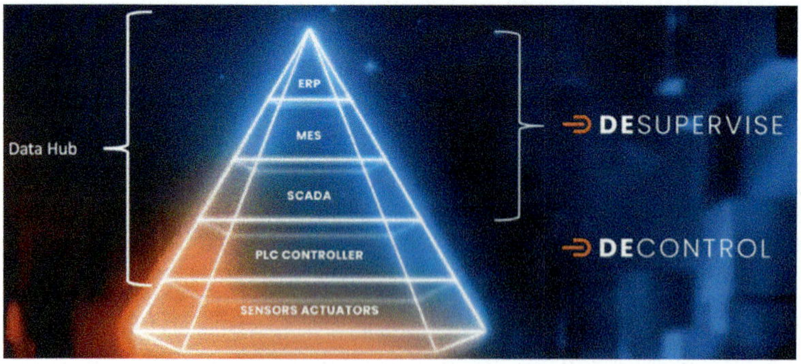

Bild 3: Intelligent verknüpft: Automatisierung ohne Brüche (Quelle: Demig Prozessautomatisierung GmbH)

z. B. die Anlagenkonfiguration, geschrieben wird. Die in dieser Konfiguration gespeicherten Daten beschreiben das gesamte System der Anlage, z. B. einen Ofen. Damit das System zuverlässig läuft, wird diese Konfiguration sowohl auf dem Prozessregler als auch im Leitsystem gesichert und kann selbst bei Ausfall einer Komponente innerhalb weniger Minuten durch das Leitsystem wieder hergestellt werden. Diese Basis einer Demig Anlagen-Systemkonfiguration (**Bild 2**) ermöglicht ganzheitliche Prozesskontrolle und Effizienzsteigerung – auch mithilfe der integrierten Software-SPS. Ein weiterer wesentlicher Aspekt ist die normgerechte Dokumentation. Das Demig-System erfasst sicherheits- und qualitätsrelevante Daten automatisiert und regelkonform (**Bild 3**) – etwa nach ISO 9001, IATF 16949 oder der branchenspezifischen AMS2750. Auch die Erfüllung anwendungsbezogener Normen wie ISO 20431, AIAG CQI-9 oder NADCAP-Prozesse werden zuverlässig unterstützt. Mit Blick auf die Investitionssicherheit überzeugen die Demig-Systeme durch ihre Abwärtskompatibilität bis 1992. Damit lassen sich auch ältere Anlagen unkompliziert modernisieren, ohne dass teure Komplettumbauten notwendig sind. Dank des modularen Aufbaus ist das System zudem hochgradig skalierbar und flexibel anpassbar, sowohl in Bezug auf die Hardware als auch auf die Software (**Bild 4**). Neue Funktionen oder Erweiterungen lassen sich bedarfsgerecht und ohne Brüche in die bestehende Infrastruktur integrieren. Ergänzend dazu setzt Demig auf gezielten Know-how-Transfer: Schulungen, Trainings und Beratungsleistungen befähigen Mitarbeitende, die Systeme optimal zu nutzen und eigenes Prozess-

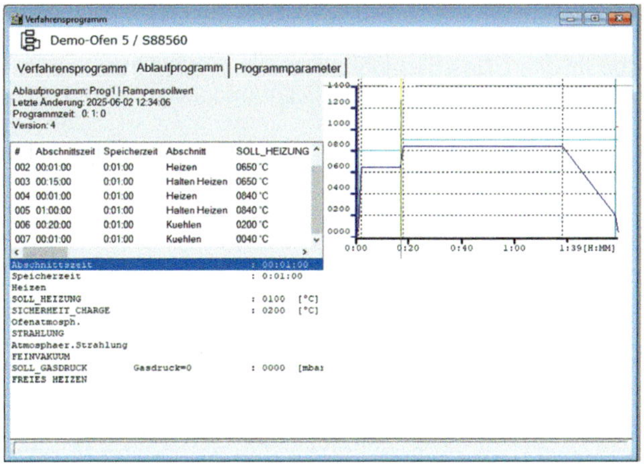

Bild 4: Anpassen leicht gemacht: der smarte Begleiter für die Ofenanzeige (Quelle: Demig Prozessautomatisierung GmbH)

3

wissen nachhaltig aufzubauen. Das ist ein entscheidender Faktor, um trotz Fachkräftemangels langfristig konkurrenzfähig zu bleiben.

3.2.2 Daten als Fundament für KI und digitale Intelligenz

Der Erstellung und Verwaltung der eben beschriebenen Prozessdaten dient das digitale Ökosystem. Es ist nur so leistungsfähig wie seine Datenbasis. Um Maschinenlernen, Prozessoptimierung und vorausschauende Wartung zu ermöglichen, müssen Daten strukturiert, vollständig und kontinuierlich verfügbar sein. Genau hier setzen die Kernkomponenten von Demig an, und zwar bei Hardware und Software, die modular aufeinander aufbauen. Digitale Automatisierungsökosysteme sind einfach, wenn sie komplexe Automatisierungsprozesse auf eine intuitive und benutzerfreundliche Weise zugänglich machen, und zwar ganz ohne unnötigen technischen Ballast. Nutzer können mit einem einzelnen Regler starten und das System nach Bedarf erweitern – ganz ohne Umbau der gesamten Anlagen.

Demig hat dazu bestens aufeinander abgestimmte Bausteine entwickelt und über die Jahre hinweg gepflegt. Sie sind so gestaltet, dass sie von den Nutzern über die verschiedenen Bereiche hinweg, also vom Instandhalter, Anlagenbediener, Logistiker bis zum Leiter eines Wärmebehandlungsbetriebes, leicht zur Effizienzsteigerung genutzt werden können. Dabei gliedern sich die Tools entsprechend den organisatorischen Ebenen. Die Bezeichnung der Produktsegmente folgt dabei der Anwenderstruktur.

Auf dem Shopfloor übernimmt DE|Control mit den Reglern und seiner Software die präzise Steuerung der Wärmebehandlungsanlagen und erfasst parallel dazu sämtliche Prozessdaten direkt im Prozess, inklusive Benutzerinteraktionen, Temperaturverläufen, Energieverbräuchen und Chargeninformationen. Die Ebenen darüber nutzen Softwaresysteme des Produktsegmentes DE|Supervise. Diese Softwaresysteme aggregieren die Daten aus der gesamten Produktion und auch aus Steuerungssystemen anderer Typen und überführen sie verschlüsselt z. B. in zentrale Datenbanken oder übergeordnete ERP-Systeme. Damit wird die Datenkonsistenz gewährleistet: eine Grundvoraussetzung für KI-basierte Auswertungen und intelligente Entscheidungsunterstützung. Ob Predictive Maintenance, automatische Qualitätsprüfung oder intelligente Energieoptimierung – ohne diese integrierte Datenbasis bleiben KI-Initiativen Stückwerk. Diese Daten müssen selbstverständlich geschützt sein – zum einen, weil sie Teil des wertvollen Erfahrungs-Know-hows der Wärmebehandlungsbetriebe sind, und zum anderen, weil sie vor Cyberangriffen gesichert werden müssen. Das bedeutet: Das Ökosystem muss so konzipiert sein, dass es unbefugten Zugriff verhindert und regelmäßig aktualisiert wird. Hierzu bieten sich bevorzugt „On-Premise-Lösungen" an. Gerade im Umgang mit digitalen Systemen in der Produktion ist ein Umdenken erforderlich – die konsequente Einführung eines lückenlosen Benutzermanagements oder regelmäßige Software-Updates beispielsweise.

3.2.3 Komplexe Technik, einfach bedienbar

Trotz der technologischen Tiefe bleiben Demig-Systeme einfach in der Anwendung. Das ist besonders in Zeiten schrumpfender Fachkräfte-Ressourcen und hoher Fluktuation wichtig. Mit den benutzerfreundlichen Tools DE|Supervise PCS, MES – Prosys (Prozessleitsystem) und DE|Supervise ERP – AMS (Auftragsmanagementsystem) werden auch komplexe Produktionsprozesse über klar strukturierte Oberflächen steuerbar: sicher, nachvollziehbar und ohne IT-Fachwissen nutzbar. Ein besonderer Mehrwert ist dabei die symbolische Programmierung (DSP). Sie erlaubt es, Programme für Prozessabläufe ohne klassische SPS-Logik zu erstellen. Dabei definiert der Maschinenhersteller oder -verantwortliche die sinnvollen möglichen Prozessschritte und -abfolgen und lässt nur genau diese auch zur Ausführung zu. Damit werden Fehler auf der Anwenderseite vermieden. Anwender können selbstständig neue Rezepte konfigurieren oder bestehende Programme anpassen. Ein digitaler Assistent führt dabei Schritt für Schritt durch den Prozess. Das reduziert Abhängigkeiten von externem Know-how und stärkt die Autonomie im Betrieb.

Ein verbreiteter Irrtum in diesem Zusammenhang ist, dass Prozessleitsysteme (PCS – Prozess Control System) und SCADA-Systeme (Supervisory Control and Data Acquisition) dasselbe leisten. Zwar bieten beide eine Visualisierung und Überwachung von Produktionsprozessen, doch ein Prozessleitsystem geht funktional weit über ein klassisches SCADA hinaus. Während ein SCADA-System primär Echtzeitdaten einer Anlage visualisiert, ermöglicht ein echtes Prozessleitsystem wie das DE|Supervise Prosys von Demig die maschinenübergreifende Datenerfassung, zentrale Prozessdokumentation und umfassende Analyse über ganze Produktionslinien hinweg. Ein SCADA-System eignet sich vor allem für die Anzeige, Bedienung und Überwachung einzelner Anlagen oder Komponenten und bietet grundlegende Funktionen wie Alarmierung, Datenaufzeichnung und einfache Steuerbefehle. Ein Prozessleitsystem hingegen integriert darüberhinausgehende Funktionen wie Rezepturmanagement, Prozesskontrolle und automatisierte Bedienung wie Starten und Stoppen von Prozessen, Chargenverfolgung, Energiemanagement, Analysewerkzeuge und eine tiefere Systemintegration – etwa mit MES- oder ERP-Systemen. Es wird damit eben ein ganzheitliches Ökosystem, eine zentrale Plattform für Prozesstransparenz, Effizienz und strategische Steuerung.

So kann Fachpersonal beispielsweise

- Energieverbräuche verschiedener Anlagen miteinander vergleichen,
- automatisiert Berichte generieren oder
- historische Trends auswerten, ohne zwischen verschiedenen Systemen oder Einzelanlagen springen zu müssen.

Ein weiterer zentraler Vorteil: Prozessleitsysteme schaffen die notwendige Struktur z. B. für ein übergreifendes Energiemanagement, etwa durch

- gebündelte Verbrauchsdaten aller Wärmebehandlungsöfen,

- die Zuordnung von Energieverbräuchen zu konkreten Produktionsaufträgen bzw. Chargen,
- die Visualisierung von Effizienzkennzahlen sowie
- die direkte Anbindung an übergeordnete Managementsysteme.

Damit werden nicht nur operative Entscheidungen unterstützt, sondern auch strategische Ziele wie CO_2-Reduktion, Nachhaltigkeit oder ein ISO-50001-konformes Energiemanagement aktiv gefördert. Ein Prozessleitsystem ist ein zentraler Baustein für digitale Souveränität, transparente Produktionsprozesse und eine zukunftssichere, effiziente Fertigung.

3.2.4 Vom Konzept zur Praxis: Alltagstauglichkeit als Erfolgsfaktor

Digitale Lösungen entfalten ihren echten Wert erst im laufenden Betrieb. Entscheidend ist nicht nur, wie leistungsfähig ein System auf dem Papier ist – sondern wie robust, sicher und einfach in der Praxis anwendbar. Gerade in der industriellen Wärmebehandlung, wo es auf Präzision, Verfügbarkeit und Normkonformität ankommt, müssen Automatisierungssysteme dauerhaft zuverlässig und gleichzeitig flexibel anpassbar sein. Ein durchdachtes Ökosystem überzeugt deshalb nicht nur mit Technik, sondern mit praktischer Handhabbarkeit im industriellen Alltag: Updates, Datensicherheit, rollenbasierter Zugriff, globale Verfügbarkeit – all das sind zentrale Faktoren, wenn Digitalisierung nicht zur einmaligen Investition, sondern zum nachhaltigen Erfolgsmodell werden soll. In der Industrie ist Wartung bei Maschinen Standard – bei Software jedoch häufig noch ein Stiefkind. Dabei sind regelmäßige Updates und Patches essenziell: Sie schließen nicht nur Sicherheitslücken, sondern sorgen auch für Stabilität, Fehlerfreiheit und insbesondere für kontinuierliche Weiterentwicklung. Neue Features, optimierte Algorithmen und gesetzliche Anpassungen fließen auf diesem Weg in die Systeme ein.

3

Wie bereits erwähnt: Softwarewartung wird in der Industrie häufig vernachlässigt. Genau hier setzt das Prinzip von „Software as a Service on Premise“ (SaaS) an: Es stellt sicher, dass Softwarelösungen kontinuierlich betreut, gewartet und weiterentwickelt werden. Regelmäßige Updates und Patches sind dabei kein lästiges Beiwerk, sondern essenziell für den Schutz vor Cyberangriffen, die Stabilität im Betrieb, die Leistungsfähigkeit sowie die Kompatibilität mit neuer Hardware. Darüber hinaus ermöglichen sie die Integration neuer Funktionen und Features, sichern die Einhaltung gesetzlicher Vorgaben und verhindern kostspielige Ausfälle. Kurz: SaaS schafft die Grundlage für eine zukunftssichere, effiziente und normkonforme Produktion.

Mit dem SaaS-Konzept von Demig erhalten Unternehmen eine leistungsstarke, skalierbare und sichere Lösung für die digitale Produktion. Statt hoher Anfangsinvestitionen profitieren Kunden von einem flexiblen Abo-Modell – ideal auch für kleine und mittelständische Betriebe, die moderne Automatisierungslösungen ohne komplexe Infrastruktur einführen möchten. Die Software ist schnell einsatzbereit, wächst mit den Anforderungen und bleibt durch automatische Updates stets auf dem neuesten Stand – ohne zusätzlichen Wartungsaufwand im Betrieb. So entsteht ein durchgängiges Ökosystem bei planbaren Kosten für maximale Transparenz, Sicherheit und Effizienz. Ergänzt wird die Lösung durch Schulungen, Support und Predictive Maintenance – für eine hohe Anlagenverfügbarkeit und nachhaltige Produktivität. Das ist ein entscheidender Faktor für langfristige Investitionssicherheit und Normkonformität – besonders in sicherheitskritischen Branchen wie der Wärmebehandlung. Ein Beispiel hierfür ist der „Demig Web-Satellite“ (**Bild 5**). Mit ihm stellt Demig eine webbasierte Lösung im Softwaresystem DE|Supervise bereit, die weltweit den Zugriff auf Anlagenzustände, Prozessdaten und Reports ermöglicht und dies abgesichert nach aktuellen Sicherheitsstandards und Unternehmensrichtlinien. Die Benutzeroberfläche lässt sich individuell gestalten: Vom einfachen KPI-Dashboard bis hin zur Ofenvisualisierung oder Energie- und Wartungsmanagement können Rollen und Inhalte exakt auf den jeweiligen Anwender zugeschnitten werden (**Bild 6**). Ob Produktionsleiter in Deutschland, Instandhalter in Polen oder Qualitätsingenieur in den USA: Alle Beteiligten greifen in Echtzeit auf dieselbe Datenbasis zu und können so fundierte, schnelle Entscheidungen treffen. Um Updates effizient umzusetzen, hat sich SaaS auch in der industriellen Automatisierung als etabliertes Betriebsmodell bewährt. Es bietet Unternehmen flexible Nutzungsmöglichkeiten bei kalkulierbaren Kosten – ohne hohe Anfangsinvestitionen oder aufwendige Einführung. In

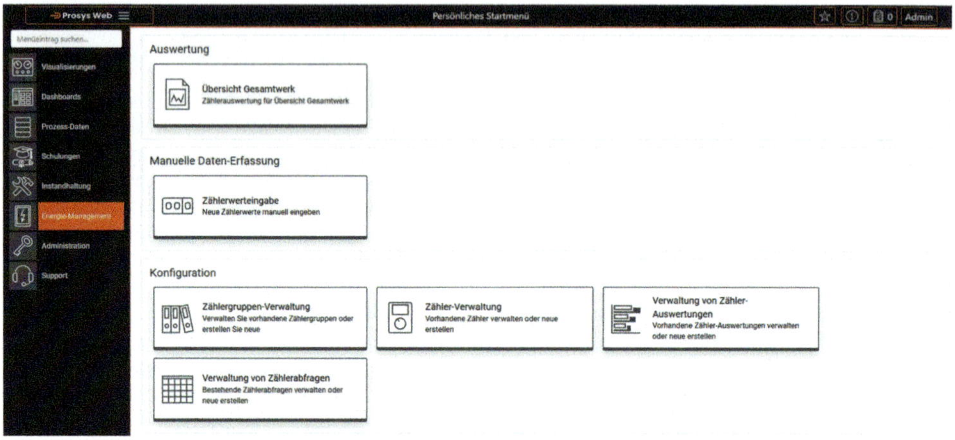

Bild 5: Webbasierter Überblick, wo immer der Anwender ist: der Demig Web-Satellite (Quelle: Demig Prozessautomatisierung GmbH)

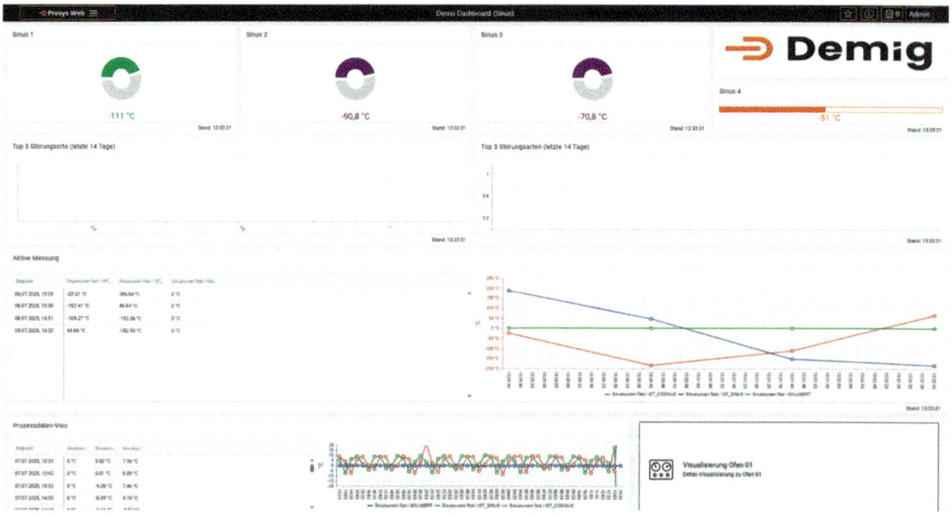

Bild 6: Immer den Überblick behalten mit klaren Dashboards und intuitiver Bedienung (Quelle: Demig Prozessautomatisierung GmbH)

Verbindung mit Schulungen, Fernwartung, Predictive Maintenance und individuellem Service entsteht eine solide Basis für den zuverlässigen Betrieb.

3.2.5 Vorteile im Überblick:

- Schneller Einstieg ohne lokale Installationen
- Flexible Skalierung je nach Unternehmens- oder Standortgröße
- Automatisierte Wartung und Sicherheitsupdates durch den Anbieter
- Hohe Verfügbarkeit
- Nahtlose Integration in bestehende Systeme via APIs
- Zentrale Datenanalyse und KI-Anbindung.

3.2.6 Fazit

Industrieunternehmen brauchen heute mehr als einzelne Tools oder isolierte Softwarelösungen. Sie benötigen ein stabiles, offenes und zukunftsfähiges Prozessautomatisierungsökosystem, das Menschen, Maschinen, Daten und Prozesse nahtlos miteinander verbindet. Demig bietet genau das – und ist damit nicht nur Technologieanbieter, sondern ein echter Digitalisierungspartner für Wärmebehandlungsbetriebe. Diese stehen unter hohem Druck: Steigende Anforderungen an Normkonformität, Effizienz und IT-Sicherheit treffen auf Fachkräftemangel und wachsende Kosten. In diesem Umfeld bieten digitale Automatisierungsökosysteme einen entscheidenden Vorteil. Sie vernetzen präzise Steuerung, durchgängige Prozessdokumentation, normgerechte Datenerfassung und KI-fähige Analysefunktionen zu einem skalierbaren Gesamtsystem. Das Ergebnis: höhere Effizienz, transparente Prozesse, einfache Bedienung und langfristige Investitionssicherheit. Durch modulare Erweiterbarkeit, Abwärtskompatibilität und nutzerfreundliche Tools lassen sich auch bestehende Anlagen zukunftsfähig integrieren. Ergänzt durch Software-Wartung, Schulungen und SaaS-Betriebsmodelle macht Demig die digitale Transformation zum industriellen Alltag – intuitiv, sicher und effizient.

4. Messen und Regeln

4.1 Regelung kohlenstoffbasierter Schutz- und Reaktionsatmosphären

Helmut Egger

Kohlenstoffbasierte Schutz- und Reaktionsatmosphären werden noch immer in einer großen Anzahl von Wärmebehandlungsbetrieben in kontinuierlichen sowie diskontinuierlichen Ofenlinien verwendet. Sie werden vor allem deshalb bevorzugt, weil sie sich sehr gut an die angewendeten Prozesse und Anforderungen der zu wärmebehandelnden Materialien anpassen lassen. In den letzten Jahren haben sich die Versorgungsbedingungen der benötigten Ausgangsgase (z. B. Erdgas, Propan) in ihrer Zusammensetzung geändert, wobei häufig Schwankungen mit daraus resultierenden Prozessstörungen auftraten, mit denen sich dieser Artikel befassen soll.

Die vorherrschenden Versionen kohlenstoffbasierter Atmosphären (**Tabelle 1**) variieren zwischen kontinuierlichen und diskontinuierlichen Ofenlinien. In Lohnwärmebehandlungsunternehmen werden vorwiegend Chargenofenanlagen betrieben, während in Härtereien in großen Betrieben (z. B. der Automotive-Industrie) kontinuierliche Wärmebehandlungslinien mittlerer bis großer Kapazität dominieren. Besonders in der zweitgenannten Kategorie werden bevorzugt Gasatmosphären verwendet, die in Generatoren (**Bild 1**) erzeugt werden. Erdgas ist nach wie vor – aus kommerziellen Gründen – das mit Abstand meistverwendete

Tabelle 1: Kohlenstoffbasierte Schutz- und Reaktionsgasatmosphären (Quelle: Egger Consulting)

Ausgangs-Medium #1	Ausgangs-Medium #2	Anmerkung	≈ CO %-Anteil im Erdogas	≈ H_2 %-Anteil im Erdogas	≈ N_2 %-Anteil im Erdogas
Erdgas	**Luft**	**dominierende Variante**	**20**	**40**	**40**
Propan	Luft	seltene Variante	22,9	31,6	45,5
Methanol	Stickstoff	im Wettbewerb zu erdgas & Luft	20	40	40
Methanol	Stickstoff	-“-, mit niedrigerem Stickstoffanteil	23	46	31
Methanol	Stickstoff	-“-, mit niedrigerem Stickstoffanteil	26	52	22
Methanol	Stickstoff	-“-, mit niedrigerem Stickstoffanteil	30	60	10
Methanol	–	höherer C-Transfer-Koeffizient bei Gasaufkohlung	33,33	66,67	0
Erdgas	Kohlendioxyd	höchster C-Transfer-Koeffizient bei Gasaufkohlung	50	50	0
Propan	Kohlendioxyd	höherer C-Transfer-Koeffizient bei Gasaufkohlung	60	40	0

Bild 1: Endogas-Generatoren – „stand-alone“ und integrierte Version (Quelle: Egger Consulting)

Ausgangsgas. Kommerzielle Dienstleister (Lohnwärmebehandlungsbetriebe), in denen vorwiegend Chargenofenlinien verwendet werden, bevorzugen die flexiblere Stickstoff-Methanol-Atmosphäre. Die Atmosphärenzusammensetzung wird meist so angepasst, dass sie im Wesentlichen einer Atmosphärenzusammensetzung wie bei endothermen Gasen – aus Erdgas hergestellt – gleicht, die aus etwa 20 % Kohlenmonoxid, 40 % Wasserstoff und 40 % Stickstoff besteht (um nur die Hauptbestandteile zu nennen). Beide Atmosphärenarten erfordern entsprechende Kohlenstoffpotenzial-Regelsysteme, bei denen sogenannte Anreicherungs- bzw. Konditionierungssubstanzen eingesetzt werden müssen, um die gewünschte Anpassung des Kohlenstoffpotenzials je nach Prozessanforderung zu ermöglichen. Diese Substanzen basieren auf Kohlenwasserstoffen und müssen während der thermischen Zersetzung „frei verfügbaren Kohlenstoff“ freisetzen, der für eine erneute Reaktion und Anpassung des Kohlenstoffpotenzials zur Verfügung steht. Die wichtigsten Bestandteile der Atmosphäre sind CO (Kohlenmonoxid) und H_2 (Wasserstoff), die beide reaktiver Natur sind und mit den potenziellen, anwendungsspezifischen „Schadstoffen“ O_2 (Sauerstoff) und H_2O (Wasserdampf) reagieren und diese absorbieren können, um eine unerwünschte Oxidation und/oder Entkohlung (**Bild 2**) auf der Oberfläche der zur Wärmebehandlung vorgesehenen Werkstücke zu verhindern.

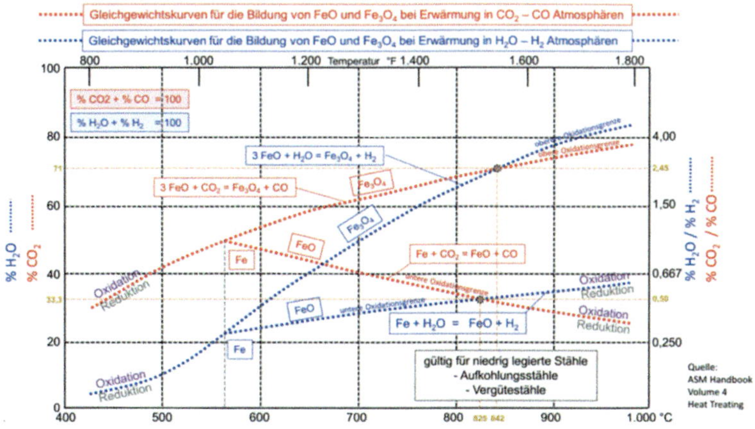

Bild 2: Gleichgewichtskurven für die Bildung von FeO and Fe_3O_4 (Quelle: Egger Consulting)

4.1.1 Berechnungsgestützte und geregelte Erzeugung von Endogas

Endogas wird – neben anderen Anwendungen – in der Automobilindustrie häufig zum Glühen, Einsatzhärten und Neutralhärten von Automotive-Bauteilen aus niedrig legiertem Stahl verwendet. Es wird in Gasgeneratoren (Bild 1) endotherm erzeugt, wobei eine Mischung aus Erdgas und Luft in eine heiße Reaktionsretorte eingeführt wird, die mit einem Katalysator auf Ni-Basis gefüllt ist. Das eingeleitete, durchfluss- und mengengeregelte Erdgas-Luft-Gemisch reagiert im Betriebstemperaturbereich zwischen 1.000 und 1.050 °C thermochemisch und verwandelt sich in ein Gasgemisch – Endogas bzw. auch RX-Gas genannt – mit den in **Bild 3** und **Bild 4** angegebenen Gasbestandteilen.

Für das erzeugte Endogas (Gasgemisch) sowie für alle darin enthaltenen Einzelgase gelten alle Naturgesetze für ideale Gase. Dies ermöglicht eine genaue Berechnung und Vorhersage der Analyse des endothermen Gases, sofern die Zusammensetzung des Ausgangsgases und die Luftbedingungen bekannt sind und angemessen berücksichtigt werden. Es

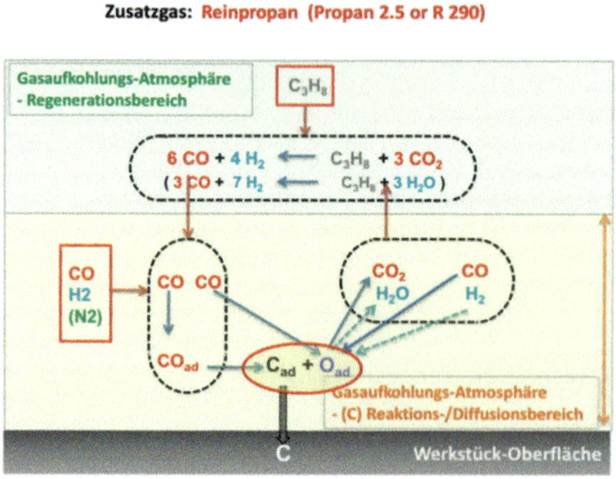

Bild 3: Gasreaktionen in Gasaufkohlungs-Atmosphären (Quelle: Egger Consulting)

1.	A	B	C		D		E	F	G
2.	Ausgangsgas (AG) - Analyse:		terranets-bw Mittelwert Jan. 2021				Ergebnis der Endogas - Analyse		
3.	Gasanteil	Formel	Ant.%		Luftbeiwert +/- %		Gasanteil	Formel	Ant.%
4.			mol %		Bereich -2 bis +4				mol %
5.			V %		2,6900				V %
6.	Methan	CH_4	95,5382				Kohlenmonoxid	CO	20,1018
7.	Ethan	C_2H_6	3,2171						
8.	Propan	C_3H_8	0,1469				Kohlendioxid	CO_2	0,2700
9.	Butan	C_4H_{10}	0,0625						
10.	Pentan	C_5H_{12}	0,0085				Wasserstoff	H_2	40,2484
11.	Höhere KW	C_6H_{14} +	0,0196						
12.	Kohlendioxid *	CO_2	0,5900				Wasserdampf	H_2O	0,2700
13.	Methan *	CH_4	0,5900						
14.	Stickstoff	N_2	0,4172				Stickstoff **	N_2	39,1098
15.	Wasserstoff	H_2	0,0000				** inkl. Edelgase		
16.	Sauerstoff	O_2	0,0000				Gesamt:	mol (V) %	100,0000
17.	Gesamt:	mol (V) %	100,0000						
18.	* CH_4 Reaktionsanteil / Abzug von Methananteil						Luftbedarf	m^3/m^3 AG	2,4450
19.	Berücksichtigung der Luftfeuchtigkeit						Endogasmenge	m_n^3/h	20,0000
20.	Wasserdampf	H_2O	0,0000		H_2O Gehalt g/m^3		AG - Menge	m_n^3/h	2,4450
21.	Methan *	CH_4	0,0000		8,000		Luftmenge	m_n^3/h	9,6330

Bild 4: Berechnungsprogramm für zu erwartende Endogas-Analyse (Quelle: Egger Consulting)

wurde ein Berechnungstool für die Endogas-Gasanalyse (Bild 4) entwickelt, das alle damit verbundenen Naturgesetze und thermochemischen Reaktionsformeln berücksichtigt, einschließlich der Auswirkungen der Luftbedingungen (relative Luftfeuchtigkeit). Basierend auf den Eingangsanalysedaten des Ausgangsgases sowie der relativen Luftfeuchtigkeit, des erforderlichen Endogas-Volumenstroms und dem gewünschten Taupunkt (TP °C) und/oder Kohlendioxidgehalt (CO_2) ermittelt das Berechnungstool dann die Sollwerte für die erforderlichen Durchflussmengen von Ausgangsgas und Luft. Ein weiteres Berechnungsprogramm steht zur Berechnung des erforderlichen mV-Sollwerts für das auf einer λ-Sauerstoffsonde basierende automatische Analysenregelungssystem zur Verfügung.

Das Berechnungstool ist auch für künftig erforderliche Konditionierungsmaßnahmen (**Bild 5**) – im Falle einer Beimischung von Wasserstoff in erheblichem Anteil – gerüstet. Durch Zugabe des Konditionierungsgases Kohlendioxid (CO_2) kann die resultierende endotherme Gasanalyse tatsächlich auf einen gewünschten, konstanten CO-Gehalt kalibriert werden, wodurch der Verdünnungseffekt ausgeglichen wird, der sonst in Kauf genommen werden müsste und der künftig durch den regionalen Erdgaslieferanten durch den beigemischten „grünen" Wasserstoff verursacht wird.

Wenn ein Endogas-Generator nur einer Ofenlinie oder einem einzelnen Prozess mit identischen Betriebsbedingungen (z. B. nur für einen bestimmten, begrenzten Kohlenstoffpegel – Regelbereich) zugeordnet ist, kann das Mengenverhältnis von Ausgangsgas und Luft so gewählt werden, dass es diesen einzelnen Zweck optimal erfüllt. Je niedriger der erforderliche Kohlenstoffpegel *Cp* im Ofen bzw. im angewandten Prozess sein muss, desto moderater kann das Mischungsverhältnis von Ausgangsgas und Luft durch leicht ansteigenden Luftüberschuss gewählt werden. Das Risiko von Störungen im Katalysatorrohr – etwa Rußbildung im Katalysator durch Luftmangel – kann damit sehr stark reduziert und ein kurzzeitiger Generatorausfall – erforderlich durch eine notwendige Regenerierung (Ausbrennen) – vermieden werden. Aus **Bild 6** können die entsprechend vorgeschlagenen Einstellkategorien mit Bezug auf bestimmte Kohlenstoffpegel-Arbeitsbereiche entnommen werden.

4

Konditionierungsmaßnahmen in der Endogaserzeugung

Ausgangsgas: Erdgas aus dem *bw-terranets Netz* (01-2021) - eingeleitet in den Endogasgenerator

Methan	CH_4	95,5382		76,4305		70,6981
Ethan	C_2H_6	3,2171		2,5737		2,3807
Propan	C_3H_8	0,1469	ÄNDERUNG DES	0,1175	GEGENMASSNAHME	0,1087
Propen	C_3H_6	0,0000	AUSGANGSGASES	0,0000	KONDITIONIERUNG MIT	0,0000
Butan	C_4H_{10}	0,0625	20% WASSERSTOFF	0,0500	7,5 % Kohlenstoff-Dioxyd	0,0463
Buten	C_4H_8	0,0000	künftige Beimischung	0,0000	künftige Beimischung	0,0000
Pentan	C_5H_{12}	0,0085	gemäß	0,0068	Kalibrationsmaßnahme	0,0063
Höhere KW	C_6H_{14+}	0,0196	Vorausschau	0,0157	zur Etablierung eines	0,0145
Kohlenstoff-Dioxyd	CO_2	0,5900	DVGW G 260 (2021)	0,4720	CO : H2 Verhältnis von 1 : 2	7,9366
Stickstoff	N_2	0,4172		0,3338	im erzeugten Endogas	0,3088
Wasserstoff	H_2	0,0000		20,0000		18,5000
Sauerstoff	O_2	0,0000		0,0000		0,0000
Gesamt	**mol (V) %**	100,0000		100,0000		100,0000

Resultierende Endogas-Zusammensetzung

Endogasanalyse	Formel	Inhalt %		Inhalt %		Inhalt %
		mol %	ÄNDERUNG IN DER	mol %	ÄNDERUNG IN DER	mol %
		V %	ENDOGASANALYSE	V %	ENDOGASANALYSE	V %
Kohlenstoff-Monoxyd	CO	20,2723	minus 1 %	19,2988	plus 2,3 %	21,6266
Kohlenstoff-Dioxyd	CO_2	0,1850		0,1850		0,1850
Wasserstoff	H_2	39,8503	plus 2,7 %	42,5739	plus 0,9 %	43,4304
Wasserdampf	H_2O	0,5653		0,6346		0,5743
Stickstoff **	N_2	39,1271	minus 1,8 %	37,3077	minus 3,1 %	34,1835
Total	mol (V) %	100,0000	** inkl. Edelgasanteile	100,0000		100,0000

Endogasvolumen V_n^3 - erzeugt aus 1 m_n^3 Erdgas inkl. H_2 Zusatz und CO_2 Kalibrierung

Erdgas+ H_2O & CO_2 Zusatz	m_n^3	1,00		1,00		1,00
Luft	m^3	2,44		1,97		1,64
Ausgangsgas : Luft	Verhältnis	1 : 2,44		1 : 1,97		1 : 1,64
Endogas	m^3	5,06		4,25		3,85

Bild 5: Mögliche Konditionierungsmasnahmen in der Endogas-Erzeugung (Quelle: Egger Consulting)

Beispiel:

Aufgabenstellung: Der Endogas-Generator soll eine Grundatmosphäre für eine kontinuierliche Ofenlinie oder mehrere Chargenöfen bereitstellen, in denen die jeweiligen Prozesse (z. B. Neutralhärten) einen Kohlenstoff-Pegelbereich zwischen 0,4 und 0,70 % C haben.

Empfohlene Generatoreinstellung:

- Sollwert für λ-Sonden EMV Regelbereich: 1.160 mV
- Max. erlaubte Abweichung: +/– 5 mV
- CO_2-Sollwertanteil im Endogas: 0,185–0,23 %
- Taupunktbereich: 0 bis + 3,5 °C.

Es wird generell empfohlen, Infrarot-Gasanalysatoren für die kontinuierliche Überwachung sowohl der Ofenatmosphäre als auch des erzeugten Endogases direkt am jeweiligen Generator zu nutzen. So ist es jederzeit möglich, sowohl die Ofenatmosphäre als auch das erzeugte Endogas auf einwandfreie Zusammensetzung zu überprüfen und aufzuzeichnen bzw. den gefahrenen Prozessen zuordenbar zu dokumentieren. Erst damit sind die – in der Automotive-Industrie weit verbreitet anzuwendenden – CQI-9-Qualitätsstandards auch vollumfänglich erfüllt.

Aus Tabelle 1 können die Schutzgasatmosphären-Bedingungen für CO-/CO_{2-} und H_2-/H_2O-basierte Atmosphären entnommen werden, die für den jeweiligen Anwendungsfall eine passende Atmosphärenzusammensetzung aufweisen. Je nach Prozesstemperatur ergeben sich unterschiedlich anzustrebende Verhältnisse zwischen den reduzierend wirkenden Atmosphärenbestandteilen H_2 und CO und den oxidierend und entkohlend wirkenden Atmosphärenbestandteilen H_2O und CO_2. Die ausgewiesenen Gleichgewichtskurven stellen die jeweiligen Betriebsbedingungen dar, wo eine neutrale Wärmebehandlung ohne jegliche Werkstoffänderung (etwa durch Oxidation und Randentkohlung) stattfinden kann.

Zwischen 825 und 842 °C (Tabelle 1) liegt die Schutzwirkung von CO und H_2 auf etwa gleichem Niveau. Diese Gleichgewichtskurven dienen der Festlegung der Atmosphärenrahmenbedingungen, bei welchen beispielsweise Bauteile – aus niedrig legierten Vergüte- und Einsatzstählen gefertigt – oxidationsfrei und kohlungsneutral wärmebehandelt werden können.

4.1.2 Analysenregelung in Endogas-Generatoren

Zwischen den Messwerten *EMV-mv* über die λ-Sonde, dem CO_2-Gehalt (in %) sowie dem Taupunkt *Tp* (in °C) besteht eine Abhängigkeit zueinander gemäß **Bild 7**.

Von den drei verfügbaren Regelungsoptionen – über $\lambda(O_2)$-Sonde, CO_2-Gasdetektor oder Taupunkt(*Tp*-)Sensor, in Verbindung mit elektronischen PID-Regelgeräten und Massendurchflusseregelgeräten – wird heutzutage jene mit λ-Sonde bevorzugt verwendet.

Wie bereits zuvor erwähnt, ist es empfehlenswert, das erzeugte Endogas mithilfe eines Infrarot-Gasanalysegerätes zu messen bzw. zu überwachen. Von Interesse sind hier alle wichtigen Gasbestandteile wie CO, CO_2, H_2 sowie der Restmethangehalt (CH_4). Die Werte für CO und CO_2 sind hierbei besonders wichtig und Ausgangspunkt für die Berechnung des Kohlenstoffpotenzials (*Cp*) in der Ofenatmosphäre, wohingegen der Restmethangehalt (CH_4) – idealerweise unterhalb von 0,30 % – einen Indikator für den Zustand des Katalysators darstellt und Rückschlüsse darüber zulässt, ob ein betriebsgefährdender Verrußungsprozess innerhalb des Katalysatorrohres ggf. bereits eingesetzt hat.

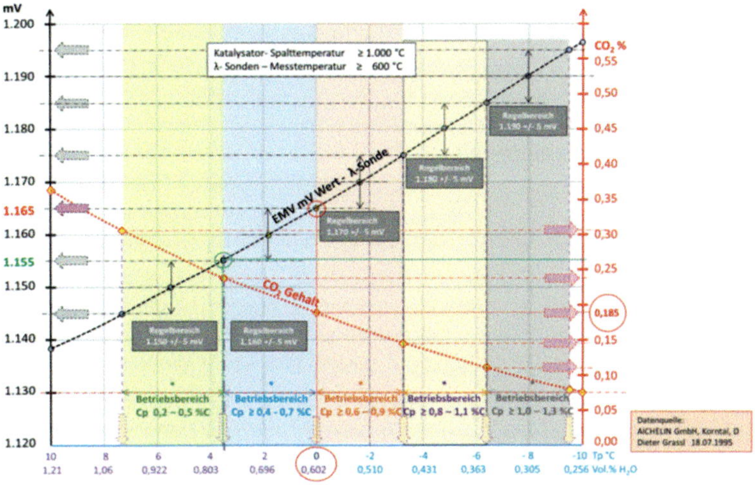

Bild 6: Endogas-Generator – Daten-Analysediagramm (Quelle: Egger Consulting)

4.1.3 Gasreaktionen in Gasaufkohlungsatmosphären

Eine Reihe von komplexen Gasreaktionen (Bild 3) findet bei diesem thermochemischen Prozess statt, bei dem Kohlenstoff aus der Spaltung von CO in atomarer Form in die Oberfläche der Werkstücke eindiffundiert. Dabei wird vorübergehend Sauerstoff (O) freigesetzt, der umgehend mit Kohlenstoffmonoxid (CO) und Wasserstoff (H_2) reagiert und hierbei in Kohlenstoffdioxid (CO_2) und Wasserdampf (H_2O) umgewandelt wird.

Der an die Werkstücke permanent abgegebene Kohlenstoff (C) muss durch geregelte Zugabe von Zusatzgas – vorwiegend Propan (C_3H_8) – über eine entsprechende Gasreaktion ersetzt werden. Die zuvor im Reaktionsbereich mit den Werkstücken gebildeten – für

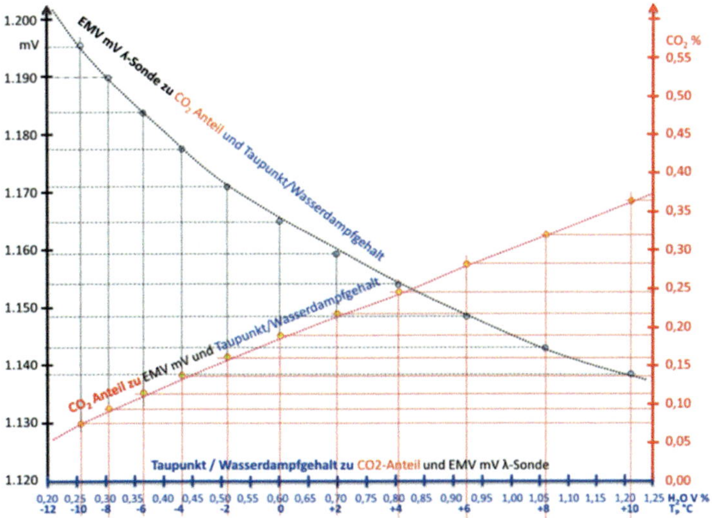

Bild 7: Zusammenhang zwischen *EMV mV*, Taupunkt und CO_2-Gehalt (Quelle: Egger Consulting)

4

den Prozess „schädlichen" – Gaskomponenten CO_2 und H_2O werden nun wieder zurückreduziert zu Kohlenstoffmonoxid (CO) und Wasserstoff (H_2).

Die ebenfalls in der Endogas-Gasatmosphäre vorhandenen Gasbestandteile Stickstoff (N_2), sowie die in sehr geringen Mengen präsenten Edelgase Helium (He), Neon (Ne), Argon (Ar), Krypton (Kr), Xenon (Xe) und Radon (Rn) sind allesamt inerte Gase, nehmen also nicht am Reaktionsgeschehen teil.

4.1.4 In Endogas-Generatoren verwendete Ausgangsgase

Erdgas war stets und ist weiterhin – aus Wirtschaftlichkeitsgründen und wegen weit verbreiteter Verfügbarkeit – das dominierende Ausgangsgas seit Einführung von Endogas-Generatoren. In den letzten Jahren hat sich jedoch im deutschen Erdgasversorgungsnetz ein grundlegender, auch noch nicht abgeschlossener Wandel vollzogen.

Die Erdgaszusammensetzung unterliegt häufigen Schwankungen, bedingt durch den Wegfall der Erdgaseinspeisung über Pipelines aus Russland. Es galt, diesen ehemals sehr hohen Versorgungsanteil (über 50 %) durch eine Vielzahl individueller Bezugsquellen (aus aller Welt) über Flüssiggaseinspeisung von LNG (Liquid Natural Gas) zu ersetzen. Zusätzlich wird zunehmend lokal Bioerdgas eingespeist sowie die zukünftige Einspeisung – von aus regenerativen Energiequellen hergestelltem – grünen Wasserstoff vorbereitet.

Hinzu kommt noch, dass gegebenenfalls das Erdgas durch die überregionalen Versorger konditioniert werden muss, um die zugesicherten und der Verrechnung zugrundeliegenden Eigenschaften (Heizwert, Wobbe-Index) sicher zu stellen. Das Mittel der Wahl hierbei ist die Beimengung von LPG (Liquified Petroleum Gas), das lediglich den vagen Anforderungen der DIN 51622 genügen muss.

Es stellt sich also die Frage, ob Erdgas unter den o. g. Umständen weiterhin auch für die Nutzung als Ausgangsgas in Endogas-Generatoren zuverlässig Verwendung finden kann.

Die Antwort dazu ist JA, wenngleich gewisse Anpassungsmaßnahme an vorhandenen Endogas-Generatoren gegebenenfalls notwendig werden können. Dies kann die Bereiche Analysenregelung und Gasüberwachung ebenso betreffen, wie auch notwendige Konditionierungsmaßnahmen, wie im Folgenden erläutert wird.

Über eine neuartige Konditionierungsmaßnahme (Bild 4) mit Beimischung von CO_2 als Konditionierungsgas zum Ausgangsgas kann auch dann eine gesicherte, konstant bleibende Endogas-Zusammensetzung erzielt werden, wenn künftig ein höherer Anteil von grünem Wasserstoff dem Erdgas beigemischt werden wird. Mit dieser Maßnahme wird es möglich, die Endogas-Analyse mit einem konstant bleibenden CO-Wert zu regeln, der für die Kohlenstoffpegel-Regelung von großem Vorteil ist

Propan als alternatives Ausgangsgas wurde und wird in seltenen Fällen verwendet, wenn ein Erdgas-Netzanschluss nicht vorhanden ist oder – basierend auf ungeklärten Problemen mit Erdgas – schlussgefolgert wurde, die Probleme durch Umstellung auf Propan vermeiden zu können. Letzteres hat sich leider als Trugschluss erwiesen.

Selbst wenn – aus Sicht der Verwendung als Prozessgas – „minderwertiges" Propan nach DIN 51622 verwendet wird, ist diese Variante im Vergleich zu Erdgas als Ausgangsgas nicht wirtschaftlich. Über die möglichen Probleme mit diesem Gas wird im Detail auf das Folgekapitel „Zusatzgase für die Kohlenstoffpegel-Regelung" verwiesen.

Ein prozesssicherer Betrieb könnte nur bei Verwendung einer Propan-Qualität mit gesichertem Reinheitsgrad und konstanter Analyse gewährleistet werden.

Ein solches Gas ist verfügbar (Propan 2.5 / R 290), kostet jedoch ein Vielfaches gegenüber Industriepropan nach DIN 51622 und scheidet daher aus wirtschaftlichen Gründen als Alternative zu Erdgas als Ausgangsgas für die Endogas-Herstellung aus, da die Verbrauchsmengen durchaus erheblich wären.

4.1.5 Zusatzgas(e) für die Kohlenstoff-Pegel-(*Cp*-)Regelung

Erdgas war in der Vergangenheit das dominierende Zusatzgas für die Kohlenstoffpegel-(*Cp*-)Regelung in CO und H_2 basierten Schutz- und Reaktionsgasatmosphären für Thermoprozesse im Betriebstemperaturbereich vorzugsweise oberhalb von 900 °C. Nur oberhalb dieser Temperatur vollzieht Erdgas eine nahezu vollständige thermische Spaltung mit nur geringem, in der Schutz- und Reaktionsgasatmosphäre verbleibendem Anteil an nicht gespaltenem Restmethan (CH_4) in der Größenordnung von unter 1 %. Bei niedrigeren Prozesstemperaturen ab etwa 820 °C – wie das vielfach der Fall bei kohlungsneutralen Vergüte-Prozessen ist – liegt der verbleibende Restmethangehalt in der Schutzgasatmosphäre meist deutlich höher, und zwar bei Werten von bis zu 3 %. Dies stellt dann sowohl für den Prozess ein Problem durch Verfälschung des errechneten Kohlenstoff-Pegels dar als auch generell durch die Gefahr von Rußausscheidung im Ofeninneren, vorwiegend an Bauteilen aus hitzebeständigen CrNi-Stählen mit der einhergehenden Gefahr von Bauteilschäden durch „Metal Dusting"-Effekte.

Die Zusammensetzung von Erdgas wird – verbindlich für die überregionalen Versorger – im DVGW-Arbeitsblatt G 260 (A) in der jeweils gültigen Fassung geregelt. Nach ausführlicher Befassung mit diesem Regelwerk kann festgestellt werden, dass im Lichte der eingetretenen Änderungen in der nationalen Versorgungslage unter Berücksichtigung von unterschiedlichsten Flüssigerdgas-(LNG-)Bezugsquellen und weiterer möglicher Einspeisequellen (Bioerdgas, Wasserstoff) die Regulierung nun eher ins Ungefähre abdriftete und – notgedrungen – deutlich mehr Spielraum in der Bandbreite der Erdgasanalyse zulässt, s. dazu **Tabelle 2**.

4

Da jedoch aus verrechnungstechnischen Gründen Erdgas nicht einfach in verbrauchten Volumensmengeneinheiten abgerechnet werden kann, sondern dies zu einem verbindlich festgelegten Heizwert (*Ho* – oberer Heizwert) in enger Bandbreite erfolgen muss, wird es notwendig, eintretende Unterschreitungen zu dem erlaubten Toleranzbandbereich fallweise durch Konditionierung des Erdgases (durch den überregionalen Versorger) mit einem höheren Kohlenwasserstoffgas auszugleichen. Das Mittel der Wahl wird in dem DVGW-Arbeitsblatt G 260 (A) vorgeschrieben bzw. spezifiziert mit LPG/Propan nach DIN 51622.

Auch diese Norm muss offensichtlich die aktuellen Marktbedingungen berücksichtigen und bleibt daher ebenfalls im Ungefähren, was die erlaubte Zusammensetzung von LPG bzw. Propan anbelangt. LPG bzw. Industriepropan wird in Raffinerien unterschiedlichster Technologieausprägung weltweit hergestellt. Je nach Sorgfalt und Wahl der Betriebsbedingungen können dabei stark unterschiedliche Reinheits- bzw. Qualitätsgrade die Folge sein. So ist beispielsweise in der DIN 51622 (2020-09) im Falle von Propan lediglich vorgegeben, dass der mehrheitliche Anteil Propan (C_3H_8) sein muss, beim Rest (also bis über 49 %) kann es sich um Alkene, wie z. B. Propen (C_3H_6) und Buten (C_4H_8), oder auch Diene, wie Propadien (C_3H_4) oder 1,3-Butadien (C_4H_6), handeln. Das sind ungesättigte Kohlenwasserstoffverbindungen mit vorliegenden Doppelbindungen zwischen einzelnen C-Atomen (**Bild 8**). Diese Gasbestandteile zeichnet alle eine schwache Bindungsstruktur im Molekülverbund zwischen den C- und H-Atomen aus, weshalb diese thermisch bereits bei niedrigen Temperaturen spalten, mit der Folge, dass die durch Doppelbindung aneinanderhaftenden C-Atome sich nicht notwendigerweise trennen und sich daher als molekularer Feststoff in

Tabelle 2: Auszug aus DVGW-Arbeitsblatt G 260 (A) 2021-09 (Quelle: Egger Consulting)

Bestandteile	Erdgas – H	Erdgas – L	Anmerkung/Quelle
Methan CH_4	% nicht mehr im Detail reguliert typisch ≥ mol 95 %	% nicht mehr im Detail reguliert typisch ≥ mol 90 %	
Sauerstoff O_2	< 1 mol %	< 3 mol %	Bestandteil im Erdgas
Schwefel (Gesamtgehalt) S_2	< 10 mg/m³	< 10 mg/m³	aus Biogas
Kohlenstoffdioxid CO_2	< 4 mol %	< 4 mol %	aus Biogas und Bestandteil im Erdgas
Kohlenstoffmonoxid CO	> 0,1 mol %	> 0,1 mol %	
Ammoniak NH_3	< 10 mg/m³	< 10 mg/m³	aus Biogas
Siliziumverbindungen Si	< 0,3 mg/m³	< 0,3 mg/m³	aus Biogas
Propan C_3H_8 Propen (Alken) C_3H_6	< 5 mol %	< 5 mol %	Bestandteil im Erdgas und Konditioniergas entsprechend DIN 51622
Butan C_4H_{10} Buten (Alken) C_4H_8	< 1,5 mol %	< 1,5 mol %	Bestandteil im Erdgas und Konditioniergas entsprechend DIN 51622
1,3 Butadien Gesamt-Diene	< 0,10 mol % < 0,50 mol %	< 0,10 mol % < 0,50 mol %	Bestandteil im Erdgas und Konditioniergas entsprechend DIN 51622
Wasserstoff H_2	% zur Zeit nicht reguliert typisch ≤ mol 20 % individuelle/regionale Festlegungen	% zur Zeit nicht reguliert typisch ≤ mol 20 % individuelle/regionale Festlegungen	aus „Green Electic Power to Gas"–H_2 – Erzeugung zur Zeit nicht reguliert

Form von Ruß im Ofeninneren ausscheiden bzw. niederschlagen. Die Gefahr einer signifikanten Prozessstörung ist dadurch jederzeit gegeben.

Aus Gründen fehlender Sicherheit bezüglich der Zusammensetzung von Erdgas – wie zuvor beschrieben – ist von einer weiteren Verwendung als Zusatzgas für die Kohlenstoff-

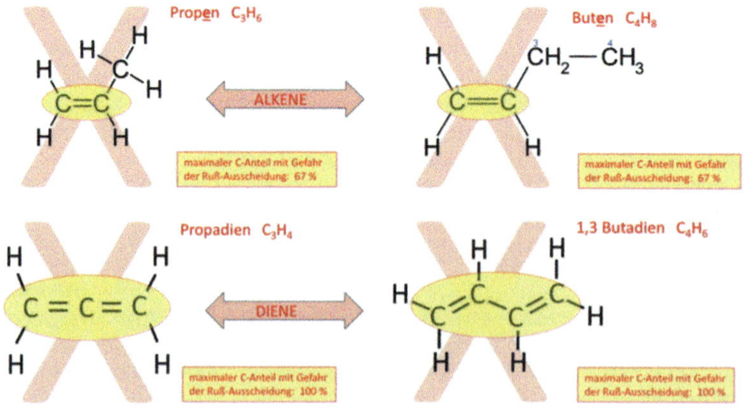

Bild 8: Besonders problematische, potenzielle Bestandteile in Erdgas und Propan (Quelle: Egger Consulting)

pegel-Regelung dringend abzuraten, vor allem dann, wenn die CQI-9-Qualitätsstandards zwingend zur Einhaltung vorgegeben sind.

4.1.6 Zusatzgasalternative Propan

Propan wird in der Automotive-Industrie bereits seit Längerem der Vorzug gegenüber Erdgas als Zusatzgas – insbesondere bei Prozesstemperaturen im Bereich von 820 bis 900 °C – gegeben, da Propan – vorausgesetzt, es liegt in der erforderlichen Qualität bzw. Reinheit vor – deutlich bessere Spalteigenschaften bei niedrigeren Prozesstemperarturen hat, als dies bei Erdgas der Fall ist. Leider ist aber der Einsatz einer kostengünstigen Propanvariante nach DIN 51622 weit verbreitet, sodass es auch hier immer wieder zu ähnlichen Problemen kommt wie beim Einsatz von Erdgas. Näheres dazu wurde bereits zuvor im Detail erläutert. Von dieser Industrie-Propan-Variante muss daher ebenfalls dringend abgeraten werden.

Das Mittel der Wahl ist daher – aus Sicht der Wärmebehandlung weitestgehend unkontaminiertes – Propangas (C_3H_8):

- Propan 2.5 / R 290
- Reinheitsgrad: 99,5 %
- max. Anteil an anderen Kohlenwasserstoffen: ≤ 50 ppm; da die Verbrauchsmengen hier – im Unterschied etwa zur Verwendung als Ausgangsgas zur Herstellung von Endogas – relativ gering sind, sollten die relativ hohen spezifischen Bezugskosten für Reinpropan keine nennenswerten Auswirkungen auf die Gesamtbetriebskosten haben.

4

4.1.7 Fazit

- Erdgas als Ausgangsgas dürfte für die Herstellung von Endogas weiterhin – auch bei künftiger Beimischung von grünem Wasserstoff bis etwa 20 % Volumenanteil – bei nur geringem verbleibendem Restrisiko verwendbar sein.
- Von der Verwendung von Erdgas als Zusatzgas für die Kohlenstoffpotenzial-Regelung wird aus den beschriebenen Gründen dringend abgeraten.
- Propan in qualitätsgesicherter Zusammensetzung (Propan 2.5 oder R 290) ist das Mittel der Wahl als Zusatzgas für eine prozesssichere Kohlenstoffpotenzial-Regelung in Endogas-Schutz- und Reaktionsgasatmosphären.

Literatur

[1] ASM Handbook Volume 4. Überarbeitete Auflage, ISBN 0-87170-379-3, Kapitel Ofenatmosphären, Fig. 1 und 3

[2] Egger, H.: Endogas-Erzeugung aus Erdgas mit zunehmend volatiler Gasbeschaffenheit. In: Prozesswärme, 4 (2021)

[3] DVGW Deutscher Verein des Gas- und Wasserfaches e.V.: Technische Regel – Arbeitsblatt DVGW G 260 (A). Ausgabe September 2021, Erdgas-Qualität

[4] DIN 51622. Ausgabe 2020-09, Flüssiggase – Propan, Propen, Butan, Buten und deren Gemische mit einem maximalen Schwefelgehalt von 30 mg/kg – Anforderungen

4.2 Verschleißreduzierung und Kosteneinsparung bei Vergütungsatmosphären durch reduzierten CO-Gehalt

Gerd Waning

Aus verschiedensten Gründen, zum Teil historisch bedingt, zum Teil aus Unkenntnis, werden Vergütungsatmosphären mit ähnlicher Zusammensetzung gefahren, wie sie auch für Aufkohlungsprozesse verwendet werden. Dieser Beitrag erklärt, wie es ohne Qualitätseinbußen möglich ist, derlei Prozesse bei reduzierten CO-Gehalten in der Atmosphäre zu fahren und hierbei auch noch Kosten bei Betriebsmitteln und der Instandhaltung einzusparen.

4.2.1 Einleitung

An vielen Wärmebehandlungsanlagen kommt es häufig zu vorzeitigem Verschleiß von Muffeln und Bändern. Durch so entstandene Schäden können erhebliche Folgekosten durch Ausschuss und verringerte Anlagenverfügbarkeit auftreten. Ebenso schlagen die Kosten für Ersatzmuffeln in erheblichem Maße zu Buche.

Ursächlich hierfür ist häufig eine Form der Hochtemperaturkorrosion in Ofenatmosphären, in die Kohlenstoffträger wie Propan, Erdgas oder Endogas eingebracht werden. Hierdurch kann es lokal zu einer intensiven Aufkohlung der temperaturbeständigen Werkstoffe kommen, welche die Hochtemperaturkorrosion beschleunigen.

Üblicherweise lässt der Oberflächenzustand des Glühgutes selbst hierbei nur vordergründig einen Rückschluss auf die Atmosphäre zu, da dieses der Atmosphäre nur kurzzeitig ausgesetzt ist, während die Ofenbauteile, wie z. B. die Muffel, permanent dem Einfluss dieser Atmosphäre unterliegen.

4.2.2 Hochtemperaturkorrosion

Grundsätzlich sind verschiedene Formen der Hochtemperaturkorrosion bekannt, wie Flüssigmetallkorrosion, Schwefelkorrosion, Aufstickung und Oxidation. An dieser Stelle wird der Fokus auf den Aspekt der Aufkohlung, die der Korrosionsform des Metal Dusting vorausgeht, gelegt.

4.2.2.1 Metal Dusting

Metal Dusting tritt als Folge einer starken Aufkohlung von Stahl-, Nickel- oder Kobaltwerkstoffen in Atmosphären auf, in denen die Werkstoffe einer Kohlenstoffaktivität von $a_c > 1$ ausgesetzt waren. Besonders stark tritt Metal Dusting im Temperaturbereich zwischen 400 und 800 °C auf. Es erreicht sein Maximum bei einer Temperatur von ca. 575 °C. Der Werkstoff zerfällt hierbei in ein pulveriges Gemisch aus Kohlenstoff und Metallpartikeln.

In extremen Fällen kommt es durch derartige Niederschläge (**Bild 1**) zu Ofenblockaden.

Die korrodierten Elemente weisen häufig Grübchen und Löcher in der Metalloberfläche auf (**Bild 2**). Der Mechanismus, der zu Metal Dusting führt, kann vereinfacht wie folgt beschrieben werden:

Bild 1: Niedergeschlagener Metal-Dusting-Staub aus einem Bandsinterofen

Bild 2: Bauteil eines Schutzgaserzeugers aus Inconel 601, durch Metal Dusting zerstört

Durch eine Umgebungsatmosphäre mit hoher Kohlenstoffaktivität kommt es insbesondere an Oberflächen mit Ungänzen in der Passivschicht zur Aufkohlung das Materiales. Es bilden sich zunächst stabile Carbide $M_{23}C_7$, M_7C_3 und MC (M = Fe, Ni, Cr). Wird die Löslichkeitsgrenze für Kohlenstoff überschritten, bilden sich metastabile Carbide M_3C an der Oberfläche und an den Korngrenzen als Zwischenprodukt. Bei weiterer Kohlenstoffzufuhr zerfallen die Carbide an der Oberfläche zu 3 M + C. An der Oberfläche der Metallpartikel scheidet sich anschließend weiterer Kohlenstoff aus der Atmosphäre ab.

4.2.3 Aufkohlung

Hier unterscheidet man i. d. R. zwei wesentliche Mechanismen.

4.2.3.1 Aufkohlung bei Anwesenheit von Kohlenmonoxid in der Schutzgasatmosphäre

Zur Aufkohlung von Stahlwerkstoffen kommt es stets dann, wenn die Kohlenstoffaktivität in der umgebenden Ofenatmosphäre höher liegt als die im Stahl.

Thermodynamisch beschreiben lässt sich die Kohlenstoffaktivität einer kohlenmonoxidhaltigen Ofenatmosphäre vereinfacht u. a. mittels folgender Gleichungen, die auf das Boudouard- bzw. das Wassergasgleichgewicht als Hauptreaktionen zurückzuführen sind:

$$2\,CO \Leftrightarrow C_{ad} + CO_2 \qquad \lg(a_C) = \lg\,(p^2CO/pCO_2) + 8.817/T - 9{,}071 \qquad (1)$$

(Boudouard-Gleichgewicht)

$$CO + H_2 \Leftrightarrow C_{ad} + H_2O \qquad \lg(a_C) = \lg\,(pCO^*pH_2/pH_2O) + 7.100/T - 7{,}49 \qquad (2)$$

(Wassergasgleichgewicht)

In der Praxis macht man sich häufig eine weitere thermodynamische Gesetzmäßigkeit zu Nutze:

$$CO \Leftrightarrow C_{ad} + 1/2O_2 \qquad \lg(a_C) = \lg\,(pCO/(pO_2)^{1/2}) - 5.927/T - 4{,}545 \qquad (3)$$

(Sauerstoffgleichgewicht)

Aus den ermittelten Werten lässt sich wiederum ein temperaturabhängiger *C*-Pegel als Kenngröße für den Kohlenstoffgehalt der Atmosphäre ableiten:

$$\lg a_c = 2.300/T - 2{,}21 + 0{,}15 * Cp + \lg\ (Cp) \qquad (4)$$

Nachdem ein *C*-Pegel einer Ofenatmosphäre ermittelt wurde, lässt sich abschätzen, ob sich die Schutzgasatmosphäre aufkohlend, entkohlend oder neutral in Bezug auf die eingesetzten Stahlwerkstoffe (z. B. Band, Muffel, Werkstück) verhält.

Eine Verschiebung des Gleichgewichtes und damit eine Veränderung des *C*-Pegels ist durch Zugabe bestimmter, sauerstoffhaltiger Gase wie Luft oder CO_2 in Richtung eines niedrigeren *C*-Pegels einfach realisierbar. Selbstverständlich steigt die Reaktionsgeschwindigkeit mit zunehmender Prozesstemperatur, so wie bei anderen chemischen Reaktionen auch. Auch die Diffusionsgeschwindigkeit des Kohlenstoffes im Stahl beschleunigt sich mit dem Temperaturanstieg. Der Einfluss von Legierungselementen wie Cr oder Ni wurde an dieser Stelle vernachlässigt; hier wird auf die weiterführende Literatur verwiesen.

4.2.3.2 Rußgrenze durch die Boudouard-Reaktion

Bei einer der bereits o. a. Hauptreaktionen in der Gasphase gilt es, einen weiteren Aspekt zu berücksichtigen, worauf an dieser Stelle eingegangen werden soll:

Die Temperaturabhängigkeit der Reaktion

$$2CO \Leftrightarrow C + CO_2 \qquad (5)$$

führt dazu, dass CO-haltige Gase bei ihrer Abkühlung in Abhängigkeit vom jeweiligen CO_2-Gehalt in CO_2 und Ruß zerfallen (**Bild 3**).

Mit der Kenntnis dieser Zerfallsreaktion und der Hauptreaktion, die zum *C*-Pegel führt, lässt sich nunmehr eine Beziehung zwischen dem *C*-Pegel und der Rußbildung darstellen, da beide von den Größen Temperatur, CO-Gehalt und CO_2-Gehalt abhängig sind (**Bild 4**). Hieraus wird ersichtlich, dass eine Atmosphäre mit einem bestimmten *C*-Pegel unabhängig von

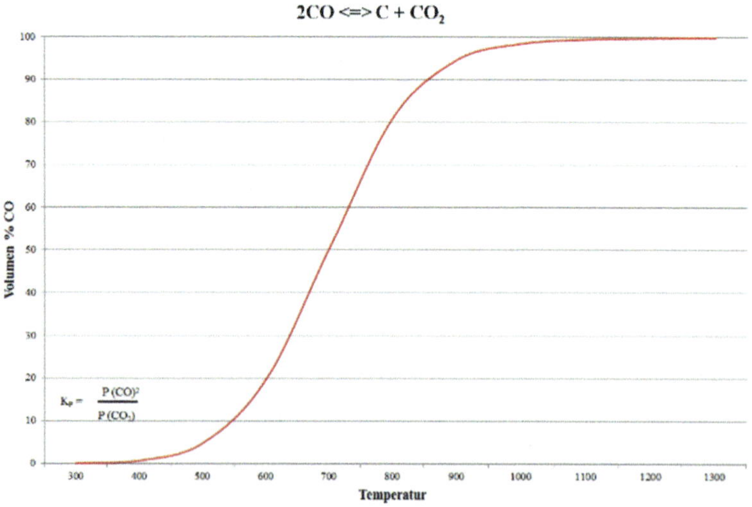

Bild 3: Temperaturabhängiger Zerfall von CO

Bild 4: Rußgrenze in Abhängigkeit vom *C*-Pegel

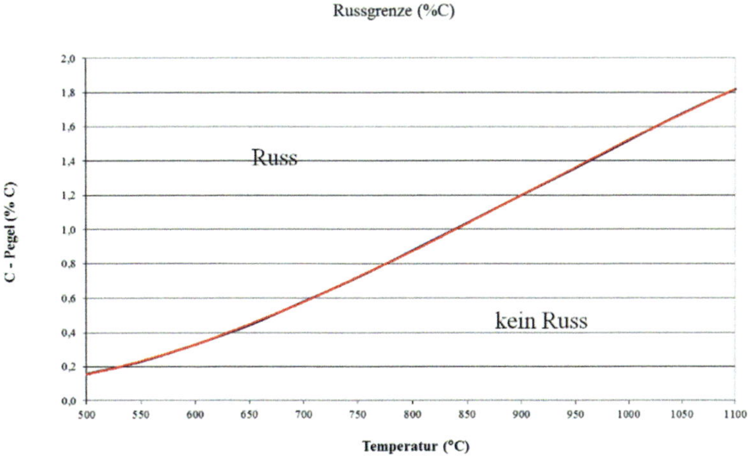

ihrem CO-Gehalt zu Ruß und CO_2 zerfallen kann. Dies kann z. B. bei Abkühlung der Atmosphäre in Ofenbereichen mit niedrigerer Temperatur erfolgen. Abhängig von der Temperatur, der ein Gas im Ofen ausgesetzt ist, ergeben sich somit unterschiedlichste Effekte auf das Bauteil (Entkohlung/Neutral/Aufkohlung) sowie auf die Rußbildung.

4.2.3.3 Aufkohlung in Abwesenheit von Kohlenmonoxid in der Schutzgasatmosphäre

Hierbei liegen häufig Schutzgasatmosphären auf der Basis von H_2 und N_2 vor, denen gelegentlich geringe Mengen eines sogenannten Fettungsgases wie Propan (C_3H_8) zugefügt werden. Das Fettungsgas wird zugegeben, um ein Entkohlen der Bauteile zu verhindern oder zu minimieren. Eine analoge Wirkung entfalten in manchen Prozessen die durch die Bauteile in den Ofen eingeschleppten Schmiermittelrückstände oder Binder. Bekannt sind ebenso Atmosphären auf der Basis von CH_4 und N_2.

Bei solchen Begasungsregimen kann es ebenso zu einer erheblichen Aufkohlung des Bandes, sofern vorhanden, und der Muffel kommen.

Während es sich bei dem im Kapitel 4.2.3.1 „Aufkohlung bei Anwesenheit von Kohlenmonoxid in der Schutzgasatmosphäre" behandelten Aufkohlungsmechanismus um einen thermodynamisch wohlbekannten und gut beschreibbaren Gleichgewichtsvorgang handelt, liegen bei solcherlei Atmosphärensystemen die Verhältnisse anders. Die vollständige Abwesenheit von Sauerstoff bzw. sauerstoffhaltiger Gase (bei Betrachtung des idealen, geschlossenen Systems), mit denen sich ein Gleichgewicht einstellen könnte, führt dazu, dass die Zersetzung des beigefügten Fettungsmediums pyrolytisch, beispielsweise nach dem folgenden Muster, verläuft:

$$C_3H_8 \Leftrightarrow C_2H_4 + CH_4 \quad (6)$$

$$C_2H_4 \Leftrightarrow 2\,H_2 + 2C \quad (7)$$

$$C_3H_8 \Leftrightarrow C_2H_2 + CH_4 + H_2 \quad (8)$$

$$C_2H_2 \Leftrightarrow 2\,H_2 + 2C \quad (9)$$

Der detaillierte Reaktionsmechanismus enthält 227 Spezies und 827 zumeist reversible Elementarreaktionen. Hier wird jedoch nur ein vereinfachter Reaktionsmechanismus unter Berücksichtigung der metastabilen Zwischenzustände betrachtet. Es bleibt zu erwähnen, dass ggf. im Ofenraum herumvagabundierende kohlenwasserstoffhaltige Reststoffe in ähnlicher Weise in den thermochemischen Prozessablauf durch Einbringung weiteren Kohlenstoffes eingehen können.

Insbesondere C_2H_2 und C_2H_4 reagieren mit katalytischer Unterstützung der Bauteiloberfläche unter weiterer Bildung von Methan und Wasserstoff und führen direkt zur Aufkohlung der Metalloberflächen.

Anmerkung: Im Bereich der Niederdruckaufkohlung von Einsatzstählen wird bereits erfolgreich seit Jahren mit dieser Technologie mit den Aufkohlungsgasen Acetylen oder Propan gearbeitet. Um jedoch die Karbidbildung zu vermeiden, wird in diesen Prozessen immer nur kurzzeitig das Aufkohlungsgas aufgegeben, woran sich jeweils eine Diffusionsphase anschließt. In neuerer Zeit werden auch Aufkohlungsprozesse nach diesem Schema, allerdings bei Atmosphärendruck, gefahren und erprobt.

4.2.3.4 Schlussfolgerungen zu den unterschiedlichen Aufkohlungsmechanismen

Bei der Aufkohlung mit CO-haltigen Gasen ist ein bestimmtes Kohlenstoffpotenzial definierbar. In bestimmten Prozessphasen kann es auch hier zu einer verstärkten Aufkohlung kommen. CO-haltige Atmosphären ermöglichen im Gegensatz zu CO-freien Atmosphären jedoch auch eine definierbare Entkohlung der kontaktierenden Stahlwerkstoffe. Ggf. eingebrachte kohlenstoffhaltige Reststoffe aus Zersetzungsvorgängen gehen aufgrund der hohen Prozesstemperaturen nahezu unmittelbar in die thermochemische Gesamtbilanz mit ein und lassen sich durch die entsprechenden Messmethoden berücksichtigen, da sie in eine Gasphase umgewandelt werden, sofern die Rußgrenze lokal nicht überschritten wird.

Eine Aufkohlung bei der Verfahrensweise ohne CO-haltige Gase unter Zugabe eines Fettungsgases ist im geschlossenen System nahezu immer gegeben. Die Intensität der Aufkohlungswirkung richtet sich im Wesentlichen nach den tatsächlich dem Prozess zugeführten Fettungsgasmengen, deren Kohlenstoffgehalt zum Teil in die Stahloberflächen eindiffundiert und zum Teil auch als Ruß ausfallen kann. Eine Entkohlung ist in geschlossenen Systemen nicht möglich. Ggf. eingebrachte Reststoffe aus vorgelagerten Kohlenwasserstoffzersetzungsvorgängen gehen in diesen Vorgang mit ein.

In realen, offenen Durchlaufofenanlagen kann es in Abhängigkeit von den tatsächlichen Strömungsvorgängen zu einer Überlagerung beider Aufkohlungsmechanismen kommen. Beispielsweise bei einer Begasung mit einer N_2-H_2-C_3H_8-Atmosphäre, die am Anfang der Kühlstrecke eingebracht wird: Im vorderen Bereich des Ofens, nahe der Kohlenwasserstoffzersetzung, kommt es dann zur Gleichgewichtsauf- oder Abkohlung, da hier sauerstoffhaltige Rückstände aus der Kohlenwasserstoffzersetzung wirken können und zur CO-Bildung beitragen. Im nachgelagerten Bereich der Hochtemperaturzone kommt es zur direkten Aufkohlung (bzw. Rückkohlung, wenn zuvor entkohlt wurde).

Um die Aufkohlung von Ofenbauteilen, die zu Metal Dusting führt, zu minimieren, ist es demzufolge erforderlich, auf eine Ofenatmosphäre zurückzugreifen, die so wenig Aufkohlungswirkung wie möglich zeigt.

Hierzu bietet sich eine Reduzierung des CO-Gehaltes der Atmosphäre als einer der einfach zu beeinflussenden Parameter an.

4.2.4 Wirtschaftlichkeitsbetrachtung

Anhand eines konkreten Falles wurden die wirtschaftlichen Aspekte einer solchen Maßnahme für die jährlichen Begasungskosten der Öfen bei einem Schraubenhersteller betrachtet. Für die fünf im Betrieb befindlichen Ofenanlagen sollten nicht allein die Auswirkungen der Kostenentwicklung für den Austausch eines Methanolanteils durch vermehrten Stickstoffeinsatz betrachtet werden, sondern auch weitere Varianten einander gegenübergestellt werden. Hierzu zählten:

4.2.4.1 Stickstoff - Methanol mit 20 % CO (Status quo)

Die *Cp*-Regelung erfolgt mittels einer herkömmlicher O_2-Sonde und einem CO-Festwert von 20 %. Änderungen und Investitionen sind logischerweise nicht erforderlich. Bei diesem bewährten Prozess hat man es mit einem bekannten, hohen Anteil von Ruß in den Anlagen zu tun. Die Kohlenstoffzufuhr beträgt 1,6 kg/h bei einem Schutzgasbedarf von 15 Nm^3/h.

4.2.4.2 Stickstoff - Methanol mit 5 % CO

Die *Cp*-Regelung erfolgt ebenso mittels einer herkömmlicher O_2-Sonde, jedoch mit einem CO-Festwert von 5 %. Änderungen und Investitionen sind ebenfalls nicht erforderlich. Bei diesem ähnlichen Prozess kann man von einem geringeren Rußausfall ausgehen. Die Kohlenstoffzufuhr verringert sich auf 0,4 kg/h bei einem Schutzgasbedarf von 15 Nm^3/h.

4.2.4.3 Endogas, 20 % CO

4

Die *Cp*-Regelung erfolgt analog zu Kapitel 4.2.4.1. Für die Installation von Endogasgeneratoren und deren Infrastruktur sind erhebliche Investitionen erforderlich. Hinsichtlich der in Europa auch zukünftig zu erwartenden Varianz der Qualität durch unterschiedliche Quellen ergeben sich derzeit noch nicht abschätzbare Risiken hinsichtlich der Konstanz der Gaszusammensetzung. Rohrleitungen für ein toxisches Gas sind zu installieren und zu warten. Bei diesem Prozess ist mit einem ähnlichen Rußanfall wie in Kapitel 4.2.4.1 zu rechnen. Ansonsten kann man von einem ähnlich funktionierenden Prozess wie in Kapitel 4.2.4.1 ausgehen. Die Kohlenstoffzufuhr beträgt 1,6 kg/h bei einem Schutzgasbedarf von 15 Nm^3/h.

4.2.4.4 Endogas, 5 % CO

Hinsichtlich der Investitionskosten und Risiken gilt dasselbe wie bei der Variante in Kapitel 4.2.4.3, jedoch mit geringerem zu erwartenden Rußanfall. Auch die Prozessregelung sollte mit dem vorhandenen Equipment realisierbar sein. Die Kohlenstoffzufuhr verringert sich auf 0,4 kg/h bei einem Schutzgasbedarf von 15 Nm^3/h.

4.2.4.5 N_2/H_2 mit 10 % H_2 (und CO)

Hierbei ließe sich die vorhandene Regelung nicht mehr nutzen, da bei diesem Prozess keine nennenswerte Menge an CO als Basisrechengröße für eine Regelung vorhanden ist. Alternativ wäre es möglich, mittels einer externen Sauerstoffsonde zu arbeiten, die in der Lage dazu ist, besonders niedrige Sauerstoffpartialdrücke zu messen (> 1.200 mV). Die Hauptbestandteile eines solchen Systems sind eine extern in einem separaten Ofen untergebrachte Sauerstoffsonde zur präzisen Bestimmung des Sauerstoffpartialdruckes der Ofenatmosphäre sowie ein CO-Analysator. Zusammen mit der Prozesstemperatur wird aus den

Messsignalen der Kohlenstoffpegel der Ofenanlage bestimmt. Zur Aufrechterhaltung eines minimalen CO-Gehaltes wird der Atmosphäre bei Bedarf reines Kohlenmonoxid geregelt zudosiert. Je nach Ofenstabilität könnte mit einem CO-Gehalt von 0,15–0,25 % gearbeitet werden, wodurch sich die Kohlenstoffzufuhr auf Werte von unter 0,5 kg/h reduzieren ließe.

Sofern noch nicht vorhanden, wäre eine zusätzliche H_2- und CO-Versorgung zu installieren. Ein großer Vorteil dieses Systems bestünde in einem minimalen Rußanfall. Allerdings würde diese Prozessvariante, verglichen mit der bestehenden Variante, eine erhebliche Veränderung bedeuten.

Tabelle 1: In der Wirtschaftlichkeitsberechnung berücksichtigte Kostenfaktoren

Berechnungsgrößen
Wasserstoff
Methanol
Kühlwasser
Raumkosten
Erdgas zur Endogaserzeugung
Energieverbrauch zur Endogaserzeugung
Wärmeverlust des Generators
Endogaserwärmung im Ofen
Methanolerwärmung, Verdampfung, Spaltung
Erwärmung H_2 auf Ofentemperatur
Erwärmung N_2 auf Ofentemperatur
Elektr. Gebläseenergie

4.2.4.6 Kostenfaktoren

Die in **Tabelle 1** aufgelisteten Berechnungsgrößen wurden in der Wirtschaftlichkeitsbetrachtung berücksichtigt.

4.2.4.7 Berechnungsergebnisse

Der Vergleich der einzelnen Verfahrensvarianten aus wirtschaftlicher Sicht erfolgte unter Außerachtlassung von Wartungs- und Investitionskosten auf einer jährlichen Basis (**Tabelle 2**). Die Betriebsdauer für die zugrunde gelegten fünf Öfen wurde mit 8.424 h/a bei 860 °C Ofentemperatur angesetzt. Der Schutzgasbedarf pro Ofen wurde mit 15 m^2 pro Stunde angesetzt. Etwaige Abfackelungsverluste des Endogasgenerators wurden ebenso nicht berücksichtigt.

4.2.4.8 Ergebnisse

Als kostengünstigste Variante konnte die Betriebsweise mit 5 % Methanol ermittelt werden, als teuerste die mit Stickstoff und 5 % Wasserstoff. Es muss jedoch klar festgehalten werden, dass solch ein Berechnungsergebnis stark variieren kann und stets von den im Einzelfall zugrundeliegenden Kostenfaktoren abhängig ist. Beispielsweise können die Kühlwasserkosten bei dem Betrieb von Endogasgeneratoren ganz erheblichen Schwankungen unterworfen sein, wenn z. B. in einem Fall nahezu kostenfreies Brunnenwasser verwendet werden kann, in einem anderen Fall aber auf teures Stadtwasser zurückgegriffen werden muss. Die hier gezeigte Berechnung basiert auf der Kostenbasis eines bestimmten Kunden, wie sie tatsächlich vorgefunden wurde.

4.2.5 Fazit

Neben den abschließend dargestellten erheblichen wirtschaftlichen Vorteilen ergeben sich durch eine Reduzierung des Kohlenmonoxidgehaltes einer Vergütungatmosphäre verschiedenste prozesstechnische Vorteile. Gegebenenfalls lassen sich Vorgaben hinsichtlich

Tabelle 2: Ergebnis der Wirtschaftlichkeitsbetrachtung fünf unterschiedlicher Begasungsvarianten

Beispielrechnung (€)	5 Plattenbandöfen zu je 15 Nm^3/h Schutzgasbedarf (jährliche Kosten, ohne Abschreibung und Wartung)				
	Methanol 20% CO	Methanol 5% CO	Endogas 20% CO	Endogas 5% CO	N_2/H_2 10% H_2
Stickstoff	59.136,48	125.665,02	0,00	110.880,90	133.057,08
Wasserstoff	0,00	0,00	0,00	0,00	126.360,00
Erdgas zur Spaltung	0,00	0,00	138.996,00	34.749,00	0,00
Energieverbrauch zur Endogaserzeugung	0,00	0,00	36.806,14	9.201,54	0,00
Wärmeverlust Generator	0,00	0,00	9.266,40	9.201,54	
Endogaserwärmung im Ofen	0,00	0,00	66.153,56	16.538,39	0,00
Methanol	158.895,81	39.723,95	0,00	0,00	0,00
Methanolerwärmung + Verdampfung + Spaltung	90.297,98	22.574,49	0,00	0,00	0,00
Erwärmung H_2 auf Ofentemperatur	0,00	0,00	0,00	0,00	5.396,87
Erwärmung N_2 auf Ofentemperatur	23.030,33	48.939,44	0,00	43.181,86	51.818,23
elektr. Gebläseenergie	0,00	0,00	2.965,25	2.965,25	0,00
Kühlwasser	0,00	0,00	53.913,60	53.913,60	0,00
Raumkosten	0,00	0,00	1.600,00	1.600,00	0,00
Summe Stoff- und Energiekosten	331.360,59	236.902,91	308.100,95	253.740,13	316.632,18

4

der Bauteilqualität unter dem Aspekt der Randoxidation nur so befriedigend und reproduzierbar darstellen. Die dargestellten Verbesserungen hinsichtlich der Ofensauberkeit und dem geringeren Verschleiß zeigen sich selbstverständlich erst nach längerer Betriebszeit, konnten jedoch auch in konkreten Anwendungsfällen bestätigt werden. Hierzu zählt auch die längere Verwendbarkeit des verwendeten Härteöles.

Um die Aufkohlung von Ofenbauteilen, die zu Metal Dusting führt, zu minimieren, ist es erforderlich, auf eine Ofenatmosphäre zurückzugreifen, die so wenig Aufkohlungswirkung wie möglich zeigt.

Hierzu bietet sich eine Reduzierung des CO-Gehaltes der Atmosphäre als ein einfach zu beeinflussender Parameter an.

Unter Umweltgesichtspunkten kann selbstverständlich ebenso festgehalten werden, dass es sich bei einer Betriebsweise mit geringerem Kohlenmonoxidgehalt auch um eine Variante handelt, bei der mit erheblich geringeren CO_2-Emissionen gearbeitet wird. Während dem Prozess mit 20 % Kohlenmonoxid stündlich 1,6 kg/h Kohlenstoff zugeführt wurde, verringerte sich diese Menge auf 0,4 kg/h, wenn die CO-Menge auf ein Viertel reduziert wurde.

5. Induktive und konduktive Erwärmung

5.1 Homogene Temperaturen dank Feldführern bei der konduktiven Erwärmung gekrümmter Werkstücke

Dr. Christine Tränkner, Dipl.-Ing. (FH) André Wieder

Bei der konduktiven Erwärmung gebogener Werkstücke mit Wechselstrom kommt es aufgrund unterschiedlicher Stromdichten an Innen- und Außenradien zu sehr inhomogenen Temperaturverteilungen. Technische Lösungen zur Homogenisierung der Temperatur an den Biegungen lassen sich teils wegen fehlender Prozessstabilität nur schwer in der industriellen Serienfertigung umsetzen oder liefern nicht die gewünschte Qualität in ausreichend kurzen Taktzeiten. Speziell entwickelte wassergekühlte Feldführer jedoch beeinflussen gezielt die generierte Leistung in den Biegungen, können damit Temperaturunterschiede zwischen Innen- und Außenradius verringern und eignen sich auch für den industriellen Einsatz.

5.1.1 Herausforderungen bei der konduktiven Wärmebehandlung gekrümmter Werkstücke

Die konduktive Erwärmung bietet energetische sowie prozesstechnische und wirtschaftliche Vorteile und wird insbesondere bei der Erwärmung von metallischen Stangen, Rohren, Bändern oder Drähten eingesetzt. Das Werkstück wird dabei mithilfe von Kontakten in den Stromkreis eingebunden und aufgrund seines elektrischen Widerstandes gemäß dem Jouleschen Gesetz erhitzt. Die thermische Leistungsdichte $\dot{q} = \rho \cdot j^2$ (d. h. die pro Zeit- und Volumeneinheit entstehende Wärmemenge) ist abhängig vom spezifischen Widerstand ρ und der Stromdichte j. Die Wärme entsteht, wie auch beim induktiven Erhitzen, direkt im Werkstück. Man spricht daher auch von der „direkten Widerstandserwärmung“. Im Vergleich mit indirekten Methoden ergibt sich häufig ein besserer Wirkungsgrad. Vorteilhaft (beispielsweise für die Schmiedeerwärmung) ist, dass das Werkstück schnell über die gesamte Länge auf die gewünschte Temperatur gebracht werden kann. Der „spezifische Energieverbrauch“ kann dabei geringer sein als bei der induktiven Erwärmung ([2], S. 321). Ideal ist das Verfahren für lange, gerade Werkstücke ohne Querschnittsänderung. Werden dagegen gekrümmte Werkstücke auf diese Weise wärmebehandelt, kommt es an den Biegungen aufgrund unterschiedlicher Stromdichten in diesen Bereichen zu einer inhomogenen Temperaturverteilung. Die Innenradien werden dabei überhitzt („höhere Stromdichte“), während die Außenradien kälter bleiben als die geraden

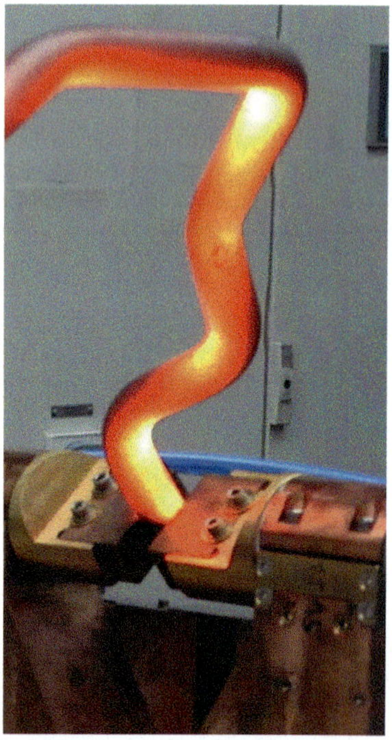

Bild 1: Inhomogene Erwärmung eines gebogenen Stabes bei der konduktiven Erwärmung

Abschnitte („verringerte Stromdichte"). In **Bild 1** sieht man diesen Effekt deutlich. Unterschiedliche Biegeradien führen dabei zu einer unterschiedlich starken Überhitzung. Für temperaturkritische Anwendungen wie z. B. das Härten oder Anlassen von Stählen ist das nicht akzeptabel.

5.1.2 Ansätze zur Homogenisierung der Temperatur an den Biegungen

Um die beschriebenen Temperaturunterschiede zu vermeiden bzw. zu verringern, werden unterschiedliche Methoden verwendet. So können beispielsweise die Innenradien mittels Druckluft oder einem flüssigen Medium aktiv gekühlt werden. Nachteilig ist hier die fehlende Prozessstabilität, da die Werkstückoberfläche (Zunder, Walzhaut usw.) den Abkühlungseffekt beeinflusst und ein gleichmäßiges, reproduzierbares Abkühlen erschwert. Eine geregelte Abkühlung über eine Temperaturmessung ist aus den gleichen Gründen ebenfalls schwierig. Das aktive Abführen der thermischen Energie verschlechtert einerseits den Wirkungsgrad, andererseits müssen – um Temperaturunterschiede zu den geraden Partien zu vermeiden – kleinere Heizraten und damit längere Prozesszeiten in Kauf genommen werden (vgl. [3], S. 2). In der Praxis werden üblicherweise Temperaturunterschiede von ca. 100 bis 150 °C akzeptiert, um die Bauteile überhaupt in einer wirtschaftlichen Geschwindigkeit auf Härtetemperatur zu erwärmen. Dabei muss im kältesten Bereich des Werkstücks die Ac3-Härtetemperatur von ca. 900 bis 950 °C erreicht werden, um den gewünschten Härteeffekt zu erzielen. Entsprechend sind in den überhitzten Bereichen des Stabes Temperaturen von bis zu 1100 °C und teilweise darüber hinaus zu finden. Bedingt durch diese Überhitzungen entstehen metallurgische Nachteile wie übermäßige Verzunderung, Randentkohlung und Grobkornbildung, was zur Qualitätsminderung beim Bauteil führt. Somit muss immer zwischen Qualität und Taktzeit abgewogen werden.

1975 wurde in der Offenlegungsschrift 24 17 107 des deutschen Patentamtes über das „Verfahren zur konduktiven Erhitzung gekrümmter oder geknickter Werkstücke mittels Wechselstrom" [4] eine neue Methode vorgeschlagen: die Stromdichteverteilung in den Innenradien durch passive Feldführungselemente aus magnetisch leitfähigem Material, beispielsweise Eisenblechen, zu beeinflussen und damit den Energieeintrag anzupassen. Diese Bleche selbst sollen durch einen Sprühkopf mit Kühlluft oder „einem nebligen Wasser-Luft-Gemisch gekühlt" werden (ebd., S. 6), da sie sich sonst zu stark erwärmen. Die direkte Kühlung beeinflusst allerdings die Prozessstabilität, da nicht verhindert werden kann, dass die Werkstückoberfläche ebenfalls durch das Kühlmedium erreicht wird. Nachteilig ist auch die mangelnde Flexibilität bezüglich der Positionierung der Feldführer, die nicht einzeln bewegt werden können.

Gesucht wurde nun ein Verfahren, das neben der benötigten Prozessstabilität und Handhabbarkeit auch in der industriellen Praxis bei sehr kurzen Taktzeiten einsetzbar ist.

In den Patenten BE 488 851 A [5] und FR 990735 A [6] wird ebenfalls eine Idee beschrieben, wie die Temperatur in geraden Werkstücken, die konduktiv erwärmt werden, in einzelnen Bereichen beeinflusst werden kann: Durch die Erhöhung des induktiven Widerstandes in einzelnen Segmenten wird die Stromdichte herabgesetzt und damit die umgesetzte Leistung verringert. Hierfür werden, je nach Anwendung, U-förmige Blechpakete oder massive Formkerne genutzt.

5.1.3 Weiterentwickelte Methode für die industrielle Serienfertigung

Die Entwicklung der Firma ITG Induktionsanlagen GmbH baut darauf auf und bietet dabei signifikante Verbesserungen. Sie besteht aus Feldführungselementen, die aus zu einem

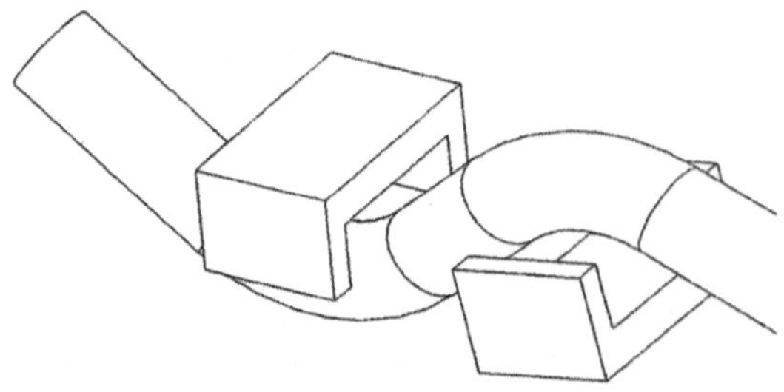

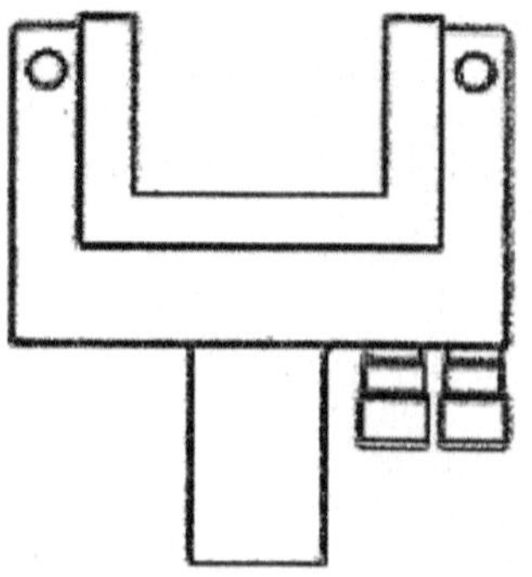

Bild 2: Positionierung der Feldführungselemente in den Biegungen (links); wassergekühltes Feldführungselement in kompakter U-Form (rechts) [3]

Paket verschweißten Blechen in U-Form in einer gekühlten Fassung aus Kupfer bestehen (vgl. **Bild 2**). Diese sind einzeln positionierbar und können dadurch flexibel auch für unterschiedliche Werkstückgeometrien genutzt werden. Durch die in die Feldführer integrierte Kühlung ergibt sich eine kompakte Bauweise; das Werkstück kann so nicht durch eine direkte Kühlung beeinflusst werden. Dies sorgt für eine gute Prozessstabilität. Vorteilhaft ist, dass die Feldführer je nach Biege- und Krümmungsradius unterschiedlich nahe am Werkstück angebracht werden können. Da die Verschiebung der Stromdichte abhängig ist von der „Eintauchtiefe" des Rohres in den Feldführer, kann so der Energieeintrag gezielt beeinflusst werden. Durch die homogen erwärmten Biegungen wird der Querverzug und damit die nötige Nacharbeit minimiert. Weil zur Temperaturhomogenisierung keine Ausgleichszeiten notwendig sind, kann die Erwärmung schneller stattfinden; damit sind kürzere Taktzeiten möglich. Im industriellen Umfeld werden die Feldführungselemente bereits erfolgreich eingesetzt.

5.1.4 Anlagentechnik und Verwendung der Feldführer

Herausforderungen können sich bei mehrfach gebogenen Rohren ergeben, bei denen verschiedene Feldführer auf dichtem Raum positioniert werden müssen. Hier gibt es die Möglichkeit, statt der Elemente in U-Form zusammengesetzte Feldführer zu verwenden, die aus zwei Einzelelementen in L- bzw. I-Form bestehen und an schwer zugänglichen Stellen flexibler angeordnet werden können. Der Luftspalt zwischen den beiden Einzelteilen kann dabei mehrere Millimeter betragen, ohne dass die Wirkung groß beeinträchtigt wird. Die wiederholgenaue Positionierung der Feldführungselemente ist sehr wichtig für die Prozessstabilität. Hier gibt es beispielsweise die Möglichkeit, nach Einlegen des Werkstücks in die Maschine die Feldführer gemeinsam auf einer Trägerplatte passgenau heranzufahren. Bei größeren thermischen Maß- oder Formänderungen des Werkstücks können die Feldführungselemente auch nachgeführt werden. Die Temperaturen können während oder nach Abschluss der Erwärmung mittels Wärmebildkamera überprüft werden. Diese Technik hat sich in der Serienfertigung schon bewährt.

5.1.5 Beispiel: Erwärmen eines gebogenen Rohres auf Härtetemperatur

Im Folgenden sollen einige Ergebnisse aus der Entwicklung und Erprobung der Feldführer aufgeführt werden. **Bild 3a** zeigt die Erwärmung eines gebogenen Rohres ohne jegliche Kühlung oder Homogenisierung der Temperatur – deutlich ist die stärkere Erwärmung des

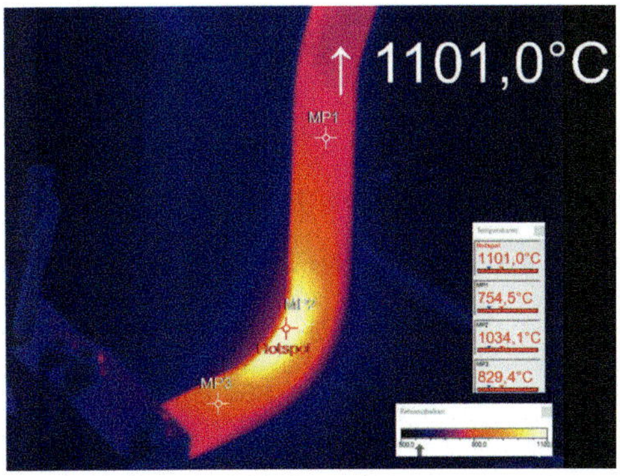

Bild 3a: Konduktive Erwärmung ohne Kühlung

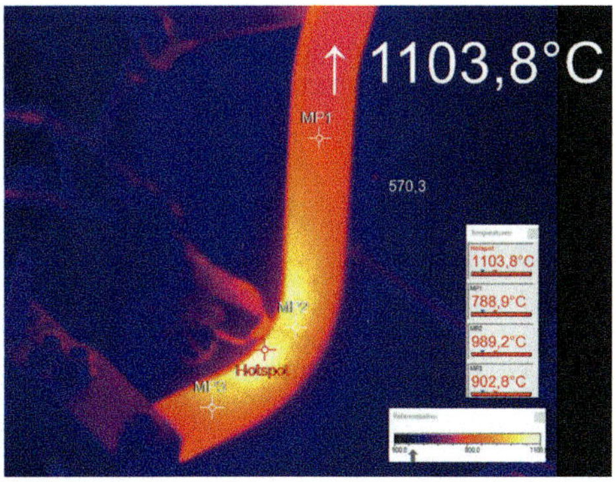

Bild 3b: Konduktive Erwärmung mit aktiver Kühlung

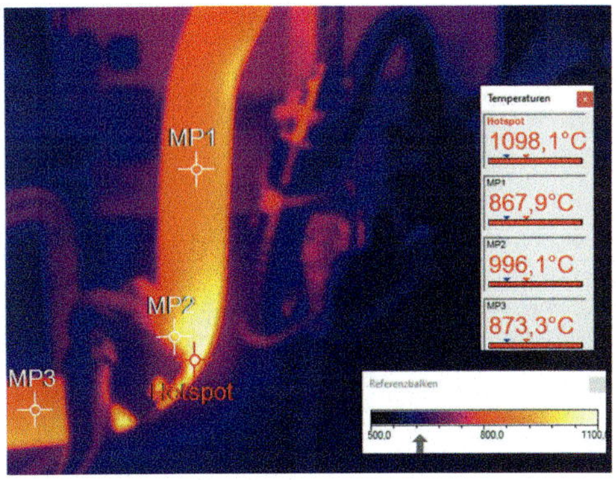

Bild 3c: Konduktive Erwärmung mit Feldführer

Bild 4: Homogene Temperaturverteilung eines gebogenen Stabes kurz vor dem Abschrecken während eines Härteprozesses

Innenradius' zu sehen. Bei **Bild 3b** wurde bei ansonsten gleichem Versuchsaufbau die Biegung mit Druckluft gekühlt. Dadurch erreicht man eine insgesamt größere Homogenität, der Hotspot liegt jedoch weiterhin im Innenradius. Deutliche Temperaturunterschiede von mehreren hundert Grad findet man auch hier. **Bild 3c** wurde nach dem Anbringen eines Feldführelementes aufgenommen; hier wurde noch ein Prototyp verwendet. Die Position der Kamera musste verändert werden, da die Biegung ansonsten kaum sichtbar gewesen wäre. Der Hotspot liegt nun am Außenradius, d. h., in diesem Falle müsste die Ausrichtung des Feldführers zur Biegung noch angepasst werden. **Bild 4** zeigt nun die homogene Temperaturverteilung in einem mehrfach gekrümmten Stab während eines Härteprozesses. Hier wurden die Feldführungselemente gerade entfernt, um das Werkstück in die Abschreckeinheit zu transportieren.

5.1.6 Fazit

Beim konduktiven Erwärmen gekrümmter Metallstangen oder -rohre kommt es an den Biegungen zu signifikanten Temperaturunterschieden. Die Firma ITG Induktionsanlagen GmbH hat ein Verfahren entwickelt, um mit kompakten, gekühlten Feldführungselementen die

Stromdichte und damit den Wärmeeintrag in den Biegungen so zu beeinflussen, dass eine homogene Erwärmung auch bei kurzen Taktzeiten möglich ist. Gute Ergebnisse aus der industriellen Serienfertigung bestätigen den Nutzen dieser Technik.

Literatur

[1] Pfeifer, H.; Nacke, B.; Beneke, F. (Hrsg.): Praxisbuch Thermoprozesstechnik. Band I, 2. Auflage, Vulkan-Verlag GmbH, Essen, 2010

[2] Schulze, D.; Lüdtke, U.: Theoretische Grundlagen der Elektroprozesstechnik, Teil 1. Vulkan-Verlag GmbH, Essen, 2023

[3] Seidler, T.; von Naumann, M.; Wieder, A.: Verfahren zur konduktiven Erwärmung von gekrümmten metallischen Werkstücken, Anlage zur Ausführung des Verfahrens sowie Feldführungselement zur Ausführung des Verfahrens oder als Teil der Anlage. Patent EP 3 742 865 B1, Anmeldetag: 23.05.2019

[4] Seulen, G.: Verfahren zur konduktiven Erhitzung gekrümmter oder geknickter Werkstücke mittels Wechselstrom. Offenlegungsschrift 24 17 107, DE000002417107A, Anmeldetag: 08.04.1974

[5] Seulen, G.: Procédé de chauffage de pièces métalliques. Patent BE 488 851A, 1949

[6] Seulen, G.: Procédé de chauffage de pièces métalliques. Patent FR 990 735 A, Anmeldetag: 03.05.1949

5.2 Intelligent eingesetzte Umrichtertechnologie für das ultratiefe Randschichthärten großer Wellen

Dr. Christine Tränkner, Martin Bauer

Mithilfe des Randschichthärtens ist es möglich, Werkstücke mit einer harten, verschleißfesten Oberfläche bei gleichzeitig zähem Kern herzustellen. Üblicherweise betragen die erzielten Randschichthärtungstiefen einige Millimeter. Werden sehr große Bauteile mit einer höheren Härtetiefe benötigt, muss die Induktionstechnik entsprechend angepasst werden. Im Folgenden soll am Beispiel eines Retrofits einer bestehenden Anlage gezeigt werden, wie die Energieversorgung für das induktive Randschichthärten bei sehr großen Walzen mit entsprechend hohen Randschichthärtungstiefen von mehreren Zentimetern umgesetzt wurde.

5.2.1 Einleitung

Sehr große Walzen und Wellen mit Massen von mehreren Tonnen werden beispielsweise in der Metallverarbeitung zur Blechherstellung verwendet. Die dabei auftretenden hohen Belastungen führen zu einem starken Verschleiß der Oberfläche. Um diesen zu minimieren, werden verschiedene Verfahren eingesetzt. Neben der Möglichkeit, Schichten aus Keramik oder Hartmetall als Verschleißschutz aufzubringen, kann auch eine Oberflächenhärtung einen schnellen Materialabtrag verhindern. Hierbei wird das Gefüge im Randbereich umgewandelt, ohne die chemische Zusammensetzung zu verändern. Dabei wird die Randschicht des Werkstücks bis in eine bestimmte Tiefe auf Härtetemperatur erwärmt und dann schnell abgeschreckt, sodass sich in diesem Bereich ein hartes Martensitgefüge bildet. Da die Erwärmung sehr rasch erfolgt, spricht man auch vom „Kurzzeitaustenitisieren". Eine schnelle und effiziente Methode ist dabei das induktive Härten. Hier wird mithilfe einer von Wechselstrom durchflossenen Spule im Werkstück ein veränderliches magnetisches Feld und somit elektrischer Wechselstrom induziert. Bei elektrisch leitfähigen Materialien kommt es so zur Erwärmung durch Wirbelstromverluste, bei ferromagnetischen ebenso durch Hystereseverluste. Die benötigte Energie wird – bei auf die jeweilige Erwärmaufgabe angepasster Frequenz – durch passende Umrichter bereitgestellt. Diese werden für die spezielle Anwendung optimiert und an die Kundenwünsche angepasst. Dies soll im Folgenden am Beispiel einer Anlage des Unternehmens SIJ Ravne Systems d.o.o. zum induktiven Oberflächenhärten von Walzen und Wellen beleuchtet werden, bei der für sehr große Werkstücke auch entsprechend große Randschichthärtungstiefen (kurz: SHD, surface hardening depth) gewünscht wurden. Für kleine Teile mit Abmessungen im Zentimeterbereich werden üblicherweise Härtetiefen von einigen Millimetern erzielt, während hier bis zu 40 mm gewünscht waren. Dies stellt eine besondere Herausforderung an die Auslegung der Anlagentechnik dar. Das Unternehmen ITG Induktionsanlagen sah sich dieser Herausforderung durch den Einsatz modernster Umrichtertechnologie gewachsen.

5

5.2.2 Optimierte mechanische Eigenschaften durch das induktive Randschichthärten

Der Vorteil des Randschichthärtens liegt darin, dass nur die Oberfläche des Werkstücks gehärtet wird, während der Kern seine Zähigkeit behält. Man erhält dadurch eine verschleißfeste Oberfläche, vermeidet aber, dass das Teil als Ganzes zu spröde wird. Die Einhärtungstiefe hängt von der gewünschten Anwendung ab und kann durch die Prozessführung in gewissen, vom jeweiligen Material abhängigen Grenzen beeinflusst werden. Ein wichtiger Parameter ist dabei die Frequenz der am Induktor anliegenden Wechselspannung.

Über

$$\delta = \sqrt{\frac{\rho}{\pi \cdot \mu \cdot f}} \tag{1}$$

mit δ = Eindringtiefe in mm, μ = Permeabilitat in H/m und f = Frequenz in Hz ist die Eindringtiefe des Magnetfeldes in die Probe direkt mit der Frequenz verknüpft. Als Eindringtiefe oder Stromeindringmaß bezeichnet man die Länge, auf der die im Werkstück induzierte Stromdichte um den Faktor e^{-1} abgesunken ist. In diesem Bereich wird demnach der Hauptteil der induzierten Energie, etwa 86 %, umgesetzt und in Wärme umgewandelt [1], [2]. Für hohe Eindring- und damit Härtetiefen sind also niedrige Frequenzen notwendig; mit hohen Frequenzen können dünne Randhärteschichten erzeugt werden.

Während bei komplexen Werkstückgeometrien häufig speziell an die Form angepasste Induktoren entwickelt werden, werden bei achsensymmetrischen Teilen oft einfache Rundspulen verwendet. Diese werden im sogenannten Vorschubverfahren am Werkstück entlang bewegt; alternativ kann auch das Werkstück bewegt werden. Auch die dabei verwendete Geschwindigkeit und damit Dauer der Erwärmung hat einen Einfluss auf die Dicke der wärmebehandelten Schicht. Für jedes Bauteil kann der Erwärmprozess durch die Wahl der geeigneten Parameter angepasst werden. Die Auslegung von Induktor und Anlage erfolgen damit auf Basis der zu behandelnden Werkstücke.

5.2.3 Retrofit für bestehende Anlage

Im vorliegenden Fall wurde von SIJ Ravne Systems – Teil der SIJ Slovenian Steel Group d.d. [3] – die Modernisierung einer bestehenden Induktionshärteanlage, die zum Härten von großen Arbeitswalzen für Kaltwalzwerke, Zwischenwalzen, Stützwalzen und anderen Rollen genutzt wird, angefragt. Dabei sollte der Prozess optimiert sowie seine Stabilität verbessert werden. Außerdem wurden aufgrund gestiegener Kundenanforderungen höhere Härtetemperaturen und Härtetiefen benötigt. Die zu erwärmenden Werkstücke haben dabei eine Masse von bis zu 8 t bei einer Länge von maximal 5,5 m und einem Durchmesser von bis zu 800 mm [4]. Verarbeitet werden beispielsweise chromhaltige Werkzeugstähle wie 86CrMoV7 und Stahlqualitäten wie 100Cr6 oder 42CrMo4. Die gewünschten Oberflächenhärten variieren je nach Anwendung und können beispielsweise zwischen 45 und 50 HRC liegen oder auch 65 bis 68 HRC betragen. Die SHD hängt von der jeweiligen Stahlsorte sowie vom Walzentyp ab und liegt für kleinere Walzen im Bereich von 6 bis 12 mm, bei großen Walzen werden bis zu 40 mm angefordert. Es werden somit unterschiedliche Eindringtiefen δ und damit auch ein breites Frequenzspektrum benötigt.

5.2.4 Erweiterter Frequenzarbeitsbereich für unterschiedliche Eindringtiefen

Für die Lösung dieser komplexen Aufgabe hat man sich für den Einsatz von zwei Weitbereich-Umrichtern entschieden. Die beiden Umrichter decken zusammen einen Frequenzbereich von 30 bis 1200 Hz ab. Mit dieser Konstellation hat der Kunde die Möglichkeit, mit einer Energie und eingestellter Frequenz kleine und mittlere Härtetiefen direkt zu erzielen. Für die ultratiefen Härtungen können die beiden Umrichter parallel mit unterschiedlichen Frequenzen und Leistungen gefahren werden. Dabei wärmt die tiefe Frequenz vor, auch um im Werkstück die Gefahr von Härterissen bei zu schnellem Aufheizen zu verringern. Mit der

Bild 1: Zweifrequenzanlage mit gekoppelten, simultan betriebenen Induktoren für eine niedrige und eine hohe Frequenz

zweiten Frequenz wird im Anschluss die endgültige Austenitisierungstemperatur erreicht. Verwendet werden einwindige Induktoren mit an die Werkstücke angepassten Durchmessern und Höhen. Für große Teildurchmesser sind diese beispielsweise ca. 200 mm hoch. Mit den zur Verfügung stehenden Frequenzen erreicht man im Stahl bei 400 °C Eindringtiefen zwischen 2 und 10 mm, bei höheren Temperaturen auch wesentlich mehr. **Bild 1** zeigt einen Härteprozess für ultratiefes Randschichthärten bei einer großen Welle.

Die Resonanzfrequenzen im Schwingkreis ergeben sich aus Induktivität und Kapazität im Lastkreis. Da die Induktivität meist durch die Anwendung (Induktor und Werkstück) vorgegeben ist, kann die Frequenz nur durch die geeignete Wahl der Kapazität beeinflusst werden. Über eine Kondensatorstufenschaltung können Kondensatoren hinzu- oder abgeschaltet werden. Dies kann entweder mechanisch über Schaltmuttern erfolgen oder elektrisch über Schütze eingestellt werden. Bei der hier betrachteten Anlage wurde die Anpassung über Schütze realisiert. Im Gegensatz zur mechanischen Umschaltung ist hierbei keine Elektrofachkraft nötig. Möglich ist auch das Umschalten durch die Maschinensteuerung mittels Hinterlegung im Rezept. In diesem werden die für jedes Werkstück verwendeten Prozessparameter gespeichert und nach Auswahl des entsprechenden Bauteiles abgerufen.

5.2.5 Besonderheiten der Anlage

Für den hier vorgestellten Erwärmprozess müssen, um die gewünschten großen Einhärtetiefen bei gleichzeitig hoher Produktivität zu erreichen, die beiden enggekoppelten Induktoren simultan betrieben werden. Dabei besteht die Gefahr, dass sich die entstehenden Magnetfelder gegenseitig beeinflussen. Dies konnte bei dem hier realisierten Aufbau durch angepasste Abstände der Induktoren vermieden werden. Hinsichtlich der Umrichter fiel die Wahl bei den verwendeten niedrigen Frequenzen auf moderne IGBT-Wechselrichter. Der

Einsatz von SiC-Wechselrichtern eignet sich bei höheren Frequenzen. Der geringe Blindstromanteil und das damit einhergehende gute Netzverhalten ergeben sich durch die Verwendung von ungesteuerten Gleichrichtern mit 6-Puls-Brückenschaltung (B6U) im Umrichter, wobei auch 12-Puls-Umrichter möglich gewesen wären. Hierbei treten nur geringe Oberwellenanteile auf. Dabei ist die Anlage hocheffizient: Bei einer Frequenz von 50 Hz ergibt sich, unabhängig von der abgerufenen Leistung, für die Stromversorgung vom Netz bis einschließlich zum Wechselrichter ein Wirkungsgrad von 95 %. Der Gesamtwirkungsgrad wird dann beispielsweise noch durch Verbindungen wie Stromschienen oder Kabel oder den Wirkungsgrad des Induktors beeinflusst.

Damit ein möglichst breites Teilespektrum gefahren werden kann, gibt es verschiedene, an die jeweilige Geometrie angepasste Induktorsätze. Nach einem Wechsel muss nur das jeweilige Rezept, in dem die zugehörigen Prozessparameter gespeichert sind, geladen werden. Die Frequenz- und Leistungsanpassung geschieht über das Umschalten von Trafo und Kondensatorstufen. Um die Prozesssicherheit zu gewährleisten, erfasst die SPS-Steuerung kontinuierlich die Betriebsdaten der Anlage, wie beispielsweise Bauteiltemperaturen im Umrichter oder Durchlaufmengen aus den Kühlkreisläufen. Störungen werden dem Bediener direkt auf dem HMI-Bedienpanel angezeigt. Damit es bei irregulären Werten nicht sofort zu einer Notabschaltung mit Anlagenstillstand und damit zu einem Produktschaden am hochwertigen Bauteil kommt, wird ohne Unterbrechung des Prozesses zuerst eine Warnung ausgegeben. Somit kann der Bediener hier direkt eingreifen oder sich nach erfolgter Wärmebehandlung um die Störung kümmern. Es sind zudem Überwachungen eingebaut, die den Zwischenkreisgleichstrom, die Ausgangsspannung sowie den Ausgangsstrom begrenzen und die Anlage beim Überschreiten der Toleranz abschalten, um Schäden am Gerät zu verhindern. Die übergeordnete Steuerung wird mittels Profinet-Schnittstelle angebunden. In **Bild 2a** ist das Übersichtsbild des Bedienpanels dargestellt. Von hier aus erreicht man über das Touchpad schnell und einfach die Anzeigen (z. B. Bauteiltemperaturen) oder

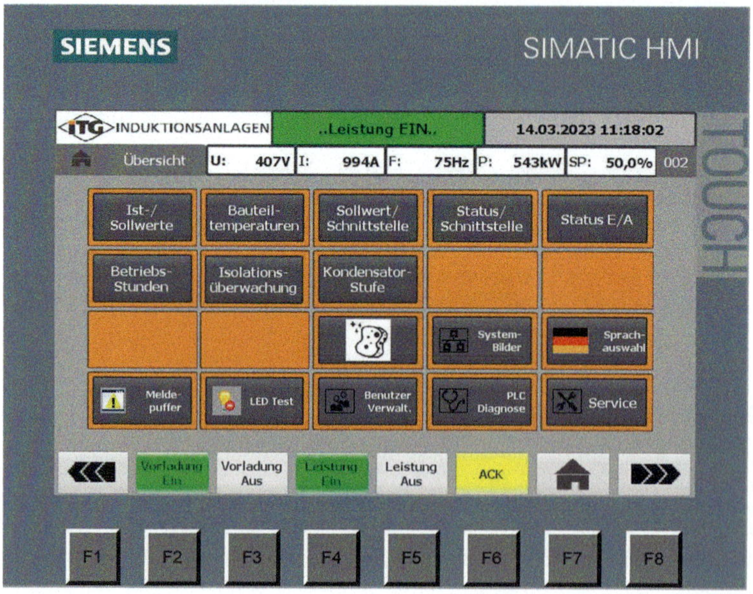

Bild 2a: Übersichtsbild des Bedienpanels: Von hier aus kann beispielsweise auf die Soll-/Ist-Werte oder die Bauteiltemperaturen zugegriffen werden.

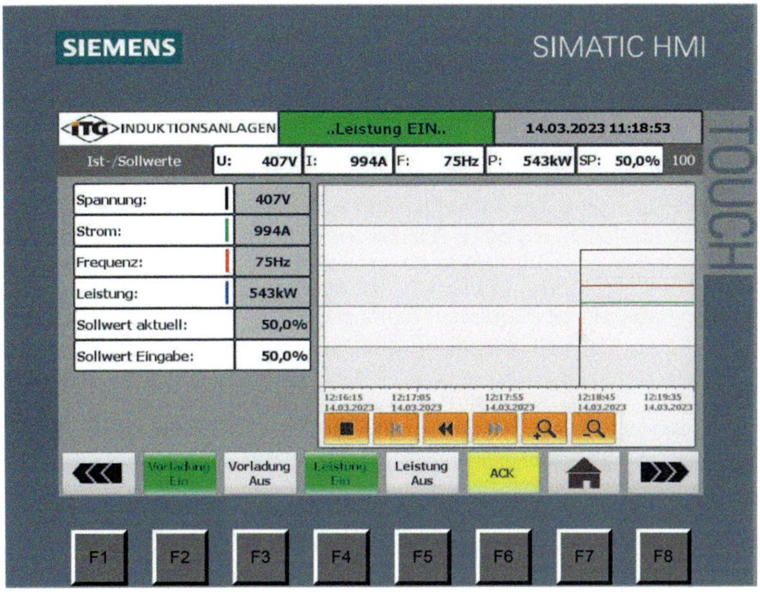

Bild 2b: Oberfläche des Bedienpanels mit Anzeige der Soll-/Ist-Werte

Steuerbildschirme (z. B. Kondensatorstufenanpassung). **Bild 2b** zeigt beispielhaft Soll- und Ist-Werte für einen laufenden Prozess.

Insgesamt bietet die Anlage eine hohe Verfügbarkeit. So wurden in den ersten zehn Monaten nach der Inbetriebnahme über 800 Walzen gehärtet.

5.2.6 Verbesserte Eigenschaften der gehärteten Wellen

In **Bild 3** ist der Härteverlauf für eine mittelgroße Walze dargestellt. Deutlich wird, dass die Einhärtetiefe stark von der gewählten Frequenz beeinflusst wird, während die Härtewerte

5

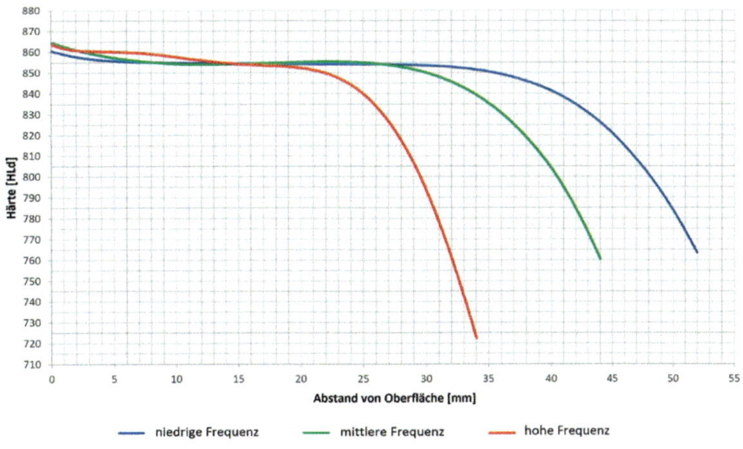

Bild 3: Härteverlauf bei einer mittelgroßen Welle bei verschiedenen verwendeten Frequenzen

der Randschicht kaum streuen. Von SIJ Ravne Systems wird eine homogene Härte der Einhärtezone innerhalb von +/- 0,5 HRC garantiert [5]. Das Gefüge besteht beispielsweise bei den Arbeitswalzen aus niedrig angelassenem Martensit mit verschiedenen Sekundärkarbiden.

5.2.7 Zusammenfassung

Um die gewünschte hohe Produktivität in Kombination mit den benötigten großen Einhärtetiefen zu erreichen, wurde im Rahmen eines Retrofits die Umrichtertechnik an einer bestehenden Induktionshärteanlage für große Walzen modernisiert. Zwei gekoppelte, simultan betriebene Induktoren erwärmen die Walzen dabei auf Härtetemperatur. Die Umrichter können dabei ein sehr breites Frequenzspektrum abbilden. Besonderes Augenmerk wurde beim Umbau der Anlage auch auf Bedienerfreundlichkeit und Prozesssicherheit gelegt.

Literatur

[1] Benkowsky, G.: Induktionserwärmung. Verlag Technik GmbH, Berlin, 1990

[2] RWE Energie Aktiengesellschaft (Hrsg.): Induktive Erwärmung: Physikalische Grundlagen und technische Anwendungen. 1991

[3] Unternehmenshomepage [https://sij.ravnesystems.com/en/], zuletzt abgerufen am 11.04.2023

[4] Katalog „Heat Treatment". Online, [https://sij.ravnesystems.com/assets/ Uploads/HEAT-TREATMENT.pdf], zuletzt abgerufen am 18.11.2022

[5] Katalog „Forged Rolls". Online, [https://sij.ravnesystems.com/assets/Uploads/ FORGED-ROLLS.pdf], zuletzt abgerufen am 14.03.2023

5.3 Ultra-Große Lager: Ein komplettes Sortiment an Schlupflos-Induktionserwärmungslösungen

Livia Scarpellini, Marco Bono, Mario Cesano, Dario Ciscato, Beatrice Pillin

Die Ultra-Large-Bearing(ULB)-Industrie kann die Produktionsleistung durch den Einsatz von Induktionserwärmung für eine breite Palette thermischer Prozesse steigern. Dieser Beitrag präsentiert die technologischen, wirtschaftlichen und prozessbezogenen Optimierungen, die durch den Einsatz der Induktionserwärmungstechnologie sowohl beim Härten als auch beim Anlassen erzielt werden können. Zwei verschiedene Lösungen sind für das Härten von Laufringen (schlupflos) verfügbar: ein Hochleistungs-Induktions-Einzelprozess (Schuss-Härtung) für kleine bis mittelgroße Ringe und das Induktions-Schlupflos-Vorschub-Härten für große Lager. Der Induktions-Anlassprozess mit extrem niedriger Frequenz wird in diesem Beitrag beschrieben und mit einem traditionellen Ofen verglichen. Diese Technologien werden vorgestellt und verglichen, um Anwendungsbereiche, spezifische Merkmale, metallurgische Ergebnisse sowie Prozess- und Kosteneffizienzen aufzuzeigen.

5.3.1 Einleitung

Die ULB-Industrie wächst exponentiell und Prozessoptimierung sowie Kostenreduzierung sind herausfordernde und wichtige Themen. Die primäre Energieeinsparung wird ein wirksamer Weg sein, um sowohl die Prozesskosten als auch den globalen CO_2-Fußabdruck zu reduzieren. Die Induktionserwärmungstechnologie kann auf eine Vielzahl von Produktionsprozessen angewendet werden (Schmelzen, Schmieden, Erwärmen, Härten usw.) und erhöht die globale Energieeffizienz mit einer vollständigen Reihe neuer Funktionen und Möglichkeiten, die mit dieser Technologie verbunden sind.

Der besondere Aspekt der Induktionserwärmung ist die direkte Erzeugung der Wärme im Werkstück: Das von der Spule erzeugte Magnetfeld induziert Ströme, die durch das Stück fließen und das Material dann durch ohmsche (Joule-)Erwärmung erhitzen. Die Tiefe der Erwärmung hängt von der Frequenz des Wechselstroms ab: Je höher die Frequenz, desto dünner oder flacher die Erwärmung. Spannung, Frequenz, Prozesszeit, Kühlmedium, magnetische Flusskonzentratoren und die Form des Induktors können so ausgelegt werden, dass nur ein bestimmter Teil des Werkstücks auf vollautomatische, stabile und softwaregesteuerte Weise erwärmt wird [1–3].

Der sehr hohe Wirkungsgrad, der mit einem Induktionserwärmungssystem erzielt werden kann (bis zu 95 %), ist ein wichtiger Punkt für die Kostenreduzierung und die Strategie zur Energieeinsparung. Im Vergleich zu traditionellen Erwärmungstechnologien, die auf Wärmeübertragung (Elektro- oder Gasöfen) mit geringem Wirkungsgrad und Gasemissionen basieren, ist die Induktionserwärmung die führende grüne Technologie für die Wärmebehandlung.

Der vorliegende Beitrag konzentriert sich auf die Härtungs- und Anlassverfahren für die ULB-Industrie.

5.3.2 Härten

Der Härtungsprozess ist für die ULB-Produktion erforderlich, um die Oberflächenverschleißfestigkeit in den Laufbahnbereichen zu erhöhen. Andere nicht-funktionale Zonen können gehärtet werden, um Kratzer oder Verformungen zu vermeiden. Die erforderliche Härte, Randschichttiefe und Verteilung können je nach Lageranwendung variieren. Das Härten kann auf der traditionellen thermo-chemischen Diffusionsaufkohlung oder auf dem elektromagnetischen Induktionserwärmungsverfahren basieren.

Das traditionelle Aufkohlungsverfahren von Stählen wird üblicherweise zum Härten von Lagern verwendet. Bei dieser Art von Lagern wird die gehärtete Schicht durch Anreicherung der Lager-Oberflächenschicht mit Kohlenstoff erzielt. Dieser Diffusionsprozess wird gefolgt von einem Abschrecken, um die angereicherte Schicht zu härten, und einem Anlassverfahren, um die Sprödigkeit zu reduzieren. Die Aufkohlung erfolgt in Öfen bei hohen Temperaturen und mit langen Haltezeiten. Dies führt zu einem energie- und zeitaufwendigen Prozess und fördert die Kornvergröberung im Material. Darüber hinaus müssen Lager nur auf den Laufbahnen aufgekohlt werden, was bedeutet, dass das Teil vor der Aufkohlung durch Auftragen eines chemischen Maskiermittels auf die Bereiche, in denen eine Härtung unerwünscht ist, vorbereitet werden muss. Unter der Annahme einer ausreichenden Abkühlgeschwindigkeit wird der Härtegradient hauptsächlich durch die Verteilung der Kohlenstoffkonzentration im Stahl bestimmt. Bei diesem Prozess entstehen große Verformungen durch die martensitische Umwandlung und die schnelle Abkühlung des gesamten Teils. Die Aufkohlung von großformatigen Lagern erfordert sehr lange Haltezeiten im Ofen, um eine aufgekohlte Schicht von mehreren Millimetern zu erzielen. Dies macht den Prozess kostspielig und es treten große Verformungen im Teil auf. Aus diesen Gründen werden großformatige Schwenklager in Stählen mit niedrigem Kohlenstoffgehalt nicht bevorzugt. Stähle mit mittlerem Kohlenstoffgehalt sind eine geeignete Option – dank ihrer guten Formbarkeit, großen Härtbarkeit, guten Widerstandsfähigkeit und Zähigkeit, wenn eine ordnungsgemäße Wärmebehandlung durchgeführt wird [4], [5].

5.3.2.1 Induktionshärten

Induktionshärten ist ein schnelles und energiesparendes Verfahren im Vergleich zum Aufkohlen. Es ist eine äußerst vielseitige Heizmethode, die eine lokale oder gleichmäßige Oberflächenhärtung bis zur Durchhärtung des Bauteils durchführen kann.

Im Gegensatz zum Aufkohlen erfordert die Induktionshärtetechnologie nicht das Erhitzen des gesamten Teils. Nur die Oberflächenschicht wird behandelt, wodurch das Kernmaterial und seine Mikrostruktur unverändert bleiben und weniger Verzug entsteht. Die Erwärmung erfolgt, indem ein leitfähiges Materialteil in das Magnetfeld gebracht wird, das durch einen

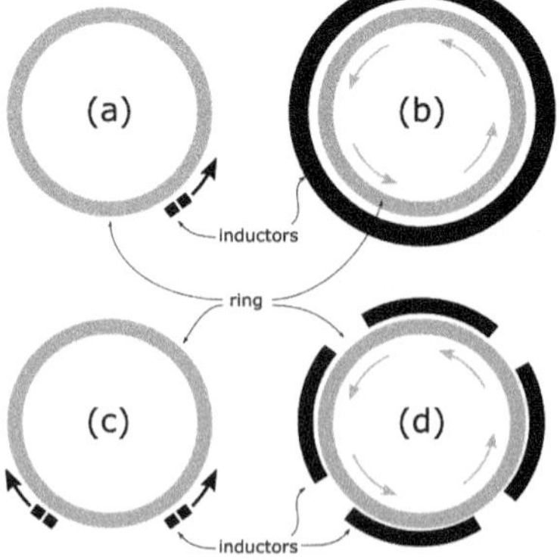

Bild 1: Verschiedene Induktionshärtekonstruktionen für ULB: (a) Vorschub, (b) Einzelschuss mit festem Induktor, (c) Schlupflos-Vorschub und (d) Einzelschuss mit mehreren Induktoren

Hochfrequenzwechselstrom erzeugt wird, der durch einen Induktor, normalerweise eine wassergekühlte Kupferwicklung, fließt. Wie in der Einleitung dieser Arbeit beschrieben, sind die spezifischen Merkmale des Induktionserwärmungssystems die Schlüssel zur Erzielung des gewünschten thermischen Profils, der Härtequalität und der Prozesswiederholbarkeit.

Bei der spezifischen ULB-Härteanwendung können verschiedene Induktionserwärmungslösungen eingesetzt werden. Wie in den Skizzen in **Bild 1** beschrieben wird, kann das aktive Induktionssystem ein kleines Ringsegment mit einer relativen Vorschubbewegung wie in (a) und (c) erhitzen oder, wie in (b) und (d), den Ring mit einem breiteren Induktor umschließen. Die Schlupflos-Vorschub- und die Einzelschussverfahren werden in den folgenden Abschnitten beschrieben. Die Systemwahl hängt von den Elementdurchmessern, Prozessanforderungen und Spezifikationen ab.

5.3.2.2 Schlupflos-Vorschub-Härten

Wie in Bild 1(c) gezeigt, ist der nahtlose Prozess eine Vorschubheiztechnik. Die Induktoren sind haarnadelförmig und erhitzen einen kleinen Bogenabschnitt des Umfangs. Die erforderliche spezifische Leistung ist höher, da jeder Materialabschnitt eine kürzere Heizzeit hat, aber die erhitzte Oberfläche viel kleiner ist als im Einzelschussfall; aus diesem Grund ist die erforderliche Leistung relativ gering, aber mit einer längeren Prozesszeit verbunden, die mit der Vorschubgeschwindigkeit und dem Ringumfang zusammenhängt.

Die Betriebsbedingungen der Lager erfordern, dass die Ringoberfläche eine gehärtete Schicht mit gleichmäßigen mechanischen Eigenschaften über den gesamten Umfang aufweist. Eine weiche Zone mit geringerer Oberflächenhärte und mechanischer Festigkeit würde zu einer Region führen, die stärkerem Materialverschleiß ausgesetzt ist und somit zu einem privilegierten Bereich für die Rissbildung und -ausbreitung. Ein einfacher Vorschub-Induktionshärteprozess, wie in Bild 1(a) gezeigt, würde eine weiche Zone erzeugen, wenn sich der Vorschubweg vom Ende mit dem Vorschubstartpunkt überschneidet. Diese Überlappung würde bereits abgeschrecktes Material anlassen [6]. Die schlupflose Vorschubtechnologie ermöglicht einen Induktionshärteprozess ohne weiche Zonen. Es wurden zwei Heizköpfe verwendet, wobei jeder Kopf mit zwei Spulen ausgestattet ist, die als Vorheiz- und Heizspule bezeichnet werden. Jede Spule hat eine unabhängige Stromversorgung, die eingestellt werden kann, um die für den Prozess benötigte richtige Leistung abzustimmen. Die beiden Köpfe können sich um das Lager bewegen, wobei jeder die Hälfte des Rings abdeckt. An jedem Kopf ist die Heizspule an einem mechanischen System montiert, das sich dreht, um die Position der Spule einzustellen. Sowohl die Vorheiz- als auch die Heizspule sind parallel zur Lagerlaufbahnfläche eingestellt. Während sich die Heizköpfe um die Laufbahnringfläche bewegen, wird diese Parallelität durch eine rotierende Hülse beibehalten, die an den Heizköpfen installiert ist. Jeder Kopf ist mit einer Hauptkühlbrause und einer Hilfskühlbrause ausgestattet. Kupferwicklungen und Kühlbrausen müssen speziell für die Geometrie der Lagerlaufbahn ausgelegt sein. Kühlbrausen sind am Heizkopf montiert, um den Teil der Laufbahn zu kühlen, der gerade von den Induktionsspulen erhitzt wurde. Die Hauptkühlbrause ist für die schnelle Abkühlung verantwortlich, die für eine martensitische Umwandlung notwendig ist, während die Hilfskühlbrause dabei hilft, die Phasenumwandlung abzuschließen und das Material auf Raumtemperatur abzukühlen. In **Bild 2** ist eine Skizze der Positionen der Heizköpfe in verschiedenen Prozessschritten dargestellt.

Die Erwärmung beginnt an einem Punkt des Rings, wobei die beiden Heizköpfe nahe beieinander liegen und sich der Ring oszilliert, um eine heiße Zone zu erzeugen, die als Startzone bezeichnet wird (Bild 2a). Während dieser Oszillationen ist es das Ziel, eine gleichmäßige Temperatur im erhitzten Material zu erreichen. Zu diesem Zeitpunkt sind die Heizköpfe nahe beieinander, wobei die Heizspule eines Kopfes neben der Heizspule des anderen

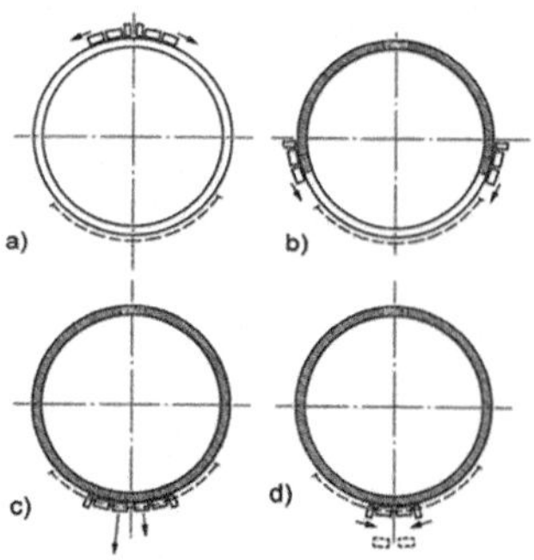

Bild 2: Skizzen des schlupflosen Prozessschritts

Kopfes liegt. Die Induktoren bilden somit eine kompakte Spulengruppe. Wenn die Startzone auf Härtetemperatur erwärmt ist, stoppt der Ring seine Oszillation und bewegt sich in entgegengesetzte Richtungen, um sich zu öffnen. Der erhitzte Teil in der Startzone sollte groß genug sein, damit die Heizköpfe sich in eine Position öffnen können, in der die Hauptbrause eingeschaltet werden kann, um das Material zu kühlen. Jeder Heizkopf bewegt sich um den Ring herum und deckt die Hälfte des Lagerumfangs ab, während er im Vorschubmodus heizt (Bild 2b).

Die Vorschubgeschwindigkeit und die Heizleistung werden so eingestellt, dass die gewünschte Härtetiefe erzielt wird. Während des Vorschubes kühlen die auf den Heizköpfen montierten Brausen die von den Spulenpaaren erhitzte Oberfläche. Die Heizköpfe bewegen sich in Richtung des Lagerbereichs gegenüber der Startzone, der Endzone genannt wird. Wenn sich die beiden Heizköpfe einander nähern, sind die Vorheizspulen der beiden Paare nebeneinander (Bild 2c).

In diesem Moment ist die Stromversorgung der Vorwärmspule des ersten Heizkopfes unterbrochen und die Spule wird von der Lagerfläche wegbewegt. Das Wegbewegen einer Spule schafft Platz, damit sich die beiden Heizköpfe annähern können. Wenn die Vorwärmspule des zweiten Heizkopfes und die Heizspule des ersten Heizkopfes nebeneinander liegen, wird die Stromversorgung der zweiten Vorwärmspule erneut abgeschaltet und die Spule von der Schienenoberfläche wegbewegt. Der Teil der Schienenoberfläche, von dem die zweite Vorwärmspule wegbewegt wurde, wird dann von den beiden sich annähernden Heizspulen erwärmt (Bild 2d).

Wenn die beiden Heizspulen nebeneinander liegen, wird ihre Stromversorgung abgeschaltet und der Drehtisch wird schnell bewegt, damit eine der Hilfsduschen den zuletzt erwärmten Schienenteil bedeckt. Ein konstanter Luftspalt zwischen den Spulen und der Lagerfläche ist entscheidend für ein gleichmäßiges Erwärmungsmuster. Während der Vorschub-Erwärmung treten im Lager aufgrund der fortschreitenden Phasentransformation, die auf der Ringoberfläche stattfindet, Verformungen auf. Die für den Härteprozess notwendige schnelle Abkühlung und die martensitische Umwandlung erzeugen Spannungen in der gehärteten Schicht. Diese Spannungen führen zu einer Lagerverformung, die den Luftspalt während des Prozesses verändern würde. Um dies zu vermeiden, wurde ein mechanisches Nachführsystem eingesetzt. Jeder Heizkopf ist mit einer Nachführeinrichtung ausgestattet, die in Kontakt mit der Ringoberfläche gehalten wird. Während sich der Heizkopf um das Lager dreht, misst das Nachführsystem Änderungen der Lagerflächenposition und kompensiert die Spulenposition, um den Luftspalt konstant zu halten [7], [8].

Die Start- und Endzone sind die kritischsten Bereiche für den Prozess. In diesen Zonen ist eine leichte Überhitzung der Oberfläche notwendig, um eine Härtetiefe zu erzielen, die mit den Vorschubzonen vergleichbar ist. Das Erreichen der maximal zulässigen Oberflächentemperatur (1.050 °C) kann zu Kornvergröberung führen. Aus diesem Grund ist die Analyse der Mikrostruktur erforderlich. Trotz der erforderlichen Überhitzung bleibt die Mikrostruktur feinkörnig. In **Bild 3** und **Bild 4** wurde die Korngröße an der Oberfläche im am stärksten erwärmten Teil der Endzone gemessen. Es wurde eine durchschnittliche Korngröße von 8,5 gemäß dem Standard ASTM E112 erzielt. Diese Korngröße ist vergleichbar mit der in den Vorschubzonen gemessenen Korngröße. An den Positionen der Endzone, wo sich die Vorschubgeschwindigkeit ändert, wurde die Mikrostruktur überprüft, um die gute Umwandlung

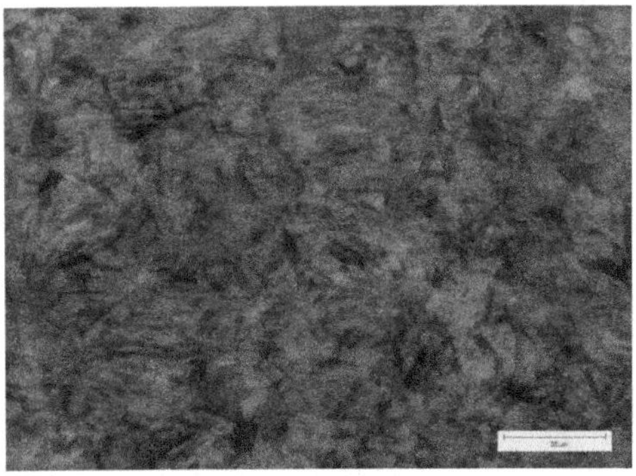

Bild 3: Martensitische Struktur in der Endzone mit einer Korngröße von 8 gemäß dem Standard ASTM E112, 500X

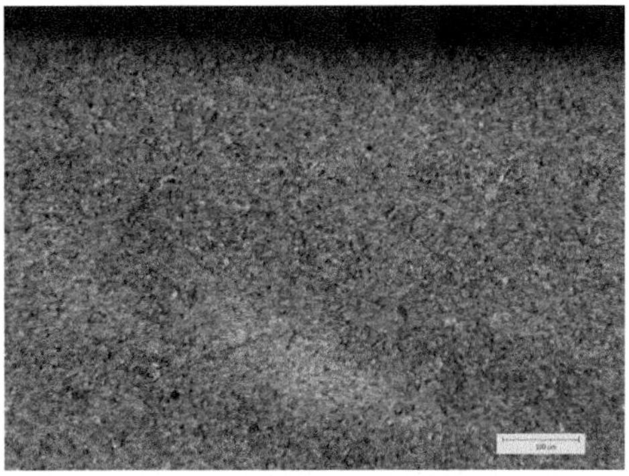

Bild 4: Martensitische Mikrostruktur in der Endzone ohne Bainit oder nicht-martensitische Umwandlungsprodukte, 100X

des Austenits in Martensit zu verifizieren. Bild 4 zeigt ein Beispiel der martensitischen Mikrostruktur ohne sichtbaren Bainit oder nicht-martensitische Umwandlungsprodukte.

5.3.2.3 Härten mit einem einzigen Schuss

Die in Bild 1(b) und (d) gezeigten Skizzen stellen lediglich einige mögliche Schemata eines Härtesystems mit einem einzigen Schuss dar. Bei diesem Verfahren erwärmt ein kreisförmiger Induktor oder eine Gruppe von Induktoren den Ringbereich gleichzeitig, während das Teil in Rotation gehalten wird. Ziel der Rotation ist die Homogenisierung des Temperaturprofils. Das Abschrecksystem (oberhalb, unterhalb und/oder zwischen den Induktoren platziert) wird verwendet, um das gesamte Teil am Ende des Erwärmungsprozesses gleichzeitig abzukühlen. Mit dem Einzelschusssystem wird die gesamte Oberfläche gleichzeitig erwärmt und dann abgeschreckt: das resultierende Härteprofil und die metallurgische Struktur sind entlang des Umfangs ohne Nahtstelle gleichmäßig.

Das Design eines solchen Induktorsystems kann je nach gewünschtem Härtungsmuster und den Fähigkeiten der Maschine variieren. Verschiedene Designschemata sind möglich, wie in **Bild 5** gezeigt, mit einer breiten Palette von Merkmalen, die mit der Wahl der Konstruktionsparameter verbunden sind und berücksichtigt werden müssen: Art und Verteilung der Flusskonzentratoren, Modulgeometrie und Teilung, Kupferabdeckung, Luftspalt zwischen dem Ring und dem Induktor, Stromrichtung und elektrische Verbindungen. In **Bild 6** sind zwei Beispiele für behandelte Elemente bei Verwendung unterschiedlicher Induktor-Designs dargestellt: ein Hair-Pin-Induktor für die linke Komponente und ein voller Kreisring-Induktor für das Element auf der rechten Seite. Beide Induktoren wurden so konstruiert, dass sie die erforderlichen Härtungsprofile erreichen.

5

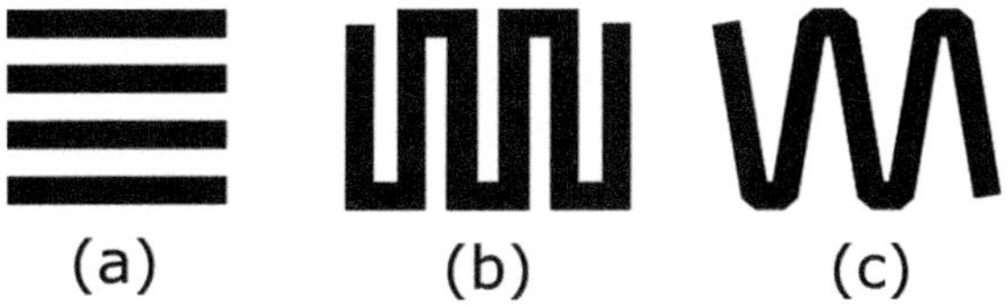

Bild 5: Einige mögliche Designs von Einzelschuss-Induktoren: (a) volle Kreisringe, (b) „Hair-Pins"-förmige Module und (c) „W"-förmige Module

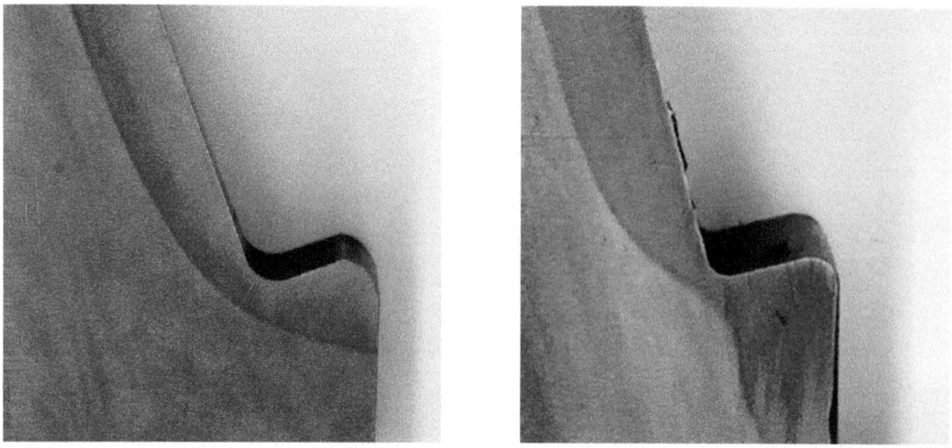

Bild 6: Geätzte Abschnitte des Rippensegments eines Rings, erhalten mit verschiedenen Einzelschuss-Verfahren

Ein einziger voller Kreisring-Induktor kann so konstruiert werden, dass er den Spurbereich mit 100-%-Abdeckung gleichmäßig erwärmt, wie in Bild 1(b) gezeigt ist. Nicht alle Ringe können mit diesem System behandelt werden: Der Induktor und die Abschreckduschen müssen so konstruiert werden, dass thermische Ausdehnungen des Rings und geometrische Interferenzen zwischen den Teilen berücksichtigt werden. Der Induktor benötigt einen einzigen Wandler, der in der Lage ist, die erforderliche Leistung an das System zu liefern. Aus diesem Grund ist die Wahl einer solchen Bauart durch die verfügbare Leistung auf die Elementgröße beschränkt. Je größer der Ring ist, desto höher ist die erforderliche Leistung unter Berücksichtigung einer minimalen Leistungsdichte pro Fläche.

Um gewisse Einschränkungen zu vermeiden, ist ein System wie in Bild 1(d) dargestellt vorzuziehen. Dieses Design umfasst mehrere separate Induktoren, die jeweils einen Abschnitt des Rings unabhängig voneinander erwärmen. Einerseits erfordert ein Mehrfachinduktordesign die Auslegung und Steuerung eines komplexeren elektromechanischen Systems. Andererseits ist jeder Induktor sowohl aus mechanischer als auch aus elektrischer Sicht unabhängig, was zu einer höheren Produktionsflexibilität führt (d. h. die Anzahl der Induktoren kann je nach Elementdurchmesser geändert werden) sowie zu der Möglichkeit, die relative Position von Element und Induktor während des Heizprozesses mit einem aktiven Nachführsystem anzupassen und einer geringeren Leistung, die für jeden unabhängigen Generator benötigt wird, sowie einer höheren Modularität.

Ein Nachteil von Einzelschuss-Systemen ist die benötigte hohe Leistung: die spezifische Leistungsdichte, die für die Wärmebehandlung erforderlich ist, wird auf eine große Fläche angewendet. Dies bedeutet hohe Leistungspegel für kurze Zeit. Auf der anderen Seite ist die kürzere Zykluszeit im Minutenbereich im Vergleich zu den Stunden, die für den Vorschubprozess benötigt werden, ein wesentlicher Vorteil dieser Art von Design.

Wo das Grundmaterial und das Elementdesign die Erwärmung zulassen, kann die vom System benötigte maximale Leistung durch einen Vorwärmschritt reduziert werden. Ein Niedrigleistungs- und ein Langzeitschritt werden verwendet, um die Temperatur bis zu einem Wert unterhalb der Materialumwandlung (500–700 °C) zu erhöhen.

Tabelle 1: Qualitativer Vergleich der Härteprozesse

Härtung Technologie	Traditioneller Ofen	Induktion Schuss-Härtung	Induktion Schlupflos-Vorschub
Benötigte Leistung	HOCH	HOCH	NIEDRIG
Prozesszeit	Tage	Minuten	Stunden
Produktivität	Chargen	Inline	Inline
Durchmesserbereich	bis zu 8 m	< 2 m	1–8 m
Energieverbrauch	HOCH	NIEDRIG	NIEDRIG
Systemkomplexität	NIEDRIG	HOCH	HOCH
Prozesskomplexität	NIEDRIG	HOCH	HOCH
Basismaterialkosten	HOCH	NIEDRIG	NIEDRIG
Systemflexibilität	HOCH	NIEDRIG	MITTEL
Verformungen	HOCH	NIEDRIG	NIEDRIG
Umweltfreundlich	NEIN	JA	JA

5.3.2.4 Induktionshärten - Zusammenfassung

Die Schuss- und die Vorschub-Schlupflosverfahren wurden in den vorhergehenden Abschnitten beschrieben. Für beide Verfahren hat die Anwendung der Induktionserwärmung positive Auswirkungen auf den ULB-Härteprozess (**Tabelle 1**).

Die Prozess-Einrichtung und -Entwicklung für ein Schuss- oder ein Vorschub-System erfordert ein feines Design und Studium des Induktors. Die Prozesszeit, Leistung, Frequenzen, das Abschrecksystem (Duschen, Abschreckmitteltyp und -fluss) und die Prozessrezepte müssen zusammen mit dem Induktor-Design analysiert und studiert werden, um Überhitzung, Entkohlung, Kornwachstum und Risse zu vermeiden. Aus denselben Gründen erfordert die Schlupflos-Prozessentwicklung die Definition der drei Hauptrezeptschritte (Start- und Endzone sowie der Vorschubbereich) mit geringerem Aufwand beim Induktor-Design, das speziell mit der Bahnform verbunden, aber unabhängig vom Elementdurchmesser ist.

Die Wahl der Technologie ist dann mit einem Kompromiss zwischen Leistungs- und mechanischen Einschränkungen verbunden: Elemente mit Durchmessern von mehr als 2.000 mm erfordern zu hohe Leistungen, um richtig erwärmt zu werden. Auf der anderen Seite erlauben Durchmesser von weniger als 1.000 mm nicht die Montage von Vorschub-Induktionsheizköpfen. Daher ist man gezwungen, ein Schuss-System zu verwenden.

5.3.3 Anlassverfahren

Das Anlassen ist erforderlich, um die Sprödigkeit des gehärteten Teils zu reduzieren. Die Temperatur muss auf einige hundert Grad erhöht werden, um innere Spannungen abzubauen. Der traditionelle Prozess basiert auf Gas- oder Elektroöfen, in denen die Teile für die Prozesszeit bei der erforderlichen Temperatur gehalten werden, normalerweise im Bereich von Stunden. Bei extrem großen Lagern beruht der traditionelle Anlassprozess auf riesigen Öfen. Diese Systeme haben eine riesige thermische Trägheit mit langen Heiz- und Kühlzeiten und werden manchmal 24 Stunden am Tag bei Temperatur gehalten, was zu einer sehr geringen Energieeffizienz führt und die Anlassoperation in Chargen anzuordnen zwingt [9].

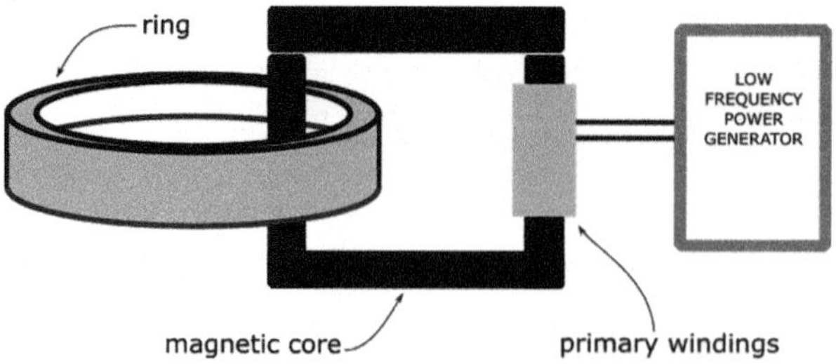

Bild 7: Ultra-Niederfrequenz-Anlassprozess – Systemkonzeptbeschreibung

Die elektromagnetische Erwärmung kann auch in diesem Bereich erfolgreich angewendet werden, indem eine über 60 Jahre alte Technologie genutzt wird, die beim Schrumpfen weit verbreitet ist [10]. Die Idee ist, den Ring als kurzgeschlossene Sekundärwicklung eines Transformators zu verwenden. Wie in **Bild 7** gezeigt, wird der Ring in ein geteiltes magnetisches Joch gelegt. Eine Primärwicklung, die aus mehreren Windungen besteht, induziert ein Magnetfeld im Joch; der induzierte Strom im Ring erzeugt die Wärme durch ohmsche Erwärmung. Mehrere Heizmodule können um denselben Ring platziert werden, um die Leistungsübertragung zu erhöhen. Die Modularität des Systems verleiht dem System Flexibilität für verschiedene Ringdurchmesser und Heizzeiten.

Die Effizienz ist ein Schlüsselfaktor bei der Umstellung von traditionellen Öfen auf Induktionserwärmung (**Tabelle 2**): Laut mehreren Studien [9] liegt die Effizienz eines statischen Ofens bei etwa 20 % bei voller Auslastung; diese Zahl sinkt bei Teilauslastung und eine durchschnittliche Effizienz von etwa 10 % bei Teilauslastung ist eher üblich.

Ultra-Niederfrequenz-Wechselrichter und die Integration mit standardmäßigen industriellen Prozesssteuerungen haben zu einer statischen Induktions-Anlasstechnologie geführt, die vollständig in die Produktionslinie integriert werden kann. Die Nutzung der Induktionserwärmung verändert die Physik des Phänomens: Die Heizrampe hängt nicht mehr vom Fourierschen Gesetz der Wärmeübertragung ab; die Wärme wird jetzt im Material selbst erzeugt und dies verändert die Gleichung, die die Oberflächentemperatur beschreibt, von exponentiell zu linear; dies führt zu viel kürzeren Heizzeiten mit gesteuerten Rampen.

Das Ultra-Niederfrequenz-Induktionshärtesystem, das in **Bild 8** dargestellt ist, hat einen Gesamtwirkungsgrad von 59 % und reduziert somit den Energieverbrauch um mehr als 50 %; bei teilweiser Ofenauslastung können die Energieeinsparungen enorm sein.

Tabelle 2: Vergleich

Traditioneller Ofenverbrauch [%]	20 %	10 %
Ringmasse [kg]	4.000	4.000
Temperiertemperatur [°C]	200	200
Enthalpieänderung [kW/h]	121,6	121,6
Jährliche Produktion [#]	2.000	2.000
Durchschnittliche Energiekosten [€/MW]	150	150
Jährliche Energiekosten für den Ofen (*) [MW/h]	182,400	364,800
Jährliche Induktionsenergie-Rechnung [MW/h]	61,831	61,831
Jährliche Ersparnis [k/€]	120,6	303,0

(*) Basierend auf einem Wirkungsgrad von 20 %

Bild 8: Maschine für ULB-Härtung mit Ultra-Niederfrequenz (SAET S.p.A.); Vier-Modul-Design mit verstellbarer Position für verschiedene Ringdurchmesser

Bild 9: Temperaturverlauf an verschiedenen Kontrollpunkten an der Oberfläche für ULB-Härtung; Aufheiz- und Haltezeit durch Rezept definiert (SAET S.p.A.); ULF-Vier-Modul-Maschine

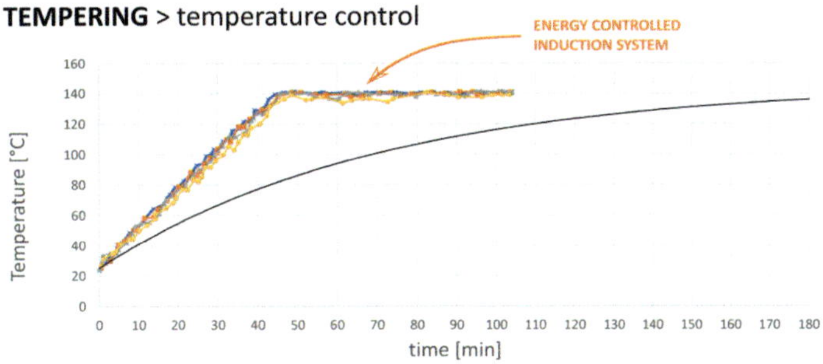

Prozess: Bei einem Härtungsprozess ist die Temperaturverteilung ein Schlüsselparameter; in einem 100 m² großen Ofen für eine ULB wurde die Temperaturhomogenität im gesamten Bereich auf +/- 4 °C geschätzt. Das Induktionshärtesystem wurde entwickelt, um vergleichbare Leistungen zu erzielen und es wurden umfangreiche 3D-Modellierungen durchgeführt, um parasitäre Effekte auf Ringbereiche näher an den Wicklungen abzuschätzen.

Zur Validierung des Prozesses wurde eine vollständige Reihe von Messungen an einem 4.000 kg schweren Ring durchgeführt: Fünf magnetische Thermoelemente wurden an der Laufbahn in verschiedenen Bereichen angebracht (**Bild 9**).

Einschränkungen: Die Temperaturverteilung ist nicht vollständig geometrieunabhängig; bei dieser Art von Anwendung ist die Stromverteilung auf einem Querschnitt des Lagers ziemlich homogen, aber leider ist die Verlustleistung auf Oberflächen im Hinblick auf die Temperaturverteilung auf einem engen Bereich relevant. Einige spezifische Formen mit engen Querschnitten könnten lange Aufheizzeiten oder eine erweiterte Temperaturschwankung über den Querschnitt hinweg (+/- 10 °C) erfordern.

5.3.4 Schlussfolgerungen

Einige Anwendungen der Induktionserwärmungstechnologie, die speziell für ULB-Wärmebehandlungsprozesse entwickelt wurden, wurden in diesem Beitrag vorgestellt.

Die Vorteile in Bezug auf Energieeinsparungen und technologische Möglichkeiten wurden sowohl für Härtungs- als auch für Anlassprozesse beschrieben, mit einigen Vergleichen zu traditionellen Technologien.

Zwei Induktionshärtungsprozesse (Schuss und Schlupflos) wurden beschrieben und verglichen. Vor- und Nachteile beider Ansätze wurden diskutiert und mit einigen erzielten industriellen Ergebnissen präsentiert.

Das Ultra-Niederfrequenz-Härtungssystem wurde vorgestellt, mit der Beschreibung des Systems und einigen Ergebnissen, die während der Inbetriebnahme der Industriemaschine erzielt wurden.

Literatur

[1] Mühlbauer, A.: Geschichte der Induktionserwärmung und -schmelzung, Vulkan-Verlag GmbH, 2008

[2] Lupi, S.: Fundamentals of Electroheat: Electrical Technologies for Process Heating. Springer Verlag, New York, USA 2016

[3] Rudnev, V.; Totten, G. E.: ASM Handbook. Volume 4C: Induction Heating and Heat Treatment. ASM International, Ohio, USA, 2014

[4] Bhadeshia, H. K. D. H.: Steels for bearings. In: Prog. Mater. Sci., 57 (2012) 2

[5] Lai, J.; Ovize, P.; Kuijpers, H.; Bacchettto, A.; Ioannides, S.: Case Depth and Static Capacity of Surface Induction-Hardened Rings. In: J. ASTM Int., 6 (2009) 10

[6] Scarpellini, L.; Cesano, M.; Pillin, B.; Ciscato, D.: Seamless induction hardening of a 42CrMo4 slewing bearing. In: Proceedings of the 30th Heat Treating Society Conference, Detroit, Michigan, USA, Okt. 2019, S. 50–55

[7] Cesano, M.; Ardissone, F.: Method and device for localized induction hardening of circular mechanical component parts of large-size, in particular rings for rolling bearings. EP2542707B1, 11. Mai 2016

[8] Cesano, M.; Cena, C.: Device and method for performing a localized induction hardening treatment on mechanical components, specifically thrust blocks for large-sized rolling bearings. US20100243643A1, 30. September 2010

[9] Källen, M.: Energy Efficiency Opportunities within the Heat Treatment Industry. 2012, doi: 20.500.12380/159604

[10] Semiatin, S. L.: Elements of Induction Heating: Design, Control, and Applications. In: ASM International, 1988

6. Vakuumwärmebehandlung

6.1 Charakteristika von Vakuumkammeröfen

Björn Eric Zieger

Moderne Vakuumofentechnik bietet eine breite Auswahl an Anlagengrößen und Anlagenausführungen und ermöglicht somit auch ein weitgefächertes Prozess-Spektrum. Neben dem typischen Härten, Anlassen und Glühen verschiedenster Chargen und Komponenten sind auch Sinterprozesse, Vakuumlöten, Niederdruckaufkohlen und Tiefkühlen realisierbar. Wesentliche Anlagenmerkmale und Features werden im Folgenden vorgestellt.

Einkammervakuumöfen sind in drei unterschiedliche Bauformen einzuteilen. Horizontal von vorne zu beladende Vakuumkammerofenanlagen werden den meisten Aufgabenstellungen gerecht.

Vakuumhubwagenöfen (**Bild 1**), deren Beladung von unten erfolgt, werden häufig bei der Wärmebehandlung von größeren rotationssymmetrischen Bauteilen eingesetzt. Solche Anlagen sind u. a. bei der Wärmebehandlung von Triebwerkskomponenten für die Luft- und Raumfahrt sowie bei der Herstellung von großen Press-Matrizen im Einsatz. Neben den höheren Investitionskosten ist hier auch eine größere Hallenhöhe notwendig.

Vakuumschachtofenanlagen werden von oben beladen. Sie werden u. a. bei der Wärmebehandlung von Bauteilen eingesetzt, bei denen das Durchmesser-/Länge-Verhältnis sehr groß ist, wie z. B. bei Räumnadeln und Extruder-Schnecken.

Je nach Bauform werden solche Einkammervakuumöfen für maximale Chargengewichte von 50 bis zu 15.000 kg und mehr gebaut.

Bild 1: Vertikaler Vakuumhubwagenofen, Beispielfoto (Quelle: IVA Schmetz GmbH)

6.1.1 Prinzipieller Aufbau von Vakuumkammeröfen

Der Vakuumofen ist elektrisch beheizt und mit einem doppelwandigen, wassergekühlten Ofenkessel konzipiert. Die thermische Belastung auf das metallische Ofengehäuse wird hierdurch reduziert, da nach außen nur die Leerverluste an das Kühlwasser im Doppelmantel abgegeben werden.

Im Ofengehäuse befinden sich die Heizkammer mit Isolationsauskleidung und Heizungselementen sowie die Schnellkühleinrichtung mit Wärmetauscher. Das Hochleistungsradialgebläse mit Elektromotor ist ebenfalls innerhalb des Rezipienten installiert.

6.1.2 Vakuumsysteme

In Vakuumöfen wird üblicherweise ein Unterdruck zwischen 10^{-1} und 10^{-6} mbar erzeugt. Für das Härten von Stahl mit metallisch blankem Oberflächenergebnis ist in den häufigsten Fällen ein Vakuum im 10^{-2}-mbar-Bereich vollkommen ausreichend. Hier wird ein mechanischer Vakuumpumpenstand, bestehend aus einer Drehschieberpumpe und einer Wälzkolbenpumpe (Rootspumpe) eingesetzt. Um höhere Vakua zu erreichen, können mehrstufige mechanische Pumpenstände oder Hochvakuumpumpen wie Öldiffusionspumpen, Turbomolekularpumpen oder Kryo-Pumpen, eingesetzt werden. Mit diesen Pumpen ist es möglich, den Rezipienten bis in den 10^{-6}-mbar-Bereich zu evakuieren.

6.1.3 Heizeinsätze

Die Heizkammern der meisten Vakuumöfen für die Wärmebehandlung sind mit einer Grafitisolation ausgekleidet. Die Heizkammern können rechteckig oder rund ausgeführt sein. Die Prozesstemperaturen zur Wärmebehandlung von metallischen Werkstücken liegen meist bei bis zu 1.250 °C. Die Auslegung der Vakuumofenanlagen liegt i. d. R. bei einer maximalen Arbeitstemperatur von 1.350 °C. Darüber hinaus gibt es aber auch Löt- und Sinterprozesse, die Temperaturen von bis zu ca. 1.600 °C verlangen. Die Arbeitstemperatur bestimmt auch die Stärke der Heizkammerisolation. So kommt standardseitig eine 40-mm-Isolation in Grafitheizeinsätzen von Vakuumöfen zum Einsatz. Im Temperaturbereich von bis zu 1.600 °C werden dann schon Isolationsstärken von 60 mm eingesetzt.

Verstärkte Isolation in modifizierten Heizkammern wird aber auch vermehrt zugunsten einer höheren Energieeffizienz verwendet. Dies trägt zu einer Energieverbrauchsreduzierung von ca. 10 % und mehr bei.

Für spezielle Applikationen oder Forderung von kohlenwasserstofffreier Atmosphäre werden ganzmetallisch isolierte Heizkammern **(Bild 2)** verwendet. Bei diesen Heizeinsätzen werden als Isolation mehrere hintereinanderliegende Strahlschilde aus Molybdän und hitzebeständigen Edelstählen eingesetzt. Die Höhe der Anwendungstemperatur bestimmt

Bild 2: Molybdän-Heizkammer eines horizontalen Vakuumofens, Beispielfoto (Quelle: IVA Schmetz GmbH)

6

die Anzahl an Strahlschilden und deren Werkstoff. Diese Heizkammern können rechteckig oder rund ausgeführt werden.

6.1.4 Heizungssysteme und Ausführungen

Wichtig bei der Beheizung im Vakuumofen ist die homogene Erwärmung der Charge, um Verzug bereits in der ersten Phase des Wärmebehandlungsprozesses zu vermeiden. Die Heizleistung einer Vakuumofenanlage liegt je nach Nutzraumgröße im Bereich von etwa nur 50 bis zu über 1.000 kW. In der Praxis hat sich für die meisten Prozesse und Anlagenausführungen ein einzoniger Heizungsaufbau durchgesetzt. Für Bauteile oder Chargen mit sehr unsymmetrischem Aufbau können Heizsysteme auch in mehreren separaten Zonen (z. B. Decke, Seiten, Boden) ausgeführt werden. Darüber hinaus können neben einer zusätzlichen Tür- und Rückwandheizung auch noch Mittenheizungen oder Kombinationen von separaten Heizzonen konfiguriert werden.

Während im hohen Temperaturbereich das Heizen nur unter Strahlungserwärmung – also in Vakuumatmosphäre – erfolgt, kann im unteren Bereich der Vorteil einer Heißgaskonvektion unterstützen.

6.1.5 Abkühlmechanismen in Einkammervakuumöfen

Gemäß den Erfordernissen der Wärmebehandlungscharge und der Werkstoffe können unterschiedliche Mechanismen für die Kühlung eingesetzt werden.

Mit einem geregelten Abheizen wird die Chargentemperatur gemäß einem Gradienten langsam gesenkt.

Bei einer Vakuumkühlung erfolgt die Chargenkühlung über die Leerverluste an den wassergekühlten Ofenkessel. Die Wärmeenergie wird sehr langsam an die kalte Innenseite des wassergekühlten Doppelmantels abgegeben. Zur Beschleunigung können ggf. Heizraumklappen geöffnet werden.

Eine Gaskühlung mit stehendem Schutzgas forciert den Wärmetransfer von der Charge über das Schutzgas an den wassergekühlten Ofenkessel. Zur weiteren Forcierung können ggf. auch hier Heizraumklappen geöffnet werden.

Die Hochdruckgasabschreckung erzeugt höchste Chargen-Kühlgeschwindigkeiten. Sie spielt verfahrenstechnisch für Glüh- und Lötanwendungen eine eher untergeordnete Rolle, ist aber beim Härten von Stahlbauteilen zumeist eine Notwendigkeit.

6.1.6 Hochdruckgasabschreckung

Bei der Hochdruckgasabschreckung wird die Ofenanlage mit Schutzgas (beim Härten von Stählen meistens Stickstoff, bei Titanlegierungen häufig Argon) auf einen vorwählbaren Druck geflutet. Ofenanlagen werden hier auf maximal Abschreckdrücke im Bereich von 1,5 bis 15 bar (abs.) ausgelegt. Die Kühlventilatoreinheit, bestehend aus Drehstrommotor und Ventilatorrad, erzeugt einen Kühlgasfluss, der die komplette Charge durch geöffnete Heizraumklappen oder Düsen durchströmt. Hierbei wird die Wärmeenergie der Charge erfasst und über das Schutzgas an den Gas-/Wasser-Wärmetauscher abgeführt.

Das Härten von Bauteilen aus mittel- und hochlegierten Stählen stellt verschiedenste Anforderungen an die Gasabschreckung.

Die Stahlqualität und deren Legierungszusammensetzung sowie auch die Bauteildicke und die Chargengröße bestimmen die notwendige Mindestabkühlgeschwindigkeit, wobei diese nicht höher als metallurgisch erforderlich sein sollte. Die Geschwindigkeit der Chargenabschreckung wird anlagentechnisch neben dem Kühlgasdruck z. B. auch vom Gasvolumenstrom (Leistung des Motors und Auslegung des Ventilatorlaufrades), dem Gas-/Wasser-Wärmetauscher (konstruktive Ausführung und Leistung) sowie der peripheren Kühlwasserversorgung (Menge und Temperatur etc.) bestimmt. Generell wird eine gleichmäßige und verzugsarme Wärmebehandlung angestrebt. Hier stellen die Bauteilgeometrie und der Chargenaufbau entsprechende Anforderungen. Die Homogenität der Chargenkühlung wird anlagentechnisch beeinflusst durch die Verteilung der Kühlgasdurchströmung. Diese wird bestimmt durch die Anordnung und die Ausführung der Ein- und Austrittsöffnungen wie Gasverteilerplatten oder Düsen.

6.1.7 Einkammervakuumofen mit eckiger Heizkammer und geradliniger Kühlgasdurchströmung mit den Systemen 2R bzw. 2x2R

Für einen Großteil der typischen Wärmebehandlungsprozesse in Einkammervakuumöfen hat sich das Durchströmprinzip in einer rechteckigen Heizkammer durchgesetzt. Für eine hohe Kühlgleichmäßigkeit sind große, in ihrer Abmessung der Charge entsprechende Gaseintritts- und Gasaustrittsöffnungen erforderlich. Diese Gaseintritts- und Gasaustrittsöffnungen sind mit speziellen Gasverteilerelementen ausgestattet.

Die Homogenität der Gasabschreckung kann mit den Systemen 2R (**Bild 3**) oder 2x2R, der programmierbaren vertikalen und/oder der horizontalen Reversion der Kühlgasdurchströ-

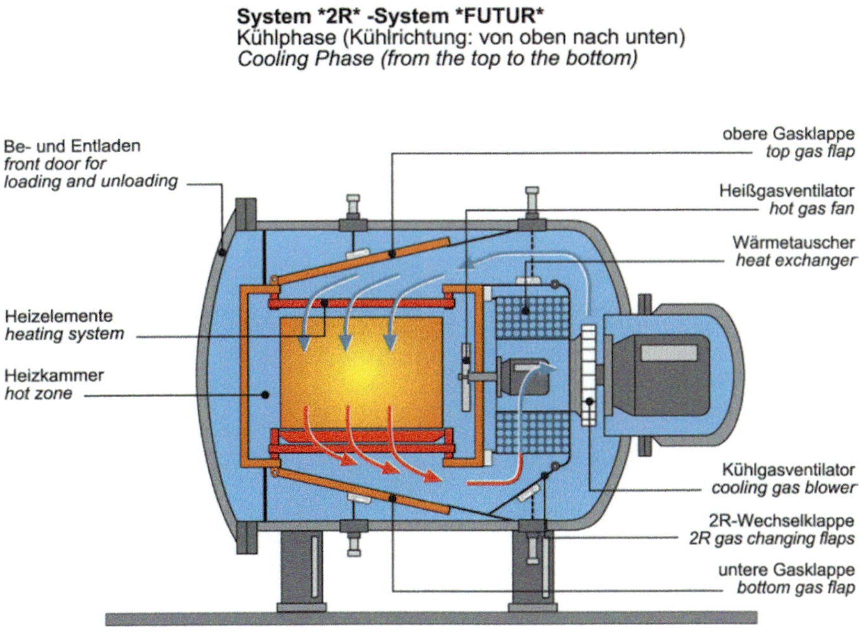

Bild 3: Prinzip der vertikalen Reversion der Kühlgasdurchströmung in einer rechteckigen Heizkammer mit dem System 2R (Quelle: IVA Schmetz GmbH)

mung, erhöht werden. Dabei ist das mehrfache Wechseln der Kühlrichtung von der horizontalen auf die vertikale Durchströmung ebenso gut möglich wie die Realisierung einer rotierenden Kühlung. Der Gasrichtungswechsel kann zeit- oder temperaturgesteuert, entsprechend den Anforderungen des Wärmebehandlungsgutes, durchgeführt werden. Die Abschreckung erfolgt gleichmäßiger, Härtestreuung innerhalb der Charge und Verzug am Einzelbauteil werden weitgehend reduziert. Darüber hinaus wird auch eine schnellere und damit ökonomischere Chargenkühlung realisiert.

6.1.8 Vakuumofen mit rechteckigen Heizeinsatz und separater Abschreckzone mit dem System 2PLUS

Eine Möglichkeit zur Steigerung der Abschreckgeschwindigkeit in einem Vakuumkammerofen mit rechteckiger Heizkammer bietet eine separate Abschreckzone ohne Vakuum und druckdichte Trennung im gemeinsamen Ofenkessel. Dieser Ofentyp mit dem System 2PLUS (**Bild 4**) beinhaltet in Verbindung mit einer separaten, kalten Abschreckzone die wesentlichen Merkmale einer Einkammerausführung. Die Charge wird in der Heizkammer auf Temperatur gebracht (wahlweise unter Konvektion oder Strahlung) und nach dem Erreichen der Härtetemperatur automatisch in die Abschreckzone transportiert und dort abgeschreckt. Hierzu öffnet ein Strahlungsschieber die thermische Barriere zwischen heißer und kalter Zone. Die Charge wird innerhalb weniger Sekunden in die Kühlzone gefahren. Der Verschlussschieber schottet anschließend wieder die kalte gegen die heiße Kammer ab. Die Abschreckzone ist auf optimale Kühlgasströmung ausgelegt, d. h., die Kühlkammer ist an den Chargennutzraum angepasst. Der Kühlgasstrom muss dann nur die Energie der Charge und der Chargentransporteinheit aufnehmen und abführen. Hier werden nicht

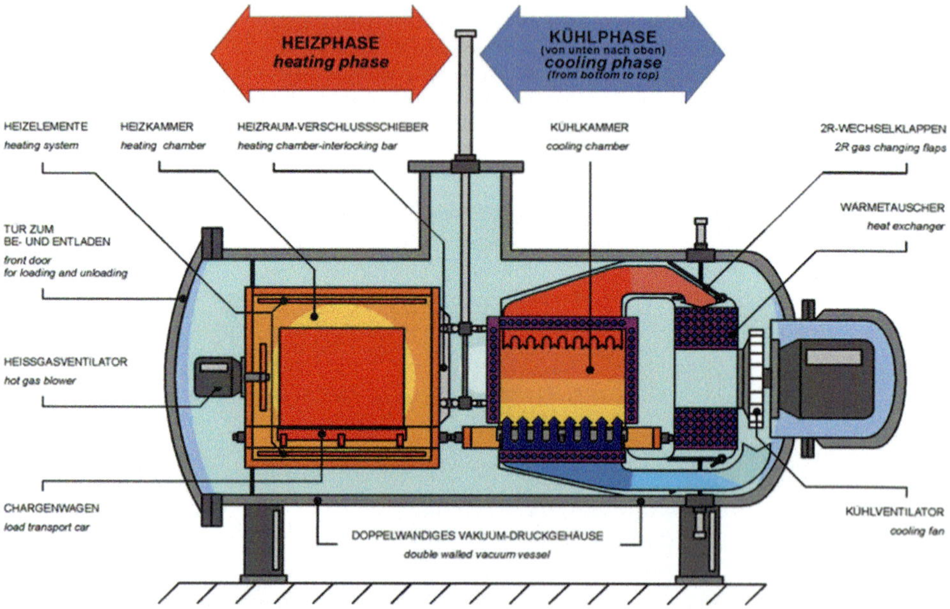

Bild 4: Prinzip eines Vakuumofens mit separater Abschreckzone mit dem System 2 PLUS (Quelle: IVA Schmetz GmbH)

wie bei einem Einkammerofen auch die in der Heizkammer befindliche Isolation, Heizstäbe und Gasleiteinrichtungen aktiv mit abgekühlt. Die schnelle Abschreckung der Charge erfolgt zudem nicht nur über den Kühlgasstrom, sondern auch über die intensive Strahlung von zusätzlich angeordneten Wärmetauschern.

Im Vakuumkammerofen mit kalter Abschreckzone lassen sich die Kühlzeiten gegenüber einem Einkammerofen bei analogem Abschreckdruck verdoppeln. Betriebskosten und Laufzeiten werden eingespart, da der Heizraum während der Betriebszeit nicht aktiv abgekühlt werden muss. Dies zahlt sich insbesondere bei Vergütungsprozessen mit mehrmaligem Anlassen aus. Erst zum Prozessende, bei Entnahme der Charge, ist es erforderlich, die Heizkammer mit abzukühlen. Darüber hinaus bieten diese Vakuumöfen mit separater Abschreckzone auch die Möglichkeit, Chargenthermoelemente mitzuführen und die Bauteiltemperaturen über den kompletten Zyklus – Heiz-, Transport- und Kühlvorgang –, zu messen und zu dokumentieren. Die exakte Temperaturführung realisiert auch hier ein Höchstmaß an Reproduzierbarkeit. Dieser Anlagentyp wird dort eingesetzt, wo zu härtende Bauteile und Chargen eine geradlinige Kühlgasdurchströmung bei tendenziell höherer Abschreckgeschwindigkeit fordern.

6.1.9 Einkammervakuumofen mit runder Heizkammer und optimiertem Kühlgaseinströmung mit dem System RD PLUS

Bei einigen Vakuumprozessen wie dem Härten von sehr großen und massiven Bauteilen, wie z. B. Druckgussformen, und der damit erforderlichen hohen Abschreckgeschwindigkeit kann eine runde Heizkammer mit allseitig radialer Düsenkühlung vorteilhaft sein. Der Kühlgasstrom trifft hier rundum auf die Oberfläche der Charge und wird nach hinten zum Wärmetauscher abgeführt. Das optimierte Düsenkühlsystem RD PLUS (**Bild 5**) gleicht mit einer intensivierten Kühlung an der Unterseite die strömungstechnischen und thermischen Widerstände der Chargentragkonstruktion, Chargier-Grundroste etc. aus. Eine höhere Anzahl von Kühldüsen im Bodenbereich und damit eine maximierte Kühlgasanströmung von dort realisieren hierbei den notwendigen Ausgleich. Die Kühldüsen im Bodenbereich sind so positioniert, dass die Gasströmung zwischen den Tragbänken zur Charge kanali-

Bild 5: Runde Heizkammer mit optimierter Düsenkühlung mit dem System RD PLUS, Beispielfoto (Quelle: IVA Schmetz GmbH)

6

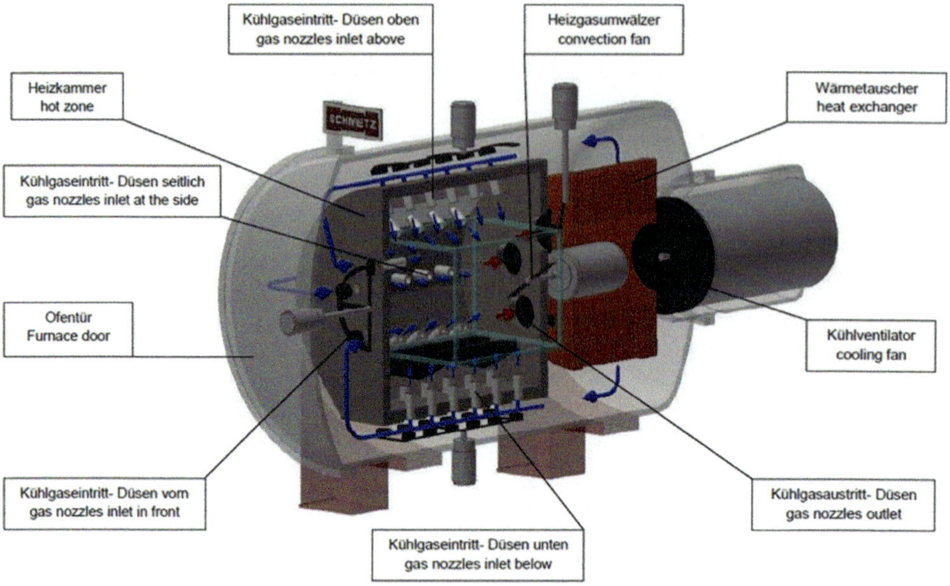

Bild 6: Prinzip eines Vakuumofens mit runder Heizkammer und optimierter Düsenkühlung mit dem System RD PLUS (Quelle: IVA Schmetz GmbH)

siert wird. Im hinteren Bereich der Heizkammer sind entsprechend dimensionierte Kühlgasaustrittsöffnungen angeordnet. Durch diese Austrittsöffnungen wird der Gasstrom zum integrierten Gas-Wasser-Wärmetauscher abgesaugt (**Bild 6**). Darüber hinaus kann der Kühlventilator auch noch eine zusätzliche Kühlgasdurchströmung der Heizkammer von vorne nach hinten erzeugen. Hierfür sind im Heizkammertürbereich optional zu öffnende Kühlgaseintrittsdüsen installiert.

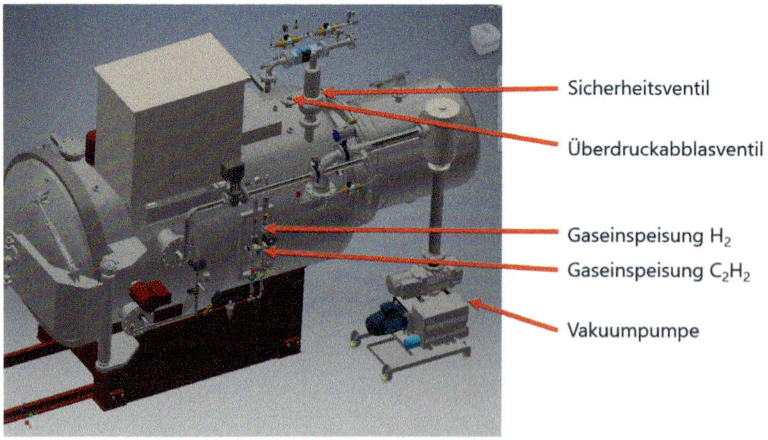

Bild 7: Prinzip des Aufbaus eines Vakuumofens mit einem Niederdruckaufkohlen-System CARB PLUS (Quelle: IVA Schmetz GmbH)

Vakuumöfen mit dem Düsenkühlsystem RD PLUS und dessen Weiterentwicklungen sind mit Nutzraumabmessung von 900 x 1.200 x 900 mm (B x L x H) und nun von bis zu ca. 2.000 x 3.000 x 1.500 mm (B x L x H) und Nutzgewichten von bis zu 15.000 kg im internationalen Einsatz.

6.1.10 Niederdruckaufkohlen im Einkammervakuumofen mit dem System CARB PLUS

Das Niederdruckaufkohlen (LPC, Low Pressure Caburizing) im Vakuumofen mit dem System CARB PLUS (**Bild 7**) erfolgt unter Einsatz von Acetylen. Dieses Einsatzhärteverfahren wird bei geringen Drücken bei Temperaturen im Bereich von bis zu 1.070 °C durchgeführt.

Es werden gleichmäßige Aufkohlungs- und Einsatzhärtetiefeergebnisse auch in dicht gepackten Chargen an allen Bauteilpositionen – auch in engsten Ausbohrungen und an komplexen Geometrien – realisiert.

Die steuerbare Überdruckgasabschreckung minimiert auch hier Bauteilverzüge. Bei diesem vollautomatischen, mannlosen Verfahren mit Aufkohlen, Abschrecken und Anlassen werden blanke Oberflächenergebnisse ohne Randoxidation reproduziert.

6.1.11 Tieftemperaturbehandlung im Vakuumofen mit dem System COOL PLUS

Das Tiefkühlen bei der Vergütung von Werkzeugen ist ein etabliertes Verfahren. Hierbei werden durch die Tieftemperaturbehandlung Werkzeugeigenschaften durch gezielte Optimierung von metallurgischen Strukturen, wie z. B. die Reduzierung von Restaustenit, signifikant verbessert.

Die Integration der Tiefkühleinrichtung in den Vakuumhärteofen realisiert einen mannlosen, vollautomatischen Vergüteprozess. Da die Werkstücke zwischen dem Tiefkühlen und dem anschließenden Anlassen (**Bild 8**) nicht in Kontakt mit der Umgebungsatmosphäre treten, kann verfahrensbedingt keine Oberflächenkorrosion auftreten. Hierdurch wird die für die Vakuumwärmebehandlung typische metallisch blanke Oberflächenqualität sichergestellt.

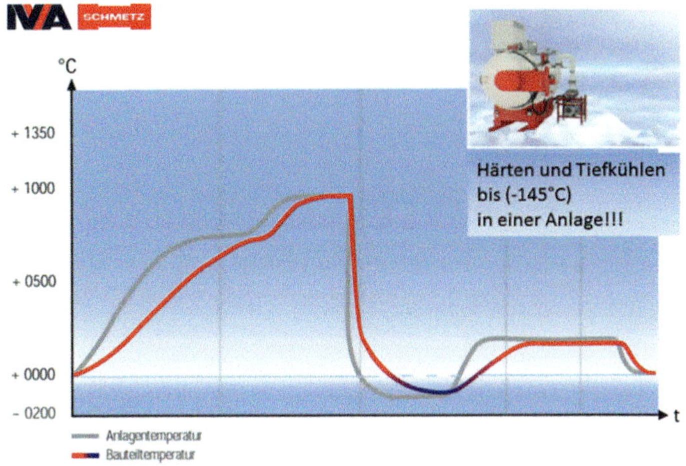

Bild 8: Prinzip des Prozesses im Vakuumofen mit integrierter Tiefkühlung mit dem System COOL PLUS (Quelle: IVA Schmetz GmbH)

6

Durch den auch mit Chargen-Thermoelementen gesteuerten und selektierbaren Tiefkühlgradienten wird ein sanftes, kontrolliertes Tiefkühlen ermöglicht. Folglich wird die Gefahr der Spannungsrissbildung durch schroffes Tiefkühlen, wie dies bei anderen Tieftemperaturverfahren der Fall ist, erheblich eingeschränkt.

Das Funktionsprinzip bilden die Eindüsung und Vergasung von flüssigem Stickstoff durch ein Düsensystem in die grafitisolierte Heizkammer des Vakuumkammerofens. Bei der Vergasung tritt eine etwa 700-fache Stickstoffvolumenvergrößerung ein. Im Chargennutzraum wird der noch tiefkalte, gasförmige Stickstoff über die Konvektionseinrichtung gleichmäßig verteilt. Hierbei wird der Charge die Wärmeenergie entzogen und dann weiter abgeführt.

6.1.12 Fazit

Moderne Vakuumkammeröfen können mit ihrer Vielzahl an möglichen Features und Charakteristika für unterschiedlichste Wärmebehandlungsprozesse und Chargen konzipiert werden. Kleine Anlagen mit Nutzraumabmessungen von z. B. 300 x 300 x 300 mm (B x L x H) werden neben Zwecken der Forschung und Entwicklung auch für kleinere Chargen oder Kleinstkomponenten, wie z. B. dem Härten von Chirurgieinstrumenten, flexibel eingesetzt.

In großen Einkammervakuumöfen werden z. B. 4.000 kg Plattenwärmetauscherchargen in Großserie gelötet. In Anlagen mit Chargennutzgewichten von bis zu 15.000 kg werden entsprechende Großkomponenten, wie z. B. Formwerkzeuge, mittels Hochdruckgasabschreckung gehärtet.

6.2 Erforschung des Potenzials des Niederdruckkarbonitrierens

Grzegorz Głuchowski, Dr. Robert Pietrasik, Michal Bazel

Angesichts des wachsenden Interesses am Niederdruckkarbonitrieren und des ständig steigenden Bedarfs der Kunden an der Verbesserung der Nachhaltigkeit ihrer Unternehmen war es notwendig, die verfügbaren technologischen Möglichkeiten für das Niederdruckkarbonitrieren in Vakuumöfen zu untersuchen. Im Jahr 2021 schloss Seco/Warwick eine Partnerschaftsvereinbarung mit dem gewerblichen Wärmebehandlungsunternehmen Hart-TECH in Łódź, Polen. Gegenstand der Vereinbarung war die gemeinsame Erforschung und Entwicklung einer thermochemischen Wärmebehandlungstechnologie für das Niederdruckkarbonitrieren in einem industriellen Vakuumofen. Der erste Schritt bestand darin, eine Reihe von Stahlsorten auszuwählen, die sich für Niederdruckkarbonitrierverfahren eignen, die erforderlichen Projektmeilensteine zu definieren und Prozessparameter für erste Test-Wärmebehandlungsverfahren festzulegen.

Das Karbonitrieren selbst ist seit vielen Jahren ein etabliertes thermochemisches Verfahren. Seine Niederdruckvariante, die auf der Niederdruckaufkohlung basiert, hat in den letzten Jahren an Popularität gewonnen. Die wichtigsten Prinzipien des Niederdruckkarbonitrierens (Vakuum) werden von Daniel H. Herring wie folgt erläutert: „Das Vakuumkarbonitrieren ist ein modifizierter Aufkohlungsprozess, keine Form des Nitrierens. Diese Modifikation besteht darin, dass Ammoniak in die Aufkohlungsatmosphäre eingebracht wird, um der aufgekohlten Hülse während der Herstellung Stickstoff hinzuzufügen. Karbonitrieren ist eine Oberflächenhärtungsbehandlung, bei der Kohlenstoff und Stickstoff in den Stahl oberhalb der Austenitisierungstemperatur (Ac3) eingebracht werden. Beim Abschrecken wird ein martensitischer Zustand erreicht. Die Härte des Einsatzes hängt von der Kohlenstoff- und Stickstoffkonzentration des Einsatzes ab." [1]

Das Hauptanwendungsgebiet des Karbonitrierens ist die Wärmebehandlung von unlegierten Stählen wie Automatenstählen sowie Maschinen- und Baustählen. Die erzielte Einhärte-

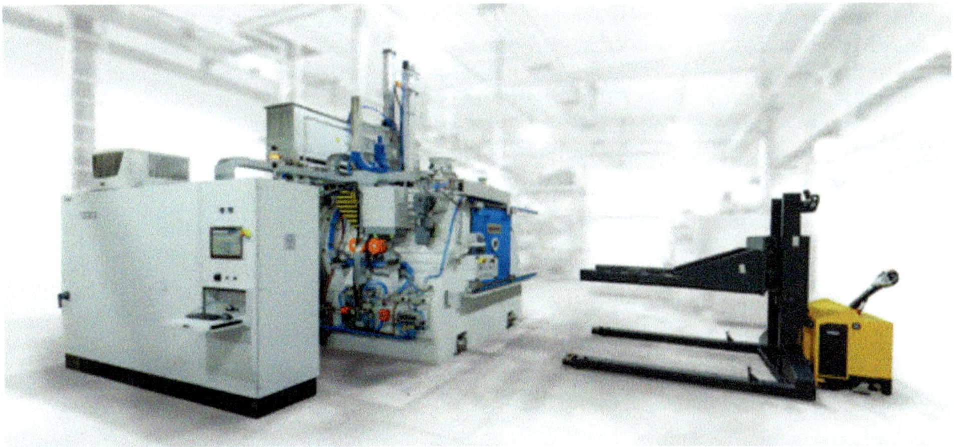

Bild 1: Doppelkammervakuumofen mit Ölabschreckung, Typ Case Master Evolution (Quelle: Seco/Warwick)

tiefe ist geringer als bei aufgekohlten Teilen (karbonitrierte Schichten liegen typischerweise im Bereich von 0,2 bis 0,5 mm), da die Kohlenstoffdiffusion durch Stickstoff behindert wird und die Prozesstemperaturen beim Karbonitrieren i. d. R. niedriger sind und kürzer dauern. Das Ergebnis des Verfahrens ist eine Verbesserung der mechanischen Eigenschaften der Teile durch die Bildung einer Epsilon-Nitrid-Phase, die die Gehäusehärte erhöht und die Verschleiß- und Abriebfestigkeit verbessert [2].

Die gängigste Methode in der Industrie ist das Karbonitrieren in atmosphärischen Aufkohlungsöfen, entweder in Chargenöfen (z. B. mit abgedichteter Abschreckung) oder in Durchlauföfen (z. B. mit Gitterband). Ein gemeinsames Merkmal ist, dass bei dieser Lösung eine entflammbare und explosive endotherme Atmosphäre verwendet wird und der Prozess selbst nicht nur zur Emission von Kohlendioxid, sondern auch von anderen toxischen Gasen (CO, SO_x, NO_x) führt. Darüber hinaus müssen die atmosphärischen Aufkohlungsöfen mit offenen Flammenvorhängen und Nachbrennern arbeiten, die eine echte und ständige Brandgefahr darstellen. Vakuumöfen und Niederdrucktechnologien hingegen wirken all diesen Nachteilen entgegen und bieten sicherere, schnellere und sauberere Prozesse, die den aktuellen Erwartungen von Behörden und Industrie entsprechen. Die Wärmebehandlung im Vakuum gilt als grundlegende Zukunftstechnologie und bietet die Möglichkeit, von den neuesten Errungenschaften auf dem Gebiet der Material- und Produktentwicklung sowie der Fertigungsmethoden zu profitieren [3]. Trotz strengerer Vorschriften zur Begrenzung der Auswirkungen der globalen Erwärmung und zur Verringerung der CO_2-Emissionen, wie z. B. der UN-Resolution „Transforming Our World: the 2030 Agenda for Sustainable Development" [4], und des Pariser Abkommens [5] sowie der Tatsache, dass immer mehr Länder [6] und Unternehmen erklären, ihren CO_2-Fußabdruck deutlich zu verringern oder kohlenstoffneutral zu sein, ist es unmöglich, eine kohlenstofffreie Produktion zu erreichen. Es ist unmöglich, kohlenstoffneutrale thermochemische Wärmebehandlungsprozesse, wie das Aufkohlen oder das Karbonitrieren, in herkömmlichen Atmosphärenöfen zu erreichen. Deshalb wenden sich immer mehr Kunden an Seco/Warwick, um umweltfreundliche vakuumbasierte Lösungen für eine nachhaltige Wärmebehandlung zu erhalten.

6.2.1 Prüfverfahren für das Niederdruckkarbonitrieren

Zu Beginn der Forschungsarbeiten sollten die Unterschiede in den grundlegenden metallografischen Ergebnissen (Mikrohärte und Mikrostruktur) zwischen den Proben nach dem Niederdruckaufkohlen (LPC) und den Proben nach dem Niederdruckkarbonitrieren (LPCN) bewertet werden. Es wurden Rezepte für zwei Wärmebehandlungsverfahren der Niederdruckaufkohlung definiert und durchgeführt; ein Unterschied zwischen den Verfahren bestand in der Zugabe von gasförmigem Ammoniak in einem der Verfahren, um eine Stickstoffanreicherung der Oberflächenschicht zu erreichen. Die folgenden Stahlsorten wurden in beiden Verfahren als Referenzproben verwendet:

- Baustähle zum Vergüten: 36MnB4 und 42CrMo4
- Baustähle für die Aufkohlung: 21TiMnCr12 und 16MnCr5
- Kohlenstoffstähle: S355J2 und C15
- Automatenstähle: 11SMn30 und 11SMnPb37.

Die nachstehenden Diagramme zeigen die Ergebnisse für die Oberflächenhärte und die effektive Einsatztiefe nach dem ersten Testverfahren.

Die Ergebnisse für die Oberflächenhärte (**Bild 2**) und die effektive Einsatztiefe (**Bild 3**) haben deutlich gezeigt, dass die Anwendung eines Niederdruckkarbonitrierverfahrens das Erreichen einer höheren Oberflächenhärte und einer dickeren effektiven Einsatztiefe

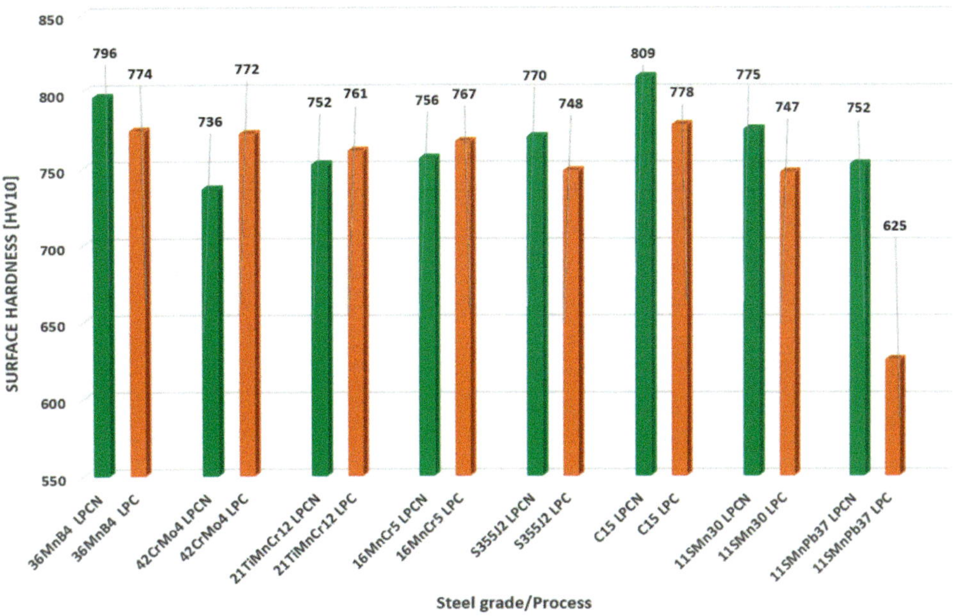

Bild 2: Vergleich der Oberflächenhärte nach dem LPCN-/LPC-Verfahren (Quelle: Seco/Warwick)

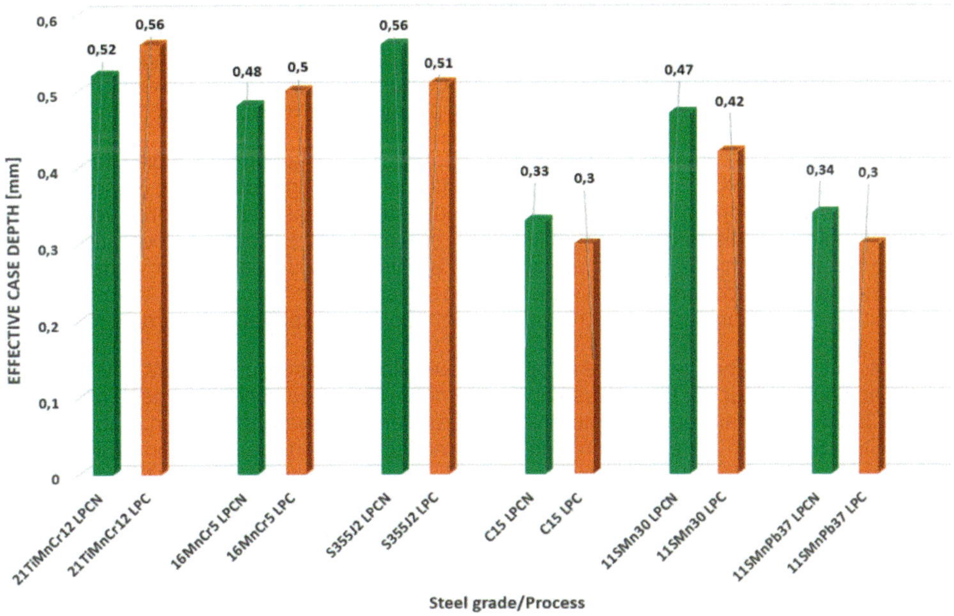

Bild 3: Vergleich der effektiven Gehäusetiefe nach dem LPCN-/LPC-Verfahren (Quelle: Seco/Warwick)

6

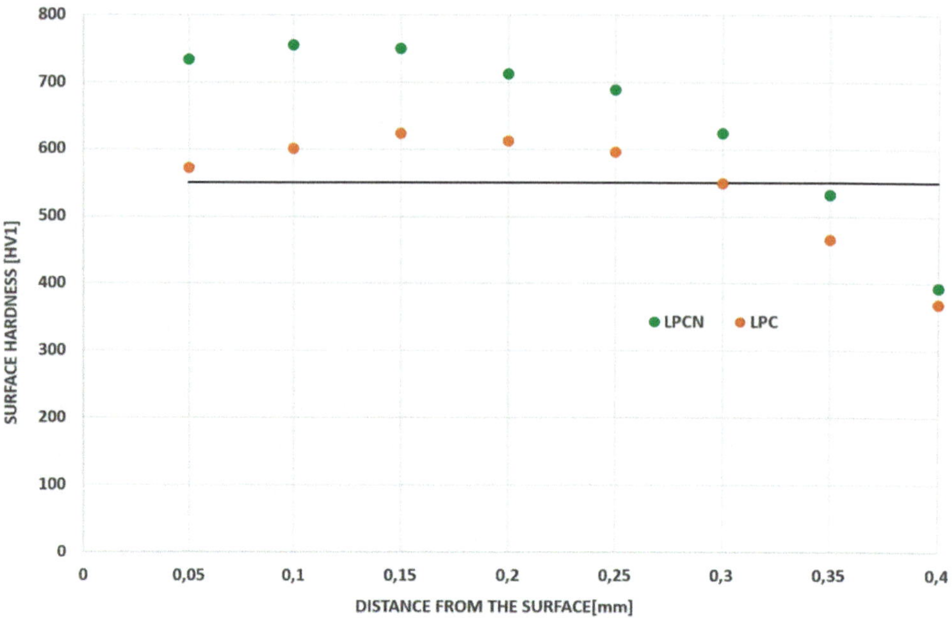

Bild 4: Vergleich der Oberflächenhärte nach dem LPCN-/LPC-Verfahren für 11SMnPb37 (Quelle: Seco/Warwick)

ermöglicht, insbesondere bei Automatenstählen. Insbesondere bei 11SMnPb37 wurden eine Erhöhung der effektiven Einsatztiefe um 13 % (Bild 2) und eine Erhöhung der Oberflächenhärte um 150 HV (**Bild 4**) beobachtet. Bei den anderen Stahlsorten war der Anstieg entweder minimal oder es wurde überhaupt kein positiver Effekt beobachtet.

Auf der Grundlage der gesammelten Informationen wurde festgelegt, dass die Stahlsorten 11SMn30 und 11SMnPb37 als Ausgangsmaterial für die nachfolgenden Tests dienen sollen. Diese Stahlsorten werden in der Industrie am häufigsten für das Karbonitrieren verwendet und die nach dem Niederdruckaufkohlungsprozess erzielten Ergebnisse bestätigten die Vorteile dieses Verfahrens. Andere ausgewählte Stahlsorten wurden von der weiteren Verarbeitung ausgeschlossen, da die erzielten Ergebnisse keine signifikante Verbesserung zeigten und diese Stahlsorten in der Industrie nicht so häufig verwendet werden.

In weiteren Untersuchungen wurde untersucht, wie sich die Änderung der Wärmebehandlungsbedingungen auf die Ergebnisse auswirkt, indem fünf zusätzliche Prozesse mit unterschiedlichen Temperaturen und Ammoniakdosierungen durchgeführt wurden. Dabei wurde ein breites Spektrum an effektiven Temperaturen für die Verarbeitung von 11SMnPb37-Stahl ermittelt und die optimalen Parameter für die Ammoniakdosierung bestätigt. Mit diesen Informationen wurden die Schlüsselparameter für ein effizientes Niederdruckkarbonitrierverfahren ermittelt, das günstige Ergebnisse und wirtschaftliche Effizienz gewährleistet. Nach der Durchführung der Testprozesse wurden auch die Stickstoff- und Kohlenstoffkonzentrationen in der Gehäusetiefe gemessen. Der ermittelte Wert für Stickstoff im Untergrund betrug ca. 0,3 % und die Gesamttiefe der Hülse ca. 0,50 mm. Bemerkenswert ist, dass nach der Niederdruckaufkohlung eine größere Eindringtiefe des Kohlenstoffs beobachtet wurde, was bedeutet, dass die Zugabe von Stickstoff die Diffusion des Kohlenstoffs in das wärmebehandelte Material erleichterte, wodurch gleichzeitig eine größere Einsatztiefe erreicht werden konnte.

6.2.2 Tribologische Tests

Tribologische Tests an der Kugel-Scheiben-Reibpaarung wurden durchgeführt, um die Verschleißfestigkeitseigenschaften der Scheibenoberfläche zu bewerten. Die genormten Scheiben für diese Tests wurden aus 11SMnPb30 hergestellt und entweder einer Niederdruckaufkohlung oder einem Niederdruckkarbonitrierverfahren unterzogen. Als genormte Gegenprobe für die Scheiben diente eine Kugel aus 100Cr6-Stahl. Die Kontaktfläche der beiden Werkstoffe des Paares war trocken und für jedes Paar wurden drei Tests unter den folgenden Bedingungen durchgeführt:

- Geschwindigkeit: 0,1 m/s
- Zeit: 1.000 s
- Reibradius: 0,009 m
- Drehgeschwindigkeit: 106 U/min
- Anzahl der Umdrehungen: 17.684
- Reibweg: 1.000 m.

Die Ergebnisse der tribologischen Tests waren wie folgt:

- Für den normalisierten Scheibenverschleiß wies die Indikatorprobe nach der Niederdruckaufkohlung einen mehr als fünfzigmal höheren Wert auf als die Probe nach dem Niederdruckkarbonitrieren. Das bedeutet, dass die mit LPCN behandelte Probe einen deutlich geringeren Verschleiß aufwies als die mit LPC behandelte Probe (**Bild 5**).

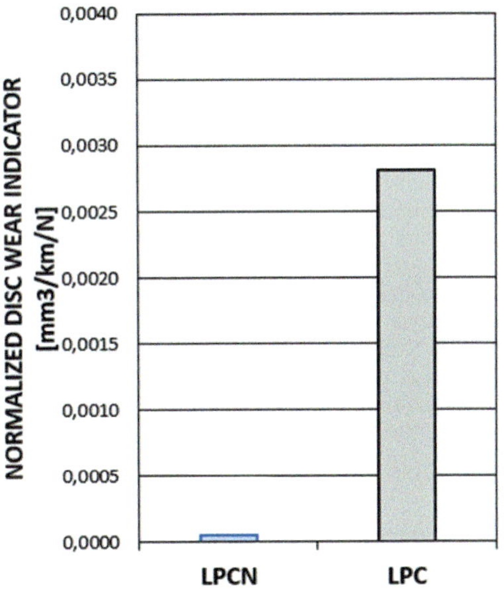

Bild 5: Vergleich der Verschleißindikatoren nach LPCN und LPC (Quelle: Seco/Warwick)

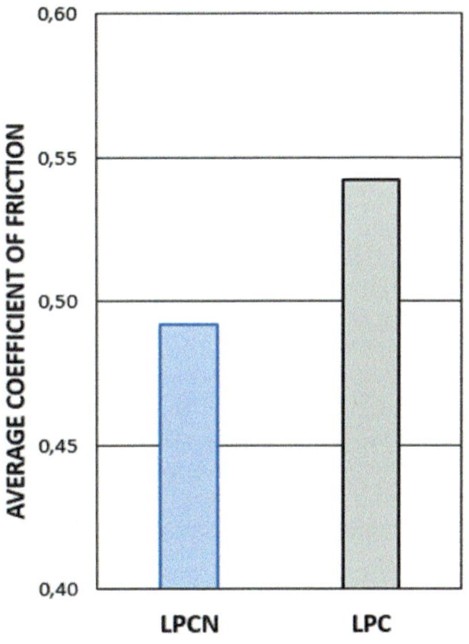

Bild 6: Vergleich der durchschnittlichen Reibungskoeffizienten (Quelle: Seco/Warwick)

- Der durchschnittliche Reibungskoeffizient war bei der Scheibe nach der LPCN-Behandlung niedriger, was bedeutet, dass die Bremskraft in diesem Fall ebenfalls geringer war (**Bild 6**).
- Die maximale Verschleißtiefe ist bei der mit LPCN behandelten Bandscheibe deutlich geringer, was darauf hindeutet, dass die Bandscheibe nach der LPCN-Behandlung eine deutlich bessere Verschleißfestigkeit aufweist (**Bild 7**).

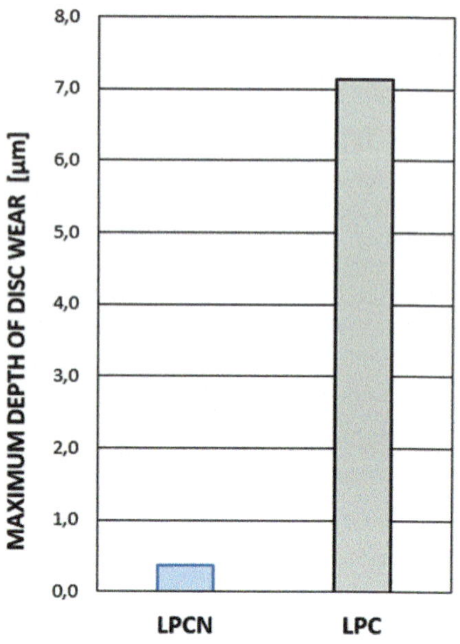

Bild 7: Vergleich der maximalen Abnutzungstiefe der Scheiben (Quelle: Seco/Warwick)

6.2.3 Zusammenfassung

Diese Forschung hat dazu beigetragen, optimale Prozessparameter für das Niederdruckkarbonitrieren in Seco/Warwick-Vakuumöfen zu definieren. Mithilfe des hochentwickelten Softwaretools SimVAC definiert der Anwender Rezeptursegmente für das Niederdruckkarbonitrieren in Abhängigkeit von der gewünschten effektiven Einsatztiefe. Nach Beendigung des Niederdruckaufkohlungssegments wird die Charge auf eine Austenitisierungstemperatur abgekühlt, bei der automatisch eine definierte Menge Ammoniak für einen bestimmten Zeitraum in den Ofen injiziert wird. Anschließend wird die wärmebehandelte Charge unter Schutzbedingungen in die Abschreckkammer überführt und in Öl abgeschreckt. Der gesamte Wärmebehandlungsprozess läuft automatisch nach einem vorprogrammierten Rezept ab, was eine vollständige Wiederholbarkeit gewährleistet und kann aus der Ferne überwacht werden.

Die durchgeführten Tests und Untersuchungen haben bestätigt, dass das Niederdruckkarbonitrieren in Vakuumöfen eine praktikable und umweltfreundliche Alternative zum herkömmlichen Karbonitrieren in Atmosphärenöfen darstellt. Die Gesamtergebnisse haben eindeutig gezeigt, dass das Niederdruckkarbonitrieren bei bestimmten Stahlsorten die Härtbarkeit, die Mikrohärte und das Gefüge verbessert und eine Erhöhung der effektiven Einsatztiefe und der Abriebfestigkeit bewirkt. Freischneidende Stahlsorten haben sich als am empfänglichsten für die positiven Auswirkungen der Niederdruck-Karbonitrier-Technologie erwiesen. Die Verbesserung der metallografischen Ergebnisse und die Verbesserung der tribologischen Eigenschaften waren bei diesen Stahlsorten am bedeutendsten. Abgesehen davon kann das Niederdruckkarbonitrieren auch für kohlenstoffarme Stähle und für niedrig legierte Stähle, die für die Aufkohlung vorgesehen sind, als praktikables Verfahren angesehen werden. Die Versuche haben auch bestätigt, dass bei mittel- und hochlegierten Stählen der Einsatz des Niederdruckkarbonitrierens keine sichtbaren Verbesserungen der metallographischen Eigenschaften bewirkt.

Seco/Warwick und verschiedene Industriepartner haben andere mögliche Prozesskonfigurationen für bestimmte Stahlsorten untersucht, um die Niederdruckkarbonitriertechnologie weiter zu verbessern und zu optimieren. Zu den aktuellen Untersuchungen gehören Arbeiten zur Verbesserung und Anpassung der Stickstoffkonzentration innerhalb der effektiven Härtetiefe sowie der Stickstoffdiffusionsprofile zur Erfüllung spezifischer Industrieanforderungen. Erfahrung und technologisches Fachwissen ermöglichen es Seco/Warwick, umfassende und konstante Unterstützung bei der Mitentwicklung verschiedener Niederdruckaufkohlungs- und -karbonitrierverfahren zu bieten. Bestätigte technologische Ergebnisse werden in der Zukunft garantiert mit voller Reproduzierbarkeit erreicht. Mit der ständigen Weiterentwicklung vakuumbasierter Technologien ist Seco/Warwick führend auf dem Weg in eine nachhaltige Zukunft der Wärmebehandlungstechnologie.

Literatur

[1] Herring, D. H.: Vacuum Heat Treatment. Principles|Practices|Applications. In: BNP Media II, 2012, S. 377

[2] Herring, D. H.: Vacuum Heat Treatment. Principles|Practices|Applications. In: BNP Media II, 2012, S. 377

[3] Herring, D. H.: Vacuum Heat Treatment. Principles|Practices|Applications. In: BNP Media II, 2012, S. 444

[4] Resolution adopted by the General Assembly on 25. September 2015: Transforming our world: the 2030 Agenda for Sustainable Development. Online, [https:// www.un.org/en/development/desa/population/migration/generalassembly/ docs/globalcompact/A_RES_70_1_E.pdf], zuletzt abgerufen am 15.01.2024

[5] The Paris Agreement. Online, [https://unfccc.int/process-and-meetings/the-parisagreement], zuletzt abgerufen am 15.01.2024

[6] Carbon Neutrality Coalition. Online, [https://carbon-neutrality.global], zuletzt abgerufen am 15.01.2024

6.3 Wasserstoff & Niederdruckaufkohlung: Innovative Vakuumtechnik

Dr. Klaus Buchner, Dominic Kraus

Mit der Ergänzung des Angebots um Vakuumanlagentechnik bietet Aichelin umfassende Lösungen für die Wärmebehandlung und setzt erstmals auch Niederdruckaufkohlungsanlagen und klassische Vakuumhärteöfen (**Bild 1**) ein. Diese ermöglichen eine präzise Steuerung metallurgischer Prozesse und eröffnen neue Möglichkeiten für anspruchsvolle Anwendungen.

Um den Anforderungen an die Abschrecktechnologie gerecht zu werden, werden sowohl Öl- als auch Hochdruckgasabschreckung angeboten. Für CO_2-neutrale Betriebsmedien wurde neben den klassischen Gasen Stickstoff und Helium auch die Hochdruckgasabschreckung mit Wasserstoff wiederentdeckt. Wie in [1] dargestellt, lassen sich die Einflussfaktoren des Hochdruckgasabschreckens in drei Kategorien unterteilen: physikalische Einflussgrößen, anlagenspezifische Faktoren einschließlich Peripherie sowie die Eigenschaften der Charge und deren Zusammensetzung.

Bild 2 zeigt die Abhängigkeit des Wärmeübergangskoeffizienten von dem Druck und der Gasart. Aufgrund seiner physikalischen Eigenschaften erweist sich Wasserstoff dabei als ideales Abschreckmedium. Bedingt durch die niedrigere Gasdichte ergibt sich für Helium und Wasserstoff ein signifikant geringerer Leistungsbedarf für die Gasumwälzung, sodass auch Drücke bis 20 bar wirtschaftlich darstellbar sind.

Bei der Betrachtung von Wasserstoff-Stickstoff-Gemischen hinsichtlich der Abschreckwirkung zeigt sich, dass Gemische einen höheren Wärmeübergangskoeffizienten als reiner Wasserstoff aufweisen, wobei dies im Wesentlichen von den Strömungsverhältnissen abhängt [2]. Die bessere Abschreckwirkung von Wasserstoff in Verbindung mit höheren Drücken im Vergleich zu Stickstoff ermöglicht eine martensitische Härtung auch von Bau-

Bild 1: Moderner Vakuumhärteofen mit Wasserstoffabschreckung und LPC-System (Quelle: Aichelin Group)

Bild 2: Einfluss von Abschreckdruck und Gasart auf den mittleren Wärmeübergangskoeffizienten (Quelle: s. [1])

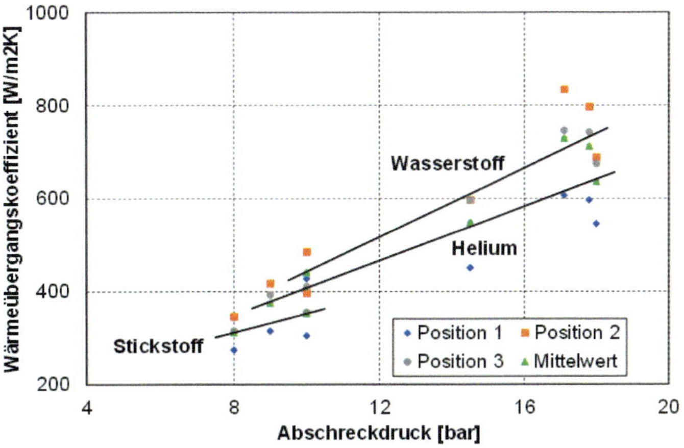

Tabelle 1: Durchhärtbare Querschnitte verschiedener Kaltarbeitsstähle (Quelle: s. [3])

Werkstoff	6 bar N_2	6 bar H_2	20 bar H_2
1.2080	70 bis 80	110	200
1.2842	< 15	30	50
1.2067	-	5	15

teilen mit größeren Querschnitten. In **Tabelle 1** sind exemplarisch drei Kaltarbeitsstähle hinsichtlich des durchhärtbaren Querschnitts für unterschiedliche Medien bzw. Abschreckdrücke aufgeführt [3]. Damit wird eine Alternative zur Ölabschreckung bei gleichzeitig verbessertem Formänderungsverhalten der wärmebehandelten Bauteile aufgezeigt. Alternativ dazu ermöglicht die gesteigerte Abkühlrate den Einsatz kostengünstigerer Werkstoffe. Niedriglegierte Einsatzstähle wie 1.7131 oder 1.7147, die sich in Kombination mit Ölabschreckung bewährt haben, erreichen bei angepasster Hochdruckgasabschreckung mit Wasserstoff vergleichbare Härtewerte.

6.3.1 Potenziale der Niederdruckaufkohlung

Nachfolgend soll mit Blick auf die Anwendungstechnik anhand des konkreten Beispiels der Niederdruckaufkohlung (LPC) das Potenzial der Vakuumtechnik in diesem Technologiebereich hinsichtlich der Prozesszeitverkürzung durch Erhöhung der Prozesstemperatur näher beleuchtet werden.

Das Einsatzhärten als solches kann als Kombination von Aufkohlen und dem anschließenden martensitischen Härten angesehen werden. Im industriellen Maßstab dominiert heute die Gasaufkohlung. Dabei erlaubt die Gleichgewichtsatmosphäre eine In-Situ-Messung und damit eine gezielte Regelung des Kohlenstoffpegels. Dieser Prozess wird durch die Niederdruckaufkohlung ergänzt. Hierbei wird Acetylen (Ethin) als kohlenstoffspendendes Gas eingesetzt. Der Prozess erfolgt bei Drücken unterhalb von 10 mbar, wobei sich Puls- und Diffusionsphasen abwechseln. Die spezifischen Vor- und Nachteile der Gas- und Niederdruckaufkohlung sind in den Ausführungen von Edenhofer [4] und Liedtke [5] beschrieben.

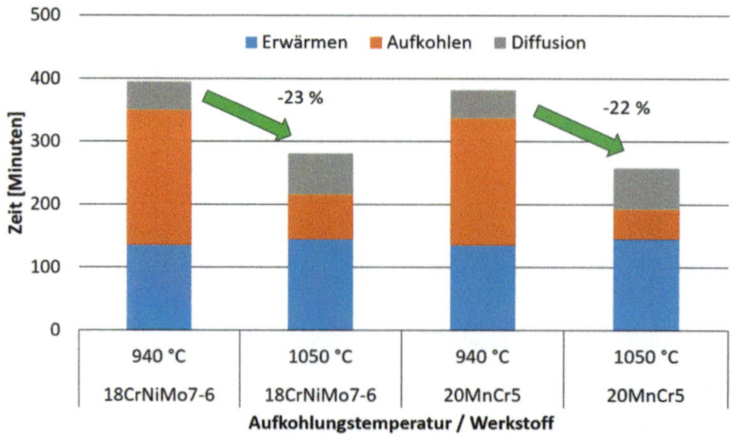

Bild 3: Prozesszeiten beim Niederdruckaufkohlen von Pulsatorrädern (Quelle: s. [6])

Wesentlich für beide Prozesse ist die Temperaturabhängigkeit des Stoffübergangs- und des Diffusionskoeffizienten, der durch einen Arrhenius-Ansatz beschrieben werden kann. Somit erlaubt eine Erhöhung der Prozesstemperatur eine signifikante Verkürzung der Prozessdauer. **Bild 3** zeigt dies exemplarisch für zwei unterschiedliche Prozesstemperaturen bei unterschiedlichen Legierungssystemen für den Niederdruckaufholungsprozess.

Trotz erhöhtem Energieaufwand für das Aufheizen der Charge dominiert die Energieeinsparung durch die verkürzte Prozesszeit, was einen Beitrag zur Optimierung der Energieeffizienz und damit zur Reduktion des CO_2-Fußabdrucks darstellt. An dieser Stelle sei auf den Aspekt der Feinkornstabilität hingewiesen; etwaige Problematiken des Kornwachstumes durch die erhöhte Prozesstemperatur gilt es zu berücksichtigen. Die Arbeit von Steinbacher und Stenico [6] zeigt jedoch, dass dies durch geeignete Maßnahmen kompensiert werden kann, sodass die Prozesssicherheit der Hochtemperaturaufkohlung gegeben ist. Die Untersuchungen zur Zahnfuß- und zur Grübchentragfähigkeit zeigen keine negativen Auswirkungen. Zur Unterdrückung der Stickstoffeffusion wurde in der Diffusionsphase ein Mindestdruck von 200 mbar gehalten.

Die beiden ausgewählten Werkstoffe (20MnCr5 und 18CrNi-Mo7-6) unterscheiden sich in ihrer Härtbarkeit. Dadurch ermöglichen sie in Kombination mit verschiedenen Abschreckmedien – insbesondere auch der Hochdruckgasabschreckung mit Wasserstoff sowie der Gasabschreckung mit Stickstoff oder Helium und der klassischen Ölabschreckung – ein breites Spektrum an Bauteilgeometrien.

6.3.2 Fazit

Um sowohl den ökonomischen als auch den ökologischen Anforderungen der Zeit gerecht zu werden, ist mehr denn je eine ganzheitliche Betrachtung der verschiedenen technologischen Ansätze im Bereich der Wärmebehandlung erforderlich. Je nach Anwendungsfall kann sich entweder die atmosphärische Wärmebehandlung oder die Vakuumanlagentechnik als vorteilhaft erweisen. Speziell im Bereich der Abschrecktechnik müssen bewährte Technologien aufgrund veränderter Randbedingungen sowie zur Nutzung neuer technologischer Potenziale neu bewertet werden. Die Hochdruckgasabschreckung mit Wasserstoff in modernen Vakuumanlagen kann dabei einen zukunftsweisenden Ansatz darstellen.

Literatur

[1] Hoffmann, F.; Gondesen, B.; Lohrmann, M.; Lübben, T.; Mayr, P.: Möglichkeiten und Grenzen des Gasabschreckens. In: HTM Journal of Heat Treatment and Materials, 53 (1998) 2, S. 81–86

[2] Laumen, C.; Holm, T.; Lübben, T.; Hoffmann, F.; Mayr, P: Hochdruck-Gasabschrecken mit Wasserstoff. In: HTM Journal of Heat Treatment and Materials, 53 (1998) 2, S. 72–80

[3] Altena, H.: Hochdruck-Wasserstoffabschreckung. In: HTM Journal of Heat Treatment and Materials, 50 (1995) 1, S. 27–30

[4] Edenhofer, B.: Einsatzhärten – Ein Prozess mit neuen Entwicklungen und Perspektiven. In: HTM Journal of Heat Treatment and Materials, 56 (2001) 1, S. 14–22

[5] Liedtke, D.: Stand des Einsatzhärtens aus industrieller Sicht. In: HTM Journal of Heat Treatment and Materials, 64 (2009) 6, S. 323–337

[6] Steinbacher, M.; Stenico, A.: Prozesssicheres Hochtemperatur-Aufkohlen feinkornstabilisierter Stähle in Atmosphären- und Vakuumöfen. Abschlussbericht zum AiF-Forschungsvorhaben 86ZN, Bremen, Juli 2006

6.4 Zuverlässige Vakuumversorgung beim Plasmanitrieren

Dr. Fabian Fahlbusch

Die Wärmebehandlung von Stählen und anderen Metallen, hauptsächlich für die Luftfahrt- und Automobilindustrie sowie für die Kraftwerktechnik, ist die Kernkompetenz der HWL Löttechnik GmbH in Berlin. Das Unternehmen bietet nahezu alle Verfahren der Wärmebehandlung an – vom Induktionshärten und -glühen über das Vakuumhärten, -glühen und -löten bis hin zu sämtlichen Arten des Einsatzhärtens. Beim Plasmanitrieren nutzt HWL Vakuumtechnik von Busch Vacuum Solutions, die im Nitrierofen zum Einsatz kommt.

Eine immer größere Bedeutung in der Wärmebehandlung gewinnt bei HWL das Plasmanitrieren. In diesem Verfahren verfügt das Unternehmen über jahrzehntelange Erfahrung. Heute sorgen moderne Anlagentechnik und -steuerung dafür, dass Aufbau und Zusammensetzung der Verbindungs- und Diffusionsschichten kontinuierlich gesteuert und überwacht werden können. Gearbeitet wird mit einem gepulsten Gleichstromplasma, durch das ein gleichmäßiges Wärmebehandlungsergebnis erzielt wird. Dieses thermochemische Verfahren hat den Vorteil, dass die Wärmebehandlung bei vergleichbar niedrigen Temperaturen von 520 bis 580 °C durchgeführt werden kann. Damit das Plasma elektrisch leitfähig ist, müssen freie Ladungsträger für die Stromübertragung vorhanden sein. Bei Atmosphärendruck wären wirtschaftlich unrealistische Temperaturen notwendig, um die elektrische Leitfähigkeit des Plasmas zu erzeugen. Bei HWL arbeitet man bei Drücken von 2,5 hPa (mbar), was in der Praxis eine Wärmebehandlung bei unter 600 °C ermöglicht. Die im Vergleich zu anderen Wärmebehandlungsmethoden niedrige Temperatur wirkt sich äußerst positiv auf das Verzugsverhalten der Bauteile aus. Außerdem bietet dieses Verfahren den Vorteil, dass einzelne Bereiche von Bauteilen, die nicht nitriert werden sollen, mechanisch abgedeckt

Bild 1: Vakuumsystem bestehend aus einer R5 Drehschieber-Vakuumpumpe und einem PUMA Vakuum-Booster von Busch Vacuum Solutions

und somit gezielt vom Nitrierprozess ausgeschlossen werden können. Dadurch werden die Eigenschaften der Randschichten unter den maskierten Bereichen nicht verändert.

Vor dem eigentlichen Plasmanitrierprozess werden die zu behandelnden Bauteile exakt auf den Montagevorrichtungen platziert. Hier zahlt sich die langjährige Erfahrung von HWL aus, denn die Teile müssen optimal im Ofen angeordnet sein, um die gewünschten Oberflächeneigenschaften zu erzielen. Nach dem Chargiervorgang und dem Schließen des Nitrierofens wird dieser auf den notwendigen Prozessdruck evakuiert und über die Wandheizung erwärmt. Nach diesem Erwärmungsprozess werden die Bauteile unter Stickstoff-Atmosphäre einer Glimmentladung ausgesetzt. In dieser Glimmentladung bildet sich ein Plasma, der Stickstoff dissoziiert dabei, ionisiert und brennt an der Oberfläche der Bauteile ab. Die exakte Behandlungstemperatur und die Nitrierdauer sind abhängig von Material, Größe und Beschaffenheit der Bauteile sowie der zu erzielenden Nitrierhärtetiefe. Nach dem Nitriervorgang wird der Ofen mit den Bauteilen abgekühlt. Der ganze Prozess dauert zwischen 17 und 30 Stunden. Während dieser Zeit ist die Vakuumanlage in Betrieb (**Bild 1**).

Nachdem HWL bereits bei anderen Wärmebehandlungsanlagen gute Erfahrungen mit Vakuumpumpen von Busch Vacuum Solutions gemacht hat, wurde ein neuer Nitrierofen angeschafft, dessen Vakuumsystem ebenfalls von Busch stammt. Es besteht aus einer ölgeschmierten R5 Drehschieber-Vakuumpumpe als Vorpumpe und einem PUMA Vakuum-Booster. Dieses Vakuumsystem erreicht einen Enddruck von < 1 x 10-2 hPa (mbar); der eigentliche Arbeitsdruck während des Prozesses liegt bei 2,5 hPa (mbar). Damit wird das optimale Saugvermögen des Vakuumsystems genutzt, das in diesem Arbeitsbereich am höchsten ist (**Bild 2**).

Zu Beginn des Prozesses evakuiert die R5 Drehschieber-Vakuumpumpe den Ofen von Atmosphärendruck auf ein Grobvakuum von 100 hPa (mbar). Erst jetzt wird der PUMA Vakuum-Booster zugeschaltet, der als Verstärker das Saugvermögen des Vakuumsystems erheblich erhöht, sodass der Prozessdruck schnell erreicht und sicher gehalten werden kann.

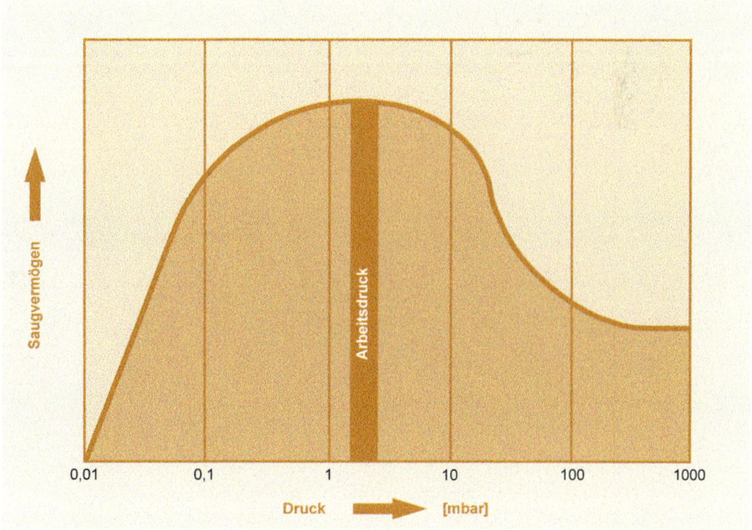

Bild 2: Saugvermögen des Vakuumsystems (Quelle: Busch Vacuum Solutions)

Durch die Kombination des Vakuumsystems mit einer Drehschieber-Vakuumpumpe und deren genau auf den Prozess abgestimmten Steuerung ist es möglich, mit geringstmöglichem Energieaufwand ein Maximum an Saugvermögen zu erreichen.

Das präzise Einhalten des Arbeitsdrucks und des Saugvermögens ist ein Garant dafür, nachvollziehbare Prozesse zu fahren und zu dokumentieren. Dadurch lassen sich die exakt gewünschten Produkteigenschaften erzielen. Bei HWL werden meist hochlegierte Edelstähle plasmanitriert, aber auch Baustähle oder Sintermetalle werden durch dieses Verfahren wärmebehandelt. Der kontinuierliche Betrieb wird nur durch die Rüst- bzw. Bestückungszeiten unterbrochen. Die absolute Zuverlässigkeit der Vakuumtechnik hat höchste Priorität, denn ein Ausfall des Vakuumsystems während des Prozesses kann die komplette Charge hochwertiger und kostspieliger Präzisionsbauteile unbrauchbar machen. Die Wartung des Vakuumsystems beschränkt sich auf ein Minimum. Neben der täglichen Sichtkontrolle des Ölstands wird alle zwei Jahre das Öl in der R5 Drehschieber-Vakuumpumpe sowie im Getriebe des PUMA Vakuum-Boosters gewechselt.

7. Qualitätsmanagement

7.1 Normen für die Wärmebehandlung und deren Bedeutung

Arnold Horsch

Normung bezeichnet die Formulierung, Herausgabe und Anwendung von Regeln, Leitlinien oder Merkmalen durch eine anerkannte Organisation und deren Normengremien. Sie sollen auf den gesicherten Ergebnissen von Wissenschaft, Technik und Erfahrung basieren und auf die Förderung optimaler Vorteile für die Gesellschaft abzielen. Die Festlegungen werden mit Konsens erstellt und von einer anerkannten Institution angenommen [1].

Wie bei fast allen industriellen Produkten und Verfahren ist auch in der Wärmebehandlung eine Normung und Standardisierung von Verfahren und Beschreibungen für alle teilnehmenden Parteien von großem Vorteil.

7.1.1 Historie

Die allererste DIN-Norm aus dem Jahr 1918 betraf ein Produkt, das den meisten Anwendern weniger geläufig sein dürfte: den Kegelstift. Es handelt sich um ein Verbindungselement aus dem Maschinenbau. Durch Normung und Standardisierung können für beinahe jedes Gerät problemlos passende Ersatzteile beschafft werden.

Mit der Industrialisierung entstand eine arbeitsteilige Produktion. Damit Teile aus verschiedenen Fabriken auch zusammenpassten, waren Normen erforderlich. Noch vor dem DIN gab es nationale Normensetzer in Ländern, in denen die Industrialisierung bereits weiter fortgeschritten war: 1901 wurde in London die British Standards Institution gegründet.

Auch in Deutschland gab es zu dieser Zeit bereits Normen, aber nur für einzelne Werke oder Branchen. Entscheidender Auslöser für eine branchenübergreifende Standardisierung war der Erste Weltkrieg. Mit der Gründung des heutigen Deutschen Instituts für Normung (DIN) überließ der Staat die Normierung weitgehend der Privatwirtschaft. Das zeigt schon der Taufname der Organisation, die damals „Normenausschuss der Deutschen Industrie“ (Nadi) hieß. Bis heute ist das DIN privatwirtschaftlich organisiert. Neben Vertretern aus Unternehmen und Verbänden gehören den 118 vom DIN koordinierten Normenausschüssen allerdings zahlreiche Behördenvertreter an. Es gilt: DIN-Normen sind keine Gesetze, verbindlich werden sie erst, wenn ein Gesetz oder eine Verordnung auf eine Norm verweist [2].

In Deutschland wurden die ersten Normen für die Wärmebehandlung spät herausgegeben. Die Geschichte des Normenausschusses Werkstofftechnologie (NWT) beginnt vor mehr als 60 Jahren. Bereits im Jahre 1955 hat das DIN „versucht[,] einheitliche Richtlinien für die Härteangabe in Zeichnungen herauszugeben“, woraus schließlich die Norm „DIN 6773, Wärmebehandlung von Eisenwerkstoffen – Darstellung und Angaben wärmebehandelter Teile in Zeichnungen, Ausgabe Oktober 1967“ entstand, die, wie es in den Erläuterungen zur DIN 6773 heißt, in überarbeiteter Form und in vier Teile aufgeteilt zwischen Oktober 1976 und Mai 1977 herausgegeben wurde.

Um das Ergebnis der Arbeiten für Normen mit wärmebehandlungstechnischen Inhalten im Sinne der Normung im DIN vor der Öffentlichkeit zu vertreten, wurde angeregt, einen Fachnormenausschuss Wärmebehandlungstechnik zu gründen. Das wurde am 20. November 1972 in Bonn realisiert. In Anwesenheit von Nikolaus Ludwig, Dr.-Ing. Helmut Reihlen und Rudolf Muschalla vom DIN sowie Vertretern aus der stahlverarbeitenden Industrie, aus Forschungsinstituten (AWT, MPI, DIW), den Fachnormenausschüssen Materialprüfung, Nichteisenmetalle, Pulvermetallurgie und Maschinenbau, dem Verein Deutscher Gießereifachleute und dem Verband Deutscher Gesenkschmieden wurden die aufzugreifenden Normungsarbeiten beschlossen und die vier Arbeitsausschüsse

- Wärmebehandlungsverfahren,
- Wärmebehandlungsangaben in Fertigungsunterlagen,
- Prüfung der Wirkungen der Wärmebehandlung und
- Anforderungen an Wärmehehandlungsanlagen und -mittel

gegründet, die sich 1973 konstituieren sollten [3].

Zu Obleuten der Arbeitsausschüsse wurden Dr. Ruth Chatterjee, Paul Birk und Dr. Dieter Liedtke gewählt. Auch hier diente die Normung zur Klärung technischer Prozesse und von Begriffsbestimmungen.

7.1.2 Welche Vorteile bieten Normen für Unternehmen?

Die Frage, welche Vorteile Normen für Unternehmen bieten, ist von großer Bedeutung. Einige der wichtigsten Punkte sind:

- Gleichwertige, erfolgsversprechende Chancen am Markt
- Verbesserte Prozesse und Produktqualität durch aktuellen Stand der Technik
- Es wird geliefert, was bestellt wurde
- Einheitliche Sprache durch genormte Begriffe
- Vereinfachte Handhabung von Reklamationen
- Internationale Wahrnehmung und Anerkennung.

7.1.3 Normung in der Wärmebehandlung

In Deutschland ist die Normung der Wärmebehandlung im NA 145-02 FB – Fachbereich Wärmebehandlungstechnik angesiedelt. Dieser hat vier Untergremien. Dies sind [4]:

- NA 145-02-01 AA Terminologie; Obmann: Thomas Waldenmaier, Bosch
- NA 145-02-02 AA Wärmebehandlungsverfahren/Wärmebehandlungsangaben; Obmann: Thomas Waldenmaier, Bosch
- NA 145-02-03 AA Prüfung nach der Wärmebehandlung; Obmann: Arnold Horsch, Arnold Horsch e.K.
- NA 145-02-04 AA Anforderungen an Wärmebehandlungsanlagen und -mittel; z. Zt. kein Obmann, Ausschuss nicht aktiv.

In allen Ausschüssen zusammen sind insgesamt neun Personen aktiv.

Gibt man beim Beuth Verlag „Wärmebehandlung Metall“ als Suchbegriff ein, erhielt man am 15.07.2023 2.033 Einträge, in denen dieser Begriff enthalten ist, also eine Vielzahl an Normen, die zu beachten sind. Diese Normen umfassen alle Metalle und nicht nur die Eisenwerkstoffe, wobei hier nicht immer spezifisch auf die Wärmebehandlung eingegangen wird, sondern das Thema in diesen Normen auch nur Erwähnung finden kann.

7.1.3.1 Welche Arten von Normen gibt es in der Wärmebehandlung?

Es stellt sich die Frage, welche Arten von Normen es für die Wärmebehandlung gibt. Da wären zunächst die Normen, die direkt zum Thema Wärmebehandlung erstellt wurden und die

- die Wärmebehandlungsverfahren beschreiben,
- die festlegen, wie Angaben zur Wärmebehandlung gemacht werden,
- die Begriffe festlegen und
- die Prüfung von Bauteilen beschreiben.

Die vier Gremien im NA 145-02-FB sind z. Zt. für 18 DIN-EN-ISO-Normen zuständig. Bedenkt man, dass der Begriff „Wärmebehandlung Metall“ 2.033 Treffer im Verzeichnis der Normen hat, sind dies nicht sehr viele Normen, die hier bearbeitet werden.

Zu beachten ist, dass es viele Normen außerhalb des NA 145-02 FB – Fachbereich Wärmebehandlungstechnik gibt, die ebenfalls Normungen zur Wärmebehandlung und deren Ergebnisse festlegen, z. B. verschiedene Produktnormen. Daher ist es sehr wichtig, neben den Prozessen und Prüfungen der Wärmebehandlung auch immer die einzelnen Produktnormen der zu behandelnde Bauteile zu beachten.

Es gibt nicht nur DIN-EN-ISO-Normen in diesem Bereich, sondern auch eine Vielzahl an Werksnormen und anderen Normen wie z. B. ASME, ASTM, SAE-Standards, AIAG (CQI9) und weitere nationale und internationale Normen. Hier muss der Hersteller von Produkten und Prozessen genau darüber informiert sein, welche Normen für ihn gültig sind.

Auch die Werkstoffprüfnormen (Härteprüfung, Zugversuch usw.) und weitere Normen können mitgelten.

7.1.3.2 Normen zur Wärmebehandlung

In **Tabelle 1** sind alle Normen, die im NA 145-02 FB – Fachbereich Wärmebehandlungstechnik bearbeitet werden, aufgelistet.

7.1.3.3 Welche Normungssysteme gibt es?

Weltweit gibt es mehrere wichtige Normungssysteme für die Wärmebehandlung, die Prüfung wärmebehandelter Bauteile und die zu erzielenden Wärmebehandlungsergebnisse.

In der Welt außerhalb der USA:

- ISO, EN
- Nationale Normen wie:
  - DIN, GOST, JIS und weitere.

Im amerikanischen Raum:

- ASTM: American Society for Testing and Materials
- AWS: American Welding Society
- SAE: International: Society of Automotive Engineers
- u. v. m.

Tabelle 1: Normen, die im NA 145-02 FB – Fachbereich Wärmebehandlungstechnik bearbeitet werden

Norm Nr. - Ausgabe	Titel
E DIN 17029: 2025, Entwurf	Zeichnungstoleranzen und Prüfverfahren wärmebehandelter Bauteile
DIN 17014-3:1976-05	Wärmebehandlung von Eisenwerkstoffen; Kurzangabe von Wärmebehandlungen
DIN 17022-1: 1994-10	Wärmebehandlung von Eisenwerkstoffen – Verfahren der Wärmebehandlung – Teil 1: Härten, Bainitisieren, Anlassen und Vergüten von Bauteilen
DIN 17022-2: 1986-06	Wärmebehandlung von Eisenwerkstoffen; Verfahren der Wärmebehandlung; Härten und Anlassen von Werkzeugen
DIN 17022-3:1989-04	Wärmebehandlung von Eisenwerkstoffen; Verfahren der Wärmebehandlung; Einsatzhärten
DIN 17022-4:1998-01	Wärmebehandlung von Eisenwerkstoffen – Verfahren der Wärmebehandlung – Teil 4: Nitrieren und Nitrocarburieren
DIN 17022-5:2000-03	Wärmebehandlung von Eisenwerkstoffen – Verfahren der Wärmebehandlung – Teil 5: Randschichthärten
DIN 17023:2000-03	Wärmebehandlung von Eisenwerkstoffen – Wärmebehandlungs-Anweisung (WBA) – Vordruck
DIN 17052-1:2013-09	Wärmebehandlungsöfen – Teil 1: Anforderungen an die Temperaturgleichmäßigkeit
DIN 17052-2:2018-08	Wärmebehandlungsöfen – Teil 2: Anforderungen an die Ergebnisgleichmäßigkeit nach einer Wärmebehandlung
DIN 17052-3, in Planung	Wärmebehandlung von Eisenwerkstoffe – Teil 3: Ermittlung der Abschreckgleichmäßigkeit
DIN 30901:2016-12	Wärmebehandlung von Eisenwerkstoffen – Ermittlung der Tiefe und Ausbildung der Randoxidation
DIN 30902:2016-12	Wärmebehandlung von Eisenwerkstoffen – Lichtmikroskopische Bestimmung der Dicke und Porosität der Verbindungsschichten nitrierter und nitrokarburierter Werkstücke
DIN ISO 15787:2018-08	Technische Produktdokumentation – Wärmebehandelte Teile aus Eisenwerkstoffen – Darstellung und Angaben
DIN EN ISO 642:2024-11	Stahl – Stirnabschreckversuch (Jominy-Versuch)
DIN EN ISO 4885:2018-07	Eisenwerkstoffe – Wärmebehandlung – Begriffe
DIN EN ISO 18203:2022-07	Stahl – Bestimmung der Dicke gehärteter Randschichten
ISO 16573-1:2020-08	Stahl – Messverfahren zur Bewertung der Wasserstoffversprödungsbeständigkeit von hochfesten Stählen – Teil 1: Dauerbelastungsprüfung
ISO 20431	Heat Treatment Control of Quality

Außerdem:

- Werksnormen
- Militärnormen
- Bahnstandards
- Automobilnormen, z. B. AIAG, CQI 9
- Viele Produktnormen, z. B.:
  - DIN EN ISO 683 Teil 1–5: für eine Wärmebehandlung bestimmte Stähle, legierte Stähle und Automatenstähle
  - DIN EN SO 898-1:20213-05: mechanische Eigenschaften von Verbindungselementen aus Kohlenstoffstahl und legiertem Stahl usw.
- u. v. m.

Alle diese Normen sollten beachtet werden.

7.1.4 Anwendung von Normen

Die erste Frage, die sich stellt, ist: Müssen Normen angewendet werden? Hier ist die Aussage sehr einfach: Nein, die Anwendung von Normen ist freiwillig. Ausnahme: Der Gesetzgeber schreibt dies vor.

Bei der Frage, wie Normen angewendet werden sollen, sind die wichtigsten Hinweise:

- Lesen Sie die aktuelle Norm von der ersten bis zur letzten Seite; auch die Anhänge müssen gelesen werden.
- Tun Sie, was Sie lesen.

Sollten Sie nicht wissen, welche Normen für Ihren Prozess und dessen Ergebnisse anzuwenden sind:

- Fragen Sie Ihren internen oder externen Auftraggeber.
- Recherchieren Sie sorgfältig in den zugänglichen Datenbanken.
- Führen Sie nicht einfach die Wärmebehandlung durch, ohne die wichtigsten Vorschriften zu kennen.
- Führen Sie keine Wärmebehandlung durch, deren vorgegebene Ergebnisse nach den aktuellen Regeln der Technik nicht prüfbar sind, z. B. ist eine Toleranz von 60^{+2} HRC nicht prüfbar.

7.1.5 Beispiele für Fehlanwendungen

7.1.5.1 Prüfung der Oberflächen und Kernhärte

Ist in der Zeichnung oder im Auftrag eine Toleranz vorgegeben, die sich nach den Härteprüfnormen und den Normen der allgemeinen Messtechnik nicht sicher prüfen lässt, sollte die Wärmebehandlung nicht durchgeführt werden. Zwischen den Parteien sollten Toleranzen festgelegt werden, die sich realistisch prüfen lassen; ein paar Beispiele sind in **Tabelle 2** gezeigt.

Diese Toleranzen erscheinen groß für den teilweise bisher genutzten Toleranzrahmen, sind aber realistisch in Bezug auf die erlaubten Abweichungen der Härteprüfmaschinen und die

Tabelle 2: Beispiele für Toleranzen

Vorgegebene Härtetoleranz, nicht sicher prüfbar	Toleranzfeld	Prüfbares Toleranzfeld	Messunsicherheit	Richtige Toleranzangabe
60^{+2} HRC	2 HRC	2 HRC	≈ 0,8–1,0 HRC	+6 HRC
750^{+50} HV 1	50 HV	50 HV	≈ 20 HV	+150 HV
250^{+25} HB 2,5/187,5	25 HB	25 HB	≈ 3 HB	+50 HB

zu berücksichtigenden Messunsicherheiten [7]. Im ersten Beispiel, 60^{+2} HRC, hat man bereits die Situation, dass die Messunsicherheit MU bereits das Toleranzfeld von 2 HRC komplett belegt. Für eine sichere und normgerechte Prüfung bleibt kein Spielraum mehr.

7.1.5.2 Bestimmung der CHD - DIN EN ISO 18203:2022-07 [5]

Die Bestimmung der CHD ist eine in weiten Teilen der Industrie übliche Prüfung. Sie sollte nach DIN EN ISO 18203:2022-07 durchgeführt werden.

Hier stehen als wichtigste Randbedingungen folgende Angaben:

- Im Kapitel 3.1 „Einsatzhärtungstiefe":
  - Einsatzhärtungstiefe CHD (Case Hardening Depth), senkrechter Abstand von der Oberfläche bis zu der Schicht, die eine Vickers-Härte HV (Vickers Hardness) von 550 HV nach ISO 6507-1 oder eine äquivalente Knoop-Härte nach ISO 4545-1 aufweist.
  - Anmerkung 1 zum Begriff: Bei Stählen mit einer Vickers-Härte von mehr als 450 HV in einem Abstand von der Oberfläche gleich der dreifachen Einsatzhärtungstiefe (ermittelt mit einem Härtegrenzwert von 550 HV) kann für die Bestimmung der Einsatzhärtungstiefe nach Vereinbarung zwischen den interessierten Parteien ein Härtegrenzwert von über 550 HV, in Stufen von 25 Einheiten, ausgewählt werden.
  - Anmerkung 2 zum Begriff: Im Allgemeinen umfasst Einsatzhärten Aufkohlen oder Karbonitrieren, gefolgt von einer Härtung (s. ISO 4885).
- Im Kapitel 7.1 „Auswahl und Vorbereitung der Proben":
  - Falls nicht anders vereinbart, muss die Messung an einem Querschliff des Werkstücks in dem folgenden festgelegten Zustand durchgeführt werden: Ein Abschnitt senkrecht zur Längsachse des Produkts oder, falls das Produkt keine Längsachse aufweist, ein Abschnitt senkrecht zur Oberfläche an einer zwischen den beteiligten Parteien zu vereinbarenden Stelle.
- Im Kapitel 8.1 „Härteprüfverfahren":
  - Entlang einer oder mehrerer paralleler Linien werden senkrecht zur Oberfläche und innerhalb eines Bereichs (W) von 1,5 mm Breite Härteprüfeindrücke aufgebracht. Der Abstand zwischen diesen Linien muss der Anforderung aus ISO 6507-1 entsprechen. Der Abstand zwischen zwei benachbarten Eindrücken (Δd) darf nicht weniger als das Dreifache ihrer Diagonalen betragen. Die Differenz zwischen den aufeinanderfolgenden Abständen der einzelnen Eindrücke von der Oberfläche (z. B. $d2-d1$) darf nicht mehr als 0,1 mm betragen und die kumulativen Abstände von der Oberfläche müssen mit einer Genauigkeit von ±25 µm gemessen werden. Die Diagonalen der Eindrücke müssen mit der in ISO 6507-2 festgelegten Genauigkeit gemessen werden. Die Mitte des ersten Eindrucks muss sich in einem Abstand $d1$ zur Oberfläche befinden, der

mindestens dem 2,5-Fachen seiner Diagonalen entspricht. Prüfungen mit Vickers- oder Knoop-Eindrücken werden bei Drücken von 0,9807 bis 9,807 N durchgeführt. Messungen der Eindrücke müssen mit einem qualitativ hochwertigen Lichtmikroskop und bei geeigneter Beleuchtung durchgeführt werden – mit oder ohne Kamerasystem (und möglich mittels Bildanalyse), bei ausreichend großer Vergrößerung, sodass die Diagonalen zwischen 25 und 75 % der Okular- oder Bildschirmbreite oder -höhe betragen und die Spitzen der Eindrücke ohne Verzerrung fokussiert werden können. Aufgrund der Abstände zwischen den Eindrücken werden für diese Aufgabe üblicherweise Prüfkräfte von 0,9807 bis 2,942 N verwendet. Die Messungen werden auf der vorbereiteten Oberfläche in zwei oder mehreren Bereichen vorgenommen, deren Lage zwischen den beteiligten Parteien vereinbart werden muss. Für jeden Bereich werden die Ergebnisse aufgezeichnet, um die Kurve zu erhalten, auf der die Härteschwankungen je nach Abstand von der Oberfläche dargestellt werden.

- Im Kapitel 10 „Prüfbericht":
  - Der Prüfbericht muss mindestens die folgenden Angaben enthalten, sofern nicht anders zwischen den beteiligten Partnern vereinbart: Verweisung auf dieses Dokument, d. h. ISO 18203:2016; das Randschichthärtungsverfahren; das Messverfahren (Härteprüfverfahren, Prüfkraft, ob das mikrophotographische Verfahren eingesetzt wurde, Ätzmittel, Vergrößerung, Abstand zwischen parallelen Linien usw.); das Ergebnis der Einsatzhärtungstiefe (CHD), Randschichthärtungstiefe (SHD), Nitrierhärtetiefe (NHD), Gesamtdicke der Randschichthärtungstiefe (THD) oder Verbindungsschichtdicke (CLT).

Tatsächliche Prüfung in > 90 % aller CHD-Messungen:

- Es wird ein Härteverlauf durchgeführt.
- Die Probe wird irgendwie aus dem Bauteil herausgetrennt.
- Die Prüfung erfolgt am abschließend wärmebehandelten und angelassenen Bauteil.
- Die CHD wird im Protokoll angegeben.
- Keine Abweichung von der Norm wird im Prüfbericht angegeben.

Was ist falsch?

- Die Probe wird falsch geprüft; in der Norm steht eindeutig: „Im Allgemeinen umfasst Einsatzhärten Aufkohlen oder Carbonitrieren gefolgt von Härtung". Nach DIN EN ISO 4885 ist dies ein Prozess ohne Anlassen. Hier muss also im Prüfbericht vermerkt werden, dass die CHD nach dem Anlassen geprüft wurde.
- Es wird nur einer und nicht mind. zwei Härteverläufe durchgeführt; dies muss im Prüfbericht vermerkt werden.
- Es wird nicht beachtet, dass die Proben quer zur Längsachse herausgetrennt werden sollen.
  - Ist die Probe nicht in Längsrichtung entnommen, muss dies im Prüfbericht vermerkt werden.
  - Es werden so gut wie nie Vereinbarungen zur Lage der Prüfstelle getroffen. In der Zeichnung sollte, nach DIN ISO 15787:2018-08 [8], Kapitel 5.5, eine Prüfstelle normgerecht eingetragen sein.

7.1.5.3. Bestimmung der Kernfestigkeit - Kernhärte

Kernfestigkeit

An dem folgenden Beispiel ist gut zu erkennen, wo die Tücken im Detail stecken. Steht in der Zeichnung als Wärmebehandlungsangabe „34 CrMo 4 + QT“, ohne weitere Angaben von Festigkeitswerten, ist dies eine komplette Zeichnungsvorschrift, mit der alles geklärt ist, wenn man denn weiß, wo nachgesehen werden muss: in der entsprechenden Norm:

- DIN EN ISO 683-2:2018-09: Für eine Wärmebehandlung bestimmte Stähle, legierte Stähle und Automatenstähle – Teil 2: Legierte Vergütungsstähle [6]; für den relevanten Querschnitt steht in
  - Tabelle 8, welche Festigkeitswerte, Dehnung und Kerbschlagwerte erreicht werden müssen, wenn in der Zeichnung keine weiteren Angaben gemacht sind.
  - Bild 2, wie die Proben zu entnehmen sind.
- Nach DIN ISO 15787:2018-08, Technische Produktdokumentation – Wärmebehandelte Teile aus Eisenwerkstoffen – Darstellung und Angaben [8]:
  - In Kapitel 5.12 „Festigkeitsangaben“: Festigkeitswerte werden nur angegeben, wenn es a) notwendig ist und b) wenn Form und Maße des Teils es zulassen, dass eine wärmebehandelte Probe zum Prüfen der Festigkeit verwendet werden kann. Wenn notwendig, sind Ort und Lage anzugeben, wo diese entnommen werden kann. Die Angabe einer Kernhärte ist in diesen Fällen nicht erforderlich. Den Festigkeitswerten sind Toleranzen zuzuordnen, deren Schreibweise in derselben Art erfolgen kann wie die für Härtewerte oder Härtetiefen. Die Toleranzen sollten so groß sein wie funktionell zulässig.
  - Die korrekte Schreibweise wäre z. B.: vergütet, Rm = (800 + 100/0) MPA, $R_{p0,2}$ ≥ 650 MPA, A ≥ 12 %.

Kernhärte

- Nach DIN ISO 15787:2018-08, Technische Produktdokumentation – Wärmebehandelte Teile aus Eisenwerkstoffen – Darstellung und Angaben [8]:
  - Die Kernhärte ist in der Zeichnung anzugeben, wenn eine Festlegung zur Prüfung getroffen wurde. Die Kernhärte ist anzugeben als Rockwellhärte nach ISO 6508-1, als Vickers-Härte nach ISO 6507-1 oder als Brinellhärte nach ISO 6506-1. In der vorherigen Ausgabe dieser Norm von 2010 stand auch noch die folgende Anmerkung: Zur Prüfung ist eine Zerstörung oder Beschädigung des Werkstücks unumgänglich. Gegebenenfalls kann die Prüfung an einer zu diesem Zweck zusammen mit den Werkstücken wärmebehandelten Kontrollprobe vorgenommen werden.
  - Die korrekte Schreibweise der Kernhärte wäre z. B.: Kernhärte (350^{+80}) HBW 2,5/187,5. Hier ist kein Wärmebehandlungsverfahren spezifiziert, es bleibt also dem Wärmebehandler überlassen, welches Wärmebehandlungsverfahren eingesetzt wird oder es wird „vergütet“ angegeben, dann ist folgende Schreibweiserichtig: vergütet, ($350^{+50/0}$) HBW 2,5/187,5.

Wird keine weitere Angabe außer +QT auf der Zeichnung angegeben, müsste konsequenterweise die DIN EN ISO 683-2:2018-09 [6] gelten. Danach wird ein Zugversuch durchgeführt.

Prüfstelle der Kernfestigkeit/Kernhärte

Es gibt keine allgemeine Abgabe, wo die Kernfestigkeit/Kernhärte zu prüfen ist. Hier bleiben nur die Angaben der

- DIN EN ISO 683-2:2018-09 (**Bild 1**): Für eine Wärmebehandlung bestimmte Stähle, legierte Stähle und Automatenstähle – Teil 2: Legierte Vergütungsstähle [6].
- DIN EN ISO 377:2013-11: Stahl und Stahlerzeugnisse, Lage und Vorbereitung von Probenabschnitten und Proben für die mechanische Prüfungen [9].

Wünscht der Auftraggeber die Prüfung in der Mitte des Querschnittes, muss dies auf der Zeichnung vermerkt sein. In anderen Fällen sollten die Angaben der zuvor genannten Normen Beachtung finden.

Fehler, die bei der Kernfestigkeit/Kernhärte entstehen

Wird z. B. eine Welle aus einem vorvergüteten Bolzen 34CrMo4, der ein Ausgangsmaß von Ø = 50 mm hat, hergestellt, kann es sein, dass der Konstrukteur mit falschen Festigkeitswerten rechnet. Ist der größte Durchmesser nach der Bearbeitung 38 mm und der kleinste Durchmesser 10 mm, geht der Konstrukteur wahrscheinlich davon aus, dass das Bauteil auch am kleinsten Durchmesser die Festigkeit erreicht. Man kann aber nicht davon ausgehen, dass er in diesem Bereich die Festigkeit sicher erreicht. Dies kann wesentlich von der Ausgangsqualität des Vormaterials abhängen.

Der Wärmebehandler wird diesen Bolzen / diese Stange nach DIN EN ISO 683-2 [6] behandeln.

Hier steht in der Tabelle 8:

- $d \leq 16$ mm, Rm = 1.000 bis 1.200 Mpa, $R_{p0,2}$ min., 800 Mpa
- 40 mm < $d \leq 100$ mm, Rm = 800 bis 950 Mpa, $R_{p0,2}$ min., 550 Mpa.

Die Festigkeit wird unter Umständen nicht erreicht und der wichtige Wert der $R_{p0,2}$ wird mit großer Sicherheit bei Ø = 10 mm unterschritten. Es kann zum Bauteilversagen kommen.

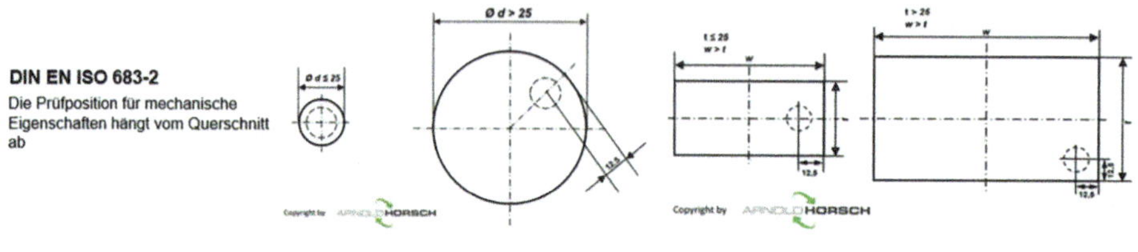

Bild 1: DIN EN ISO 683-2

7.1.6 Zusammenfassung

Falsche, nicht normgerechte Vorgaben in der Wärmebehandlung eröffnen ein breites Feld an Fehlern und Missverständnissen zwischen dem Auftraggeber und dem Wärmebehandler. Die Praxis hat gezeigt, dass unzureichende Vorgaben und ein fehlendes Pflichtenheft bzw. HTO (Wärmebehandlungsanweisung) immer wieder zu Ausschuss und zu Streitigkeiten führen.

Die Beachtung der Normen für die Wärmebehandlung ist wichtig und zur Vermeidung von Reklamationen unerlässlich. Es müssen auch solche Normen beachtet werden, die im ersten Moment für den Wärmebehandler nicht im Fokus liegen, wie die Werkstoff- und Bauteil-normen. Nur mit Sorgfalt und Kenntnis der gültigen Regelwerke, können Reklamationen und Bauteilversagen verhindert werden.

Auch ist zu beachten, dass Normen und Regelwerke nicht alle Eventualitäten abbilden können. Die Anwender wie Konstrukteure, Wärmebehandler, Einkäufer usw. müssen bereits bei Auftragserteilung klare Regeln für die Fertigung und Abnahme festlegen, um Fehler zu vermeiden.

Literatur

[1] DIN EN 45020:2007-03: Normung und damit zusammenhängende Tätigkeiten. Allgemeine Begriffe

[2] Schäder, B.: 100 Jahre Deutsches Institut für Normung – Hier wird für Ordnung gesorgt. Stuttgarter-zeitung.de, zuletzt abgerufen am 22.12.2017

[3] Liedtke, D.: 40 Jahre Normenausschuss Werkstofftechnologie (NWT) im DIN

[4] [https://www.din.de/de/mitwirken/normenausschuesse/nwt/nationale-gremien], zuletzt abgerufen am 08.07.2024

[5] DIN EN ISO 18203:2022-07

[6] DIN EN ISO 683-2:2018-09: Für eine Wärmebehandlung bestimmte Stähle, legierte Stähle und Automatenstähle – Teil 2: Legierte Vergütungsstähle

[7] Neue DIN 17029 E: Entwurf, Zeichnungstoleranzen und Prüfverfahren wärmebehandelter Bauteile

[8] DIN ISO 15787:2018-08: Technische Produktdokumentation – Wärmebehandelte Teile aus Eisenwerkstoffen – Darstellung und Angaben

[9] DIN EN ISO 377:2013-11: Stahl und Stahlerzeugnisse, Lage und Vorbereitung von Probenabschnitten und Proben für die mechanische Prüfungen

7.2 Zerstörungsfreie Härtebestimmung mithilfe des magnetischen Barkhausen-Effekts

Dr. Christian Radek, Jennifer Striewisch

Harte Stoffe ritzen weiche – diese Erkenntnis bildet die Grundlage aller klassischen Härteprüfverfahren. Ob Rockwell, Vickers, Brinell oder moderne Methoden wie UCI: Sie alle hinterlassen Spuren am Werkstück. Doch es geht auch anders. Berührungslos und zerstörungsfrei nutzt das QASS-μmagnetic-Messsystem den magnetischen Barkhausen-Effekt, um die Härte zu bestimmen. Diese Methode eröffnet völlig neue zukunftsweisende Möglichkeiten für die Qualitätskontrolle in der Wärmebehandlung und Fertigung, wie zum Beispiel die 100-%-Prüfung in der Serie.

Zur Qualitätssicherung von wärmebehandelten Bauteilen stehen grundsätzlich drei verschiedene Zertifizierungsverfahren zur Verfügung:

- Konformitätszertifikat (Certificate of Conformity – CoC): Die Zertifizierung erfolgt ohne zerstörende Prüfung auf der Basis von Erfahrungswerten etablierter Prozesse. Prozessschwankungen oder schleichende Prozessveränderungen bleiben dabei unentdeckt.
- Chargenzertifikat (Certificate of Batch – CoB): Eine repräsentative Stichprobe wird zerstörend geprüft. Dies verursacht zusätzliche Kosten durch Probennahme und Materialausschuss. Zudem besteht das Risiko nicht-repräsentativer Stichproben, beispielsweise durch unzureichende Berücksichtigung verschiedener Ofenpositionen beim Einsatzhärten oder des zeitlichen Ablaufs beim Induktivhärten.
- Analysenzertifikat (Certificate of Analysis – CoA): Jedes Werkstück wird individuell zerstörend geprüft. Da die Prüfung an den gehärteten, verschleißkritischen Bereichen erfolgen müsste, widerspricht dies dem Verwendungszweck der Wärmebehandlung. Alle drei Verfahren weisen erhebliche Einschränkungen auf.

7.2.1 Lösung

Die Lösung bietet der magnetische Barkhausen-Effekt, den man auch als Barkhausen-Rauschen kennt. Die Härtemessung mittels Barkhausen-Rauschen basiert auf dem Zusammenhang zwischen mechanischer Härte und magnetischen Eigenschaften. Die magnetischen Eigenschaften werden durch dieselben Faktoren bestimmt, die auch die mechanische Härte definieren.

Das QASS-μmagnetic-Messsystem (**Bild 1**) nutzt das Barkhausen-Rauschen, um in industrieller Umgebung die zeitaufwendige zerstörende Härtebestimmung durch ein einfach zu bedienendes berührungsloses und zerstörungsfreies Messverfahren zu ersetzen. Dabei stehen vor allem die folgenden Aspekte im Vordergrund: Mit einer effektiven Messdauer in der Größenordnung von einer Sekunde und einer Echtzeitverarbeitung der Messdaten ist das System in der Lage, Entscheidungen sofort zu treffen und an die Maschinensteuerung zu übermitteln, z. B. um Bauteile direkt auszuschleusen oder Prozessparameter zu optimieren. Das System ist für einen 24/7-Dauerbetrieb ausgelegt. Durch die zerstörungsfreie 100-%-Prüfung werden Prozessunregelmäßigkeiten und schleichende Prozessveränderungen frühzeitig erkannt. Das erlaubt die schnelle Reaktion auf problematische Entwicklungen und verhindert die Produktion von Ausschuss.

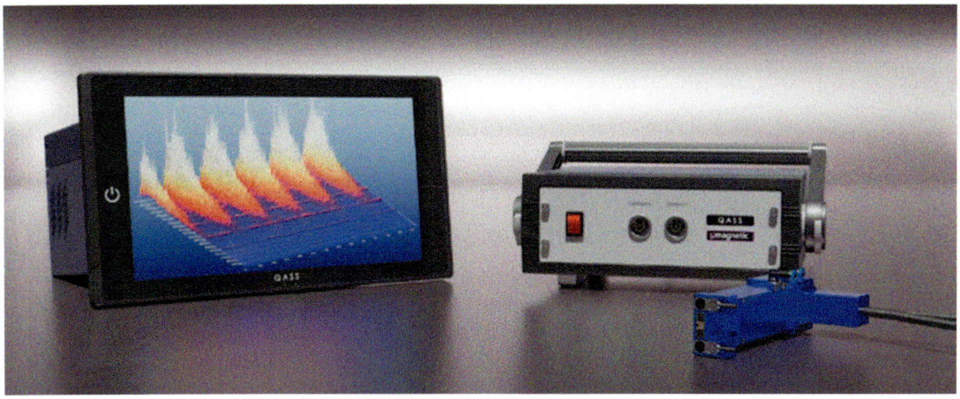

Bild 1: Das QASS-µmagnetic-Messsystem besteht aus dem Optimizer4D (links), dem Leistungsverstärker (rechts) und dem µmagnetic-Sensor (vorne). Auf die Verkabelung wurde der Übersichtlichkeit halber verzichtet. (Quelle: QASS GmbH)

Die Integration des QASS-µmagnetic-Messsystems in bestehende Prozesse ist problemlos möglich. Die Verwendung der Spektralanalyse in Kombination mit Methoden des maschinellen Lernens erlauben, dass störende elektromagnetische Einflüsse (z. B. Frequenzumrichter) erkannt und bei der Auswertung herausgefiltert werden. Das Gerät ist mit den üblichen Schnittstellen (IO-Anschlüsse, Profinet, LAN, WLAN) für eine effektive Maschinenkommunikation ausgestattet. Der modular aufgebaute Sensor (**Bild 2**) ist dafür ausgelegt, in industrieller Umgebung eingesetzt zu werden. Durch eine austauschbare Sensorspitze können unterschiedliche Geometrien (z. B. Verzahnungen, plane Oberflächen, Wellen) vermessen werden. Eine aktive Kühlung erlaubt die Verwendung von bis zu 500 °C. Im Gegensatz zur traditionellen Härtemessung erlaubt das QASS-µmagnetic-Messsystem die Messung in Bewegung (bis zu 600 mm/s). Auf diese Weise lassen sich Härtegradienten sowie Härteanomalien wie z. B. Weichfleckigkeit feststellen. Bei der berührungslosen Messung in Bewegung muss der Abstand entweder konstant gehalten oder erfasst werden, um den Einfluss des Abstandes auf den Messwert zu eliminieren. Zu diesem Zweck kann der Sensor mit induktiven Abstandssensoren ausgestattet werden. Für jedes System kann eine kunden- und anwendungsspezifische Benutzeroberfläche erstellt werden.

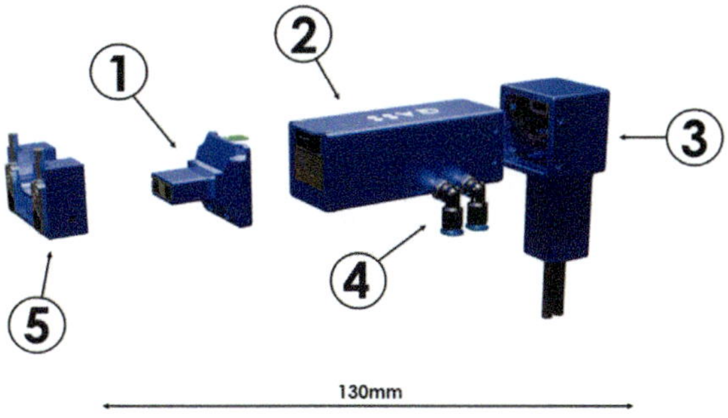

Bild 2: Der QASS-µmagnetic-Sensor ist modular aufgebaut, bestehend aus Messspitze (1), Sensorkörper (2) und aktiver Kühlung (4). Die Messspitze kann auf spezifische Bauteilgeometrien angepasst und ausgetauscht werden. Der Anschluss an den Optimizer4D und den Leistungsverstärker erfolgt über den gewinkelten Adapter (3). Induktive Abstandssensoren können optional mit einem Adapter (5) an der Messspitze angebracht werden. (Quelle: QASS GmbH)

Eine Rechteverwaltung stellt sicher, dass nur diejenigen Benutzer Zugriff haben, die dafür autorisiert sind. Über Schnittstellen lassen sich Signaleinrichtungen, Monitore und Displays anschließen, die den Betriebszustand signalisieren. Alle QASS-Systeme sind bereits für die Remote-Unterstützung durch Experten vorbereitet, sodass viele Fragen auch ohne Vor-Ort-Einsatz gelöst werden können.

7.2.2 Magnetischer Barkhausen-Effekt – Grundlage des Barkhausen-Rauschens

Die theoretischen Grundlagen gehen auf den französischen Physiker Pierre-Ernest Weiss (1865 bis 1940) und den deutschen Physiker Heinrich Georg Barkhausen (1881 bis 1956) zurück. Weiss postulierte, dass ferromagnetische Materialien kleine Bereiche gleicher magnetischer Orientierung enthalten – die Weissschen Bezirke oder Domänen. In einem nach außen hin nicht magnetisiertem Material sind diese Domänen zufällig orientiert und heben sich in ihrer magnetischen Wirkung gegenseitig auf. Bei zunehmender Magnetfeldstärke eines äußeren Feldes wachsen zunächst die Domänen, die bereits ähnlich zum äußeren Feld orientiert sind. Entscheidend ist, dass jeder Weisssche Bezirk als Einheit die Orientierung wechselt – ein sprunghafter Vorgang, der zu einer Änderung des magnetischen Flusses führt. Diese Flussänderungen induzieren in einer Messspule Stromimpulse, die Barkhausen als charakteristisches „Knacken" in einem Lautsprecher hörbar gemacht hatte – das Barkhausen-Rauschen.

Die mechanische Härte korreliert direkt mit der Defektdichte (Leerstellen, Fremdatome, Zwischengitteratome, Versetzungen) im Kristallgitter, die wiederum die magnetische Härte bestimmt. Werkstoffe mit hoher mechanischer Härte sind daher auch magnetisch hart. Im Martensit beispielsweise ist der Kohlenstoff auf Zwischengitterplätzen zwangsgelöst, was die Defektdichte deutlich erhöht. In Ferrit oder Perlit hingegen hat sich der Kohlenstoff als Karbid an Korngrenzen ausgeschieden. Dieser fundamentale Zusammenhang bildet die Grundlage für die zerstörungsfreie Härtemessung mittels Barkhausen-Rauschen-Analyse.

7.2.3 QASS-µmagnetic-Härtemessung

Ein mikromagnetisches Messsystem besteht aus einem Elektromagneten, der kontinuierlich ein periodisches magnetisches Wechselfeld mit definierter Messfrequenz erzeugt, und einer Empfängerspule zur Detektion der magnetischen Flussänderungen. Anders als bei Barkhausens ursprünglichem Experiment wird das Signal nicht hörbar gemacht, sondern als Amplituden-Zeit-Signal aufgezeichnet und digital weiterverarbeitet.

Das QASS-µmagnetic-Messsystem unterwirft dieses Amplituden-Zeit-Signal in Echtzeit einer Fast-Fourier-Transformation (FFT) – ein wesentlicher Unterschied zu konventionellen Systemen. In **Bild 3** ist das resultierende Amplituden-Zeit-Frequenz-Signal dargestellt, das die charakteristischen Barkhausen-Hügel zeigt, die jeweils einer Ummagnetisierung entsprechen. Diese Vorgehensweise ermöglicht frequenzselektive Analysen und den Einsatz von Frequenzfiltern. Mithilfe dieser Filterung können verschiedenste elektromagnetische Störquellen (Frequenzumrichter, Servomotoren usw.) aus dem industriellen Umfeld, gezielt vom Messsignal eliminiert werden.

Analysiert man auf diese Art und Weise einen harten und einen weniger harten Werkstoff, so sind die Härteunterschiede bereits in dieser 3D-Landschaft sichtbar. Harte Werkstoffe zeigen schwache und wenig intensive Signale, während weniger harte Werkstoffe im direkten Vergleich intensive Signale zeigen. **Bild 4** illustriert diesen Zusammenhang.

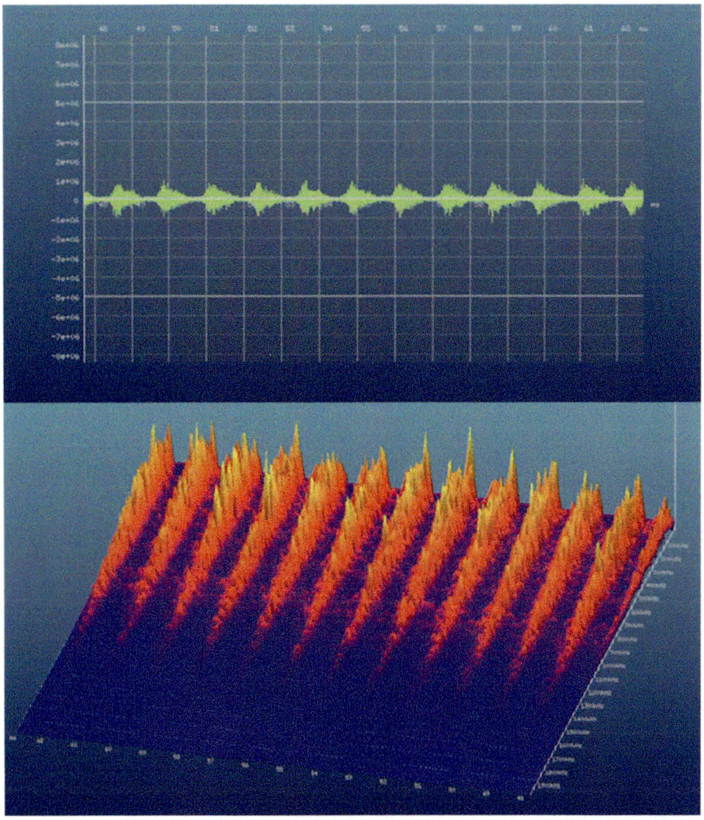

Bild 3: Das obere Bild zeigt das Amplituden-Zeit-Signal, aus dem mithilfe der Fourier-Transformation das Amplituden-Zeit-Frequenz-Signal berechnet wird, das im unteren Bild dargestellt ist. (Quelle: QASS GmbH)

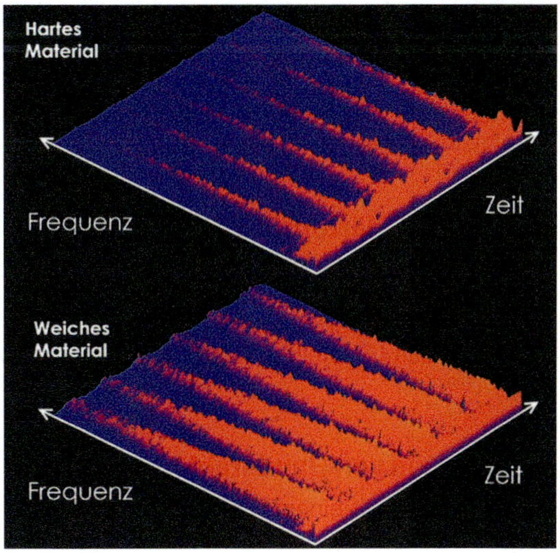

Bild 4: Bereits aus den Signallandschaften nach der FFT kann man erkennen, welches Material eine höhere (obere Abbildung) und eine weniger hohe Härte (untere Abbildung) hat. (Quelle: QASS GmbH)

Die QASS-µmagnetic-Härtemessung erfolgt als Vergleichsmessung. Dazu werden Proben gleichen Materials und gleicher Geometrie benötigt, deren Härte zerstörend bestimmt wurde. Das Material muss, abgesehen von den Abkühlgeschwindigkeiten, die gleiche Wärmebehandlung erfahren haben. Dies ist erforderlich, weil das Barkhausen-Rauschen-Verfahren von verschiedenen Faktoren beeinflusst wird, die ausgeschlossen werden müssen:

- Eigenspannungen: Mechanische Spannungen im Material, die z. B. durch die Wärmebehandlung entstehen oder abgebaut werden.
- Gefügeunterschiede: Unterschiedliche Korngrößen oder Karbidverteilungen.
- Abstand: Der Abstand hat einen großen Einfluss auf das Messsignal und muss entweder konstant oder bekannt sein (Abstandssensoren), damit dessen Einfluss eliminiert werden kann.
- Geometrische Faktoren: Wandstärken, Krümmungen, Randeffekte oder andere geometrische Einflüsse.
- Oberflächenbeschaffenheit: Rauheit und Verschmutzungen spielen keine Rolle außer sie sind ferromagnetisch (eisenhaltige Stäube).

Mit den Kalibrierdaten lässt sich eine Funktion berechnen, die den meist linearen Zusammenhang zwischen Härte und gemessener Energie herstellt. Mit dem QASS-µmagnetic-System können zerstörungsfreie Härtemessungen durchgeführt werden, die beliebig häufig wiederholt werden können und selbst bei kurzen Messzeiten von unter einer Sekunde eine hohe Reproduzierbarkeit bieten. Bei sorgfältiger Kalibrierung kann eine Umrechnungsgenauigkeit von +/- 1 HRC erreicht werden.

7.2.4 Anwendungen in der Industrie

Die etablierte Praxis der QASS-Technologie in der Automobilindustrie besteht darin, jedes Werkstück individuell zu prüfen: Getriebewellen weisen nach dem Härten häufig Verzug auf und müssen gerichtet werden. Europäische Automobilhersteller spezifizieren seit Jahren, dass dieser Richtprozess durch QASS-Körperschallsensorik überwacht werden muss, um Risse im Entstehungsmoment zu detektieren und betroffene Wellen automatisch auszuschleusen. Derselbe Anspruch, der 100-%-Überwachung der Fertigung, lässt sich mit dem QASS-µmagnetic-System nun auch für die Härtemessung in der Serie realisieren. Auch ergänzend zu den konventionellen Prüfverfahren können sich aus dem Einsatz der Technologie erhebliche Vorteile ergeben. Die automatisierte Härteüberwachung mittels Barkhausen-Rauschen wird bereits in vielen verschieden Anwendungsfeldern in der Industrie eingesetzt. Im Folgenden ist ein Auszug der Einsatzmöglichkeiten dargestellt.

7.2.5 100-%-Härteprüfung von gehärteten Gussteilen

Für einen Hersteller von verschleißarmen Gussteilen wurde eine autark arbeitende Prüfstation entwickelt. Ein Industrieroboter greift dazu ein Gussteil aus einer Schüttung (bin picking) und transportiert dieses zu einer QASS-µmagnetic-Härteprüfeinrichtung. Anschließend wird das Gussteil optisch auf Gussfehler untersucht, verwogen und nach Gewicht wegsortiert.

7.2.6 Härten von Bandstahl

In Kooperation mit einem Bandstahlhersteller wurde ein Härteprüfsystem entwickelt, das im Sekundenabstand kontinuierliche Härtemessungen an einem Stahlband, das mit einer

Geschwindigkeit von bis zu 20 m/min unter dem Sensor durchläuft, durchführt. Das Messsystem arbeitet vollautomatisch im 24/7-Betrieb, verfügt über Hinderniserkennung und nutzt Methoden des maschinellen Lernens zur automatischen Kalibrierung auf neue Stahlsorten. Dadurch wird ein lückenloser Härteverlauf über mehrere Kilometer Bandlänge dokumentiert.

7.2.7 Härtebestimmung beim Presshärten

Beim Presshärten wird das warme Blech während eines Tiefziehprozesses durch Kühlkanäle im Tiefziehwerkzeug lokal schneller abgekühlt als an anderen Stellen. Auf diese Weise erreicht man insbesondere bei Karosserieteilen Härtegradienten. Um den Erfolg des Presshärteprozesses festzustellen, wird das 400 °C heiße Material vermessen. Dazu wird die Kontaktzeit kurzgehalten und der Sensor aktiv mit Wasser gekühlt.

7.2.8 Schleifen von Walzen

Walzen erfahren beim Walzen eine Kaltverfestigung. Daher ist es notwendig, diese in regelmäßigen Abständen abzuschleifen, um die verfestigten Stellen zu entfernen. Der Endpunkt der Überarbeitung wird i. d. R. über Erfahrungswerte und optische Begutachtung ermittelt. Dabei wird häufig mehr Material abgetragen als notwendig, um sicherzustellen, dass alle Fehlstellen beseitigt wurden. Teilweise kommen dabei Hilfsmittel wie das minimalinvasive Rückprallverfahren nach Leeb zum Einsatz, die nur punktuelle Aussagen über die Härte der Walze liefern. Auch diese kleinsten Eindrücke müssen durch einen weiteren Schleifgang entfernt werden. Mit der 100-%-Härteüberwachung mittels Barkhausen-Rauschen entfällt das Abschleifen der Eindrücke und der Abtrag wird auf die reale Beschädigung beschränkt.

7.2.9 Schleifbranddetektion

Schleifbrand ist eine thermische Schädigung der Bauteiloberfläche, die durch unsachgemäße Prozessführung entsteht. Der Wärmeeintrag führt lokal zu Eigenspannungsveränderungen, evtl. zu einer Härteabnahme und im ungünstigsten Fall zur Bildung einer Neuhärtezone aus nanokristallinem, glashartem Martensit an der Bauteiloberfläche. Schleifbrand, wenn er nur schwach ausgeprägt ist, ist visuell schwer oder gar nicht erkennbar. Das klassische Verfahren zur Untersuchung von Bauteilen auf Schleifbrand ist das Nitalätzen. Dieses nasschemische Verfahren ist zeitaufwendig und arbeitet mit aggressiven Chemikalien wie Salpetersäure.

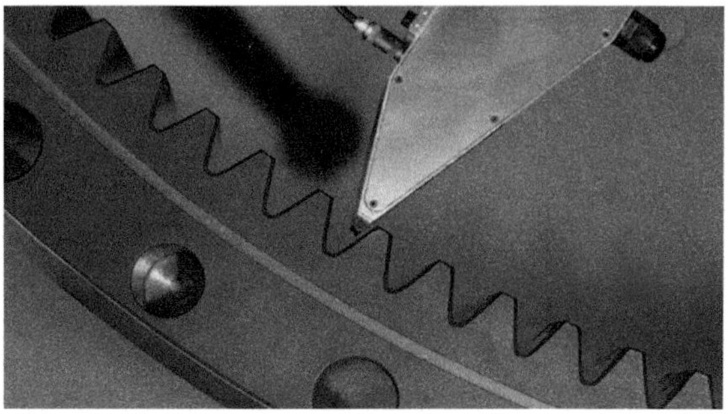

Bild 5: Berührungslose Prüfung eines Hohlrades aus einem Planetengetriebe auf Schleifbrand. Die Sensorspitze des für diese Anwendung angepassten Sensors tastet jede Zahnflanke robotergeführt und berührungslos zeilenweise ab. (Quelle: QASS GmbH)

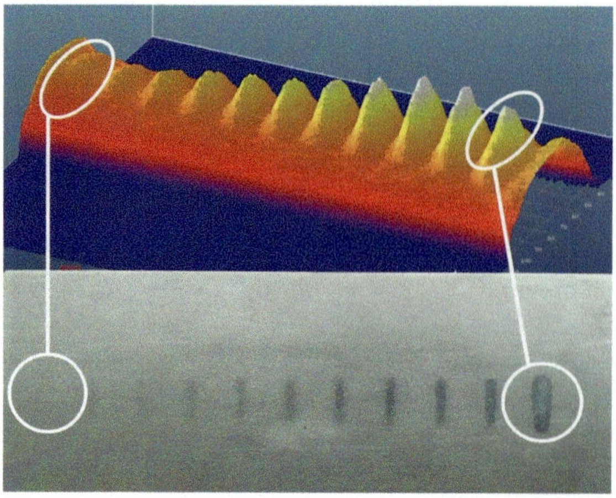

Bild 6: Das zeilenweise Abfahren des Nitalbad-Testkörpers zeigt den Energieverlauf, der sich umgekehrt proportional zur Härte verhält. Es wurden alle elf Anlasszonen gefunden. Der erste Peak (rechts) ist deutlich kleiner als der zweite, da sich bei dem ersten Peak eine Neuhärtezone aus glashartem Martensit auf der Anlasszone befindet. Der letzte Peak auf der linken Seite ist zwar unsichtbar, aber mikromagnetisch noch messbar. (Quelle: QASS GmbH)

In enger Zusammenarbeit mit einem Hersteller von Planetengetrieben für Windkraftanlagen wurde ein robotergestütztes Verfahren entwickelt, um Hohlräder auf Schleifbrand hin zu untersuchen. **Bild 5** zeigt einen für diese Anwendung optimierten Sensor knapp über der Oberfläche einer Zahnflanke, die zeilenweise berührungslos abgefahren wird.

Die Firma imq in Crimmitschau (Sachsen) stellt Nital-Ätzblöcke her, die dazu verwendet werden, die Güte von Nital-Ätzbädern zu bestimmen. Der etwa 60 mm x 20 mm x 8 mm große Stahlblock enthält elf künstlich mit einem Laser eingebrachte Anlass- und Neuhärtezonen. Durch zeilenweises Abfahren des Blockes entstand das Bild, das in **Bild 6** gezeigt wird.

7.2.10 Lagerichtige Zuführung von induktiv gehärteten Bauteilen

Die in Segmenten induktiv gehärtete Welle, die in **Bild 7** dargestellt ist, muss lagerichtig für die weitere Bearbeitung einer Maschine zugeführt werden. Die Vermessung der Welle ent-

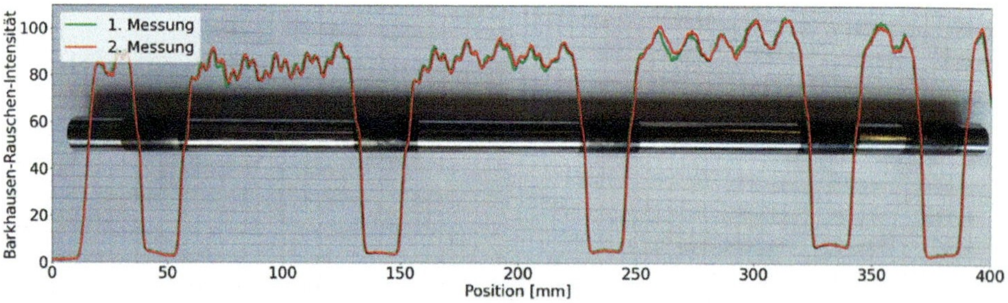

Bild 7: Die Welle zeigt fünf Zonen, die induktiv gehärtet wurden. Die mikromagnetische Vermessung, die als Überlagerung eingefügt wurde, zeigt an den induktiv gehärteten Bereichen deutliche Energieunterschiede zum Grundmaterial. Die periodischen Schwankungen sind auf die Schraubenlinie zurückzuführen, die von der zerspanenden Bearbeitung (Drehen) der Welle stammen. (Quelle: QASS GmbH)

lang der Längsachse ergibt den Grafen, der in Bild 7 über der Welle dargestellt ist. Mit dieser Information wird die Welle lagerichtig der Maschine zugeführt.

Autorenverzeichnis

Martin Bauer
ITG Induktionsanlagen GmbH
Neckarsteinacher Straße 88
69434 Hirschhorn/Neckar
Kapitel 5.2

Michal Bazel
SECO/WARWICK S.A.
ul. Sobieskiego 8
66–200 Świebodzin
Polen
Kapitel 6.2

Marco Bono
SAET S.p.A.
Via Torino, 213
10040 Leini (Turin)
Italien
Kapitel 5.3

Dr. Joachim Boßlet
Durferrit GmbH
Industriestraße 3
68169 Mannheim
Kapitel 2.6

Dr. Klaus Buchner
AICHELIN Holding GmbH
Fabriksgasse 3
2340 Mödling
Österreich
Kapitel 1.3, Kapitel 2.2, Kapitel 6.3

Mario Cesano
SAET S.p.A.
Via Torino, 213
10040 Leini (Turin)
Italien
Kapitel 5.3

Dario Ciscato
SAET S.p.A.
Via Torino, 213
10040 Leini (Turin)
Italien
Kapitel 5.3

Helmut Egger
Ehemals EGGER Consulting
Kapitel 4.1

Dr. Fabian Fahlbusch
Busch Dienste GmbH
Schauinslandstraße 1
79689 Maulburg
Kapitel 6.4

Grzegorz Głuchowski
SECO/WARWICK S.A.
ul. Sobieskiego 8
66–200 Świebodzin
Polen
Kapitel 6.2

Dr. Sebastian Goderbauer
gapzero mathematical decision support GmbH
Karl-Benz-Str. 9
40764 Langenfeld
Kapitel 3.1

Gerald Hiller
ECM GmbH
Carl Legien Straße 15
63073 Offenbach-Waldhof
Vorwort

Frank Hippenstiel
BGH Edelstahlwerke GmbH
Am Stahlwerk 1
01705 Freital
Kapitel 3.1

Arnold Horsch
Arnold Horsch e.K.
Berghauser Str. 62
42859 Remscheid
Kapitel 7.1

Julian Irretier
IBW Dr. Irretier GmbH
Mühsol 44
47533 Kleve
Kapitel 1.2

Dr. Olaf Irretier
IBW Dr. Irretier GmbH
Mühsol 44
47533 Kleve
Kapitel 1.2

Marco Jost
IBW Dr. Irretier GmbH
Marie-Curie-Str. 7
40625 Düsseldorf
Kapitel 1.2

Björn Kintscher
SMS group GmbH
Am SMS Campus 1
41069 Mönchengladbach
Kapitel 2.5

Andreas Kraus
JUMO GmbH & Co. KG
Moritz-Juchheim-Straße 1
36039 Fulda
Kapitel 2.3

Dominic Kraus
AICHELIN Holding GmbH
Fabriksgasse 3
2340 Mödling
Österreich
Kapitel 6.3

Dr. Mathias Kunz
WPX Faserkeramik GmbH
Redcarstraße 44 B
53842 Troisdorf
Kapitel 2.5

Jonas Metzger
BGH Edelstahlwerke GmbH
Am Stahlwerk 1
01705 Freital
Kapitel 3.1

Michael Müller
BGH Edelstahlwerke GmbH
Am Stahlwerk 1
01705 Freital
Kapitel 3.1

Dr. Robert Pietrasik
HART-TECH Sp. z o.o.
Niciarniana 45
92–320 Łódź
Polen
Kapitel 6.2

Beatrice Pillin
SAET S.p.A.
Via Torino, 213
10040 Leini (Turin)
Italien
Kapitel 5.3

Dr. Giuseppe Polignano
FAI-FTC S.p.A.
Via Enrico Mattei, 12
25026 Pontevico (BS)
Italien
Kapitel 1.2

Dr. Christian Radek
QASS GmbH
Schöllinger Feld 28
58300 Wetter
Kapitel 7.2

Dr.-Ing. Hans-Jürgen Rönnecke
Acritudo Systems GmbH
Am Großen Teich 20
58640 Iserlohn
Kapitel 2.4

Livia Scarpellini
SAET S.p.A.
Via Torino, 213
10040 Leini (Turin)
Italien
Kapitel 5.3

Jörg Scheyhing
ECG Energie Consulting GmbH
Wilhelm-Leonhard-Straße 10
77694 Kehl-Goldscheuer
Kapitel 2.1

Michael Schlerkmann
FAI-FTC SpA
Via Enrico Mattei, 12
25026 Pontevico (BS)
Italien
Kapitel 1.2

Uwe Schmelzing
Aalberts Surface Technologies GmbH
Walter-Freitag-Straße 25
42899 Remscheid
Vorwort, Kapitel 1.1

Björn Schuh
Demig Prozessautomatisierung GmbH
Friedrich-Wilhelm-Straße 148
57074 Siegen
Kapitel 3.2

Jennifer Striewisch
QASS GmbH
Schöllinger Feld 28
58300 Wetter
Kapitel 7.2

Dr. Christine Tränkner
ITG Induktionsanlagen GmbH
Neckarsteinacher Straße 88
69434 Hirschhorn/Neckar
Kapitel 5.1, Kapitel 5.2

Dr. Thomas Waldenmaier
Robert Bosch GmbH
Robert-Bosch-Campus 1
71272 Renningen
Kapitel 1.4

Gerd Waning
WANING Ingenieurdienstleistungen
Otto-Wulff-Straße 8
32107 Bad Salzuflen
Kapitel 4.2

Dipl.-Ing. (FH) André Wieder
ITG Induktionsanlagen GmbH
Neckarsteinacher Straße 88
69434 Hirschhorn/Neckar
Kapitel 5.1

Björn Eric Zieger
IVA Schmetz GmbH
Holzener Straße 35–39
58708 Menden
Kapitel 6.1

Inserentenverzeichnis